Algal Biorefineries

Aleš Prokop • Rakesh K. Bajpai • Mark E. Zappi
Editors

Algal Biorefineries

Volume 2: Products and Refinery Design

Springer

Editors
Aleš Prokop
Department of Chemical and Biological Engineering
Vanderbilt University
Nashville, TN, USA

Rakesh K. Bajpai
Department of Chemical Engineering
University of Louisiana at Lafayette
Lafayette, LA, USA

Mark E. Zappi
Department of Chemical Engineering
University of Louisiana at Lafayette
Lafayette, LA, USA

ISBN 978-3-319-20199-3 ISBN 978-3-319-20200-6 (eBook)
DOI 10.1007/978-3-319-20200-6

Library of Congress Control Number: 2013948473

Springer Cham Heidelberg New York Dordrecht London

Printed on acid-free paper

Springer International Publishing AG Switzerland is part of Springer Science+Business Media (www.springer.com)

Editorial and Introduction to Volume II

This volume is a continuation of the first volume (Bajpai et al. 2014) and has two parts. Consistent with the theme of volume 1, the different chapters in this volume are intended to provide its readers with a wide perspective of current state of the art as well as technical and economic issues in utilizing algae to capture energy and carbon from renewable resources. Future chapters henceforth will focus on specific issues related to biorefinery operations.

The first part deals with algae culturing and their optimization. The very first chapter sets the stage for systems biology approach to algal culture. Such approach, particularly that of computational systems biology, may serve as the identification of gene targets for enhanced production of biofuel precursors. Models based on systems biology should ensure the continued increase in biomass and production of fuel precursors from microalgal feedstock while it is expected to reduce the production cost and allow biofuels to enter the market. The second chapter discusses the employment of genetically modified algae, the latter being set within the framework of issues relevant to more general microbial systems. The next chapter by Perez-Garcia and Bashan shows how heterotrophic and mixotrophic cultivation offers an alternative to photoautotrophic cultivation as it provides more different products and an improvement of the economic feasibility of microalgae-biorefined products as compared to traditional autotrophic culture. The next four chapters review different culturing scenarios for autotrophic cultures. New designs are also introduced. Both open and closed designs are considered here. The chapter by Koller provides a very comprehensive discussion of closed reactor systems useful for special conditions which guarantee the algal culture protection against microbial contamination. Thin-layer technology (previously reviewed by Doucha and Lívanský 2014) is gaining worldwide use as it provides higher productivities and efficiencies. Following is a chapter by Revah and Morales on gas balance at algal phototrophic growth and how this can be employed for culture control. The last two chapters of this part demonstrate new products due to selenium and rare element accumulation. Selenium-rich algae may offer an interesting alternative for contamination control as other algae species and protozoal contaminants are unable to grow under higher selenium concentration.

The second part is directed to different possibilities of "unit operation" and whole biorefinery design. The first chapter by Chirdon shows how biorefineries may increase profitability by utilizing existing material streams that may have little or negative value including various nutrient-rich wastewater streams and how to convert proteinaceous waste streams from biorefineries into valuable products. Planning a biorefinery operation should be designed with consideration of inclusion of the products and coproducts to optimize the economic viability of the system. Likewise, Özçimen discusses employment of algae cake as the utilization of algal wastes is the most appropriate option for energy recovery and cost reduction. The following paper by Theegala covers separate steps such as cell disruption and oil extraction. Such topics are typically not considered in the overall biorefinery design and may improve the economics to get on par with petroleum products. In the same line, the extension of the number of products by anaerobic (or combined aerobic/anaerobic) culturing may provide a variety of products to conventional autotrophic biorefinery design such as methane and hydrogen. This paper also employs relatively simple quantitative treatment to help the optimization of product formation. The next two chapters review two different biomass treatments such as hydrothermal and cracking. The chapter by Ruiz et al. focuses on macroalgal biomass. The last two chapters focus on life cycle analysis of algal biorefinery. Cheali et al. provide systematic framework for synthesis and design of processing networks with a number of scenarios produced prior to the identification of optimal designs. The approach allows an expansion as more promising data and technologies are available. Agusdinata et al. consider environmental factors to assess potential emission reduction resulting from the adoption of algal biofuels in the US airline industry.

The editors would like to thank internal and external reviewers for their willingness to assess the submitted chapters:

Aleš Prokop, Rakesh K Bajpai, Paul Pedigo, Samuel Abramson, Vilém Zachleder, Manfred Lübken, Tomáš Branyik, William Chirdon, Melek Erol Trygun, Octavio Perez-Garcia, Larry E Erickson, David Glass, Bohumil Dušek, Martin Koller, GC Sin, Milada Vítová, Didem Őzçimen, and Jiří Masojídek.

Nashville, TN, USA — Aleš Prokop
Lafayette, LA, USA — Rakesh K. Bajpai
Lafayette, LA, USA — Mark E. Zappi

References

Bajpai RK, Prokop A, Zappi M (eds) (2014) Algal biorefineries, volume 1: cultivation of cells and products. Springer, Dordrecht, The Netherlands, 324 p

Doucha J, Lívanský K (2014) High density outdoor microalgal culture. In: Bajpai RK, Prokop A, Zappi M (eds) Algal biorefineries, volume 1: cultivation of cells and products. Springer, Dordrecht, The Netherlands, pp 147–173

Contents

Part I
Algal Growth, Products and Optimization

Microalgal Systems Biology for Biofuel Production

Seong-Joo Hong and Choul-Gyun Lee

Abstract In recent years, microalgae has received a lot of attention as potential sources of renewable energy, especially given the increase in oil prices and environmental concerns. Therefore, microalgal systems biology can shed light on complex interactions in biological systems through integration of various omics data. Genome-scale metabolic reconstruction can provide insights into cellular metabolism and species-specific adaptive features, whereas *in silico* analysis is a powerful tool for analysis of metabolic flux and identification of gene targets for enhanced production of biofuel precursors. Here, we highlight the current state of research in microalgal systems biology and evaluate the potential for future sustainable microalgae-based biofuel engineering and development of "green cell factories."

Keywords Systems biology • Metabolic reconstruction • *In silico* analysis • Flux balance analysis • Omics analysis • Biofuel • Fatty acids • Isoprenoids • β-oxidation pathway • Photosynthesis • Microalgae

1 Introduction

Systems biology, the understanding of biology at the system level through analysis of genotype-phenotype relationships, was formally defined more than a decade ago and is now an established research field (Kitano 2002; Palsson 2006a). Data from high-throughput omics technologies targeting the genome and metabolome are of central importance to systems biology research. Omics data sets can be used to construct *in silico* metabolic models of biochemical reaction networks for analysis and prediction of cellular behavior under genetic and environmental perturbations. Therefore, systems biology is a powerful tool for improving strain engineering and has been recently applied to enhancement of production of biofuel precursors from microalgae.

S.-J. Hong • C.-G. Lee (✉)
National Marine Bioenergy Research Center, Department of Biological Engineering, Inha University, 100 Inha-Ro, Nam-Gu, Incheon 402-751, Korea
e-mail: leecg@inha.ac.kr

A. Prokop et al. (eds.), *Algal Biorefineries*, DOI 10.1007/978-3-319-20200-6_1

Several attempts to construct the metabolic networks of microalgae have been reported. Central metabolic pathways in *Chlorella pyrenoidosa* were built and simulated by metabolic flux analysis under various trophic conditions (Yang et al. 2000). The first report of a constraint-based model of eukaryotic microalgae was published 5 year ago (Boyle and Morgan 2009); those authors constructed the central metabolism of the green microalga *Chlamydomonas reinhardtii* based on its genomic sequence. In 2011, Chang et al. reconstructed the metabolic network of *C. reinhardtii* based on its entire genome and quantitatively analyzed its growth efficiency under stimulation by different light sources. Additionally, the development of next-generation sequencing (NGS) has accelerated accumulation of genomics and transcriptomics data; as a result, the number of automated metabolic models of microalgae has increased rapidly (Fabris et al. 2012; Molnár et al. 2012; Radakovits et al. 2012). However, microalgal systems biology was limited by organismal diversity: microalgae employ significantly different metabolic pathways and subcellular compartments according to the taxonomic classification originating from their symbiosis and evolutionary selection.

Despite these limitations, the use of microalgae for biofuel production can help achieve sustainability through photosynthetic conversion of inorganic carbon into organic carbon compounds. Therefore, microalgal cultivation requires air (with CO_2) and light in photoautotrophic culture system. This process removes CO_2 from the atmosphere and reduces carbon emissions from fossil fuels such as natural gas, oil, and coal.

Microalgal production of biofuels and removal of CO_2 requires an increase in biomass and production of target metabolites. Herein, we focus on recent reports of biofuel production in eukaryotic microalgae to overcome the unresolved problems. In particular, we describe the application of metabolic reconstruction to elucidate the cellular network for pathway optimization and photosynthetic carbon flux, which may consequently help improve microalgal production of advanced biofuels.

2 Systems-Level Tools for Microalgal Systems Biology

Engineering bioenergy production from microalgae necessitates a systems-level approach to understanding metabolism. Thus, metabolic reconstruction models of microalgae have been used to analyze the metabolic network. Mathematical simulations using *in silico* models have been performed to identify target genes to improve bioenergy production. Additionally, massive omics datasets can identify the biological function of each gene and the metabolic and regulatory network of microalgae.

2.1 Metabolic Reconstruction

Genome-based metabolic reconstruction requires genome-scale sequencing and annotation of target microalgae. A specific database of omics information or literature is a good starting point for metabolic reconstruction. The *in silico* model of *C.*

reinhardtii was the first example of genome-scale metabolic reconstruction for eukaryotic microalgae (Chang et al. 2011), made possible by the first release of genome sequences of eukaryotic microalgae and construction of a specific online database and various studies on *C. reinhardtii* (Merchant et al. 2007). Gene-protein-reaction (GPR) associations are used to reconstruct a genome-based metabolic model. A GPR association means that a gene encoding a protein is associated with a chemical reaction(s). Therefore an *in silico* model based on GPR associations can describe cellular behavior depending on genetic properties. Generation of a genome-base metabolic network comprises 4 steps (Fig. 1): (1) genome annotation: construction of a framework; (2) metabolic reconstruction: formation of GPR associations; (3) network gap analysis: filling missing reactions in the network; and (4) model validation: comparison of experimental data with simulation results. Construction of a metabolic framework requires acquisition of genomic, proteomic, and reaction data for target microalgae. The genomic data (sequence and annotations) can be downloaded from specific databases of eukaryotic microalgae from the Joint Genome Institute (http://genome.jgi-psf.org/viridiplantae/viridiplantae.info.html). Proteomics data including protein functions and Enzyme Commission numbers are needed for translated metabolic proteins to identify GPR associations and can be obtained from websites such as BRENDA and UniProt.

Biochemical databases such as KEGG, ChlamyCyc, and AraCyc assist in connecting proteomic data to biochemical reactions in metabolic pathways. Metabolite specificity, chemical properties, stoichiometry, directionality, and localization are

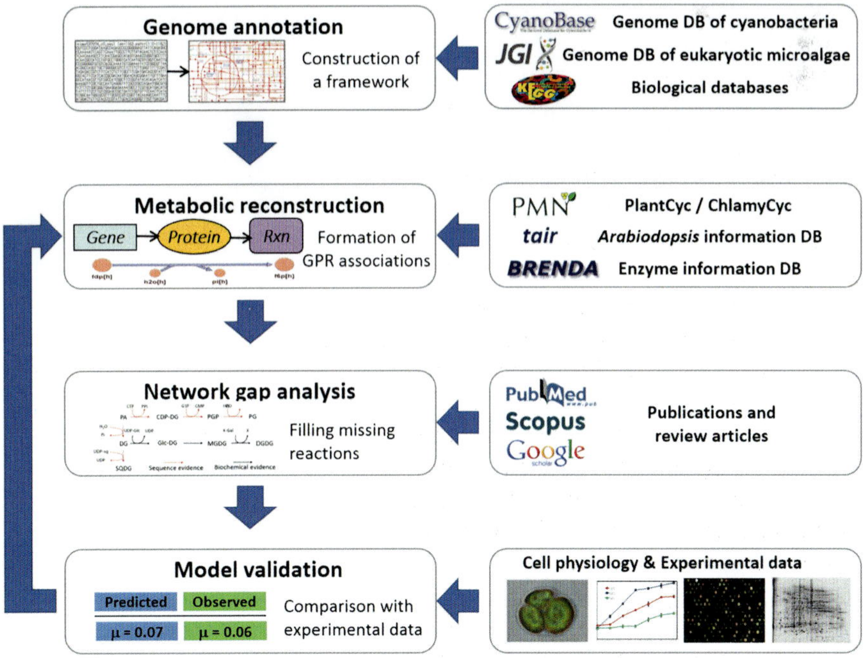

Fig. 1 The metabolic-reconstruction procedure for microalgae

combined to build a metabolic reaction. Metabolite specificity, including catalytic activity and cofactors, is identified using proteomic databases, and metabolite chemical properties are calculated from physical parameters such as pKa values to determine the charge. Reaction reversibility is determined by thermodynamic measurements based on enzyme specificity and involvement of high-energy metabolites (e.g., ATP, GTP). Additionally, cellular localization of biochemical reactions must be considered to reconstruct metabolic pathways for eukaryotic microalgae because of enzymes and metabolites in compartment-specific metabolic pathways. The chloroplast, the photosynthetic organelle in microalgae, represents the major difference between microalgae and other eukaryotic microorganisms. Therefore, understanding energy metabolism in photosynthetic organisms and information about metabolic connectivity in other eukaryotic *in silico* models are essential for microalgal metabolic reconstruction. *In silico* models of *Arabidopsis thaliana* and *C. reinhardtii* contain manually curated compartmentalization of biochemical reactions and are good examples of utilization of cellular localization data. Generally, protein localization can be identified by amino acid sequence by means of bioinformatic tools such as WoLF PSORT and PredictProtein that are established for plants.

After establishing GPR associations in the metabolic network, ATP and NADPH production are tested to confirm the ability of photosynthesis to produce energy for growth, then production of necessary biomass precursors is verified. If the models cannot produce energy or the biomass precursors, the gap in metabolic connectivity must be identified. To confirm production of essential biomass precursors, the biomass objective function (BOF) is constructed by calculating the molar content of the cell biomass. The BOF can be computed as a stoichiometry-based biomass yield, which also represents the growth rate for given substrate uptake rates (Feist and Palsson 2010).

If the *in silico* model can grow according to flux balance analysis (FBA) using BOF, its growth is compared with live cell growth to validate the physiological changes in the *in silico* model. Additionally, deletion of stand-alone genes can be used to validate GPR associations and obtain information about the effects of genetic modification on the target microorganisms.

2.2 Metabolic Flux Analysis

Metabolic networks comprise metabolic pathways that contain biochemical reactions governed by the genome. Therefore, the conversion rate from a substrate to a product, which is involved in further biochemical reactions, indicates the behavior of the metabolic network. Metabolic flux analysis is an effective method for quantitative assessment of metabolic networks because metabolic flux can be predicted by simulation and experimental approaches. Information about metabolic flux can be used to determine fundamental characteristics of cellular physiology and critical parameters of metabolic pathways (Stephanopoulos 1999) and reflects cellular regulation under different environmental conditions such as nutrient availability, pH, temperature, and light intensity, which can fluctuate in time and space (Lee et al. 2006). When cells are exposed to an environmental perturbation, cellular physiology and intracellular composition are affected by systems-level metabolic flux (Stitt 2013). Additionally, metabolic maps are so diverse and complex that for improvement of a strain, it is important to verify the possibility of genetic modification and to predict changes in metabolic fluxes (Stephanopoulos 1999).

To analyze metabolic flux, two methods can be used: prediction by computer simulation and measurement by experimental analysis. Experimental measurements can be obtained by isotopic analysis. Simulation involves calculation of intracellular fluxes using a mass balance from the stoichiometric model. Based on the biochemical reactions in the metabolic reconstruction, the stoichiometric matrix, $S(\mathrm{m} \times \mathrm{n})$, is generated to analyze metabolic characteristics of the target microalgae. Therefore, S_{ij} in the stoichiometric matrix represents the stoichiometric coefficient of the i-th metabolite in the j-th reaction. This matrix can be solved for the steady-state condition by flux balance analysis based on linear programming (Varma and Palsson 1993a, b). Equation (1) describes the problem, where $v(\mathrm{n} \times 1)$ is the vector representing reaction flux in the metabolic network.

$$S \cdot v = 0 \tag{1}$$

To calculate v for the maximum biomass, BOF is determined, optimized, and added to the matrix S using $S_{i,BOF}$ with the constraints shown in Eq. (2):

$$\alpha_i \le v_i \le \beta_i \tag{2}$$

α_i and β_i are the lower and upper boundaries for reaction flux in S. Equation (2) becomes $-\infty \le v_i \le \infty$ if the reaction is reversible (or for metabolite transport when metabolites are present in the media) and $0 \le v_i \le \infty$ if the reaction is irreversible (or for metabolite transport when metabolites are not present in the media). For computational analysis of cell growth, BOF is defined as an object function to be maximized by simulation software such as COBRA Toolbox.

The metabolic flux map from FBA can be used to predict cell growth, nutrient uptake, and product output under defined culture conditions. Cellular physiology can also be predicted accurately by single or multiple gene deletion analysis from a GPR association (Edwards et al. 2001). These predictions form the basis for genetic engineering without a trial-and-error process. Metabolic flux analysis indicates the yield of cofactors such as ATP, NADPH, and NADH in an intracellular network (Varma and Palsson 1993a).

It is useful to explain the energy flow in microalgae. The robustness of the metabolic network can be predicted by metabolic flux analysis to improve the network and to design novel metabolic pathways at the systems level (Edwards and Palsson 2000). Phenotypic phase plane (PhPP) analysis is an effective method for identifying the steady-state solution space by projection in two dimensions (Edwards et al. 2002). Steady-state flux distributions can be sorted into a number of regions, each containing similar metabolic flux patterns and characterized by equivalent shadow prices, which indicate how the BOF value could change by addition of metabolites to the network. Regions determined by the shadow prices can provide information about the limiting factor for growth of target microalgae under culture conditions. Furthermore, various methods such as MOMA, OptKnock, OptGene, GDLS, and OptFlux, can suggest gene deletion strategies to improve the microorganisms (Burgard et al. 2003; Lun et al. 2009; Rocha et al. 2010; Schellenberger et al. 2011). However, these methods have hardly been applied to microalgal models, because constraint-based metabolic reconstruction of microalgae is at the early stage of metabolic reconstruction and is limited to two strains (*C. reinhardtii* and *Synechocystis* sp. PCC6803). This limitation can be overcome by the iterative process of metabolic reconstruction for a high-quality in silico model as microalgal omics data accumulate (Thiele and Palsson 2010).

Since the whole genome-sequence of the prokaryotic microalga *Synechocystis* sp. PCC6803 was published in 1996, various *in silico* models of the *Synechocystis* central metabolism or a whole-genome metabolic network have been built and used to predict cell growth under various trophic conditions (Shastri and Morgan 2005; Hong and Lee 2007; Nogales et al. 2012) or to optimize H_2 production (Montagud et al. 2011) by FBA. Recently, optimization of ethanol production for a *Synechocystis* model by MOMA and ROOM was published. Two mutants were identified as candidates to increase ethanol productivity (Sengupta et al. 2013). Compared to cyanobacterial *in silico* models, metabolic reconstruction of eukaryotic microalgae was delayed by the belated publication of its whole-genome sequence. The *in silico* model of the central metabolism in *C. reinhardtii* was built in 2009 and was simulated under three trophic conditions by FBA (Boyle and Morgan 2009). Two years later, lipid metabolism and biofuel production were included in the whole-genome metabolic reconstruction (Chang et al. 2011). This *in silico* model was validated using cellular growth under trophic conditions and FBA-based gene deletion analysis comparisons with gene knockouts. Additionally, photosynthetic efficiency was evaluated for this constraint model using different light sources. The *in silico* model of *Chlorella* sp. FC2 IITG was also simulated by FBA and dynamic FBA to maximize biomass and neutral lipid production (Muthuraj et al. 2013) and to explore cellular physiology and changes in intracellular flux during the transition from nutrient sufficiency to nutrient deficiency.

2.3 Omics Analyses

Advances in technologies for biological component analysis have made a massive amount of omics data available. Omics technologies involve analysis of differences in DNA (genomics), RNA (transcriptomics), proteins (proteomics), and metabolites (metabolomics) within the cell using high-throughput assays and are important for systems biology to provide an understanding of metabolic- and regulatory-network changes in biological systems. Therefore, extensive omics data are needed to construct a genome-scale metabolic network with GPR association. Various additional omics technologies such as lipidomics, peptidomics, polymeromics, glycomics, interactomics, and fluxomics, have been developed with the development of analytical methods for biological components (Altuntaş and Schubert 2014; Han and Gross 2003; Winter and Krömer 2013).

Genomics, in particular, has developed exponentially in recent decades, and genome sequences are available for a variety of organisms including microalgae such as *C. reinhardtii, Nannochloropsis gaditana, Phaeodactylum tricornutum, Ostreococcus tauri, Ostreococcus luminarius, Micromonas pusilla, Coccomyxa subellipsoidea, Porphyridium purpureum,* and *Monoraphidium neglectum* (Battchikova et al. 2010; Blanc et al. 2012; Bogen et al. 2013; Derelle et al. 2006; Merchant et al. 2007; Palenik et al. 2007; Radakovits et al. 2012; Worden et al. 2009). Genomics plays a significant role in systems biology because genome annotation provides a framework for metabolic reconstruction. Modern NGS technology also permits the use of transcriptomic data to construct metabolic networks. The transcriptome of *Dunaliella tertiolecta* was sequenced and annotated to identify lipid, starch, and catabolic pathways for biofuel production (Rismani-Yazdi et al. 2011). *Botryococcus braunii* produces liquid hydrocarbons from isoprenoids, and its metabolic pathways for producing isoprenoid liquid hydrocarbons and storage metabolites such as triacylglycerol and starch were reconstructed recently based on transcriptomic data (Molnár et al. 2012). Although the expressed genes are analyzed under a specific condition, *de novo* transcriptome analysis can provide an alternative source of genome annotation. Additionally, transcriptomics can support incomplete genome annotation via experimental definition and validation. For example, 174 transcripts in the incomplete genomic data of *C. reinhardtii* were verified by reverse transcription (RT)-PCR and rapid amplification of cDNA ends (RACE) for metabolic reconstruction (Manichaikul et al. 2009). Proteomics data contain information about peptide sequences, which can also support incomplete genome annotation and network gap analysis. To accurately predict growth, a platform for metabolic reconstruction was developed based on proteomic data (Vanee et al. 2014). Metabolomics technologies produce a large compound set, which can provide information about missing metabolic pathways. Metabolome-scale metabolic reconstruction has also been attempted using the chemical transformation patterns of compound–compound pairs in enzymatic reactions with chemical fingerprints (Kotera et al. 2013).

Omics analysis not only suggests the framework for metabolic reconstruction but also explains the changes in intracellular metabolic and regulatory pathways. Integration of omics analysis and *in silico* models enables the design of more efficient microalgae for bioenergy production. Transcriptional profiling tools such as DNA microarrays and NGS determine differential gene expression by quantifying gene expression under different culture conditions. Nitrogen deficiency is an inducible factor for accumulation of storage compounds. To identify gene regulation during a nitrogen deficiency, transcriptomic analysis was performed for *C. reinhardtii, N. gaditana*, and *Chlorella variabilis* (Boyle et al. 2012; Carpinelli et al. 2014; Guarnieri et al. 2011). Proteomic profiles under nitrogen-deficient conditions were also analyzed to evaluate metabolic responses and to identify the targets for enhanced lipid production in *P. tricornutum, Chlorella vulgaris, Chlorella protothe-*

coides, and *Nannochloropsis oceanica* (Dong et al. 2013; Guarnieri et al. 2013; Li et al. 2013; Yang et al. 2014). Furthermore, qualitative and quantitative analysis of metabolomic changes revealed lipid profiles in *C. reinhardtii, N. oceanica*, and *D. tertiolecta* (Courant et al. 2013; Kim et al. 2013; Xiao et al. 2013).

Omics technologies can be applied to systems biology to improve accuracy of *in silico* models. Metabolic reconstruction points to similarities between live cells and an in *silico* model by assigning a confidence score to a GPR association with different data sources, e.g., biochemical data, sequencing data, genomics data, physiology data, and *in silico* modeling (Palsson 2006b). Additionally, metabolic networks can be integrated with massive omics data analysis to explain cellular mechanisms (Hyduke et al. 2013; Saha et al. 2014). Therefore, systems biology and omics technologies are complementary because omics data provide a framework for metabolic reconstruction and improve the accuracy of the *in silico* model, and the metabolic network from genome-scale models can help to analyze a large amount of omics data.

3 Carbon Metabolism of Microalgae

Microalgae are a major focus of attention regarding bioenergy production because they convert light energy into chemical energy. Therefore, it is important to understand carbon metabolism during microalgal bioenergy production. Under photoautotrophic conditions, carbon metabolism is different in microalgae than in other microorganisms. Chloroplasts in microalgae capture and utilize sunlight in the Calvin cycle to convert carbon dioxide, an inorganic carbon source, into glucose, an organic carbon source. Additionally, chloroplasts produce secondary metabolites with various functional groups to protect their photosystems from stressful conditions. Although land crops were considered for bioenergy production, microalgae have several important advantages: they are not edible, which means they are not competing with food crops; they grow faster and have higher lipid content than do other biofuel-producing competitors (Chisti 2007); and they can be cultivated anywhere including in rivers, on land, and in the ocean. Carbon metabolism in microalgae has been analyzed to strengthen these advantages.

Photosynthesis requires two steps: light-dependent reactions and a light-independent reaction. The light-dependent reactions generate biochemical energy (ATP and NADPH) using light energy, and the light independent reaction uses the biochemical energy to fix carbon molecules from CO_2 into 3-carbon (C3) compounds. C3 compounds are then converted to acetyl coenzyme A (CoA) via glycogenesis. Chloroplasts also contain metabolic pathways, such as fatty-acid synthesis and chlorophyll and carotenoid synthesis, because they originated from symbiosis with cyanobacteria. Fatty acids are synthetized from acetyl-CoA in the chloroplast to 16- or 18-carbon chains then released into the cytosol or endoplasmic reticulum (ER). Elongation and desaturation of fatty acids occur in the ER just like in other eukaryotic microorganisms. The enzymes for triacylglycerol (TAG) biosynthesis were identified on the surface of the ER (Radakovits et al. 2010).

Lipid droplets were observed in the chloroplasts of *C. reinhardtii* under nitrogen-deficient conditions. The TAG synthesis pathway in the chloroplast was recently identified by lipid analysis and microscopy (Fan et al. 2011); a single ω-3 fatty acid desaturase (CrFAD7; locus Cre01.g038600), which is the only known ω-3 fatty acid desaturase expressed in *C. reinhardtii*, was isolated from the chloroplast (Nguyen et al. 2013).

Although microalgae have a great potential for bioenergy production, the microalgal bioenergy has many problems that must be solved. Under stressful conditions, microalgae accumulate storage compounds such as carotenoids, lipids, and hydrocarbons (Hu et al. 2008), and they slow down their growth and consume biochemical energy for protection and maintenance of viability (Sharma et al. 2012). For example, *B. braunii* accumulates intracellular hydrocarbons up to 75 % of its dry weight during the observed slow growth (Banerjee et al. 2002). These stress responses decrease production of target biofuel materials. Therefore, a method for enhancement of content of storage metabolites while increasing biomass is needed to improve bioenergy production. One possible approach is inhibition of starch synthesis, which increases the total lipid content in *C. reinhardtii* up to 3.5-fold under severe light and nitrogen deficiency (Li et al. 2010).

Microalgal photosynthesis with sunlight is also inefficient and limits sustainable bioenergy production. The theoretical solar conversion efficiency for photosynthesis is 8–10 %, whereas the efficiency for green algae does not exceed 3 % (Melis 2009). Photosynthetic energy loss can occur through reflection, photochemical inefficiency, the energy gap between the reaction center and carbohydrate metabolism, photorespiration, and respiration (Zhu et al. 2008). Many studies have shown increased microalgal photosynthetic efficiency. *D. tertiolecta* mutants with small chlorophyll antennae exhibit a 2- to 3-fold increase in productivity relative to the wild type (Melis et al. 1998). Ribulose-1,5-bisphosphate carboxylase/oxygenase (RuBisCO), which converts CO_2 and ribulose-1,5-bisphosphate to 3-phosphoglyceric acid (PGA), is the main enzyme in the Calvin cycle. RuBisCO also catalyzes formation of 2-phosphoglycolate and PGA in the photorespiratory pathway. To increase the specificity of RuBisCO for CO_2, directed mutagenesis was performed in cyanobacteria and *C. reinhardtii* (Parry et al. 2003).

Stress factors affect the production cost of bioenergy. In general, environmental stressors such as strong light, salinity, extreme temperatures, and a nutrient deficiency induce storage metabolite production (Hu et al. 2008; Radakovits et al. 2010), which consumes resources that could otherwise be applied to production of the target bioenergy metabolites. Therefore, it is necessary to characterize the global regulatory factors induced under stress conditions. Many researchers have explored stress response mechanisms in microalgae using omics analysis (Rismani-Yazdi et al. 2011; Molnár et al. 2012; Shrestha et al. 2012; Bochenek et al. 2013; Yang et al. 2013). Lei *et al.* assessed the correlation between fatty acid composition and changes in the expression of genes in the fatty acid synthesis pathway of *Haematococcus pluvialis* under various stress conditions such as a nitrogen deficiency, high/low temperature, and salinity (Lei et al. 2012). They found that a high temperature, high salinity, and a nitrogen deficiency affect the fatty acid content and, furthermore, that acyl carrier protein (ACP), 3-ketoacyl-ACP-synthase, and acyl-ACP thioesterase in the fatty acid pathway might be the rate-limiting targets for improvement of biofuel quality and production.

The fatty acid composition affects important fuel properties of biodiesel such as oxidative stability and cold filter plugging point (CFPP), and these properties vary for different microalgal strains (Francisco et al. 2010). Oxidative stability is dependent on the degree of unsaturation, which affects stability during production, usage, and storage. The CFPP predicts viscosity of biodiesel at low temperatures. Biodiesel has a low CFPP and can gel or crystallize in cold weather; this effect is problematic for fuel systems because of fuel atomization or clogging. The CFPP decreases with the increase of the degree of unsaturation, in contrast to oxidative stability. Thus, it is necessary to control the degree of saturation depending on climatic conditions. Systems biology can identify candidate genes that control the fatty acid composition of microalgae. An *et al.* evaluated fatty acid composition and fatty acid desaturases of *Chlamydomonas* sp. from Antarctic environments under different temperatures (An et al. 2013) and showed that the composition of polyunsaturated fatty acids and saturated fatty acids is affected by temperature-dependent regulation of fatty acid desaturases. Additionally, they found that salt stress increases lipid content, whereas low salinity affects fatty acid composition.

A large amount of water is required for cultivation of microalgae; this problem makes harvesting of biomass difficult. Centrifugation, chemical flocculation, filtration, and other methods are used for microalgal biomass recovery (Li et al. 2008). However, these methods have low efficiency and can be costly or labor consuming. From an economic perspective, it is therefore necessary to develop cost-effective, efficient harvesting technologies. Recently ideal harvesting systems were demonstrated in cyanobacteria using genetic engineering. For example, the nickel lysis system, which disrupts the cell wall and releases cellular contents using Ni^{2+}, was introduced in *Synechocystis* sp. 6803 (Liu and Curtiss 2009). A secretion system with an acyl-acyl carrier protein thioesterase that could secrete free fatty acids into the culture medium was also constructed in *Synechocystis* sp. (Liu et al. 2011). A green-light-sensing system using green-light-sensing regulators (CcaS and CcaR) and the promoter for *cpcG2* was constructed in cyanobacteria (Abe et al. 2014). Genetic engineering of eukaryotic microalgae in a lysis or secretion system has not been reported yet, and the development of a cyanobacterial harvesting system for biofuel production will lay a foundation for genetic approaches in eukaryotic microalgae.

4 Engineering the Metabolic Pathways for Biofuel Production in Microalgae

There are 35,000 known species of microalgae, and it is estimated that there are many others (Borowitzka 2012). Microalgae comprise various classes including Chlorophyta (green algae), Rhodophyta (red algae), Bacillariophyceae (diatoms),

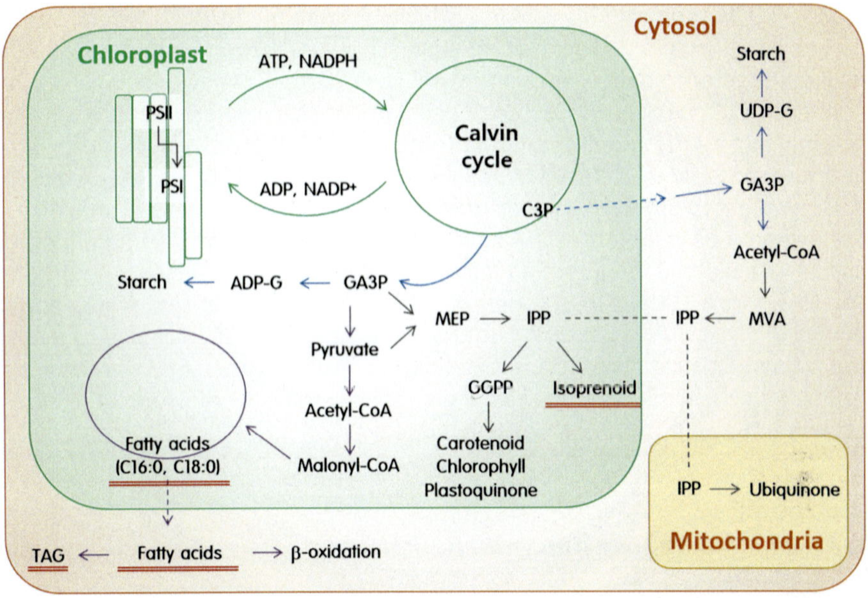

Fig. 2 Biosynthetic pathway of fatty acids and isoprenoids and the β-oxidation pathway in eukaryotic microalgae. The biosynthetic alternatives to gasoline, diesel, and jet fuel are *underlined*. *Dashed arrows* represent transport of metabolites between compartments according to *in silico* analysis. PS: photosystem; GA3P: glyceraldehyde 3-phosphate; C3P: triose phosphate; ADP-G: ADP-glucose pyrophosphorylase; UDP-G: UDP-glucose pyrophosphorylase; TAG: triacylglycerol; IPP: isopentenyl diphosphate; GGPP: geranylgeranyl diphosphate; MEP: methylerythritol 4-phosphate; MVA: mevalonic acid

Dinophyceae (dinoflagellates), and Eustigmatophyceae (yellow-green algae; Guiry 2012). Their diversity extends to the production of unique metabolites. The metabolic pathways and metabolites for biofuels production are shown in Fig. 2. To enhance production of these metabolites, we must understand and control the carbon flux through metabolic networks in microalgae.

4.1 The Fatty Acid Synthesis Pathway

The 16- or 18-carbon fatty acids synthesized in the chloroplast are elongated and unsaturated in the ER and converted into membrane-associated or storage lipids. Fatty acids and starches, which are composed of carbon, hydrogen, and oxygen, are classified as storage metabolites because they store energy for future use. Genetic engineering of microalgae was an attempt to improve the fatty acid content for biofuel production by inhibiting starch production. In 2006, Ramazanov *et al.* generated a starchless mutant of *C. pyrenoidosa* (Ramazanov and Ramazanov 2006) that exhibited an increased lipid content

(25.2–38.0 % dry weight) under nitrogen starvation. In 2010, Work *et al.* described starchless isoamylase mutants (*sta6–10*) that accumulated lipids under nitrogen-deficient conditions (Work et al. 2010). Complementation strains (*sta7–10* [c5]) can accumulate starch in a nutrient-rich medium. Another starchless mutant (BAFJ5) inhibits the small subunit of ADP-glucose pyrophosphorylase relative to the wild-type strain and accumulates lipid content up to 46.4 % of dry weight under nitrogen starvation (Li et al. 2010). Starchless mutants of *Scenedesmus obliquus* also show an 41 % increase in total fatty acid production and no change in biomass production under nitrogen-deficient conditions (de Jaeger et al. 2014). Disruption of UDP-glucose pyrophosphorylase in the diatom *P. tricornutum* causes 45-fold accumulation of TAG compared to controls (Daboussi et al. 2014). These studies demonstrated that carbon partitioning can facilitate generation of improved microalgal strains by blocking or enhancing synthesis pathways for storage metabolites.

4.2 The Isoprenoid Biosynthetic Pathway

Isoprenoids are natural compounds derived from isoprene (2-methyl-1,3-butadiene) that are produced by plants and are used as flavoring agents and pharmaceuticals (Kirby and Keasling 2009). Isoprenoids have attracted attention recently as potential alternatives to gasoline because of the branches and rings in their hydrocarbon chains (Peralta-Yahya et al. 2012). Isoprenoid synthesis in microalgae proceeds through two independent pathways: the mevalonate (MVA) pathway in the cytosol and the 2-*C*-methyl-D-erythritol 4-phosphate (MEP) pathway in chloroplasts (Vranová et al. 2013). The starting metabolite in the MEP pathway, glyceraldehyde-3-phosphate, is derived from the Calvin cycle. In the cytosol, acetyl-CoA, which is also the starting metabolite for fatty acid synthesis, is converted to 3-hydroxy-3-methylglutaryl-CoA (HMG-CoA) for mevalonate production. The isopentenyl diphosphate (IPP) that is synthetized in the chloroplast is involved in the carotenoid, chlorophyll, tocopherol, and plastoquinone synthesis pathways. In the cytosol, IPP participates in the polyprenol, sesquiterpene, and squalene synthesis pathways. The 2 isoprenoid synthesis pathways produce various hydrocarbon products in plants and algae. However, several groups of microalgae such as green algae have lost the MVA pathway (Lohr et al. 2012). Notably, *B. braunii* produces large amounts of long-chain hydrocarbons (Banerjee et al. 2002). The products can be converted to short hydrocarbons by cracking to produce biofuels (Hillen et al. 1982). However, *B. braunii* grows more slowly than other microalgae. Lindberg *et al.* generated fast-growing cyanobacteria able to produce isoprene by means of heterologous expression of the *Pueraria montana* isoprene synthase (Lindberg et al. 2010). This mutant successfully produces isoprene but in small amounts because of low carbon flux to the isoprenoid pathway. Therefore, those authors stated that the photosynthetic carbon flux must be controlled by the metabolic pathway for production of target metabolites.

4.3 Engineering the β-Oxidation Pathway

Microalgae can synthesize lipids and therefore can decrease lipid catabolism. The β-oxidation pathway is a multi-metabolic process that breaks down fatty acids with a lipase to produce energy. In algae, β-oxidation occurs in both mitochondria and peroxisomes (Winkler et al. 1988). Microalgal genomes such as *C. reinhardtii*, *C. variabilis*, and *P. tricornutum* show signs of fatty acid degradation pathways in the metabolic network (KEGG: Kyoto Encyclopedia of Genes and Genomes). Transcriptomic data from *D. tertiolecta* helped to identify all the enzymes involved in β-oxidation (Rismani-Yazdi et al. 2011). Many recent studies were an attempt to increase the lipid/fatty acid content in microorganisms and microalgae. An engineered *Escherichia coli* produces 3- to 4-fold more free fatty acids (~1.2 g/L) after elimination of 2 enzymes involved in β-oxidation (Steen et al. 2010). However, deletion of genes in β-oxidation in the yeast *Saccharomyces cerevisiae* does not enhance the lipid content (Runguphan and Keasling 2014). Trentacoste et al. (2013) reported that the lipid content in the diatom *Thalassiosira pseudonana* increases upon blockage of lipase-, phospholipase-, or acyltransferase-mediated degradation of fatty acids. Furthermore, *T. pseudonana* with gene deletions exhibits growth similar to the wild type, indicating that genetic engineering of fatty acid degradation can be used to improve lipid content in the diatoms. However, blocking the β-oxidation pathway can have a deleterious effect on cell growth and maintenance. Lipids are energy-rich storage compounds, which are used as a substrate at night. Inhibition of β-oxidation can reduce cell growth or cause cell death due to the loss of maintenance energy. Therefore, this type of genetic manipulation is not suitable for an open-pond or ocean cultivation system without continuous light.

5 Conclusion

Integration of systems biology and omics analysis has helped to improve production of lipids in microalgae. Metabolic reconstruction can be employed to generate improved strains, understand cellular networks, and optimize metabolic pathways. Advances in omics technologies should contribute to current and future successful practical applications by improving the accuracy of *in silico* models, which increase the understanding of photosynthetic carbon flux and form the basis for design, characterization, and optimization of novel pathways for biofuel production. Additionally, integration with bioprocesses can lead to optimization of culture conditions and improvement in target metabolite productivity in large-scale production. For example, *Chlorella vulgaris* has been shown to exhibit a biomass productivity of 2.11 g DCW/L/day, by optimizing light conditions in an *in silico* model of *C. reinhardtii* (Fu et al. 2012).

Various methods for microalgal cultivation and biofuel extraction are being rapidly developed. The biofuel cell factory platform involving systems biology is likely to accelerate metabolic optimization, thereby helping researchers to keep up with the developments in biofuel production processes. The development of *in silico* models based on systems biology should ensure the continued increase in biomass and production of fuel precursors from microalgal feedstock. Eventually, the development of microalgal systems biology is expected to reduce the production cost and allow biofuels to enter the market.

List of Abbreviations

ACP	Acyl carrier protein
BOF	Biomass object function
C3	Three-carbon
CFPP	Cold filter plugging point
CoA	Coenzyme A
ER	Endoplasmic reticulum
FBA	Flux balance analysis
GPR	Gene-protein-reaction
HMG-CoA	3-hydroxy-3-methylglutaryl-CoA
IPP	Isopentenyl diphosphate
MEP	2-C-methyl-D-erythritol 4-phosphate
MVA	Mevalonate
PGA	3-phosphoglyceric acid
PhPP	Phenotype phase plane analysis
RACE	Rapid amplification of cDNA ends
RubisCO	Ribulose-1,5-bisphosphate carboxylase/oxygenase
TAG	Triacylglycerol

References

Abe K, Miyake K, Nakamura M, Kojima K, Ferri S, Ikebukuro K, Sode K (2014) Engineering of a green-light inducible gene expression system in *Synechocystis* sp. PCC6803. Microb Biotechnol 7(2):177–183

Altuntaş E, Schubert US (2014) "Polymeromics": mass spectrometry based strategies in polymer science toward complete sequencing approaches: a review. Anal Chim Acta 808:56–69

An M, Mou S, Zhang X, Zheng Z, Ye N, Wang D, Zhang W, Miao J (2013) Expression of fatty acid desaturase genes and fatty acid accumulation in *Chlamydomonas* sp. ICE-L under salt stress. Bioresour Technol 149:77–83

Banerjee A, Sharma R, Chisti Y, Banerjee U (2002) *Botryococcus braunii*: a renewable source of hydrocarbons and other chemicals. Crit Rev biotech 22(3):245–279

Battchikova N, Vainonen JP, Vorontsova N, Keränen M, Carmel D, Aro E-M (2010) Dynamic changes in the proteome of *Synechocystis* 6803 in response to CO_2 limitation revealed by quantitative proteomics. J Proteome Res 9 (11):5896–5912

Blanc G, Agarkova I, Grimwood J, Kuo A, Brueggeman A, Dunigan DD, Gurnon J, Ladunga I, Lindquist E, Lucas S (2012) The genome of the polar eukaryotic microalga *Coccomyxa subellipsoidea* reveals traits of cold adaptation. Genome Biol 13(5):R39

Bochenek M, Etherington GJ, Koprivova A, Mugford ST, Bell TG, Malin G, Kopriva S (2013) Transcriptome analysis of the sulfate deficiency response in the marine microalga *Emiliania huxleyi*. New Phytol 199(3):650–662

Bogen C, Klassen V, Wichmann J, Russa ML, Doebbe A, Grundmann M, Uronen P, Kruse O, Mussgnug JH (2013) Identification of *Monoraphidium contortum* as a promising species for liquid biofuel production. Bioresour Technol 133:622–626

Borowitzka MA (2012) Species and strain selection. In: Borowitzka MA, Moheimani NR (eds) Algae for biofuels and energy, vol 5. Springer Netherlands, pp 77–89

Boyle NR, Morgan JA (2009) Flux balance analysis of primary metabolism in *Chlamydomonas reinhardtii*. BMC Syst Biol 3(1):4

Boyle NR, Page MD, Liu B, Blaby IK, Casero D, Kropat J, Cokus SJ, Hong-Hermesdorf A, Shaw J, Karpowicz SJ (2012) Three acyltransferases and nitrogen-responsive regulator are implicated in nitrogen starvation-induced triacylglycerol accumulation in *Chlamydomonas*. J Biol Chem 287(19):15811–15825

Burgard AP, Pharkya P, Maranas CD (2003) Optknock: a bilevel programming framework for identifying gene knockout strategies for microbial strain optimization. Biotechnol Bioeng 84(6):647–657

Carpinelli EC, Telatin A, Vitulo N, Forcato C, D'Angelo M, Schiavon R, Vezzi A, Giacometti GM, Morosinotto T, Valle G (2014) Chromosome scale genome assembly and transcriptome profiling of *Nannochloropsis gaditana* in nitrogen depletion. Mol Plant 7(2):323–335

Chang RL, Ghamsari L, Manichaikul A, Hom EF, Balaji S, Fu W, Shen Y, Hao T, Palsson BØ, Salehi-Ashtiani K (2011) Metabolic network reconstruction of *Chlamydomonas* offers insight into light-driven algal metabolism. Mol Syst Biol 7(1):518

Chisti Y (2007) Biodiesel from microalgae. Biotechnol Adv 25(3):294–306

Courant F, Martzolff A, Rabin G, Antignac J-P, Le Bizec B, Giraudeau P, Tea I, Akoka S, Couzinet A, Cogne G (2013) How metabolomics can contribute to bio-processes: a proof of concept study for biomarkers discovery in the context of nitrogen-starved microalgae grown in photobioreactors. Metabolomics 9(6):1286–1300

Daboussi F, Leduc S, Maréchal A, Dubois G, Guyot V, Perez-Michaut C, Amato A, Falciatore A, Juillerat A, Beurdeley M, Voytas DF, Cavarec L, Duchateau P (2014) Genome engineering empowers the diatom *Phaeodactylum tricornutum* for biotechnology. Nat Commun 5. doi:10.1038/ncomms4831

de Jaeger L, Verbeek RE, Draaisma RB, Martens DE, Springer J, Eggink G, Wijffels RH (2014) Superior triacylglycerol (TAG) accumulation in starchless mutants of *Scenedesmus obliquus*:(I) mutant generation and characterization. Biotechnol Biofuels 7:69

Derelle E, Ferraz C, Rombauts S, Rouzé P, Worden AZ, Robbens S, Partensky F, Degroeve S, Echeynié S, Cooke R (2006) Genome analysis of the smallest free-living eukaryote *Ostreococcus tauri* unveils many unique features. Proc Natl Acad Sci U S A 103(31):11647–11652

Dong H-P, Williams E, D-z W, Xie Z-X, R-c H, Jenck A, Halden R, Chen F, Place A (2013) Responses of *Nannochloropsis oceanica* IMET1 to long-term nitrogen starvation and recovery. Plant Physiol 113:214–320

Edwards JS, Palsson BO (2000) Robustness analysis of the *Escherichiacoli* metabolic network. Biotechnol Prog 16(6):927–939

Edwards JS, Ibarra RU, Palsson BO (2001) *In silico* predictions of *Escherichia coli* metabolic capabilities are consistent with experimental data. Nat Biotechnol 19(2):125–130

Edwards JS, Ramakrishna R, Palsson BO (2002) Characterizing the metabolic phenotype: a phenotype phase plane analysis. Biotechnol Bioeng 77(1):27–36

Fabris M, Matthijs M, Rombauts S, Vyverman W, Goossens A, Baart GJ (2012) The metabolic blueprint of *Phaeodactylum tricornutum* reveals a eukaryotic Entner–Doudoroff glycolytic pathway. Plant J 70(6):1004–1014

Fan J, Andre C, Xu C (2011) A chloroplast pathway for the *de novo* biosynthesis of triacylglycerol in *Chlamydomonas reinhardtii*. FEBS Lett 585(12):1985–1991

Feist AM, Palsson BO (2010) The biomass objective function. Curr Opin Microbiol 13(3):344–349

Francisco EC, Neves DB, Jacob-Lopes E, Franco TT (2010) Microalgae as feedstock for biodiesel production: carbon dioxide sequestration, lipid production and biofuel quality. J Chem Technol Biotech 85(3):395–403

Fu W, Gudmundsson O, Feist AM, Herjolfsson G, Brynjolfsson S, Palsson BØ (2012) Maximizing biomass productivity and cell density of *Chlorella vulgaris* by using light-emitting diode-based photobioreactor. J Biotechnol 161(3):242–249

Guarnieri MT, Nag A, Smolinski SL, Darzins A, Seibert M, Pienkos PT (2011) Examination of triacylglycerol biosynthetic pathways via de novo transcriptomic and proteomic analyses in an unsequenced microalga. PLoS One 6(10):e25851

Guarnieri MT, Nag A, Yang S, Pienkos PT (2013) Proteomic analysis of *Chlorella vulgaris*: potential targets for enhanced lipid accumulation. J Proteomics 93:245–253

Guiry MD (2012) How many species of algae are there? J Phycol 48(5):1057–1063

Han X, Gross RW (2003) Global analyses of cellular lipidomes directly from crude extracts of biological samples by ESI mass spectrometry a bridge to lipidomics. J Lipid Res 44(6):1071–1079

Hillen L, Pollard G, Wake L, White N (1982) Hydrocracking of the oils of *Botryococcus braunii* to transport fuels. Biotechnol Bioeng 24(1):193–205

Hong S-J, Lee C-G (2007) Evaluation of central metabolism based on a genomic database of *Synechocystis* PCC6803. Biotechnol Bioprocess Eng 12(2):165–173

Hu Q, Sommerfeld M, Jarvis E, Ghirardi M, Posewitz M, Seibert M, Darzins A (2008) Microalgal triacylglycerols as feedstocks for biofuel production: perspectives and advances. Plant J 54(4):621–639

Hyduke DR, Lewis NE, Palsson BØ (2013) Analysis of omics data with genome-scale models of metabolism. Mol BioSyst 9(2):167–174

KEGG: Kyoto Encyclopedia of genes and genomes http://www.genome.jp/kegg/

Kim S-H, Liu K-H, Lee S-Y, Hong S-J, Cho B-K, Lee H, Lee C-G, Choi H-K (2013) Effects of light intensity and nitrogen starvation on glycerolipid, glycerophospholipid, and carotenoid composition in *Dunaliella tertiolecta* culture. PLoS One 8(9):e72415

Kirby J, Keasling JD (2009) Biosynthesis of plant isoprenoids: perspectives for microbial engineering. Annu Rev Plant Biol 60:335–355

Kitano H (2002) Systems biology: a brief overview. Science 295(5560):1662–1664

Kotera M, Tabei Y, Yamanishi Y, Tokimatsu T, Goto S (2013) Supervised de novo reconstruction of metabolic pathways from metabolome-scale compound sets. Bioinformatics 29(13):i135–i144

Lee JM, Gianchandani EP, Papin JA (2006) Flux balance analysis in the era of metabolomics. Brief Bioinform 7(2):140–150

Lei A, Chen H, Shen G, Hu Z, Chen L, Wang J (2012) Expression of fatty acid synthesis genes and fatty acid accumulation in *Haematococcus pluvialis* under different stressors. Biotechnol Biofuels 5(1):1–11

Li Y, Horsman M, Wu N, Lan CQ, Dubois-Calero N (2008) Biofuels from microalgae. Biotechnol Prog 24(4):815–820

Li Y, Han D, Hu G, Sommerfeld M, Hu Q (2010) Inhibition of starch synthesis results in overproduction of lipids in *Chlamydomonas reinhardtii*. Biotechnol Bioeng 107(2):258–268

Li Y, Yuan Z, Mu J, Chen D, Feng B (2013) Proteomic analysis of lipid accumulation in *Chlorella protothecoides* cells by heterotrophic N deprivation coupling cultivation. Energy Fuel 27(7):4031–4040

Lindberg P, Park S, Melis A (2010) Engineering a platform for photosynthetic isoprene production in cyanobacteria, using *Synechocystis* as the model organism. Metab Eng 12(1):70–79

Liu X, Curtiss R (2009) Nickel-inducible lysis system in *Synechocystis* sp. PCC 6803. Proc Natl Acad Sci U S A 106(51):21550–21554

Liu X, Sheng J, Curtiss R III (2011) Fatty acid production in genetically modified cyanobacteria. Proc Natl Acad Sci U S A 108(17):6899–6904

Lohr M, Schwender J, Polle JEW (2012) Isoprenoid biosynthesis in eukaryotic phototrophs: a spotlight on algae. Plant Sci 185–186:9–22

Lun DS, Rockwell G, Guido NJ, Baym M, Kelner JA, Berger B, Galagan JE, Church GM (2009) Large-scale identification of genetic design strategies using local search. Mol Syst Biol 5(1):296

Manichaikul A, Ghamsari L, Hom EF, Lin C, Murray RR, Chang RL, Balaji S, Hao T, Shen Y, Chavali AK (2009) Metabolic network analysis integrated with transcript verification for sequenced genomes. Nat Methods 6(8):589

Melis A (2009) Solar energy conversion efficiencies in photosynthesis: minimizing the chlorophyll antennae to maximize efficiency. Plant Sci 177(4):272–280

Melis A, Neidhardt J, Benemann JR (1998) *Dunaliella salina* (Chlorophyta) with small chlorophyll antenna sizes exhibit higher photosynthetic productivities and photon use efficiencies than normally pigmented cells. J Appl Phycol 10(6):515–525

Merchant SS, Prochnik SE, Vallon O, Harris EH, Karpowicz SJ, Witman GB, Terry A, Salamov A, Fritz-Laylin LK, Maréchal-Drouard L (2007) The *Chlamydomonas* genome reveals the evolution of key animal and plant functions. Science 318(5848):245–250

Molnár I, Lopez D, Wisecaver JH, Devarenne TP, Weiss TL, Pellegrini M, Hackett JD (2012) Bio-crude transcriptomics: gene discovery and metabolic network reconstruction for the biosynthesis of the terpenome of the hydrocarbon oil-producing green alga, *Botryococcus braunii* race B (Showa)*. BMC Genomics 13(1):576

Montagud A, Zelezniak A, Navarro E, de Córdoba PF, Urchueguía JF, Patil KR (2011) Flux coupling and transcriptional regulation within the metabolic network of the photosynthetic bacterium *Synechocystis* sp. PCC6803. Biotechnol J 6(3):330–342

Muthuraj M, Palabhanvi B, Misra S, Kumar V, Sivalingavasu K, Das D (2013) Flux balance analysis of *Chlorella* sp. FC2 IITG under photoautotrophic and heterotrophic growth conditions. Photosynth Res 118(1–2):167–179

Nguyen HM, Cuiné S, Beyly-Adriano A, Légeret B, Billon E, Auroy P, Beisson F, Peltier G, Li-Beisson Y (2013) The green microalga *Chlamydomonas reinhardtii* has a single ω-3 fatty acid desaturase that localizes to the chloroplast and impacts both plastidic and extraplastidic membrane lipids. Plant Physiol 163(2):914–928

Nogales J, Gudmundsson S, Knight EM, Palsson BO, Thiele I (2012) Detailing the optimality of photosynthesis in cyanobacteria through systems biology analysis. Proc Natl Acad Sci U S A 109(7):2678–2683

Palenik B, Grimwood J, Aerts A, Rouzé P, Salamov A, Putnam N, Dupont C, Jorgensen R, Derelle E, Rombauts S (2007) The tiny eukaryote *Ostreococcus* provides genomic insights into the paradox of plankton speciation. Proc Natl Acad Sci U S A 104(18):7705–7710

Palsson BO (2006a) Introduction. In: Palsson BO (ed) Systems biology. Cambridge University Press, New York

Palsson BO (2006b) Metabolic networks. In: Palsson BO (ed) Systems biology. Cambridge University Press, New York

Parry M, Andralojc P, Mitchell RA, Madgwick P, Keys A (2003) Manipulation of Rubisco: the amount, activity, function and regulation. J Exp Bot 54(386):1321–1333

Peralta-Yahya PP, Zhang F, Del Cardayre SB, Keasling JD (2012) Microbial engineering for the production of advanced biofuels. Nature 488(7411):320–328

Radakovits R, Jinkerson RE, Darzins A, Posewitz MC (2010) Genetic engineering of algae for enhanced biofuel production. Eukaryot Cell 9(4):486–501

Radakovits R, Jinkerson RE, Fuerstenberg SI, Tae H, Settlage RE, Boore JL, Posewitz MC (2012) Draft genome sequence and genetic transformation of the oleaginous alga *Nannochloropsis gaditana*. Nat Commun 3:686

Ramazanov A, Ramazanov Z (2006) Isolation and characterization of a starchless mutant of *Chlorella pyrenoidosa* STL-PI with a high growth rate, and high protein and polyunsaturated fatty acid content. Phycol Res 54(4):255–259

Rismani-Yazdi H, Haznedaroglu BZ, Bibby K, Peccia J (2011) Transcriptome sequencing and annotation of the microalgae *Dunaliella tertiolecta*: pathway description and gene discovery for production of next-generation biofuels. BMC Genomics 12(1):148

Rocha I, Maia P, Evangelista P, Vilaça P, Soares S, Pinto JP, Nielsen J, Patil KR, Ferreira EC, Rocha M (2010) OptFlux: an open-source software platform for in silico metabolic engineering. BMC Syst Biol 4(1):45

Runguphan W, Keasling JD (2014) Metabolic engineering of *Saccharomyces cerevisiae* for production of fatty acid-derived biofuels and chemicals. Metab Eng 21:103–113

Saha R, Chowdhury A, Maranas CD (2014) Recent advances in the reconstruction of metabolic models and integration of omics data. Curr Opin Biotechnol 29:39–45

Schellenberger J, Que R, Fleming RMT, Thiele I, Orth JD, Feist AM, Zielinski DC, Bordbar A, Lewis NE, Rahmanian S, Kang J, Hyduke DR, Palsson BO (2011) Quantitative prediction of cellular metabolism with constraint-based models: the COBRA Toolbox v2.0. Nat Protoc 6(9):1290–1307

Sengupta T, Bhushan M, Wangikar PP (2013) Metabolic modeling for multi-objective optimization of ethanol production in a *Synechocystis* mutant. Photosynth Res 118(1–2):155–165

Sharma KK, Schuhmann H, Schenk PM (2012) High lipid induction in microalgae for biodiesel production. Energies 5(5):1532–1553

Shastri AA, Morgan JA (2005) Flux balance analysis of photoautotrophic metabolism. Biotechnol Prog 21(6):1617–1626

Shrestha RP, Tesson B, Norden-Krichmar T, Federowicz S, Hildebrand M, Allen AE (2012) Whole transcriptome analysis of the silicon response of the diatom *Thalassiosira pseudonana*. BMC Genomics 13(1):499

Steen EJ, Kang Y, Bokinsky G, Hu Z, Schirmer A, McClure A, Del Cardayre SB, Keasling JD (2010) Microbial production of fatty-acid-derived fuels and chemicals from plant biomass. Nature 463(7280):559–562

Stephanopoulos G (1999) Metabolic fluxes and metabolic engineering. Metab Eng 1(1):1–11

Stitt M (2013) Systems-integration of plant metabolism: means, motive and opportunity. Curr Opin Plant Biol 16(3):381–388

Thiele I, Palsson BO (2010) A protocol for generating a high-quality genome-scale metabolic reconstruction. Nat Protoc 5(1):93–121

Trentacoste EM, Shrestha RP, Smith SR, Glé C, Hartmann AC, Hildebrand M, Gerwick WH (2013) Metabolic engineering of lipid catabolism increases microalgal lipid accumulation without compromising growth. Proc Natl Acad Sci U S A 110(49):19748–19753

Vanee N, Brooks JP, Spicer V, Shamshurin D, Krokhin O, Wilkins JA, Deng Y, Fong SS (2014) Proteomics-based metabolic modeling and metabolic characterization of the cellulolytic bacterium *Thermobifida fusca*. BMC Syst Biol 8(1):86

Varma A, Palsson BO (1993a) Metabolic capabilities of *Escherichia coli*: I. Synthesis of biosynthetic precursors and cofactors. J Theor Biol 165(4):477–502

Varma A, Palsson BO (1993b) Metabolic capabilities of *Escherichia coli*: II. Optimal growth patterns. J Theor Biol 165(4):503–522

Vranová E, Coman D, Gruissem W (2013) Network analysis of the MVA and MEP pathways for isoprenoid synthesis. Annu Rev Plant Biol 64:665–700

Winkler U, Säftel W, Stabenau H (1988) β-Oxidation of fatty acids in algae: localization of thiolase and acyl-CoA oxidizing enzymes in three different organisms. Planta 175(1):91–98

Winter G, Krömer JO (2013) Fluxomics–connecting ‘omics analysis and phenotypes. Environ Microbiol 15(7):1901–1916

Worden AZ, Lee J-H, Mock T, Rouzé P, Simmons MP, Aerts AL, Allen AE, Cuvelier ML, Derelle E, Everett MV (2009) Green evolution and dynamic adaptations revealed by genomes of the marine picoeukaryotes *Micromonas*. Science 324(5924):268–272

Work VH, Radakovits R, Jinkerson RE, Meuser JE, Elliott LG, Vinyard DJ, Laurens LM, Dismukes GC, Posewitz MC (2010) Increased lipid accumulation in the *Chlamydomonas reinhardtii sta7-10* starchless isoamylase mutant and increased carbohydrate synthesis in complemented strains. Eukaryot Cell 9(8):1251–1261

Xiao Y, Zhang J, Cui J, Feng Y, Cui Q (2013) Metabolic profiles of *Nannochloropsis oceanica* IMET1 under nitrogen-deficiency stress. Bioresour Technol 130:731–738

Yang C, Hua Q, Shimizu K (2000) Energetics and carbon metabolism during growth of microalgal cells under photoautotrophic, mixotrophic and cyclic light-autotrophic/dark-heterotrophic conditions. Biochem Eng J 6(2):87–102

Yang Z-K, Niu Y-F, Ma Y-H, Xue J, Zhang M-H, Yang W-D, Liu J-S, Lu S-H, Guan Y, Li H-Y (2013) Molecular and cellular mechanisms of neutral lipid accumulation in diatom following nitrogen deprivation. Biotechnol Biofuels 6(67):1–67

Yang Z-K, Ma Y-H, Zheng J-W, Yang W-D, Liu J-S, Li H-Y (2014) Proteomics to reveal metabolic network shifts towards lipid accumulation following nitrogen deprivation in the diatom *Phaeodactylum tricornutum*. J Appl Phycol 26(1):73–82

Zhu X-G, Long SP, Ort DR (2008) What is the maximum efficiency with which photosynthesis can convert solar energy into biomass? Curr Opin Biotechnol 19(2):153–159

Government Regulation of the Uses of Genetically Modified Algae and Other Microorganisms in Biofuel and Bio-based Chemical Production

David J. Glass

Abstract Recent years have seen an increased interest in developing genetically modified algae and other microorganisms for use in biofuel and bio-based chemical production. However, this comes at a time when there is uncertainty within the industry and the academic community about how such uses will be regulated by governments in the U.S. and elsewhere in the world, as well as concerns by some observers over the adequacy of existing regulations to cover organisms created using techniques known as synthetic biology. However, a reasonable road map is emerging of a regulatory regime that can allow pilot, demonstration and commercial stage uses of modified microorganisms. In the U.S., regulations of the U.S. Environmental Protection Agency and possibly of the U.S. Department of Agriculture might govern the industrial use of microorganisms in contained photobioreactors or algae in open ponds, and these regulations generally require conducting assessments of the potential environmental risks of such large-scale uses. The EPA regulations include a mechanism by which outdoor experimentation of modified microorganisms can take place in a stepwise approach, with risks assessed as the scale of experimentation increases, which provides an accessible path to exploration of the use of modified algae in open ponds. Such risk assessments will address legitimate questions of potential ecological impact, such as the potential survival and dissemination of the production organism, the potential for heterologous genes to horizontally transfer to indigenous microorganisms, and the chance for other unintended effects on nontarget species. Numerous companies have successfully navigated these regulations, including some recent project approvals in the U.S. and elsewhere in the world.

Keywords Genetic modification • Genetically modified organism • Government regulation • Environmental impact • Risk assessment • Biofuel • Bio-based chemical

---

D.J. Glass (✉)
D. Glass Associates, Inc., 124 Bird Street, Needham, MA 02492, USA
e-mail: dglass@dglassassociates.com

A. Prokop et al. (eds.), *Algal Biorefineries*, DOI 10.1007/978-3-319-20200-6_2

1 Introduction

Genetically modified microorganisms, including microalgae and cyanobacteria, are increasingly being developed for the production of renewable fuels or bio-based chemicals. The development of biological methods of manufacturing commodity products currently made from petrochemical feedstocks promises to make an important contribution to the reduction of global carbon emissions and the movement to more sustainable industrial activities. Microbiological methods have long been used for the production of ethanol or other industrial chemicals, but the proposed use of genetically modified organisms offers potentially significant advantages over traditional methods, such as improved productivity, decreased operational costs, the ability to use a more diverse range of feedstocks, and possibly more favorable carbon footprints.

In the U.S. and most other countries around the world, manufacturing processes involving genetically modified microorganisms (GMMs) would likely trigger additional regulatory scrutiny before manufacturing could begin and products could be sold. This chapter will review the regulations that are applicable to fuel and chemical production using genetically modified algae, cyanobacteria and other microorganisms in the United States and elsewhere in the world, and which would also apply to organisms created for these purposes using synthetic biology. The chapter will also discuss the scientific concerns that have led to the imposition of these regulations, and the issues underlying the risk assessments associated with such government oversight. With proper planning and management, it should be relatively straightforward for most applicants to obtain the approvals needed to proceed with R&D or commercial use of genetically modified microorganisms in industrial biotechnology, as will be demonstrated by discussions of several cases where regulatory approvals for uses of genetically modified microorganisms have been obtained in the U.S. and elsewhere.

2 Potential Commercial Uses and Environmental Impacts of Genetically Modified Microorganisms

2.1 Strategies for Genetic Modification of Microorganisms

Much of today's commercial activity using advanced biotechnology for biofuel or bio-based chemical production focuses on the creation, selection or improvement of strains of desired microorganisms having enhanced properties for functions important for the production process. Microbiological methods for producing ethanol, fuels or other chemicals generally rely on the use of one or more selected microbial strains to catalyze biosynthesis of the desired compound, generally through a traditional fermentation process. Historically, these methods have made use of naturally-occurring or classically selected microorganisms, but in recent years the power of the new biotechnologies to develop enhanced strains is being investigated or used by numerous companies.

Much of the industrial interest in the use of microbial processes has been for the production of commodity products used as automotive fuels, primarily ethanol and diesel substitutes such as biodiesel, with some processes using modified microorganisms already in commercial use. Processes are also under development for microbial manufacture of other fuels or fuel additives, including n-butanol, isobutanol, and mixtures of alkanes or lipids that can be drop-in replacements for diesel, gasoline, or jet fuel (USDOE 2013). Biological methods are also under development or in commercial use for a variety of chemical compounds of many different uses and industrial applications. Examples include succinic acid, butadiene and its downstream products, isobutene, propanediol, and the monomeric units for numerous bioplastics such as polylactic acid, polypropylene and others (Golden and Handfield 2014).

Most commercial activity today is focused on the use of heterotrophic microorganisms, both prokaryotic and eukaryotic, and including a number of species and strains that have been used for decades in industrial production, as well as other strains not previously utilized commercially. Historically, this has involved species such as *Saccharomyces cerevisiae* and other yeasts; fungal species such as *Aspergillus* and *Trichoderma* (especially for the production of industrial or food enzymes), and bacterial species such as *Bacillus subtilis, Bacillus licheniformis, Bacillus amyloliquefaciens, Clostridium acetobutylicum*, species of *Corynebacterium* and *Lactobacillus*, and others (Adrio and Demain 2010). More recently, with the advent of genetic engineering, various strains of *Escherichia coli* and other commonly-used laboratory species have been considered for industrial use, and there has more recently been interest in utilizing less-common strains with interesting or valuable properties, such as various thermophilic microorganisms and the radiation-resistant *Deinococcus* species.

Recent years have seen an explosion in academic and industry activity in developing strategies for the use of advanced biotechnology techniques to improve such microorganisms for the production of fuels or chemicals. This is evidenced by the large and growing number of review articles that have appeared just in the last few years summarizing these strategies. For example, recent reviews of advanced biotechnology strategies for improved biofuel production include Colin et al. (2011), de Jong et al. (2012), Dellomonaco et al. (2010), Jang et al. (2012), Kung et al. (2012), Lennen and Pfleger (2012, 2013), Peralta-Yahya and Keasling (2010), and Zhang et al. (2011). Recent reviews of strategies focused on improving bio-based chemical production include He et al. (2014) (*Zymomonas* as production organism), Hong and Nielsen (2012) and Nielsen et al. (2013) (*S. cerevisiae* as production organism), Cao et al. (2011), Yu et al. (2011), and Chen et al. (2013) (*E. coli* as production organism), as well as Cao et al. (2013), summarizing methods for production of succinic acid, and more general reviews from Adrio and Demain (2010), Chen and Nielsen (2013) and Buschke et al. (2013).

The different strategies being pursued for the improvement of microorganisms for industrial purposes have been exhaustively summarized in these review articles. A brief summary is presented in Table 1. These include strategies to enhance the productivity of existing biosynthetic pathways, knocking out competing pathways

Table 1 Genetic engineering strategies for microorganisms

Overexpress key endogenous enzymes to increase yield of desired product
Modify the properties of key endogenous enzymes (e.g. directed evolution, codon optimization) to increase productivity of desired biosynthetic pathways
Introduce new enzymatic activities to enable use of different feedstocks or compounds as energy sources for the production organism
Introduce two or more genes encoding heterologous enzymes, to create entirely new biosynthetic pathways for desired products
Improve carbon flow into a desired pathway by knocking out genes encoding enzymes in competing pathways

to improve carbon flow into a desired pathway, and creation of entirely new pathways by introducing genes encoding multiple enzymes. The newer techniques of synthetic biology are often useful in implementing any of these strategies.

In spite of the commercial focus to date on heterotrophic microorganisms, photosynthetic organisms such as microalgae and cyanobacteria have also been used for commercial purposes (USDOE 2010). Species such as *Chlamydomonas, Chlorella, Haematococcus, Nannochloropsis, Dunaliela, Botryococcus, Scenedesmus* and others have been used for the production of industrially-useful compounds (Rosenberg et al. 2008; Larkum et al. 2012; USDOE 2010; Trentacoste et al. 2014). Because the processes required to grow algae, harvest the organisms and purify the product tend to be rather expensive, algal production has historically been mostly limited to specialty chemical or pharmaceutical products, characterized by low volumes and high profit margins, or products like nutritional supplements that don't require as much downstream processing as do specialty chemicals.

Similar genetic modifications are being considered for industrially-useful strains of microalgae to enable their use to produce commodity fuels and chemicals, especially to improve productivity or efficiency to overcome economic and other factors that have hindered development of the technology (USDOE 2010). Here too, progress in the field is attested by the significant number of recently-published review articles summarizing applicable advanced biotechnology strategies for strain improvement (Rosenberg et al. 2008; Radakovits et al. 2010; Jones and Mayfield 2012; Larkum et al. 2012; Rosgaard et al. 2012; Work et al. 2012; Nozzi et al. 2013; Henley et al. 2013; Enzing and Nooijen 2012). Some of the approaches being contemplated are shown in Table 2.

2.2 *Potential Environmental Impacts of Genetically Modified Microorganisms*

As the biotechnology industry grew, government regulatory frameworks developed in order to ensure the safe conduct of larger-scale industrial uses of genetically modified organisms. Early concerns about the technology focused on the potential public health impact of the creation of so-called novel life forms, but because almost

Table 2 Genetic engineering strategies for algae

Enhance photosynthesis; improve carbon fixation
Enhance pathway proteins like RuBisCO
Introduce new carbon fixation pathways
Enhance or alter lipid biosynthesis for improved diesel, jet fuel production
Enable secretion of lipids to improve harvesting and separation
Express, enhance transporter proteins
Alter cell wall composition for easier cell lysis
Metabolic engineering to enhance existing pathways
Maximize carbon flow to desired product(s)
Eliminate competing pathways
Remove toxic, harmful compounds
Introduce new pathways for desired products
Ethanol
Butanol
Improved production of hydrogen for fuel use

all industrial applications of modified microorganisms utilized well-studied species and strains known to be nonpathogenic, such concerns tended to be less prominent in the regulation of commercial activities. For larger-scale industrial uses (and agricultural applications) the emphasis of government regulation shifted to the need to assess potential environmental impacts, and to the development of appropriate risk assessment methods in support of such regulatory programs.

Microorganisms need to be grown in large scale in order to have industrial utility, especially for the production of commodity products like biofuels. For the most part, this will involve the hardware and processes that have usually been used for those native microbial species that have been exploited commercially, which might typically involve traditional industrial fermentations. These would be conducted under familiar conditions that have long been used commercially and which would be expected to afford some protection against exposure or accidental release of the microorganism. However, industrially-useful algae strains (and to some extent cyanobacteria) have traditionally been grown in open-pond reactors, where the algal cultures are exposed to the environment (USDOE 2010; Enzing and Nooijen 2012). The use of such reactors for genetically modified algae would pose much different issues for regulators conducting a risk assessment because of the inherent exposure of the production organism to the environment.

Consideration of the potential environmental risks of genetically engineered microorganisms began in earnest as the agricultural biotechnology industry was developing in the mid-1980s, with early publications such as Alexander (1985) setting forth the factors to be considered in environmental risk assessments. This concern led to more formal inquiries assessing environmental concerns in a general way, including a pioneering effort by a group of prominent ecologists and other scientists convened by the National Academy of Sciences (Tiedje et al. 1989), which generally concluded that modified microorganisms would behave in the environment similarly to nonengineered strains introduced into new environments, and that such behavior could be predicted and monitored using appropriate risk assessment tools. These early scientific reviews provided the framework for some of the earliest regulatory risk assessments of proposed field tests and other uses of GMMs, under which a number of modified microorganisms and plants were used in field experimentation beginning in the mid to late 1980s. No evidence of environmental harm was seen in any of the field tests of modified microorganisms conducted during those years (see Glass 1995 for an early review of some of these results, and Viebahn et al. 2009 for a more recent review).

As shown in Table 3, there are legitimate scientific concerns about the potential environmental effects of microorganisms having novel traits. Among the issues identified as important for a risk assessment are (a) the toxicity, infectivity or other risks inherent to the GMM itself or that might have been introduced by the genetic modifications; (b) the ability of the GMM to persist or become established in the environment; (c) the ability of the GMM to compete with or displace natural microflora at the release site; (d) the possibility that the GMM could spread or be dispersed from the release site; and (e) the possibility that genes introduced into the GMM could themselves spread through horizontal gene transfer to be taken up by and expressed in different microbial species. Other potential risks are unique to algae or cyanobacteria, such as concerns over possible impacts on native algae populations or the potential to create or exacerbate harmful algal blooms. Some of these concerns are not unique to engineered organisms, and many observers would have similar concerns about large-scale industrial uses or releases of any novel organism, whether recombinant or not (for example, see Gressel et al. 2013). Although some observers within environmental groups and the general public fear that engineered microorganisms and plants inherently have potentially serious environmental risks (e.g. Glaser and Glick 2012; Ryan 2009), especially in the context of microorganisms improved by synthetic biology (Dana et al. 2012; Snow and Smith 2012), many scientists and industry officials feel that whatever risks may exist are easily assessable and manageable, and in any event do not differ in degree from the risks posed by similar uses of naturally-occurring organisms.

Although some of the public concerns over outdoor testing of GMMs and transgenic plants began to subside (with activists' attentions shifted to food uses of modified plants), the potential environmental impacts of GMMs continued to be the

Table 3 Key issues in risk assessments of large-scale industrial uses of microorganisms, including algae and cyanobacteria

Stability of vector and introduced genes.
Possible deleterious functions encoded by transgene(s) such as toxins.
Potential for horizontal gene transfer, crossing to native species.
Potential for engineered strain to be transported outside facility, survive and compete in environment.
Potential persistence in the environment: soil or water in vicinity of site of use.
Potential disruption of natural ecosystems, such as native algae populations.
Creation or enhancement of harmful algal blooms or ecologically disruptive algal blooms.

subject of academic study, often in the context of deliberate environmental uses of GMMs in agriculture or for bioremediation. (Sayler and Ripp 2000; Davison 2005; Viebahn et al. 2009; Urgun-Demirtas et al. 2006; Singh et al. 2011). Because so much of this research has involved microorganisms intended for uses in the open environment (e.g. in agriculture), not all of it may be directly applicable to industrial uses of GMMs that are either contained in traditional fermentation processes or conducted in controlled outdoor reactors; however what has been learned from such studies may present baseline information to help assess potential adverse effects should there ever be a large-scale accidental release of industrial GMMs from production vessels.

Viebahn et al. (2009) presents a fairly recent, quite comprehensive, review of literature relating to possible survival of GMMs in the environment, as well as potential ecosystem effects and impacts on non-target species. In summarizing a large number of studies involving microorganisms that might be used in agriculture or soil remediation, these authors concluded that in most cases, GMMs did not persist in the environment nor have adverse effects on indigenous microflora or other non-target organisms, but that such effects (such as population increases in the environment) were sometimes seen in some studies. Where non-target effects were seen, they were often "transient and small compared to natural variation". In many cases, the GMM behaved similarly to its nonmodified parent, but here too there were some studies showing the opposite. These authors did not draw any broad, generalized conclusions from the literature they reviewed, and recommended that each proposed use of a GMM be assessed on its own merits.

Urgun-Demirtas et al. (2006) also review experimental results (again, largely from the perspective of bioremediation) relating to the possible environmental impacts of introduced GMMs, including a comprehensive review of studies pertaining to possible horizontal gene transfer (HGT) between introduced species and native microflora. These authors summarize several studies in which evidence for HGT has been seen between native species and from GMMs to native species, but they note that lab or microcosm studies may overestimate the extent this occurs in nature. The authors conclude that, the possibility of HGT from use of GMMs is a "crucial [issue] regarding the potential impact of [GMM] release into the field", but that studies in contained systems "have generally indicated that this may not be an insurmountable problem".

Singh et al. (2011), while largely a review of genetic engineering approaches to improve microbial bioremediation, discusses the factors that might prevent modified strains from competing with native microflora in the environment, and proposes a 6-step decision tree for performing risk assessments of modified bacteria intended for deliberate release into the environment.

The potential environmental impacts of the use of GM algae and the types of risk assessments needed to evaluate such potential impacts are similar to those discussed above, with some added concerns due to the nature of algae and cyanobacteria. Consideration of these factors has been the subject of several recently-published papers (Henley et al. 2013; Snow and Smith 2012; Gressel et al. 2013; Gressel et al. 2014; Menetrez 2012), as well as a recent workshop (Enzing and Nooijen 2012), the conclusions of which coincide with many of the points raised in these recent papers. Henley et al. (2013) presents the most comprehensive review of the potential environmental impacts of the "commodity-scale" use of GM algae, discussing such issues as the potential of a released strain to grow, persist and mutate in the environment, the possibility that GM algae could produce toxins or harmful algal blooms (HABs) or have other negative effects of aquatic ecosystems, and the possibility that introduced genes could spread by horizontal gene transfer and be expressed in indigenous microorganisms. In a shorter paper, Snow and Smith (2012) cover many of these same issues, particularly the need to assess environmental survival and persistence of an introduced strain and the potential for horizontal gene transfer. Both papers speculate on possible physical barriers or biological containment (e.g. so-called "suicide genes"; also discussed by Gressel et al. 2013) that might be effective in reducing environmental dispersal or survival of a released GM algae strain.

In two recent papers, Gressel, et al. (2013, 2014) assess the possible risks of large-scale industrial uses of both naturally-occurring and modified algal strains that have been domesticated for industrial use, and conclude that environmental risks should be assessed prior to large-scale use of either type of strain, particularly since accidental releases from production reactors are likely inevitable, even from contained photobioreactors. These authors propose mitigation strategies designed to limit the ability of production strains to survive and persist in the environment in the event of escape from production facilities. In particular, they advocate risk management strategies similar to the principles of Good Industrial Large Scale Practice, such as limiting large-scale uses to algae strains known to be nonpathogenic and to have a history of safe use.

These are the scientific issues that need to be addressed, at least at some level, before any proposed large-scale industrial use of modified organisms is to proceed. Regulatory frameworks around the world have been developed to carry out the needed risk assessments, and these should be applicable regardless of whether the organism was constructed by traditional recombinant DNA techniques or by newer techniques such as metabolic engineering or synthetic biology. However, one can have the expectation in most cases that microorganisms chosen for commercial uses at large-scale will be nonpathogenic and will not have other traits which might cause adverse environmental or health effects.

3 U.S. Framework for Regulation of Biotechnology

3.1 Overview

Regulatory frameworks have been developed in the United States and other countries to provide oversight over biotechnology and its commercial uses and to ensure that these potential environmental impacts are assessed. Because the earliest debates over biotechnology regulatory and public policy were often contentious or even confrontational, the perception developed within the industry that such government regulations were difficult to navigate, and that this, coupled with negative public opinion, placed significant barriers against the possible use of GMOs in industrial or agricultural applications, particularly those involving open environment use. Although this is not true, this misperception persists in many quarters to this day, and so it is useful to put today's regulatory frameworks into some historical perspective.

3.1.1 History

The following is a brief discussion of how the regulations that would affect uses of GMMs for fuel and chemical production evolved in the U.S. and elsewhere. More detailed summaries can be found elsewhere, including Glass (2003), Wozniak et al. (2012), Wrubel et al. (1997) and others. Biotechnology regulatory frameworks in most countries arose out of the health and safety issues that were initially raised by scientists and eventually debated by the public, shortly after recombinant DNA (rDNA) techniques were first developed in the early to mid 1970s (this early history is well documented by others, including Krimsky 1985; Glass 1991, and Glass 2003). The initial concerns were over the potential public health and safety threats that might be posed during laboratory research, if organisms having new traits were inadvertently released outside the laboratory, and this concern led to the adoption of research guidelines, which in some cases had limited applicability (e.g. the U.S. National Institutes of Health recombinant DNA guidelines, binding only on federally-funded research). As the biotechnology industry developed in the 1980s, the focus of regulatory concern shifted not only to the larger scale uses inherent in commercial application of this new technology, but also to deal with the intended use of engineered plants, animals and microorganisms for use outside the lab, in the open environment (e.g. in agriculture). In fact, the driving force for much of the regulatory action in the 1980s was concern over such "deliberate releases" to the environment, even though most governments instituted regulations that covered a wider range of commercial activities.

In the United States, the outcome of several years of public policy discussions was the adoption of a "Coordinated Framework" for biotechnology regulation in 1986 (OSTP 1986). Under this framework, it was decided that the commercial products of biotechnology would be regulated under existing laws and regulations and that it was not necessary to enact a specific law broadly covering all biotechnology activities.[1] Thus, the use of biotechnology to produce drugs, vaccines, diagnostic products, foods and food additives would be regulated by the Food and Drug Administration (FDA), using existing regulatory authority; biotechnology-derived pesticides would be governed by existing rules of the Environmental Protection Agency (EPA)[2]; and most other agricultural products would be regulated by the U.S. Department of Agriculture (USDA).

The effect of this decision was that the vast majority of biotechnology products, especially in the early years of the industry, were to be governed by the existing regulatory programs of the FDA and EPA with little or any regulatory revision; however it was also necessary to create new regulatory structures for some classes of commercial

---

[1] This is in contrast to most other countries in the world, which have generally created a single national biotechnology ("biosafety") law, often in compliance with the Cartagena Protocol on Biosafety (see below).

[2] EPA developed regulations under the U.S. pesticide law (the Federal Insecticide, Fungicide and Rodenticide act; FIFRA) to regulate proposed uses of modified and unmodified microorganisms as biopesticides. These regulations encompass risk assessments similar to those discussed in this chapter, but pesticides and other agricultural uses of microorganisms are outside this chapter's focus on fuels and chemicals. See references such as Glass (2003) or Wozniak et al. (2012) for more details on FIFRA biopesticide regulation.

product that could be anticipated to arise from biotechnology. Specifically, although it was decided that existing laws administered by the USDA could be used to regulate genetically engineered (transgenic) plants, new regulations under those laws would be needed. In addition, there were a number of potential uses for genetically modified microorganisms that could be regulated under EPA's existing statutory authority to regulate new chemicals (TSCA; discussed below), but here too a set of new rules would be needed to use this law to regulate microorganisms.

3.1.2 Applicability to Biofuels and Bio-based Chemicals

In fact, it is these EPA and USDA regulations that may govern the use of modified organisms for production of fuels or chemicals in the U.S. Many uses of modified microorganisms would be subject to regulations adopted in 1997 by the U.S. EPA under the Toxic Substances Control Act (TSCA) (Glass 2003; Bergeson et al. 2014; Wozniak et al. 2012). These regulations require notification to the agency before commercial use or importation of certain modified microorganisms, as well as agency review for proposed outdoor R&D activities of such modified organisms, e.g. open-pond growth of modified algae. The biotechnology regulations of the U.S. Department of Agriculture would, in many cases, cover uses of transgenic plants as biofuel or chemical feedstocks – these regulations will be briefly described below although they would apply to proposed uses of modified microorganisms only in rare cases.

Certain uses of modified microorganisms or algae could fall subject to FDA regulations as well. Naturally, microbial production of foods, pharmaceuticals, or other products within FDA's traditional jurisdiction would be subject to that agency's oversight, however, the nature and scientific basis for such regulation is outside the scope of this chapter and will not be discussed here. However, a common strategy for companies developing modified yeasts or other nonpathogenic microorganisms for ethanol, fuel or chemical production is to plan to use of the spent biomass that remains after the production process in animal feed. This has traditionally been done in the ethanol industry, through the production of dried distillers grains containing inactivated yeast for use in animal feed. Any proposed use of modified microorganisms in animal feed would likely require review by the animal feed division of the FDA (or equivalent bodies in other countries), although in the U.S., FDA shares some responsibility for oversight over animal feed ingredients with the Association of American Feed Control Officials (AAFCO; see below).

3.2 EPA Regulation of Industrial Uses of Modified Microorganisms

3.2.1 Overview

The use of certain genetically modified microorganisms in biofuel or bio-based chemical production may be subject to regulations promulgated by the U.S. Environmental Protection Agency under TSCA (Glass 2003; Wozniak et al. 2012; Bergeson et al.

2014). EPA uses TSCA to regulate commercial applications of genetically modified microorganisms that are not regulated by other federal agencies. TSCA (15 U.S. Code 2601) is a law requiring companies or individuals to notify EPA at least 90 days before commencing manufacture or importation of any "new" chemical, i.e., one that is not already in commerce in the United States, and which is intended to be used for a purpose not subject to federal regulation as a pesticide or under the food and drug laws. It is viewed as a "gap-filling" statute, that is meant to cover chemicals falling through the cracks of other regulatory authority, but it is also a "notification" statute, with government notice required for all new chemicals, regardless of risk, with the idea that the agency would review all the notifications and single out for further oversight those chemicals that appeared to pose potentially unacceptable risks to the environment or public health. The large majority of chemical notifications received by EPA under TSCA are cleared within the 90 day period after only brief agency review.

In the Coordinated Framework of June 1986 (OSTP 1986), EPA proposed to use TSCA in the same "gap-filling" way as it is used for chemicals, to capture those modified microorganisms to be used in commerce but that were not regulated by other federal agencies. The primary areas which were expected to became subject to the TSCA biotechnology regulations were (a) microorganisms used for production of non-food-additive industrial enzymes, other specialty chemicals, and in other bioprocesses; (b) microorganisms used as, or considered to be, pesticide intermediates; (c) microorganisms used for nonpesticidal agricultural purposes (e.g. nitrogen fixation); and (d) microorganisms used for other purposes in the environment, such as bioremediation. As the field of industrial biotechnology has developed, production of biofuels and bio-based chemicals have become the most prominent "bioprocessing" applications that might fall subject to TSCA.

Although EPA established an interim policy of TSCA regulation under the 1986 coordinated framework, because of political difficulties and interagency disputes (Glass 2003) the agency was not able to publish proposed biotechnology regulations until 1994 and was not able to finalize these regulations until 1997 (USEPA 1997b). These rules, when finally issued, amended the existing TSCA regulations by creating a new section of the Code of Federal Regulations (40 CFR Part 725), which specifies the procedures for EPA oversight over commercial use and research activities involving microorganisms subject to TSCA. The net result was to institute reporting requirements specific for microorganisms (but which paralleled the commercial notifications used for new chemicals), while also creating new requirements to provide suitable oversight over outdoor uses of those genetically modified microorganisms subject to TSCA jurisdiction.

3.2.2 Scope of the TSCA Biotechnology Regulation

The biotechnology rule requires premanufacture reporting for new organisms intended for commercial use, but it was a long-running challenge in the development of the regulations to adequately define "new organism". The final rule defines a "new organism" as an "intergeneric organism", which is defined to mean "a microorganism that is formed by the deliberate combination of genetic material originally isolated from organisms of different taxonomic genera". This is the same definition originally proposed by EPA in the Coordinated Framework and used under the agency's interim policy. The rationale for this definition, which was admittedly somewhat arbitrary, was that microorganisms that are classified within the same genus were more likely to be able to exchange genetic information in nature than microorganisms found in different genera, so that an "intergeneric" combination of genes was judged to be less likely to have occurred naturally (without human intervention) than an "intrageneric" combination. Under this formulation, microorganisms that are not intergeneric are considered not to be new, and such organisms, including naturally occurring and classically mutated or selected microorganisms, as well as GMMs modified only through gene deletions or directed evolution approaches, are exempt from reporting requirements under TSCA. Note that EPA's need to limit the TSCA regulations to "new" microorganisms in this way gives the rule a narrower scope than other U.S. federal regulations as well as the laws and regulations of other countries, in potentially excluding certain categories of modified microorganisms from the rule.

Although there has been some uncertainty in the past, it now seems clear that genetically modified algae strains would fall under EPA jurisdiction under TSCA if intergeneric and if used for a TSCA-regulated purpose. This interpretation is supported by language in EPA's 1997 Federal Register notice instituting the biotechnology rule (USEPA 1997b) which included "green and red algae" among a list of types of organisms covered by the definition of the term "microorganism", and a discussion in its Regulatory Impact Analysis accompanying the rule (USEPA 1997d), which stated that "Language in the Act has been interpreted to include living microorganisms (i.e., microscopic living cells such as bacteria, fungi, protozoa, microscopic algae, and viruses)". As discussed below, in recent years proposed uses of algae and cyanobacteria have indeed begun to be regulated under TSCA.

3.2.3 Regulation of Commercial Uses of Modified Microorganisms

Commercial uses of "new microorganisms" used for a "TSCA purpose" (that is, not regulated elsewhere in the federal government) generally require notification to EPA at least 90 days in advance of commercial use or importation. This notification takes the form of Microbial Commercial Activity Notices (MCANs). An individual MCAN is needed for each modified strain intended to be commercialized, although EPA maintains procedures to facilitate submission and review of "consolidated" MCANs covering up to six related strains of similar genetic make-up.

The information and other data that applicants need to submit in the MCAN are listed in Section 725.155 of the regulations and summarized in Table 4. Much of the required information has to do with the biological characterization of the modified microorganism and a detailed description of how it was constructed, but information is also to be submitted on the proposed use of the microorganism, the proposed production process, the containment and control procedures to be used, the likelihood for worker exposure and the steps taken to control exposure, and an assess-

Table 4 MCAN data requirements

Microorganism identity, including taxonomic identification of the "recipient" organism and the "donor" organisms that are the source of the introduced genes.
Detailed information about the construction of the microorganism.
Biological characterization of the microorganism.
Potential health effects of the microorganism, such as pathogenicity or toxicity (can be addressed from testing or from the literature).
Potential environmental effects of the microorganism (can be addressed from testing or from the literature).
Detailed information about the industrial process, including the measures that will be taken to minimize release of the microorganism from the facility.
The extent to which workers might be exposed to the microorganism.
The extent to which the microorganism might be released into the environment as a result of the process.

ment of the potential health and environmental effects of the microorganism should it be released from the facility. It is important to note that, in MCANs or other submissions to EPA under the biotechnology rule, the applicant can claim much or all of the submitted information as "confidential business information", which the agency must keep confidential and which cannot be released to the public, but the applicant must provide EPA with the justification for the confidentiality claim (in fact, this justification must be included within the MCAN filing). EPA has published a detailed "Points to Consider" document (USEPA 1997c) summarizing the required data and the format for submission, which, together with guidance from the publicly-available versions of previously-filed MCANs (i.e. the parts of prior MCANs not claimed by the applicant as confidential), can be used to help applicants prepare MCAN submissions.

In most cases, EPA review of MCANs can be expected to be fairly straightforward, and would include consideration of the potential risks and benefits of the commercial use of the modified microorganism. Most of EPA's prior reviews of MCANs have taken place within the 90-day period specified in the regulations, although EPA has the power to unilaterally extend the review period by an additional 90 days, or to ask the applicant to voluntarily suspend the review period, if the Agency decides it needs more time to complete its review or needs to request more data. MCANs for the contained use of new microorganisms in bio-based manufacturing have generally not caused any concerns or significant issues in EPA's review, and most have been routinely cleared for commercial use without any delays or difficulties; however it is possible that MCANs for algae or cyanobacteria might take slightly longer for EPA review, due to initial unfamiliarity with the species and its proposed conditions for growth and manufacture.

MCANs (like chemical PMNs) are not "approved" per se, but if no issues emerge they are cleared for commercialization if EPA takes no action and drops the MCAN from review within 90 days. However, if issues are identified, EPA has the authority to require additional data from the MCAN submitter or to limit approved uses of the microorganism in a variety of ways, including controls on workplace and/or environmental exposure and limitations on the amount of the organism that can be used commercially. Once an MCAN is dropped from review, an applicant must file a Notice of Commencement within 30 days of beginning commercial use or importation of the microorganism, a notice that requires submission only of minimal information, but which triggers recordkeeping and reporting requirements once commercialization begins. Even if EPA does not impose any restrictions on use prior to commercialization, the agency has the power to regulate microorganisms after commercialization, for example by imposing requirements for testing and submission of health and safety data, as well as taking other steps that the agency may decide is necessary to address unreasonable risks to health or the environment.

As of this writing, EPA has reviewed 63 or more MCANs (USEPA 2014; Bergeson et al. 2014), with most of the more recent ones covering microorganisms intended for use in biofuel or bio-based chemical production. These will be discussed in more detail below.

The biotechnology rule provides certain exemptions from MCAN reporting. These are the so-called "tiered exemptions" available for certain uses of modified strains of well-studied, common industrial microorganisms. First, the host, or recipient, organism must be one that is included on the list found in Section 725.420 of the regulations. This list includes many well-studied species including *E. coli* K12, *Saccharomyces cerevisiae, Bacillus subtilis,* and others, but does not include any algae or cyanobacteria species. The regulations also include a procedure whereby manufacturers may petition for inclusion of a new host on this list. In September 2012, in response to industry petitions, EPA published a notice proposing to add *Trichoderma reesei* strain QM6a and *Bacillus amyloliquefaciens* subsp. amyloliquefaciens (industrial strains) to the list of potentially exempt species, but as of this writing, the proposed rule has not yet been finalized (USEPA 2012).

Second, as specified in Section 725.421, the introduced genetic material in the microorganism must be well characterized (that is, the function of all introduced DNA is known); must be limited in size to the minimal genetic information needed; must be poorly "mobilizable", which is defined in the regulation to mean that the ability of the genetic material to be transferred and mobilized has been inactivated and that the frequency of transfer is less than 10^{-8} transfer events per recipient; and must be known to be free of harmful sequences. Given the precision possible in today's recombinant DNA techniques, these conditions should be easy to meet.

Third, in Section 725.422, the regulations specify specific containment and control procedures to minimize the possibility that the engineered microorganism might inadvertently be released from the facility. If an organism meets the first two sets of criteria (the "biological" criteria), and the applicant can certify that it will use the microorganism in strict compliance with the provisions of Section 422, the process is eligible for a "Tier I" exemption and the microorganism can be used commercially merely upon 10 days advance notice to EPA. Note, however, that in EPA's current interpretation of the Section 422 provisions, the Tier I exemption is not available if significant quantities of a live microorganism are to be transported from one facility to another (K. Moss, personal communication). For microorganisms meeting the biological criteria but which are intended for use under conditions less strict than the Section 422 procedures, the applicant can submit a petition for a "Tier II" exemption 45 days before intended manufacture. EPA would approve the Tier II request if it felt that the proposed containment and control procedures, although not identical to the Section 422 procedures, were sufficient for the organism in question. Note that the Section 422 procedures are also recommended for use with microorganisms subject to MCAN reporting.

In addition to the tiered exemptions, the TSCA regulations also provide a procedure by which companies can apply for an exemption for test marketing purposes. This requires submitting certain information to EPA 45 days in advance of the proposed activity. According to Bergeson et al. (2014), from 1997 through 2013, EPA received and approved 118 Tier I and two Tier II exemption requests, as well as one request for a test marketing exemption.

The use of genetically modified microorganisms to produce chemicals for commercial use may trigger additional regulation under TSCA. If the chemical substance synthesized by the microorganism has never before been used in commerce in the U.S., the manufacturer may have to file a traditional premanufacture notice (PMN) for the new chemical (Bergeson et al. 2012). Examples might include novel enzymes (e.g., having new activity or a novel, artificially-designed amino acid sequence) or organic chemicals that have not previously had any commercial utility. The use of a novel (i.e. intergeneric) microorganism to produce known compounds or chemical substances presents a somewhat more complicated picture. If the substance being commercialized is a single chemical compound that can be represented by a definite chemical structural diagram, it is known under TSCA as a Class 1 substance, and in most cases, if a Class 1 substance has been used in commerce and is on the TSCA Inventory, producing that substance by a novel production process, such as by a novel microorganism, would not require filing a new PMN. However, TSCA also defines another class of substances, Class 2 substances, as those that are of "unknown or variable composition, complex reaction products, [or] biological materials" (known as UVCB substances), which cannot be easily represented by a structural diagram. As described in Bergeson et al. (2012), there are several types of Class 2 substances, which are generally listed on the TSCA inventory by a definition specifying the source from which the substance has been derived. In many cases, production of a Class 2 substance by a novel microbiological method will require filing a new PMN in addition to an MCAN, since the source of the substance will differ from the source defined on the Inventory listing. For example, a substance comprising a range of alkanes or alkyl molecules for use in diesel fuel would likely require a new PMN as a UVCB substance if produced by a novel microbial method. See Bergeson et al. (2012) for more detail on scenarios that might require PMN filing in addition to MCAN filing for the production of bio-based chemicals using modified microorganisms.

Because of the statutory limitations of TSCA, most chemicals requiring PMNs under TSCA would generally not be regulated elsewhere in the federal government, but there are exceptions. Notably, microbially-produced substances intended for use as automotive fuels or fuel additives are required under the Clean Air Act to obtain registration from EPA's Office of Transportation and Air Quality under 40 CFR Part

79. This requirement is to ensure that novel fuels will not harm engines or cause undue air pollution, and would apply not only to diesel or gasoline substitutes but also to additives such as ethanol or butanol. New aviation fuels or fuels intended for use in any of the branches of the U.S. military also must undergo a registration process, which often requires establishment of, and compliance with, standards and specifications adopted by ASTM International. Furthermore, there are additional compliance requirements for manufacturers of novel fuels wishing to take advantage of the economic credits (Renewable Identification Numbers) available under the U.S. Renewable Fuel Standard. Discussion of these regulatory programs is beyond the scope of this article, but see Slating and Kesan (2012) and Danish et al. (2014) for more information.

3.2.4 Regulation of Research Uses of Modified Microorganisms

TSCA is a commercial statute, and so its jurisdiction generally does not include R&D activities. Under the parts of TSCA regulations that cover new chemicals, there is an exemption for "small quantities" of new chemicals used solely for R&D. This exemption was largely carried over into the biotechnology rule, except that EPA made the somewhat arbitrary distinction that microorganisms, because they are self-replicating, could not be considered to ever be used solely in "small quantities" unless certain restrictions were placed on how they were used. Thus, new microorganisms used solely for R&D purposes could qualify for the exemption only if they were used under suitably contained conditions (i.e., in "contained structures"). Under this definition, it is likely that most laboratory research using GMMs in biofuels or bio-based chemicals would be exempt from commercial reporting. In addition, many uses of engineered microorganisms in biofuel or bio-based chemical pilot plants or demonstration plants could also qualify for this exemption. However, R&D use of intergeneric microorganisms in the open environment, or in vessels or facilities judged not to be suitably contained, requires notification to EPA at least 60 days before the proposed use, under an application known as a TSCA Experimental Release Application (TERA; described in more detail below).

The key issue in determining if an activity qualifies for the R&D exemption is whether or not it will take place in a "contained structure". The term "structure" is defined in the biotechnology rule at Section 725.3, and includes any "building or vessel which effectively surrounds and encloses the microorganism and includes features designed to *restrict* the microorganism from leaving" (emphasis added). The key point of this definition is that the structure *minimize* (rather than *prevent*) the potential for microorganisms to escape and become established in the environment.

Under the rule, activities in contained structures would qualify for the small quantities exemption if conducted "solely for research and development" and meeting other procedural requirements. For example, the R&D must be conducted under the supervision of a technically qualified individual, who must adopt specific containment procedures. In addition, appropriate records must be kept and workers must be adequately notified of any risks. R&D meeting these requirements can be conducted with no EPA oversight or prior notice (in fact, entities determine for themselves if they are in compliance).

The rule gives EPA staff and the regulated community broad leeway in determining which structures are suitably "contained". Although there was some initial uncertainty when the rule was first proposed, it has since become clear that EPA interprets the definition broadly, so that many laboratories and greenhouses, as well as most fermentation reactor vessels, would meet the definition. Fermenters need not be indoors to meet the definition, so that large outdoor vessels can qualify if certain procedures are followed and suitable controls for the process and the facility are maintained.

It should therefore be possible for most companies to take advantage of this exemption for traditional fermentation processes taking place at pilot or demonstration plants, as long as the microorganism were used solely for research and development and neither the organism or its product are used or sold commercially. It is also likely that many uses of algae or cyanobacteria in enclosed photobioreactors would qualify as contained structures, depending on the specifics of reactor design and operation, and if the procedural requirements for the exemption are also met. However, most open-pond algae reactors would not qualify as contained structures, and would likely require prior EPA review under the TERA process.

The TERA process provides an expedited review procedure for small-scale field tests and other outdoor R&D uses of new microorganisms. Applicants proposing such uses must file a TERA with the EPA at least 60 days in advance of the proposed activity. The data requirements for TERAs are outlined in Sections 725.255 and 725.260 of the regulations, and include information about the microorganism and how it was constructed, a description of the field experimentation proposed to be conducted, along with the proposed confinement conditions and steps to be taken to monitor the possible dissemination of the organism from the test site.

EPA is required to review the submitted information and decide whether or not to approve the proposed outdoor R&D activity within 60 days, although the agency could extend the review by an additional 60 days. If EPA determines that the proposed activity does not present an unreasonable risk of injury to health or the environment, it will notify the applicant in writing that the TERA has been approved. When a TERA is approved, the applicant must carry out the testing under the conditions and limitations described in the TERA application document, and also in accordance with any requirements or conditions included in EPA's written approval. In most cases, it is likely that EPA will require applicants to conduct some form of monitoring, to detect the possible spread or dispersal of the microorganism from the test site, or to detect any other potential adverse environmental effects. EPA may require collection and submission of other data as well. EPA's approval is legally binding on the applicant, and the Agency has the additional authority to modify or revoke the approval upon receipt of evidence that raises significant questions about the potential risk of the activity.

The regulations provide some exemptions from TERA reporting for certain qualifying outdoor uses of modified microorganisms, but these exemptions are very limited, in covering only certain uses of the nitrogen-fixing bacteria *Bradyrhizobium japonicum* and *Rhizobium* (now *Sinorhizobium*) *meliloti*, species which were field tested in closely monitored experiments under EPA's interim TSCA regulatory policy. These exemptions would not be expected to apply to any potential use of modified algae or other microorganisms for fuel or chemical production.

As of this writing, there has only been limited experience with TERAs, with only 30 TERAs filed since the biotechnology rule was put into place in 1997 (USEPA 2014; Bergeson et al. 2014). These will be discussed in more detail below.

Many ecologists and public sector critics of use of GMMs in the environment have raised questions about how well the risks of such uses can be assessed. The answer to these concerns is not to prevent any outdoor uses until all risks are ruled out (e.g., as proponents of the precautionary principle would demand), but instead to allow risks to be addressed through the stepwise progression from small scale to larger scale, under a regulatory regime that not only provides oversight but also flexibility and accountability. The TERA process is well-suited for this purpose, to allow outdoor uses of GMMs to take place in a stepwise fashion under appropriate monitoring and agency oversight, to enable environmental risk assessment questions to be addressed with data from actual small-scale environmental use, thus facilitating subsequent risk assessments for larger-scale uses. Although there is no doubt that outdoor uses of genetically modified microorganisms will receive greater regulatory scrutiny than uses in contained manufacturing, EPA's TERA process should allow such uses to proceed through the normal phases of scaled testing in an orderly and responsible manner, under a level of regulatory scrutiny that is accessible to academic scientists as well as companies.

None of the projects covered by any of the previously-filed TERAs have progressed to commercial use, although EPA has approved commercial sale under TSCA of one live, modified agricultural microorganism. In September 1997, EPA approved limited commercialization of the intergeneric microorganism *Sinorhizobium meliloti* strain RMBPC-2, a modified strain with improved capacity to provide fixed nitrogen to alfalfa plants as a nutrient. Because this product was field tested under approvals granted by EPA under its pre-1997 interim biotechnology policy, EPA concluded that the commercial use of this inoculant did not pose significant environmental risks, provided it was subject to certain production limits (USEPA 1997a). Although this is the only live engineered microorganism approved for commercial use in the open environment under the EPA TSCA regulations, it does establish a precedent that EPA would be prepared to grant such approvals where warranted by the science and the data package accumulated by the applicant.

3.3 USDA Biotechnology Regulations

The U.S. Department of Agriculture (USDA) maintains regulations at 7 CFR Part 340, that have been the major U.S. government rules that have covered uses of transgenic plants in agriculture and more recently the increasing interest in using plants for other industrial purposes, such as production of pharmaceuticals, industrial products, and phytoremediation. A small number of modified agricultural microorganisms have also fallen under this regulation, and it is worth noting that there have been some in the algae community that have expressed a preference for the USDA to use this regulation to assert jurisdiction over industrial uses of modified algae, due to the commonalities between algalculture and agriculture and USDA's historical support for, and involvement with, the algae industry (Trentacoste et al. 2014; Henley et al. 2013). However, as explained below, this rule covers only outdoor uses or interstate movement of those organisms to which it applies, and so its potential applicability to contained manufacturing using GMMs is quite limited.

These rules were put into place in 1987 as an immediate outgrowth of the "Coordinated Framework" for biotechnology regulation. USDA proposed to use its existing statutory authority under a law then known as the Plant Pest Act to regulate interstate transport and field testing of genetically engineered plants intended for use in the open environment, to assess the potential environmental effects of such uses. The basis for this rule was the possibility (however remote) that such engineered plants might pose a plant pest risk, based on the presence of nucleic acid sequences arising from genera listed in the rule. These regulations were finalized in June 1987 (USDA 1987), and have been administered by a dedicated biotechnology office within USDA's Animal and Plant Health Inspection Service (APHIS).

The possible applicability of the rule to engineered microorganisms rests within its definitions. Under the rules, "regulated articles" are defined to include only "organisms that are or contain plant pests", which has been interpreted to cover only those plants (or microorganisms) engineered to contain nucleic acid sequences from certain specific microbial, plant and animal genera that contain species that were considered to be potential plant pests. The regulations included a fairly broad list of such genera, and this had the practical effect of causing most transgenic plants to be captured by the regulations: this was because the genus *Agrobacterium* was on the list, and in practice, DNA sequences from *Agrobacterium tumefaciens* were almost universally used in plant transformation procedures, and the presence of *A. tumefaciens* DNA in the resulting plant would often be enough to subject the transgenic plant to regulation under this rule. The list of known or potential plant pest species is contained in 7 CFR Part 340.2. Although the list includes several genera which might include industrially-useful species (e.g., *Pseudomonas*, *Streptomyces*), the rule would only apply if any microorganism containing nucleic acid sequences from these genera were intended to be deliberately used in the environment, which is not likely for production strains for industrial products. The rule could cover open-pond uses of algae, but the list in the regulations does not appear to include any of the genera of algae that have been suggested for industrial use. A modified algae might fall under the rule if it contained nucleic acid sequences from *Agrobacterium* or another listed genus.

The regulations give APHIS the leeway to make a determination that an organism altered or produced through genetic engineering is a plant pest or that there is reason to believe the organism is a plant pest,[3] but generally speaking, if an engineered microorganism is not from one of the genera shown on the list in Part 340.2, or does not contain any nucleic acids from any such genera, it would not *a priori* be subject to regulation under the existing rules. So, it is unlikely that USDA would use its regulatory leeway to assert authority over a proposed industrial use of a modified microorganism unless it was a fairly large-scale open-pond commercial use (e.g. of a modified algae), and only if there were some clear link, such as a possible plant pest risk, to agriculture or to a particular region or sector of U.S. agriculture. However, in view of the discretion afforded to USDA under the regulations, companies considering the use of modified algal or microbial species not having long histories of industrial use should consider informally consulting with USDA before commercial use or interstate transport of the organism. It should also be noted that several states have regulations that may affect uses of modified organisms or require state participation in USDA biotechnology reviews, particularly for activities with modified organisms conducted outside of containment.

Although the USDA rule initially required submission of permit applications for all proposed outdoor uses of organisms covered under the regulation, the regulations were substantially relaxed on two occasions (USDA 1993, 1997), with the creation of a much simpler notification process for those plant species deemed to have low potential risks. Under the current version of the regulations, transgenic varieties of most common agricultural crops and other familiar plant species meeting criteria specified in the regulations can be used in research field tests simply upon 30 days advance notice to APHIS, and the submission of only minimal information about the modified plants and the proposed field use. Such field tests must be conducted in accordance with performance standards specified in the regulations. Only uses of less-familiar transgenic plants, and presumably any modified microorganisms falling under the regulations, would now be required to undergo the longer permitting process. If an open-pond use of a modified microorganism were judged to fall under these regulations, it is likely that permits would be needed for outdoor testing, even at small scale. Commercial use would require USDA approval through the provisions under the regulation requiring applicants to petition the agency for a determination that regulated articles are determined to qualify for "nonregulated status". Such decisions by USDA clear the organism for commercial use, but in recent years have required the agency first to prepare Environmental Assessments justifying such actions.

[3] USDA now has potentially broader regulatory ability. In 2000, the Plant Pest Act, the law on which the Part 340 regulations was based, was combined with other statutes to create a new law, the Agriculture Risk Protection Act, which includes language that could give USDA the ability to regulate modified organisms based on potential invasiveness or weediness. In 2008, USDA published some possible options to amend the regulations to accomplish this, but to date the Department has never proposed any specific regulations for this purpose.

A number of modified microorganisms have received permits or have otherwise been allowed to be field tested or imported into the United States under the USDA regulations, including species such as *Pseudomonas syringae*, *Xanthomonas campestris*, *Aspergillus flavus*, and various rhizobia (Glass 2003). It is believed that these have all been for agricultural purposes and have been intended to be used under non-contained conditions.

3.4 FDA Regulation of Modified Microorganisms Used in Animal Feed

The use of spent biomass in animal feed, or to produce a substance to be used in animal feed, would be regulated in the U.S. by the Food and Drug Administration (FDA), through its Center for Veterinary Medicine (CVM). FDA does not require premarket review of most human or animal "food" per se: whole food or feed products are presumed to be safe for consumption, but FDA has enforcement powers to be sure marketed products are not adulterated. So, FDA regulation is largely directed at new substances proposed for use as human food additives or as animal feed additives. Under the Federal Food, Drug and Cosmetic Act (FFDCA), most such new substances that are intended to be components of food or to affect components of food are considered to be "food additives" and must be approved through the submission of a Food Additive Petition or, in the case of products for animal consumption, "feed additives" requiring Feed Additive Petitions. However, some substances can be used without approval of a Food or Feed Additive Petition: The FFDCA provides that "substances that are generally recognized, among experts qualified by scientific training and experience to evaluate their safety as having been adequately shown ... to be safe under the conditions of their intended use," are not considered as food additives. This created the category of substances known as GRAS: "generally recognized as safe", and many food or feed substances are used in food or feed on this basis.

Companies seeking to use spent microbial biomass in animal feed theoretically have several options to obtain clearance for such uses. One option is to file a Feed Additive Petition, which requires compilation of a significant amount of data and an often-lengthy FDA review process. The primary alternative would be to achieve GRAS status for the product, for which several routes are available. It is permissible under the law for a manufacturer to self-certify that a substance is GRAS for a specific use, if supported by appropriate publicly-available data or expert opinion, while another option would be to seek FDA's concurrence to a GRAS determination using the GRAS Notification procedure, a relatively new process instituted by FDA's Center for Veterinary Medicine in 2010 (following the successful use of a similar program within FDA's human food branch) (USFDA 2014).

However, a third option also exists. Although the law and regulations give FDA the ultimate authority to make decisions on food or feed additive petitions and

GRAS determinations, in practice CVM operates in cooperation with the Association of American Feed Control Officials (AAFCO), which is composed of state, federal, and international regulatory officials who are responsible for the enforcement of state laws regulating the safe production and labeling of animal feed. FDA CVM and AAFCO work together on animal feed regulation, particularly in the establishment of definitions to describe new feed ingredients. Each year AAFCO publishes its Official Publication which includes a model feed bill for states to adopt in regulating feed products and a list of accepted feed ingredients. Most states have adopted all or part of the model feed bill and allow feed ingredients listed in the publication to be used in their respective territories.

In many cases, it may be necessary to obtain an AAFCO ingredient definition for a new animal feed product, particularly to allow sale and use of the product in certain states within the U.S. New feed additives approved by FDA under the petition process are generally accepted as new ingredients by AAFCO, but this may not be true for products self-certified as GRAS. It is possible to work directly with AAFCO to obtain a new ingredient definition for a GRAS substance or other feed ingredient, an action to which FDA may later consent. One example of a company that has successfully obtained clearance both from FDA and AAFCO for the sale of distillers' grains containing genetically modified *S. cerevisiae* is Mascoma Corporation, which has to date obtained approvals for two such modified yeast strains (BusinessWire 2013).[4]

Regardless of the regulatory route chosen, the scientific criteria that would be considered in the regulatory risk assessments for feed use would be different from the environmental effects issues that would be considered for the programs described above, in part because of the different intended use, and in part because microorganisms used in animal feed have generally been inactivated before such uses. Therefore, these regulatory programs will not be discussed here in any additional detail.

4 International Biotechnology Regulations

Biotechnology regulations exist throughout the world, although they have developed differently than in the U.S. Many other industrialized countries or regions, particularly the European Union, Canada, Australia and Japan, implemented biotechnology laws and regulations in the early days of the growth of the industry (i.e., the 1980s and 1990s), in ways that were consistent with the regulatory approaches of these jurisdictions, resulting in some idiosyncrasies among these regulations, although there are some similarities, such as those between the U.S. and Canadian approaches. As described below, more recently, many other countries around the world have adopted biotechnology or "biosafety" laws and regulations based on the principles of an international convention adopted in 2000 – the Cartagena Protocol on Biosafety. Countries

[4]The author has consulted for Mascoma in the past, but at this writing has no financial interest in this company.

taking this route generally have a single biotechnology law that, in principle, is applicable to all research and industrial uses of genetically modified organisms, although much of the focus of the Cartagena Protocol is on agricultural applications of GMOs in the open environment, and cross-boundary movement of GMOs.

4.1 Cartagena Protocol

The Cartagena Protocol on Biosafety was adopted on January 29, 2000 as a supplementary agreement to the Convention on Biological Diversity, and took effect on September 11, 2003 (Eggers and Mackenzie 2000). Under national biosafety laws modeled on the Cartagena Protocol, government approvals are generally needed for importation of living modified organisms (LMOs), and for many industrial activities including "contained uses" or "environmental uses". Such approvals may often require a risk assessment of the LMO and its proposed use. The principles of the Protocol are often a useful guide to the biosafety policies or regulations of many governments, particularly in the developing world.

Under the Cartagena Protocol, "LMO" is defined as any living organism that possesses a novel combination of genetic material obtained through the use of modern biotechnology, with "modern biotechnology" defined to include in vitro nucleic acid techniques as well as "fusion of cells beyond the taxonomic family". Although definitions of "GMO" vary around the world, most countries have adopted a definition such as this, but it is notable that the definitions in the U.S. EPA TSCA regulations and the USDA regulations are narrower, with EPA's limited to "intergeneric" microorganisms and USDA's requiring the presence of nucleic acids from suspected plant pest species.

The Protocol is primarily intended to ensure that national authorities are notified of any proposal to introduce LMOs into their countries, particularly for the purpose of deliberate release into the environment or for use in food or feed, and further to ensure that information about uses of LMOs is provided to the public and to other countries and interested parties. A key provision of the Protocol is to require there to be "Advance Informed Agreements" (AIAs) when LMOs are shipped across national boundaries, to ensure that the recipient nation is notified of the proposed shipment, and to allow the recipient nation to conduct needed risk assessments.

In most countries, uses of microorganisms within contained manufacturing will differ from applications such as open-pond cultivation of algae, and in general will be subject to far less stringent oversight. "Contained Use" is defined in Article 3 of the Protocol as "any operation, undertaken within a facility, installation or other physical structure, which involves living modified organisms that are controlled by specific measures that effectively limit their contact with, and their impact on, the external environment". However, Article 6(2) of the Protocol provides an exemption from the AIA procedures for shipments of LMOs intended solely for contained use. Unfortunately, the definition of "contained use" in the Protocol does not distinguish between research uses and commercial uses, an ambiguity which is also found in a number of national laws, sometimes making it unclear whether there might be any permit requirements for commercial "contained uses" over and above the notification and labeling requirements under the Protocol. The Protocol can be viewed as establishing minimum requirements for applicants proposing to use LMOs in contained commercial manufacturing, such as the requirement to notify the competent national authority), with the understanding that national laws may impose additional requirements in certain countries.

In principle, under the Protocol, the required procedures to use an LMO in the open environment (e.g. in an open-pond algae reactor) would not be much different than for a proposed use in contained manufacturing, in that the recipient national government would need to be notified and would need to conduct a risk assessment. However, in the case of an intended "release" to the environment, an Advance Informed Agreement would absolutely be required (which is not the case for a proposed contained use) and the risk assessment would almost certainly be more rigorous. The Protocol provides specific guidance for the risk assessments to be conducted, with minimal information for the AIA found in Annex I and guidance for the risk assessment in Annex III. In many countries, a permit or some affirmative government permission would be needed before the LMO could be used in the open environment. Such proposals may also engender public or community interest and perhaps opposition.

4.2 European Union

The EU has adopted two directives to cover biotechnology – one covering contained uses of modified organisms, and the other covering uses of modified plants and other GMOs in the open environment (Enzing and Nooijen 2012). Each EU member state is obligated to adopt national laws corresponding to EU directives, and so all 28 EU members should have their own biotechnology laws or regulations that mirror the provisions of the two EU directives.

Uses of modified microorganisms in contained manufacturing would require national government notification, and in some cases possibly also approval, in accordance with the EU "Contained Use" Directive 2009/41/EC (European Union 2009). Article 2 of the directive defines "contained use" in a way that gives an applicant proposing to use a GMM in Europe a fair amount of leeway in determining that a system or process is "contained". Article 4 of the directive requires the user to carry out a risk assessment of the microorganism, using considerations set forth in Annex III of the directive. As a result of this assessment, the user would determine which of four containment levels is appropriate for the organism, and would be obligated to adopt appropriate containment measures in accordance with Annex IV of the directive. These requirements are similar to most other international biosafety guidelines, and most microorganisms used for fuel or chemical production would qualify to be included within the lowest level of containment. Article 6 of the directive requires users to notify the government agency designated in national legislation as having jurisdiction to enforce the contained use directive before a facility is to be used with GMMs for the first time. Annex V specifies the information required to be submitted with such notifications, and for organisms in the lowest class of risk, the necessary information is fairly minimal. The laws of individual EU nations should conform to these provisions, and in most cases there would not be any need to seek government approval for contained uses, beyond the notifications described here, although it is likely that the laws of some EU nations may require government review and approval of such proposals.

Uses of modified algae or other microorganisms in open ponds would be covered by national laws corresponding to EU Directive 2001/18/EC on "Environmental Release" (European Union 2001). Generally speaking, any outdoor activities with LMOs in Europe, including small scale field testing, would require approval from the country in which the activity is to take place. Applications for commercial use are more complicated, in that all EU member states have some say in commercial approvals granted by individual countries. Although most if not all EU members have approved numerous field tests of transgenic plants over the past two decades (most of which have been for food-producing crops), commercial approvals for food crops have proven extremely problematic, and at times have effectively been barred in Europe.

4.3 Canada

Both contained and open-pond uses of modified microorganisms may require approval from Environment Canada under the New Substances Notification regulations under the Canadian Environmental Protection Act (Darch and Shahsavarani 2012). These regulations, which in many ways resemble the U.S. TSCA biotechnology regulations, cover the use of any microorganism that is new to commercial use in Canada, and potentially cover many modified microbial strains as well as unmodified microorganisms that have not previously been used in Canada. This represents one difference from the situation in the U.S., where unmodified microorganisms (as well as some modified microbes) are not covered by the TSCA regulations. The regulations potentially cover both contained and open-environment use of microorganisms, with a greater level of scrutiny dedicated for the latter, however, as of this writing the only microorganisms approved under the program have been for enzyme manufacture or for uses not related to biofuels.

4.4 Brazil

Under the National Biosafety Law, all proposed uses of living modified organisms would require approval from the Biosafety National Technical Committee (CTNBio), followed by authorization from the applicable Ministry. CTNBio is part of the Ministry of Science and Technology and it is a multidisciplinary committee composed of representatives from many different ministries and branches of the government, which is responsible for the technical reviews of biotechnology applications. Responsibility for formulation and implementation of the National Biosecurity Policy falls to the National Biosecurity Council (CNBS), which reports to the Presidency of the Republic. Once CTNBio grants an approval for a project, the formal authorization is granted by a government ministry: either the Ministry of Agriculture for most agricultural activities; the Ministry of the Environment for nonagricultural activities taking place in the environment, or the Ministry of Health for "human and pharmaceutical uses". There have been recent approvals for contained uses of LMOs for biofuel or bio-based chemical production that have been issued by the Ministry of Health after CTNBio review (CTNBIO 2014), which are discussed below.

4.5 Japan

Japan is a signatory to the Cartagena Protocol, and it has adopted Law 97 of 2003, entitled "Law Concerning the Conservation and Sustainable Use of Biological Diversity through Regulations on the Use of Living Modified Organisms." This law placed Japanese law in conformance with the Protocol, and forms the basis for Japan's biotechnology regulatory regime (Yamanouchi 2005). Among the defined terms of Law 97 are definitions of two categories of use of LMOs. "Type 1" uses correspond to what is typically called "deliberate releases", while "Type 2" uses are "contained uses". Under this scheme, proposed uses of modified microorganisms in contained manufacturing would be regulated as Type 2 uses, although the Ministry having jurisdiction may vary, and could be either the Agriculture Ministry or the Environment Ministry. Uses of microorganisms in open ponds would be regulated more stringently as Type 1 uses, requiring submission of a greater amount of data and triggering a more intensive risk assessment.

4.6 Australia

Australia has one of the more developed biotechnology regulatory frameworks in the world, through the Gene Technology Act of 2000, which has been implemented by the Gene Technology Regulations of 2011 (Tribe 2012). Under the Gene Technology Act and its regulations, both contained and non-contained uses of LMOs would require a license from the government, through the Office of the Gene Technology Regulator (OGTR).

The Australian law uses the terminology "dealings" to refer to any proposed use of a genetically modified organism. Contained uses of microorganisms would be considered as "dealings not involving release" (DNIR). Commercial and R&D DNIRs both require government review and approval, but in general, proposals for contained uses would face a shorter, easier approval process than would a proposal for outdoor uses of GMOs. Open-pond uses would be regulated by OGTR as "dealings involving release" (DIR). These proposed uses would be subject to greater scrutiny and a more involved risk assessment than DNIR applications, but the Australian government has approved a significant number of these applications. It appears that all the approved licenses have been for transgenic crop plants or for genetically modified vaccines, and none appear to cover either modified energy crops or modified microorganisms.

4.7 China

Under China's Biosafety laws and regulations, open-pond use of modified microorganisms would likely require approval from the Agriculture Ministry. Jurisdiction over contained uses is less certain, although approval would be needed to import LMOs into China for any purpose. See Chen et al. (2006) and Gupta and Falkner (2006) for more information on the Chinese regulatory regime.

5 Successful Regulatory Applications for Industrial Uses of Genetically Modified Microorganisms

5.1 Approvals Under EPA MCANs

EPA has been receiving MCANs and other notifications of biotechnology products under its interim TSCA policy since 1987 and under the current rules since 1997, and these regulations have not proven to be a barrier to industrial biotechnology companies, including those developing biofuel products or processes. As of this writing, there are 63 MCANs listed on the EPA website (USEPA 2014) as having been filed from the 1997 inception of the regulations through December 2013. The number and frequency of these filings have increased substantially in the last 3 years, as can be seen in Fig. 1, due to a greater number of proposals for biofuel or bio-based chemical production (Bergeson et al. 2014). All but one of the MCANs listed on the EPA website were favorably reviewed by EPA, and the intended products of the microorganisms covered by these MCANs are summarized in Table 5. The breakdown of MCANs by genus is shown in Table 6.

The greatest number of MCANs cleared by EPA have been for uses of intergeneric microorganisms to manufacture industrial enzymes. Many of these, particularly in recent years, have been for enzymes intended for use in the production of cellulosic ethanol or other biofuels. In recent years, the number of MCANs for biofuel or bio-based chemical production organisms has dramatically increased, such that production of fuel ethanol has become the second largest category: notably, 16 of the 22 MCANs for this purpose have involved the use of modified strains

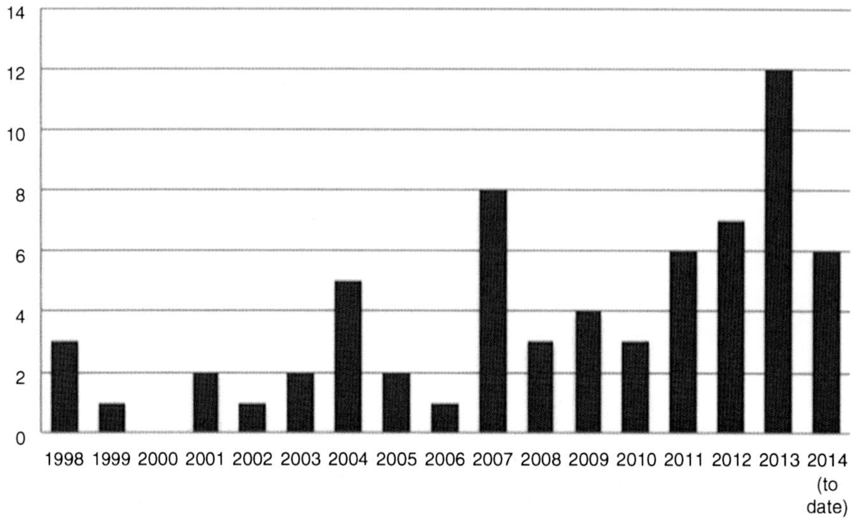

Fig. 1 MCANs Submitted to EPA, by U.S. Government Fiscal Year (through December 2013) (Source: http://www.epa.gov/biotech_rule/pubs/submiss.htm, accessed on October 23, 2014 (USEPA 2014). Data includes MCANs identified as filed through December 2013)

Table 5 Products produced by MCAN microorganisms

Industrial Enzymes[a]	27
Ethanol	22
Bio-Based and Other Specialty Chemicals	11
Research Enzymes	3
Total	**63**

Source: http://www.epa.gov/biotech_rule/pubs/submiss.htm, accessed on October 23, 2014 (USEPA 2014). Data includes MCANs identified as filed through December 2013

[a]Includes at least 6 MCANs for production of enzymes used in biofuel manufacture

of *Saccharomyces cerevisiae*. Although *S. cerevisiae* strains potentially qualify for the tiered exemptions described above, it has become common for developers of such strains to file MCANs to simplify the transfer and use of the strains to third parties, such as under partnership or licensing business models. This is largely due to EPA's policy that the Tier I exemption doesn't apply if significant quantities of live organisms are to be transferred between facilities, but also because once an MCAN is reviewed and approved by EPA, and the developer files a Notice of Commencement indicating that commercialization has begun, the strain is deemed to be placed on the TSCA Inventory and can therefore be used commercially by any

Table 6 Host organisms in MCANs filed through december 2013

Genus	Number
Trichoderma	18
Saccharomyces	16
Escherichia	5
Pichia/Komagataella	3
Zymomonas	2
Bacillus	2
Pseudomonas	2
Microalgae (unspecified species)	2
Klebsiella	1
Synechococcus	1
Unspecified (i.e. claimed as confidential)	11
Total	**63**

Source: http://www.epa.gov/biotech_rule/pubs/submiss.htm, accessed on October 23, 2014 (USEPA 2014). Data includes MCANs identified as filed through December 2013

party under any conditions. In contrast, the tiered exemptions are facility-specific, so that third party users of a modified strain for which one party has obtained an exemption would have to obtain their own approvals, through MCANs or tiered exemptions, before being allowed to use the strain.

There have been MCANs covering other species of microorganism for ethanol production, including *Zymomonas* and *E. coli*. As shown in Table 5, there have also been a considerable number of MCANs for various microbial production processes of bio-based or specialty chemicals, although the names of the applicant, the chemical and/or the production microorganism are often claimed as confidential in these filings. One series of MCANs identifies the product as an unspecified organic acid, although the identity of the submitter and the production organism have been claimed as confidential.

Among the most recent filings are two MCANs submitted by Solazyme, which are the first received and favorably reviewed by EPA under TSCA for the industrial use of modified eukaryotic algae. Although the identity of the microalgae species has been claimed as confidential in these MCANs, presumably one or both are for modified versions of the same algae species, *Prototheca moriformis*, that has been identified in online documents describing Solazyme's approvals for commercial use in Brazil (described below). Unlike many industrial uses of microalgae, Solazyme grows its algae strains in traditional contained fermentations, with the organisms growing heterotrophically, i.e. deriving their energy from chemical nutrients rather than via photosynthesis. These modified algae would be used to produce one or more chemicals, the identities of which have been claimed as confidential by the company.

MCANs have also been submitted for modified cyanobacteria. In 2012, Joule Unlimited Technologies filed the first MCAN for a modified strain of *Synechococcus* for production of ethanol, and although at this writing not yet listed on EPA's website, it is known that Algenol has also filed an MCAN for a modified cyanobacteria strain for ethanol production (P. Ahlm, personal communication).

Joule's MCAN is unique among all previously-filed MCANs in that the organisms would be grown outdoors, in durable, contained transparent photobioreactors arrayed horizontally to gather sunlight, rather than in a traditional stainless-steel fermenter.[5] In its evaluation of Joule's MCAN, EPA had no health or safety objections to use of the modified strain at Joule's Hobbs, New Mexico facility. However, because of the innovative nature of Joule's photobioreactors, EPA was not prepared to simply drop the MCAN from review, thereby granting the company unlimited rights to use the MCAN strain under any conditions. Instead, EPA and Joule entered into a voluntary consent order, which allows Joule to use the strain commercially at the Hobbs facility, while also providing EPA with further data resulting from such use. According to the EPA website, as of this writing Joule's MCAN is the only one which EPA has regulated with a consent order.

5.2 Approvals Under EPA TERAs

There has only been limited experience with TERAs since the biotechnology rule was put into place in 1997. According to EPA's website, there have been 30 TERAs submitted for field use of engineered microorganisms, almost exclusively for agricultural microorganisms, or for microorganisms to be used for bioremediation or for detection of hazardous contaminants in soil (USEPA 2014). All of these have been to propose small-scale, early-stage R&D projects, and all but three of these were approved. Two recent TERAs from 2013 were submitted by the U.S. Army Engineer Research and Development Center and the US Army Corps of Engineers to propose the use of modified strains of *Gordonia terrae* and *Rhodococcus jostii* in a field demonstration of bioaugmentation to enhance the degradation of hexahydro-1,3,5-trinitro-1,3,5-triazine (RDX) in contaminated groundwater (USEPA 2014). All the TERAs previously approved by EPA had been for use of the organisms in soils or encapsulated in devices for contaminant detection: these two TERAs appear to be the first in which release of GMMs into the groundwater was approved. EPA's approval included significant monitoring and reporting requirements.

More recently in the same year, EPA approved the first TERAs submitted for the experimental outdoor use of genetically modified algae. These are a series of applications submitted by Sapphire Energy, Inc., for open-pond testing of five intergeneric strains of the photosynthetic green algae *Scenedesmus dimorphus*. Sapphire submitted these TERAs on August 1, 2013, and EPA approved them on September 25, 2013, within the 60-day review period allotted under the regulations. The Sapphire TERAs proposed the testing of five different intergeneric strains of

[5] The author coordinated the preparation of Joule's MCAN and handled all interactions with EPA during its review of the filing, while employed by Joule Unlimited. The author also declares a financial interest in this company.

Scenedesmus dimorphus in open ponds, with the stated goals of evaluating the translatability of the genetically modified strains from the laboratory to an outdoor setting, and characterizing the potential ecological impact (dispersion and invasion) of the genetically-modified microalgae. The field trials were proposed to be conducted at the University of California San Diego Biology Field Station (BFS) in La Jolla, California, in collaboration with investigators from the university.

Sapphire's TERA included results of studies in both soil and water to show that the strains showed poor survival (i.e., zero or negative growth) in these environments, and also included a detailed description of the proposed outdoor experimentation and the procedures that will be followed to minimize and monitor the potential release of the organism from the test plots.

The studies under this TERA have been carried out. Among the main findings were that the modified algae were capable of dispersing and colonizing trap tanks up to 50 m distant from the test site, but that the rate of dispersal declines with distance; that both the modified and the wild-type algae were capable of growing in water from nearby lakes; and that the GM algae had no apparent effects on biomass, diversity or composition of native algae species found in the nearby lakes. In particular, the studies showed that the GM *Scenedesmus* is ecologically indistinguishable from the wild-type strains in its impact on native ecosystems (J. Shurin, personal communication).

5.3 Approvals Outside the United States

Although there have likely been other government approvals elsewhere in the world for use of modified microorganisms for fuel or chemical production, Internet searches for relevant information are difficult. While many countries and the Biosafety Clearinghouse that administers the Cartagena Protocol maintain detailed online records of approvals involving genetically modified crop plants, the same is generally not true for approvals for proposed uses of modified microorganisms under contained conditions; and where such online records exist, they often cover R&D as well as commercial applications, and in many cases would not distinguish industrial processes from pharmaceutical manufacture. Furthermore, within the European Union, approvals are granted by each individual EU member, meaning that there is no central location at which to search for approved industrial uses in Europe.

One exception is Brazil, where there has been considerable interest in commercialization of processes for manufacture of ethanol or bio-based chemicals, and where some records are available online. As of this writing, the advisory committee CTNBio (described above) has granted at least five approvals for industrial (i.e. non-pharmaceutical, non-food) uses of modified microorganisms in Brazil (CTNBIO 2014). These are: two approvals to Amyris Brazil S.A. for the use of modified strains of *S. cerevisiae* to produce farnesene (which Amyris uses to produce a jet fuel substitute and other products); one approval to Bio Celere Agroindustrial Ltda. for an *S. cerevisiae* strain modified to express a *Piromyces* xylA gene encoding xylose isomerase, for ethanol production; and two approvals to Solazyme Renewable Oils and Bioproducts Brazil Ltda. for the proposed use of the genetically modified microorganism *Prototheca moriformis* for the commercial production of triglycerides and bioproducts. *Prototheca moriformis* is a single-celled non-chlorophyll-containing obligatory heterotroph which reproduces asexually and does not produce spores. The latter two are notable in that they are approvals for genetically modified algae strains, which may correspond to Solazyme's two U.S. MCANs described above.

In Canada, Environment Canada maintains a website listing its risk assessment decisions and approvals of proposed uses of microorganisms subject to its New Substances Regulations described above (Environment Canada 2014). At this writing, the site lists 18 decisions, dating back to 2002; however, due to the scope of these regulations, these decisions cover proposed applications in many fields, including human and animal healthcare, and only a few relate to industrial biotechnology projects. These have included proposals for the use of modified microorganisms to produce industrial enzymes in contained manufacturing, and some proposed research projects relating to bioremediation. As noted above, these regulations administered by Environment Canada would be ones that would cover uses of modified microorganism or algae to produce fuels or chemicals.

6 Research Needs

There must be a comprehensive research base to support risk assessment if there is to be effective, science-based regulation that does not pose arbitrary barriers to commercialization. There is already a good deal of starting data regarding the environmental impacts of modified microorganisms, particularly those intended for use in agriculture or bioremediation. The available record to date gives some comfort that large-scale uses of modified microorganisms are not likely to have significant negative environmental effects. Nevertheless, there is a clear need for additional research and data, as several authors have suggested (see, for example, Snow and Smith 2012; Dana et al. 2012), particularly on microbial and algal species most likely to have industrial applicability. It may be useful to foster partnerships among industry, academia and government to focus additional research on organisms created by synthetic biology to develop this additional data.

Much of the needed research can take place at the laboratory or bench-scale, or can be conducted without the need for actual introductions of novel strains into the environment. For example, basic research into the biology, ecology and natural history of commercially-relevant wild type algae, cyanobacteria or other microbial strains would be necessary to develop data on baseline environmental behavior of such species. Basic research could also be carried out to address the important regulatory concerns discussed above, such as gene transmissibility, survival, persistence in the environment, and the genetics of algal toxin production. Laboratory, microcosm and macrocosm studies can be used to address these questions and to model the behavior of modified species in the environment. Such studies could have significant value, while recognizing the limitations of attempting to model environmental behavior in the laboratory (e.g., see Gressel et al. 2013, 2014 on the "plankton paradox").

Because of these limitations, for proposed uses of modified algae in open ponds, it will ultimately be necessary to assess environmental impacts through actual field experimentation, as has been common in agricultural research. It is appropriate to begin with small field studies designed with features to minimize potential dispersal of the test species in the environment (so-called "confinement" or even some measure of "containment"), along with soil, water and air monitoring as appropriate to detect possible dispersal or environmental persistence. If data derived from such small-scale studies provide no evidence suggesting any potential environmental harm, such studies could be followed up with larger-scale studies, in much the way new plant varieties and other new agricultural products are field tested in progressively larger field trials. In the case of genetically modified algae, even small-scale field trials would likely require some regulatory oversight, and the TERA process under EPA's TSCA regulations, discussed above, provides a very appropriate framework for doing so. Small-scale field trials of modified algae conducted under a TERA would be designed to include monitoring and other procedures to develop the data needed to support progressively larger field trials, and to ultimately support regulatory applications for approvals for commercial use in fuel or chemical production.

Biotechnology regulations around the world are structured so that conducting such studies in support of risk assessments would be the responsibility of the applicant. However, there is no reason that for-profit companies proposing larger-scale uses could not collaborate with academic or government scientists in carrying out such studies, in much the way Sapphire Energy collaborated with the University of California San Diego in carrying out the studies covered by the TERAs discussed above. Academic scientists wishing to propose and carry out their own studies of the environmental behavior of modified algae or cyanobacteria species could also take advantage of the TERA process: a number of the TERAs that EPA has reviewed and approved over the years originated from academic investigators, and the paperwork and other requirements to apply for TERA approval need not be burdensome. Finally, government agencies and their staff scientists could also become involved in carrying out such small-scale testing, as collaborators or principal investigators on such studies, or by offering field test sites at the national laboratories or other agency-controlled sites. This has often been done in the environmental remediation field, with many field tests of novel remediation technologies having been conducted at contaminated sites within facilities operated by the DOE or the military branches.

7 Conclusions

As more companies and research groups begin to contemplate or implement the use of genetically modified microorganisms, including organisms produced using synthetic biology, for biofuel or bio-based chemical production, greater attention will be turned to the need for appropriate science-based regulation and risk assessment. The scientific basis for such risk assessments is well-understood and is itself the subject of ongoing research, and regulatory frameworks exist around the world to ensure that such assessments take place. Review and approval for use of microorganisms in contained reactors (e.g. photobioreactors) should be fairly straightforward, as any potential risks would largely be mitigated by the choice of the production organism and design features of the reactor (e.g. in accordance with Good Industrial Large Scale Practice). Risk assessments of proposed open-pond uses might need to more rigorously address the key issues, but regulatory procedures like the TERA process of the U.S. EPA can ensure that risks are assessed in a stepwise manner, as field experimentation moves from small-scale to large-scale under conditions designed to minimize the potential spread of the organism from the test plot, and with appropriate monitoring and data-collection to support later experimentation or use at larger scale. Although critics may argue to the contrary, there is no reason to think that these regulations and risk assessments could not apply equally to organisms created through synthetic biology as they do for strains created by more established means of genetic manipulation. Collaborations between industry, academia and government can ensure that the technology moves forward in a responsible manner, to support the development of new processes that can address critical worldwide needs of developing novel sources of energy, while reducing carbon emissions and avoiding other detrimental environmental impacts.

List of Acronyms

EPA	U.S. Environmental Protection Agency
FDA	U.S. Food and Drug Administration
GMM	Genetically Modified Microorganism
GMO	Genetically Modified Organism
LMO	Living Modified Organism
MCAN	Microbial Commercial Activity Notice
PMN	Premanufacture Notice
TERA	TSCA Experimental Release Application
TSCA	Toxic Substances Control Act
USDA	U.S. Department of Agriculture

References

Adrio J-L, Demain AL (2010) Recombinant organisms for production of industrial products. Bioengineered 1(2):116–131

Alexander M (1985) Genetic engineering: ecological consequences. Issues Sci Technol 1(3):57–68

Bergeson LL, Auer CM, Peveler RD (2012) TSCA and the regulation of renewable chemicals. Ind Biotechnol 8(5):262–271

Bergeson LL, Auer CM, Hernandez O (2014) Creative adaptation: enhancing oversight of synthetic biology under the toxic substances control act. Ind Biotechnol. doi:10.1089/ind.2014.1532

Buschke N, Schafer R, Becker J, Wittmann C (2013) Metabolic engineering of industrial platform microorganisms for biorefinery applications–optimization of substrate spectrum and process robustness by rational and evolutive strategies. Bioresour Technol 135:544–554

BusinessWire (2013) Mascoma announces FDA favorable review of its next generation bioengineered yeast, TransFerm Yield+. http://www.businesswire.com/news/home/20130618006024/en/Mascoma-Announces-FDA-Favorable-Review-Generation-Bioengineered#.VDgUpyldW6U. Accessed 10 Oct 2014

Cao Y, Cao Y, Lin X (2011) Metabolically engineered Escherichia coli for biotechnological production of four-carbon 1,4-dicarboxylic acids. J Ind Microbiol Biotechnol 38(6):649–656

Cao Y, Zhang R, Sun C, Cheng T, Liu Y, Xian M (2013) Fermentative succinate production: an emerging technology to replace the traditional petrochemical processes. Biomed Res Int 2013:723412

Chen Y, Nielsen J (2013) Advances in metabolic pathway and strain engineering paving the way for sustainable production of chemical building blocks. Curr Opin Biotechnol 24(6):965–972

Chen C-H, Sassa Y, Suda E, Watanabe KN (2006) Biosafety system frameworks for living modified organisms in Japan and Taiwan. Plant Biotechnol 23(5):539–546. doi:10.5511/plantbiotechnology.23.539

Chen X, Zhou L, Tian K, Kumar A, Singh S, Prior BA, Wang Z (2013) Metabolic engineering of Escherichia coli: a sustainable industrial platform for bio-based chemical production. Biotechnol Adv 31(8):1200–1223

Colin VL, Rodriguez A, Cristobal HA (2011) The role of synthetic biology in the design of microbial cell factories for biofuel production. J Biomed Biotechnol 2011:601834

CTNBIO (2014) Commercial approvals: microorganisms. http://www.ctnbio.gov.br/index.php/content/view/14610.html. Accessed 22 Oct 2014

Dana GV, Kuiken T, Rejeski D, Snow AA (2012) Synthetic biology: four steps to avoid a synthetic-biology disaster. Nature 483(7387):29

Danish K, Epifani LE, Zevin A (2014) Inventory of Federal Regulations Affecting Biofuels other than the Renewable Fuel Standard. VanNess Feldman, LLP. http://bipartisanpolicy.org/sites/default/files/files/VNF_Biofuels.pdf. Accessed 16 Oct 2014

Darch H, Shahsavarani A (2012) The regulation of organisms used in agriculture under the Canadian Environmental Protection Act, 1999. In: McHughen A, Wozniak CA (eds) Regulation of agricultural biotechnology: the United States and Canada. Springer, Dordrecht, pp 137–145. doi:10.1007/978-94-007-2156-2_8

Davison J (2005) Risk mitigation of genetically modified bacteria and plants designed for bioremediation. J Ind Microbiol Biotechnol 32(11–12):639–650

de Jong B, Siewers V, Nielsen J (2012) Systems biology of yeast: enabling technology for development of cell factories for production of advanced biofuels. Curr Opin Biotechnol 23(4):624–630. doi:10.1016/j.copbio.2011.11.021

Dellomonaco C, Fava F, Gonzalez R (2010) The path to next generation biofuels: successes and challenges in the era of synthetic biology. Microb Cell Fact 9:3

Eggers B, Mackenzie R (2000) The Cartagena protocol on biosafety. J Int Econ Law 3(3):525–543. doi:10.1093/jiel/3.3.525

Environment Canada (2014) Biotechnology (living organisms) risk assessment decisions. http://www.ec.gc.ca/subsnouvelles-newsubs/default.asp?lang=En&n=8AD6A8C1-1. Accessed 22 Oct 2014

Enzing CN, Nooijen A (2012) Algae and genetic modification. Research, production and risks, COGEM

European Union (2001) Directive 2001/18/EC of the European Parliament and of the Council of 12 March 2001 on the deliberate release into the environment of genetically modified organisms. http://eur-lex.europa.eu/legal-content/EN/TXT/?uri=CELEX:32001L0018. Accessed 23 Sept 2014

European Union (2009) Directive 2009/41/EC of the European Parliament and of the Council of 6 May 2009 on the contained use of genetically modified micro-organisms. http://eur-lex.europa.eu/LexUriServ/LexUriServ.do?uri=OJ:L:2009:125:0075:0097:EN:PDF. Accessed 23 Sept 2014

Glaser A, Glick P (2012) Growing risk: addressing the invasive potential of bioenergy feedstocks. National Wildlife Federation, Washington, DC

Glass DJ (1991) Chapter 10: Impact of government regulation on commercial biotechnology. In: Ono RD (ed) Business of biotechnology, Newnes, Boston, pp 169–198, doi:http://dx.doi.org/10.1016/B978-0-7506-9119-2.50017-4

Glass DJ (1995) Biotic effects of soil microbial amendments. In: Rechcigl JE (ed) Soil amendments: impacts on biotic systems. Lewis Publishers, Boca Raton, pp 251–303

Glass DJ (2003) Regulation of the commercial uses of microorganisms. In: Encyclopedia of environmental microbiology, Wiley, New York. doi:10.1002/0471263397.env018

Golden JS, Handfield RB (2014) Why biobased? Opportunities in the emerging bioeconomy. U.S. Department of Agriculture, http://www.biopreferred.gov/files/WhyBiobased.pdf. Accessed 13 Oct 2014

Gressel J, van der Vlugt CJB, Bergmans HEN (2013) Environmental risks of large scale cultivation of microalgae: mitigation of spills. Algal Res 2(3):286–298, http://dx.doi.org/10.1016/j.algal.2013.04.002

Gressel J, van der Vlugt CJ, Bergmans HE (2014) Cultivated microalgae spills: hard to predict/easier to mitigate risks. Trends Biotechnol 32(2):65–69. doi:10.1016/j.tibtech.2013.11.003

Gupta A, Falkner R (2006) The influence of the Cartagena protocol on biosafety: comparing Mexico, China and South Africa. Global Environ Polit 6(4):23–55. doi:10.1162/glep.2006.6.4.23

He MX, Wu B, Qin H, Ruan ZY, Tan FR, Wang JL, Shui ZX, Dai LC, Zhu QL, Pan K, Tang XY, Wang WG, Hu QC (2014) Zymomonas mobilis: a novel platform for future biorefineries. Biotechnol Biofuels 7:101

Henley WJ, Litaker RW, Novoveská L, Duke CS, Quemada HD, Sayre RT (2013) Initial risk assessment of genetically modified (GM) microalgae for commodity-scale biofuel cultivation. Algal Res 2(1):66–77, http://dx.doi.org/10.1016/j.algal.2012.11.001

Hong KK, Nielsen J (2012) Metabolic engineering of Saccharomyces cerevisiae: a key cell factory platform for future biorefineries. Cell Mol Life Sci 69(16):2671–2690

Jang YS, Park JM, Choi S, Choi YJ, Seung Do Y, Cho JH, Lee SY (2012) Engineering of microorganisms for the production of biofuels and perspectives based on systems metabolic engineering approaches. Biotechnol Adv 30(5)):989–1000

Jones CS, Mayfield SP (2012) Algae biofuels: versatility for the future of bioenergy. Curr Opin Biotechnol 23(3):346–351. doi:10.1016/j.copbio.2011.10.013

Krimsky S (1985) Genetic alchemy: the social history of the recombinant DNA controversy. The MIT Press, Cambridge, MA

Kung Y, Runguphan W, Keasling JD (2012) From fields to fuels: recent advances in the microbial production of biofuels. ACS Synth Biol 1(11):498–513. doi:10.1021/sb300074k

Larkum AW, Ross IL, Kruse O, Hankamer B (2012) Selection, breeding and engineering of microalgae for bioenergy and biofuel production. Trends Biotechnol 30(4):198–205. doi:10.1016/j.tibtech.2011.11.003

Lennen RM, Pfleger BF (2012) Engineering Escherichia coli to synthesize free fatty acids. Trends Biotechnol 30(12):659–667

Lennen RM, Pfleger BF (2013) Microbial production of fatty acid-derived fuels and chemicals. Curr Opin Biotechnol 24(6):1044–1053

Menetrez MY (2012) An overview of algae biofuel production and potential environmental impact. Environ Sci Technol 46(13):7073–7085. doi:10.1021/es300917r

Nielsen J, Larsson C, van Maris A, Pronk J (2013) Metabolic engineering of yeast for production of fuels and chemicals. Curr Opin Biotechnol 24(3):398–404

Nozzi NE, Oliver JW, Atsumi S (2013) Cyanobacteria as a platform for biofuel production. Front Bioeng Biotechnol 1:7

OSTP (1986) Coordinated framework for regulation of biotechnology. Fed Regist 51:23302–23393
Peralta-Yahya PP, Keasling JD (2010) Advanced biofuel production in microbes. Biotechnol J 5(2):147–162
Radakovits R, Jinkerson RE, Darzins A, Posewitz MC (2010) Genetic engineering of algae for enhanced biofuel production. Eukaryot Cell 9(4):486–501. doi:10.1128/ec.00364-09
Rosenberg JN, Oyler GA, Wilkinson L, Betenbaugh MJ (2008) A green light for engineered algae: redirecting metabolism to fuel a biotechnology revolution. Curr Opin Biotechnol 19(5):430–436. doi:10.1016/j.copbio.2008.07.008
Rosgaard L, de Porcellinis AJ, Jacobsen JH, Frigaard NU, Sakuragi Y (2012) Bioengineering of carbon fixation, biofuels, and biochemicals in cyanobacteria and plants. J Biotechnol 162(1):134–147. doi:10.1016/j.jbiotec.2012.05.006
Ryan C (2009) Cultivating clean energy: the promise of algae biofuels. National Resources Defense Council, Washington, DC
Sayler GS, Ripp S (2000) Field applications of genetically engineered microorganisms for bioremediation processes. Curr Opin Biotechnol 11(3):286–289
Singh JS, Abhilash PC, Singh HB, Singh RP, Singh DP (2011) Genetically engineered bacteria: an emerging tool for environmental remediation and future research perspectives. Gene 480(1–2):1–9
Slating TA, Kesan JP (2012) A legal analysis of the effects of the Renewable Fuel Standard (RFS2) and Clean Air Act on the commercialization of biobutanol as a transportation fuel in the United States. GCB Bioenergy 4(2):107–118. doi:10.1111/j.1757-1707.2011.01146.x
Snow AA, Smith VH (2012) Genetically engineered algae for biofuels: a key role for ecologists. Bioscience 62(8):765–768. doi:10.1525/bio.2012.62.8.9
Tiedje JM, Colwell RK, Grossman YL, Hodson RE, Lenski RE, Mack RN, Regal PJ (1989) The planned introduction of genetically engineered organisms: ecological considerations and recommendations. Ecology 70(2):298–315
Trentacoste EM, Martinez AM, Zenk T (2014) The place of algae in agriculture: policies for algal biomass production. Photosynth Res. doi:10.1007/s11120-014-9985-8
Tribe D (2012) Gene technology regulation in Australia: a decade of a federal implementation of a statutory legal code in a context of constituent states taking divergent positions. GM Crops Food: Biotechnol Agric Food Chain 3(1):21–29
Urgun-Demirtas M, Stark B, Pagilla K (2006) Use of Genetically Engineered Microorganisms (GEMs) for the bioremediation of contaminants. Crit Rev Biotechnol 26(3):145–164. doi:10.1080/07388550600842794
USDA (1987) Introduction of genetically engineered organisms. Fed Regist 52:22892–22915
USDA (1993) Notification procedures for the introduction of certain regulated articles. Fed Regist 58:17044–17059
USDA (1997) Simplification of requirements and procedures for genetically engineered organisms. Fed Regist 62:23945–23958
USDOE (2010) National algal biofuels technology roadmap. U.S. Dept. of Energy, Office of Energy Efficiency and Renewable Energy, Washington, DC
USDOE (2013) Replacing the whole barrel to reduce U.S. dependence on oil. http://www.energy.gov/sites/prod/files/2014/04/f14/replacing_barrel_overview.pdf. Accessed 20 Oct 2014
USEPA (1997a) Fact sheet: commercialization of Sinorhizobium (Rhizobium) Meliloti, RMBPC-2. http://www.epa.gov/biotech_rule/pubs/factdft6.htm. Accessed 23 Sept 2014
USEPA (1997b) Microbial products of biotechnology; final regulation under the Toxic Substances Control Act. Fed Regist 62:17910–17958
USEPA (1997c) Points to consider in the preparation of TSCA biotechnology submissions for microorganisms. http://www.epa.gov/oppt/biotech/pubs/pdf/ptcbio.pdf. Accessed 23 Sept 2014
USEPA (1997d) Regulatory impact analysis for the regulation of microbial products of biotechnology: the regulated community. http://www.epa.gov/oppt/biotech/pubs/ria/ria013.htm. Accessed 23 Sept 2014

USEPA (2012) Microorganisms; general exemptions from reporting requirements; revisions to recipient organisms eligible for tier I and tier II exemptions. Fed Regist 77:54499–54511
USEPA (2014) TSCA Biotechnology notifications, FY 1998 to present. http://www.epa.gov/biotech_rule/pubs/submiss.htm. Accessed 23 Oct 2014
USFDA (2014) Generally Recognized as Safe (GRAS) notification program. http://www.fda.gov/AnimalVeterinary/Products/AnimalFoodFeeds/GenerallyRecognizedasSafeGRASNotifications/default.htm. Accessed 23 Sept 2014
Viebahn M, Smit E, Glandorf DM, Wernars K, Bakker PHM (2009) Effect of genetically modified bacteria on ecosystems and their potential benefits for bioremediation and biocontrol of plant diseases – a review. In: Lichtfouse E (ed) Climate change, intercropping, pest control and beneficial microorganisms, vol 2, Sustainable Agriculture Reviews. Springer, Dordrecht, pp 45–69. doi:10.1007/978-90-481-2716-0_4
Work VH, D'Adamo S, Radakovits R, Jinkerson RE, Posewitz MC (2012) Improving photosynthesis and metabolic networks for the competitive production of phototroph-derived biofuels. Curr Opin Biotechnol 23(3):290–297. doi:10.1016/j.copbio.2011.11.022
Wozniak C, McClung G, Gagliardi J, Segal M, Matthews K (2012) Regulation of genetically engineered microorganisms under FIFRA, FFDCA and TSCA. In: McHughen A, Wozniak CA (eds) Regulation of agricultural biotechnology: the United States and Canada. Springer, Dordrecht, pp 57–94. doi:10.1007/978-94-007-2156-2_4
Wrubel RP, Krimsky S, Anderson MD (1997) Regulatory oversight of genetically engineered microorganisms: has regulation inhibited innovation? Environ Manage 21(4):571–586
Yamanouchi K (2005) Regulatory considerations in the development and application of biotechnology in Japan. Rev Sci Tech 24(1):109–115
Yu C, Cao Y, Zou H, Xian M (2011) Metabolic engineering of Escherichia coli for biotechnological production of high-value organic acids and alcohols. Appl Microbiol Biotechnol 89(3):573–583
Zhang F, Rodriguez S, Keasling JD (2011) Metabolic engineering of microbial pathways for advanced biofuels production. Curr Opin Biotechnol 22(6):775–783

Microalgal Heterotrophic and Mixotrophic Culturing for Bio-refining: From Metabolic Routes to Techno-economics

Octavio Perez-Garcia and Yoav Bashan

Abstract In comparison with conventional photo-autotrophic cultivation, heterotrophic and mixotrophic cultivations of microalgae offers a feasible strategy to produce biomass and valuable chemicals through biorefinery processes. Supplementing microalgae cultures with organic carbon sources increase the biomass production and lipid/carbohydrate contents in cells. Consequently, this yields high productivity of biorefined products such as biodiesel, ethanol, starch and polyunsaturated fatty acids. Nevertheless, the addition of an organic carbon source imposes the necessity to pre-produce them and increase the cost and susceptibility of the cultures to microbial contamination. This chapter reviews the aspects related to the heterotrophic and mixotrophic cultivation of microalgae. These include advantages and limitations, metabolic routes of organic carbon assimilation, alternative carbon sources and main considerations for cultivation systems. It provides a comprehensive review of cultivated species, organic carbon sources and recently achieved productivity metrics. The chapter includes discussions regarding the main commercial products obtained using these cultivation modes and a simplified techno-economic analysis of the full biorefinery operation. The overarching aim of the chapter was to depict the main challenges for commercialization of chemical products using heterotrophic/mixotrophic cultivation of microalgae and identifies the promising research lines to achieve the same.

Keywords Heterotrophic microalgae • Mixotrophic microalgae • Metabolism • Microalgae cultivation • Bioreactor • Organic carbon source • Biorefinery • Techno-economics • Biofuels • Valuable chemicals

---

O. Perez-Garcia
Department of Civil and Environmental Engineering, The University of Auckland, 20 Symonds Street, CBD, Auckland 1010, New Zealand
e-mail: rper079@aucklanduni.ac.nz; octavio.perez@auckland.ac.nz

Y. Bashan (✉)
Environmental Microbiology Group, Northwestern Center for Biological Research (CIBNOR), Calle IPN 195, La Paz, B.C.S. 23096, Mexico

The Bashan Institute of Science, 1730 Post Oak Ct., Auburn, AL 36830, USA

Department of Entomology and Plant Pathology, Auburn University, 209 Life Sciences Building, Auburn, AL 36849, USA
e-mail: ybb0001@auburn.edu; ybashan@cibnor.mx

A. Prokop et al. (eds.), *Algal Biorefineries*, DOI 10.1007/978-3-319-20200-6_3

1 Introduction

Algae, in particular microalgae, are one of the most promising feedstock for sustainable production of biofuels and valuable chemicals. Nowadays, food supplements from microalgae comprise an important market in which compounds, such as β-carotene, astaxanthin, and polyunsaturated fatty acids (PUFAs), such as eicosapentaenoic acid (EPA), docosahexaenoic acid (DHA), and polysaccharides, such as β-glucan, dominate (Hudek et al. 2014). Research on algae is not only focusing on improving production of products, but also new algae products, such as biodiesel, bio-ethanol, and renewable chemicals, such as starch, sucrose and ethylene. In every case, the raw material for bio-refining these compounds is microalgae biomass, which is produced by culturing microalgae under controlled or semi-controlled conditions.

To achieve the largest possible microalgal productivity in a cost-effective way, selection of a microalgae mode of cultivation is of vital importance. Four major modes of microalgae cultivation can be adopted, namely photo-autotrophic, heterotrophic, photo-heterotrophic, and mixotrophic (Table 1; Wang et al. 2014). In general, microalgae are commonly grown by fixing dissolved, inorganic carbon (CO_2) and absorbing solar energy. Therefore, like most land-based plants, they perform photosynthesis and are photo-autotrophs. At the same time, some species of microalgae are also heterotrophic, using organic compounds in the growth medium as carbon and energy sources; therefore, they do not need light as an energy source (Chen 1996).

Heterotrophic growth is an aerobic process where assimilation of organic substrates generates energy through oxidative phosphorylation accompanied by oxygen consumption as the final electron acceptor. Mixotrophic cultivation is the growth mode where microalgae simultaneously uses inorganic CO_2 and organic carbon sources in the presence of light (Kang et al. 2004); therefore, photo-autotrophy and heterotrophy occur simultaneously (Wang et al. 2014). CO_2 is fixed through photosynthesis, which is influenced by illumination, while organic compounds are assimilated through aerobic respiration, which is affected by the availability of organic carbon. Several species are able to switch between photo-autotrophic and heterotrophic growth. This should not be confused with the mixotrophy regime, where both ways of uptake (organic and inorganic) occur at the same time. Mixotrophic microalgae use different sources of energy and carbon,

Table 1 Growth modes of algae (microalgae) cultivation

Growth mode	Energy source	Carbon source	Light availability requirements	Metabolism variability
Photo-autotrophic	Light	Inorganic	Obligatory	No switch between sources
Heterotrophic	Organic	Organic	No requirements	Switch between sources
Photoheterotrophic	Light	Organic	Obligatory	Switch between sources
Mixotrophic	Light and organic	Inorganic and organic	No obligatory	Simultaneous utilization

they may use organic or inorganic sources and light in different combinations. Mixotrophy makes microalgae more flexible because it may gather both carbon and energy demand by organic or inorganic sources and light simultaneously (Chen et al. 2011). In photo-heterotrophy, growth cells use light for energy, fix fixation of nitrogen and organic matter as a carbon source without CO_2 (Chen et al. 2011). Hence, because organic carbon and light are compulsory for photo-heterotrophic cultivation, it is rarely used as an approach to produce microalgal biomass to process valuable compounds (Wang et al. 2014). Therefore, photo-heterotrophic cultivation will not be discussed further.

Heterotrophic and mixotrophic cultivation of microalgae can be used in biorefinery processes. It is important to have a clear picture of where the cultivation regime fits in the overall process. Microalgae cultivation is the sum of procedures and techniques to produce the feedstock biomass that subsequently is refined to obtain valuable products. Therefore, prospecting, selecting, or developing a strain (by genetic and metabolic engineering) together with microalgae cultivation, and growth medium preparation belong to the "upstream" part of the entire process of biomass harvesting, de-watering, extraction of valuable compounds, transformation and purification belong to the "downstream" part of the process. Figure 1 presents an overview of the complete biorefinery process and highlights where the microalgae cultivation step fits in. Heterotrophic and mixotrophic cultivation of algae requires a supply of organic carbon that is provided from chemicals, such as glucose or acetate, crop flours, wastewater, food and milk industrial wastes, and lignocellulosic materials. Water supplemented with organic carbon and inorganic nutrients, such as nitrogen and phosphorus, are used to prepare the growth medium for cultivation. Medium preparation involves filtration, mixing, diluting, sterilizing, and hydrolysis of carbon sources. Appropriate strain selection supplies the inoculum (microalgae seed culture) for microalgae biomass production. Strain selection involves two main approaches: bio-prospecting and isolating strains with desirable metabolic and physiological capabilities and/or developing strains with capabilities by metabolic and genetic engineering. The cultivation step (biomass production) can be achieved by photo-autotrophic, heterotrophic, or mixotrophic growth of microalgae. The culture can be operated as batch, fed-batch, continuous, or semi-continuous modes and can be at a laboratory, pilot, or commercial scale. After cultivation, various downstream sub-processes follow to obtain the desired chemical products from the biomass. In general, these sub-processes include: (1) biomass harvesting, (2) extraction of the potential valuable chemicals, (3) separation of various microalgae raw chemicals extracted simultaneously, (4) transformation of raw materials to useful products, and (5) final purification for commercial products.

Compared with common photo-autotrophic microalgal growth, heterotrophic and mixotrophic approaches have the potential to provide larger biomass and yield of valuable organic compounds. These cultivation modes have their own technical challenges that impede large scale cultivation for producing valuable chemicals, including biofuels, in a cost-effective way (Bassi et al. 2014; Liang 2013; Perez-Garcia et al. 2011a). This chapter discusses the advantages and challenges, as well as the following topics: key aspects of cell metabolism related to growth modes,

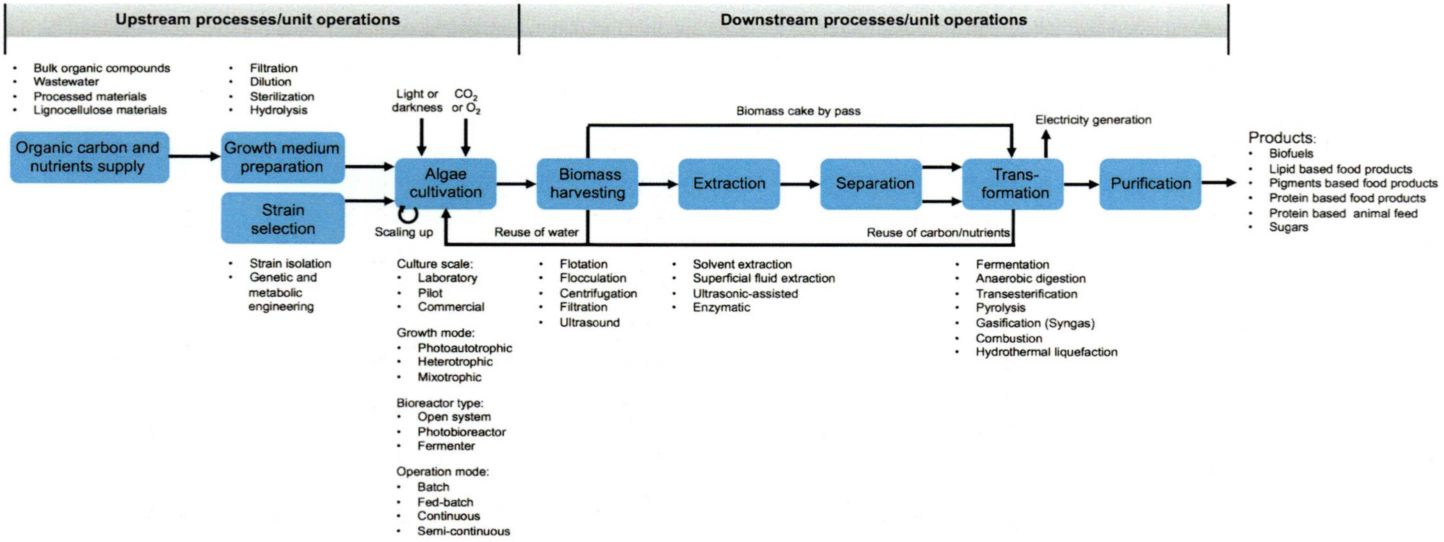

Fig. 1 Overview scheme of algae biorefinery process. Each *box* represents a specific unit operation (a part of the overall multi-step process with a specific function) achieved by a specific sub-process, *arrows* represent flow of mater or energy and *bullet points* above and below *boxes* specify sub-processes varieties or methods

potential carbon sources, and important aspects of cultivation methods. Finally, we present an analysis of important techno-economic aspects regarding heterotrophic and mixotrophic cultivation of microalgae and highlight opportunities to increase economic feasibility and sustainability of these technologies.

2 Advantages and Limitations of Heterotrophic and Mixotrophic Cultivation of Microalgae

2.1 Advantages

The advantages of heterotrophic cultivation of microalgae, in comparison with photo-autotrophic cultivation are the following: (a) Higher growth rate and biomass density (also called biomass productivity); (b) Higher lipid content per dry weight of cells (lipid productivity); (c) Higher biomass productivity per area of culture; (d) Cheaper and simpler bio-reactor design; (e) Easier scaling-up process; (f) The possibility to manipulate biomass composition by changing the culture medium's organic substrate that stimulates specific metabolic and biosynthetic pathways; and (g) Potential to remove organic carbon and several types of nitrogen and phosphorus compounds from wastewater (Brennan and Owende 2010; Chen 1996; Li et al. 2007; Lu et al. 2010; Miao and Wu 2006; Ogbonna et al. 2000; Xiong et al. 2010; Perez-Garcia et al. 2010).

Mixotrophic growth offers several advantages: (a) Higher growth rates than either heterotrophic and photo-autotrophic regimes by shortening growth cycles and producing higher biomass; (b) Prolonged exponential growth phase; (c) Reduction of lost biomass from respiration during dark hours; (d) Reduction or stopping of photo-inhibitory effect; (e) Flexibility to switch the cultivation regime to heterotrophic or photo-autotrophic regimens at will; and (f) Protection from photo-oxidative damage stimulated by accumulating oxygen in enclosed photo-bioreactors (Chojnacka and Marquez-Rocha 2004; Chojnacka and Noworyta 2004; Kröger and Müller-Langer 2011; Vonshak et al. 2000; Wang et al. 2014).

Production of microalgae under heterotrophic cultivation is a very successful route to commercialization of high-value chemicals, such as pharmaceuticals and food supplements. Economic advantages of heterotrophic and mixotrophic growth over the photo-autotrophic growth when mass-producing microalgae was summarized over 15 years ago by Chen (1996) and Borowitzka (1999); they are still valid. These microalgae biotechnology pioneers recognized that high cell population and biomass densities (between 20 and 100 g L^{-1}) can be achieved under heterotrophic cultivation in fermenters in complete darkness. Compared to the photo-autotrophic condition, heterotrophic conditions have enhanced concentrations of *Chlorella protothecoides* up to 3.4 times (Shi et al. 2002), of *C. vulgaris* up to 4.8 times (Liang et al. 2009; Perez-Garcia et al. 2010; Choix et al. 2012a;), and of *C. sorokiniana* up to 3.3 times (Zheng et al. 2012). A range of biomass between 4 and 20 g L^{-1} day^{-1} of microalgae are commonly produced by using heterotrophic cultivation (Graverholt and Eriksen 2007; Li et al. 2007; Shen et al. 2010; Xiong et al. 2010). This compares with 0.06–0.1 g L^{-1} day^{-1} in open cultivation ponds and 0.36 g L^{-1} day^{-1} in closed photo-bioreactors (Pulz 2001). Using heterotrophic and mixotrophic cultivation, researchers successfully increased the lipid content, compared to autotrophic cultivation. For example, the lipid content in a heterotrophic culture of *C. protothecoides* is 55.2 %, whereas an autotrophic culture of the same microalgae provides about 15 % (Xu et al. 2006). Cultivation of *C. vulgaris* under heterotrophic conditions, without nitrogen deprivation yielded faster growth and accumulated more lipids than under autotrophic conditions (Leyva et al. 2014).

Elimination of light in heterotrophic cultures reduces bioreactor construction costs and allows better control of growth of the microalgae and minimizes contamination by photosynthetic microorganisms. The design of photo-bioreactors for photo-autotrophic cultivation maximizes the area exposed to light irradiation to provide microalgae cells optimum photons for photosynthesis (Molina Grima et al. 1999; Rodolfi et al. 2009). While this is technologically feasible, it is difficult to reach a high biomass density of microalgae in photo-bioreactors. This happens because penetration of light in the medium is inversely proportional to cell concentration. Mutual shading of cells causes less light in the inner parts of the bioreactor. When this happens, production of biomass is very low; hence, there are very low yields of products (Borowitzka 1999; Chen et al. 2011; Molina Grima et al. 1999). Low biomass density in a reactor also increases the biomass harvesting cost. As a result, producing valuable products, specifically biofuels, from photo-autotrophic microalgae needs a long development time and huge investments before it becomes commercially viable (Liang 2013). For example, an insignificant amount of biodiesel is currently made from microalgae grown under photo-autotrophic conditions and this is not considered a commercial source. In shorter periods of time, developments to produce an abundant quantity of valuable products from microalgae should concentrate on cultivating microalgae in heterotrophic or mixotrophic growth modes, using cheap sugars or organic acids as carbon and energy sources.

Finally, cultivation of microalgae under heterotrophic and mixotrophic conditions provide environmental services. Efficiency of nutrient removal (N, P) from municipal, agricultural, or industrial wastewater by microalgae is higher under aerobic, dark heterotrophic conditions and mixotrophic conditions than under photo-autotrophic conditions (Andrade and Costa 2007; Li et al. 2011; Ogbonna et al. 2000; Perez-Garcia et al. 2010, 2011b; Zhou et al. 2013). In addition, although it is not possible to feed heterotrophic microalgae with CO_2 emissions, CO_2 sequestration is included in the overall process cycle because the initial organic substrate is produced by photosynthetic plants (Brennan and Owende 2010).

2.2 Limitations

While the metabolic and physiological features of heterotrophic and mixotrophic cultivation offer the possibility of greatly increasing cell density and productivity of microalgae, all these advantages notwithstanding, commercial production of cheap chemicals, mainly biofuel, using these cultivation regimes, are not presently cost effective (Kröger and Müller-Langer 2011; Liang 2013; Tabernero et al. 2012; Wang et al. 2014). This is caused by several major limitations: (a) Increasing energy expenses and general costs by adding an organic substrate; (b) Contamination and competition with other microorganism that grow faster than the microalgae; (c) Inability to produce light-induced metabolites into heterotrophic cultures; (d) Limited number of microalgal species that can grow heterotrophically and mixotrophycally; and (e) Indirect use of arable land for carbon source production reduces the main advantage that microalgal cultivation systems has over land-based crops (Borowitzka 1999; Chen 1996). Finally, it is important to acknowledge that heterotrophic or mixotrophic growth does not necessarily yield faster growth rates than photo-autotrophic cultivation.

Cultivation of heterotrophic and mixotrophic microalgal strains is based on high-cost carbon sources. With the essential requirement of organic compounds, the long term operation cost of heterotrophic cultivation is higher than for photo-autotrophic cultivation, even after reducing the cost of illumination in the latter case (Tabernero et al. 2012). For example, lipids from heterotrophic microalgae can partly replace fossil oils in biodiesel production, but from an economic standpoint, this concept is restricted by the costs of nutrients. For example, 80 % of the costs for producing biodiesel from *Chlorella protothecoides* comes from glucose in the growth medium (Li et al. 2007), Glucose as the carbon source yields the most lipids for many microalgae (Perez-Garcia et al. 2011a, b). Additionally, costs for nitrogen, phosphorus, vitamins, and trace metals are additional. The theoretical efficiency of heterotrophic microalgae feed with glucose is up to 75 % in the conversion to biodiesel. However, since glucose or other organic nutrients also have to be produced, in most cases from land-based crops, the efficiency of this organic nutrient production has to be added to the overall efficiency of heterotrophic or mixotrophic cultivation (Kröger and Müller-Langer 2011).

Another major disadvantage of heterotrophic and mixotrophic cultivation is their vulnerability to contamination by other microorganisms, many of them are fast growers. This may reduce the quality and quantity of products of interest. Therefore, managing axenic monoalgal culture is of vital importance for cultivation. This increases the capital and operation costs of bioreactors (Wang et al. 2014; Yan et al. 2011). Heterotrophic microalgal growth requires an oxygen supply, mixing, and nutrients, all under sterile environment. For most studies, a stirred-tank bioreactor commonly known as a fermenter is adopted. At the laboratory or pilot scale, such bioreactors are suitable. However, these fermenters are extremely expensive and economically unrealistic for production of cheap biofuels and chemicals at an industrial scale because the required volume, sterilization devices, and power consumption for mixing and aeration considerably increase capital investment and operation costs (Liang 2013). Tabernero et al. (2012) comprehensively evaluated the techno-economic feasibility of biodiesel production at the industrial scale from heterotrophic microalgae by analyzing data related to production plant construction and operation costs and lipid productivities of *C. protothecoides* in heterotrophic cultures. A non-conservative analysis revealed that an industrial plant with production capacity of 10,000 t year^{-1} of biodiesel is not economically viable unless biomass residues were refined into proteins and carbohydrates and sold. A conservative estimate showed no viability of the plant even if the residues were sold. The main economic drawback comes from the large number and volume of bioreactors (up to 465 reactors, each with working volume of 150 m^3) required to reach the targeted yearly production (Tabernero et al. 2012). Regardless of this recent grim assessment, mixotrophic cultivation should be further investigated for mass production of microalgae because it combines the features of photo-autotrophic growth, enhancing it by the partial use of organic nutrients for production of other high value chemicals (Kröger and Müller-Langer 2011).

Heterotrophic cultivation is inappropriate for most microalgae because most species are obligate autotrophs rather than facultative heterotrophs (Behrens 2005). Yet, some species are effectively grown in complete darkness and can be cultivated in conventional dark fermenters. The genera reported to grow heterotrophically include: *Amphora, Ankistrodesmus, Chlamydomonas, Chlorella, Chlorococcum, Crypthecodinium, Cyclotella, Dunaliella, Euglena, Nannochloropsis, Nitzschia, Ochromonas,* and *Tetraselmis* (Behrens 2005; Geider and Osborne 1989). Only a few microalgae species can grow mixotrophically. These include the freshwater *Brachiomonas submarina*, *Chlorella* spp., *Chlorococcum* sp., *Cyclotella cryptica*, *Euglena gracilis*, *Haematococcus pluvialis*, *Nannochloropsis* spp., *Navicula saprophila*, Nitzschia sp., *Ochromonas minima*, *Phaeodactylum tricornutum*, *Rhodomonas reticulate* and *Scenedesmus obliquus* (Bassi et al. 2014; Liang et al. 2009). Additionally, the cyanobacteria genera *Anabaena*, *Spirulina* and *Synechococcus* can be grown heterotrophically and mixotrophically (Chen et al. 1996; Kang et al. 2004). One strain of *Chlorella protothecoides* can grow on glucose, acetate, and other organic compounds in the dark, but it is unknown whether it can grow on organic carbon sources in the presence of light (Xu et al. 2006). It is important to consider that even though some species can grow on multiple carbon sources, not every microalgal species can be fed successfully with every organic nutrient for biomass production. For example, *C. vulgaris* yielded different biomass productivity in heterotrophic cultures from different carbon sources (Perez-Garcia et al. 2011a). Consequently a specific organic carbon source must fit with a specific microalga to obtain optimal growth or productivity of a compound.

To overcome the cost hurdle and make biorefineries of microalgae economically feasible, three main research areas need to be explored (Kröger and Müller-Langer 2011; Liang 2013; Tabernero et al. 2012): (a) Finding a low-value or, even better, a zero-value carbon source to support heterotrophic/mixotrophic microalgal growth; (b) Design economical bioreactors that are appropriate for industrial scale of heterotrophic/mixotrophic microalgae; and (c) Reduce downstream processing costs, especially those of harvesting and biomass transformation, for example, transesteri-

Table 2 Challenges and opportunities for the heterotrophic/mixotrophic cultivation of microalgae

Limitation	Opportunities
Carbon sources costs	Investigate new sources of cheap organic carbon, such as wastewaters, lignocellulosic material, and industrial processes waste
	Bio-prospection of strains able to assimilate cheap carbon sources
	Metabolic engineering of strains able to assimilate cheap carbon sources
	Improve methods for breakdown of lignocellulose material
Competition by fast-growing bacteria	Development of mixotrophic cultivation strategies
	Establishing cultures of microalgae able to thrive under bacteria-adverse environmental conditions
	Bio-prospection of fast-growing strains
	Metabolic engineering of fast-growing strains
	Immobilization of microalgae in polymers
Bioreactor implementation and operation costs	Cheaper materials for bioreactor vessel
	Implement alternative mixing strategies powered by a renewable energy source (hydraulic or wind)
	Implement cheap sterilization strategies
	Establish non-axenic microalgae cultures, such as open ponds
	Increase productivity of the metabolites of interest by optimizing bioreactor's operation parameters
	Risk assessment studies and regulations of GMOs in large-scale facilities
Downstream processes costs (biomass harvesting and raw product transformation)	Enhance exo-polysaccharides production to promote biomass flocculation
	Develop immobilization technique for the algae in polymeric beads/sheets
	Promote spontaneous excretion of metabolite of interest
	Selection or design of strains that excrete products
	Avoid compound extraction and separation by directly transform the biomass to products by pyrolysis, anaerobic digestion, gasification.

fication in biodiesel. Table 2 summarizes the limitations involved in heterotrophic and mixotrophic cultivation of microalgae, as well as promising research and development opportunities to address these limitations.

3 Heterotrophic and Mixotrophic Metabolism

The mode of cultivation significantly influences the metabolism and growth pattern of microalgae and determines the quality and quantity of biorefinery products. Heterotrophic or mixotrophic metabolism does not mean that every microalgae species can ingest every organic compound dissolved in water. In principle, it is impossible to generalize the properties of heterotrophically grown microalgae in relation to certain organic compounds. The review of Perez-Garcia et al. (2011b) provides an overview of the biochemistry behind use of nutrients in heterotrophic microalgae covering metabolism of major sources of carbon and nitrogen in heterotrophic and mixotrophic cultivation. The review summarizes the enzymatic pathways for assimilation of glucose, glycerol, and acetate, and the metabolism of nitrogen, including ammonium, nitrate, nitrite, urea, and organic nitrogen. This chapter, therefore, summarizes the main patterns with specific examples.

Table 4 presents metabolic performance data, such as maximum biomass concentration, growth rate, biomass productivity, and product productivity observed in studies on heterotrophic and mixotrophic growth of microalgae during mass cultivation. It shows that maximum growth rates for heterotrophic cultures for many microalgae range from 0.2 to 0.7 day^{-1}. Higher values, of up to 4.7 day^{-1}, have been reported for *Schizochytrium mangrovei* growing on food waste (Pleissner et al. 2013). Maximum growth rates observed in mixotrophic cultures ranged from 0.25 to 1.0 day^{-1} and are, in general, higher than those reported for heterotrophic cultivation. In every study where heterotrophic and mixotrophic growth were compared under identical conditions (except light exposure), mixotrophic cultures had higher growth rate and higher biomass concentration at the end of the experiment (Table 4) (Cheirsilp and Torpee 2012; Li et al. 2011; Zhou et al. 2013). For instance, *Chlorella* spp. and *Nannochloropsis* spp. grown mixotrophycally on glucose yielded higher biomass concentration than heterotrophic and photo-autotrophic cultures (Cheirsilp and Torpee 2012). An interesting observation is that the highest growth rate under mixotrophic regime (μ_{mixo}^{max}) corresponds approximately to the sum of the maximum growth rates obtained under the photo-autotrophic and heterotrophic modes ($\mu_{mixo}^{max} = \mu_{photo}^{max} + \mu_{hetero}^{max}$) (Girard et al. 2014). Similar observations are known for other microalgae and cyanobacteria species, such as *Chlorella regularis*, *C. vulgaris*, *Euglena gracilis*, *Haematococcus pluvialis*, and *Spirulina platensis* (Ogbonna et al. 2002). Nevertheless, this is only a general rule; experimental verification of growth rate is required in each case.

Apart from cell metabolism, cell structure changes, depending on growth regime. Under a photo-autotrophic regime, transmission electron microscopy showed that chloroplasts were clearly visible in photosynthetic cells (Lebsky et al. 2001). Membranes were abundantly accumulated in these chloroplasts and a number of starch granules could also be seen. In contrast, thylakoid membranes rapidly disappeared within 48 h after cells had undergone heterotrophic metabolism, suggesting degeneration of chloroplasts. Instead, the cytoplasm was almost totally filled with large lipid droplets (de-Bashan et al. 2002a). Biochemical and ultra-structural experiments suggested that chlorophyll breakdown and chloroplast degeneration was associated with lipogenesis during the heterotrophic growth of *C. protothecoides*. These cells were initially grown under photo-autotrophic conditions and were rich of chloroplasts (Xiong et al. 2010).

3.1 *Assimilation of Organic Carbon During Heterotrophic and Mixotrophic Cultivation*

3.1.1 Glucose

Glucose is the most commonly used carbon source for heterotrophic cultivation of microalgae, as is the case for many other microbial species. Far higher rates of growth and respiration are obtained with glucose than with any other substrate, such as other simple sugars, sugar alcohols, sugar phosphates, organic acids, and monohydric alcohols (Griffiths et al. 1960). This happens because glucose possesses more energy content per mol, compared with other substrates. For example, glucose produces 2.8 kJ mol^{-1} of energy compared to 0.8 kJ mol^{-1} for acetate (Boyle and Morgan 2009). The yield of 460 mg algal biomass per gram of glucose was achieved with the cyanobacteria *Spirulina platensis* (Marquez et al. 1993). This yield is comparable to yeasts and other aerobic heterotrophs. Similar results were obtained with *Chlorella protothecoides* (Li et al. 2007; Xu et al. 2006). The mass specific conversion rate from glucose to microalgal biomass varies from 40 to 64 % (Li et al. 2007; Xu et al. 2006); whereas, the lipid yield conversion has a variation ranging from 19 to 31 % of glucose (Kröger and Müller-Langer 2011).

Oxidative assimilation of glucose begins with phosphorylation of hexose, yielding glucose-6-phosphate, which is readily available for storage, cell synthesis, and respiration. An equivalent of a single phosphate bond is required per mole of glucose assimilated into glucose-6-phosphate. In that process, an additional 30 equivalents of phosphate bonds are generated by aerobic oxidation of a mole of glucose (Droop 1974). Of the several pathways used by microorganisms for aerobic glycolysis (breakdown of glucose), apparently only two, the Embden-Meyerhof pathway (EM pathway) and the Pentose Phosphate pathway (PP pathway) have been found in algae (Neilson and Lewin 1974). Under complete darkness heterotrophic growth, glucose is mainly metabolized via the PP pathway (Fig. 2), while the EM pathway is the main glycolytic process of cells in mixotrophic growth with light (Hong and Lee 2007; Lloyd 1974; Neilson and Lewin 1974; Yang et al. 2000). Both pathways are carried out in the cytosol and are functional in microalgae cells; however, the PP pathway might have a higher flux rate than the EM pathway, depending on light and presence of glucose.

The glycolytic EM pathway, Tricarboxylic Acid cycle (TCA cycle) and mitochondrial oxidative phosphorylation maintain high activities during photoautotrophic, heterotrophic, and mixotrophic growth of *C. pyrenoidosa* (chlorophyta) and *Synechocystis* sp (cyanobacteria) with glucose. This indicates a minor effect of the growth mode on these pathways in this species (Hong and Lee 2007; Yang et al. 2000). Central metabolic gene expression patterns (at the mRNA and protein levels) were analyzed for *Synechocystis* sp. growing under photo-autotrophic, heterotrophic, and mixotrophic conditions. The comparison showed that the majority of genes of the EM pathway and TCA cycle are expressed at the same levels under every trophic condition. However, expression of genes related to the Calvin cycle

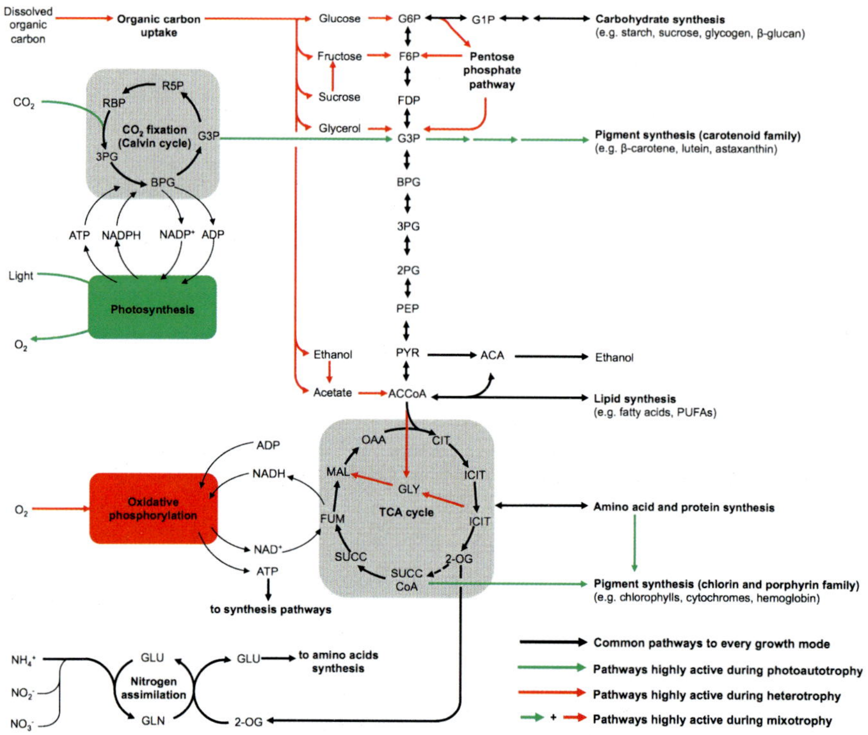

Fig. 2 Scheme of metabolic pathways for assimilation of carbon and production of energy in photoautotrophic, heterotrophic and mixotrophic microalgae. Compound abbreviations are following specified. *2-OG* 2-oxoglutarate, *2PG* 2-phospho glycerate, *3PG* 3-phospho glycerate, *ACA* acetaldehyde, *R5P* ribulose-5 phosphate, *ACCoA* acetyl-Coenzyme A, *ADP* adenosine-diphosphate, *ATP* adenosine-triphosphate, *BPG* 1,3-biphospho glycerate, *BPG* 1,3-bisphosphoglycerate, *CIT* citrate, *F6P* fructose-6 phosphate, *FDP* Fructose 1,6-biphosphate, *FUM* fumarate, *G1P* glucose-1 phosphate, *G3P* glyceraldehyde-3 phosphate, *G6P* glucose-6 phosphate, *GLN* glutamine, *GLU* glutamate, *ICIT* isocitrate, *MAL* malate, *NAD+* nicotinamide adenine dinucleotide (oxidized), *NADH* nicotinamide adenine dinucleotide (reduced), *NADP+* nicotinamide adenine dinucleotide phosphate (oxidized), *NADPH* nicotinamide adenine dinucleotide phosphate (reduced), *OAA* oxaloacetate, *OXA* oxalosuccinate, *PEP* phosphoenolpyruvate, *PYR* pyruvate, *RBP* ribulose-1,5 biphosphate, *SUCC* succinate, *SUCCCoA* succinyl-Coenzyme A

and glycogenesis (reductive EM pathway) was higher in photo-autotrophic and mixotrophic cultivation (Yang et al. 2002). The protein expression pattern under autotrophic conditions was very similar to that under mixotrophic conditions, suggesting that regulation of metabolism in these two modes occur with differences in metabolite concentrations and enzyme affinity to substrate and not at the genetic or transcriptional level (Yang et al. 2002).

Dark anaerobic cultivation of *Chlamydomonas reinhardtii*, *Chlorella vulgaris, Chlorococcum littorale*, *Oscillatoria* spp., *Spirulina* spp., and *Microcystis* spp. have been used for fermentative ethanol production (Hirano et al. 1997; John et al. 2011; Ueda et al. 1996; Ueno et al. 1998). Yet, algae cannot metabolize glucose under dark anaerobic conditions because insufficient energy is liberated during dissimilation of glucose, and low levels of the enzyme lactate dehydrogenase (EC 1.1.1.27), which is essential to complete the anaerobic fermentation process (Droop 1974; Neilson and Lewin 1974). Although these studies did not report growth under dark anaerobic conditions, they reported anaerobic endogenous respiration of storage compounds, such as starch and glycogen. Under dark anaerobic conditions, the oxidative reaction of starch is incomplete and, depending on microalgae species, H_2 gas, CO_2, ethanol, lactic acid, formic acid, and acetic acid are produced (John et al. 2011). Ueda et al. (1996) used microalgae biomass of the classes Prasinophyceae, Cryptophyceae, and Cyanophyceae as raw material for producing ethanol. The algal cells, containing a large amount of polysaccharides, were catabolized rapidly under dark anaerobic conditions to ethanol. Hirano et al. (1997) report that the intracellular conversion rate of starch to ethanol ranges from 0.2 to 0.15 g ethanol per gram of starch in *Chlamydomonas reinhardtii* under dark anaerobic conditions.

3.1.2 Carboxylic Acids (Mainly Acetate)

Uptake of dissolved carboxylic acids, such as acetic, citric, fumaric, glycolic, lactic, malic, pyruvic, and succinic, under microalgal heterotrophic cultivation, is well known for decades (Bollman and Robinson 1977). Acetate (or acetic acid) is one of the most common carbon sources for many microbial species, including microalgae (Droop 1974). The starting point for assimilating acetate is acetylation of coenzyme A by acetyl-CoA synthetase (EC 6.2.1.1). This forms acetyl coenzyme A (acetyl-CoA) in a single-step catalyzing reaction, using a single ATP molecule, (Fig. 2; Boyle and Morgan 2009; De Swaaf et al. 2003a; Droop 1974). Acetate (carried by acetyl-CoA) is generally oxidized metabolically through two pathways: the glyoxylate cycle to form malate in glyoxysomes (specialized plastids in the glyoxylate cycle) and the TCA cycle to citrate in the mitochondria, which provides carbon skeletons, energy as ATP, and energy for reduction as NADH. By definition, microalgae that grow by assimilating acetate must possess a glyoxylate cycle pathway to efficiently incorporate acetyl groups of acetyl-CoA into carbon skeletons. The operation of the glyoxylate cycle requires synthesis of isocitrate lyase (EC 4.1.3.1) and malate synthetase (EC 2.3.3.9). Both enzymes are induced when cells are transferred to media containing acetate (Boyle and Morgan 2009; Neilson and Lewin 1974). However, acetate does not always promote growth. It is toxic for many microorganisms at high concentrations. Keeping the concentration of acetate at low levels is useful for the fed-batch configuration in cultures or pH-auxostat, where pH is maintained constant (De Swaaf et al. 2003a; Zhang et al. 1999).

3.1.3 Alcohols (Mainly Glycerol)

Heterotrophic growth using glycerol as a substrate has been demonstrated for microalgae species, such as *Chlorella vulgaris*, *Nannochloropsis* spp., and *Schizochytrium limacinum* (Table 4; Chi et al. 2007; Das et al. 2011; Liang et al. 2009). Glycerol is a source of carbon and energy for microalgae growth; it is a very compatible solute for enzymes and membranes, with almost no toxic effects even at high concentrations (Richmond 1986). Large quantities of glycerol are obtained as a by-product of the biodiesel industry; thereby it is a cheap carbon substrate for heterotrophic cultivation of microalgae (Chi et al. 2007). In microalgae cells, glycerol is first phosphorylated using ATP and the glycerophosphate is then oxidized to triose phosphate. Microalgae genomes contain genes encoding for glycerol kinase (EC 2.7.1.30), sn-glycerol-3-phosphate NAD oxidoreductase (EC 1.1.1.8), and triose-phosphate (EC 5.3.1.1) (KEGG database; see Kanehisa and Goto 2000). These enzymes convert glycerol into glyceraldehyde-3-phosphate and glycerate, which are intermediates in the EM pathway of glycolysis to form pyruvate that enters the TCA cycle. In plant cells, sn-glycerol 3-phosphate, an intermediate formed during glycerol assimilation, inhibits the glycolytic activity of the EM pathway, thereby favoring the accumulation of carbohydrates from gluconeogenesis activity (Neilson and Lewin 1974). Aubert et al. (1994) suggest that the PP pathway is also inhibited when glycerol is the only carbon source (Fig. 2). Depending on the species, microalgae heterotrophic cultivation with glycerol can achieve as high a biomass concentration, growth rate, and PUFAs production as cultures using glucose as the substrate (Chi et al. 2007). Glycerol concentration in growth medium affects the biomass composition of the cultivated microalgae. For example, carbohydrate and lipid content in *C. vulgaris* cells was significantly higher in heterotrophic cultures on 2 % v/v glycerol than in cultures with 1 % v/v (Liang et al. 2009). Glycerol can also be used to cultivate microalgae mixotrophically. Mixotrophic cultures of *Nannochloropsis* spp. with glycerol reached higher biomass density and lipid production than phototrophic cultures (Das et al. 2011).

3.2 Respiration Process

In a broad sense, all organisms, including microalgae, use the same metabolic pathways for respiration. Cellular respiration is the metabolic process to produce ATP by oxidative phosphorylation. This process involves the consumption of electron donors (reduced organic or inorganic compounds) and electron acceptors (oxidized inorganic compounds). During heterotrophic and mixotrophic growth, microalgae perform aerobic dark respiration where organic compounds are used as electron donor while oxygen molecules are consumed as final electron acceptors. In microalgae, respiration plays two major roles: It serves as the exclusive source of energy for maintenance and biosynthesis under dark conditions, and it provides essential carbon skeletons for biosynthesis under any growth condition. In microalgae, dark

Table 3 Biomass composition ranges obtained in photo-autotrophic, heterotrophic, and mixotrophic cultivation of microalgae

Compound (%)	Photo-autotrophic	Heterotrophic	Mixotrophic
Protein	29–60	10–40	32–40
	(52.64±0.26)	(10.28±0.10)	(30±4.5)
Carbohydrates	10–20	15–45	10–30
	(10.62±0.14)	(15.43±0.17)	(20±5.0)
Lipids (total)	3–28	25–60	11–58
	(14.57±0.16)	(55.20±0.20)	(33±3.)
Ash	(6.36±0.05)	(5.93±0.04)	(7.00±0.26)
Others (nucleic acids and pigments)	(10.42±0.65)	(11.20±0.61)	(9.14±0.78)

Numbers in parenthesis are averages reported for *Chlorella protothecoides* growing on glucose (Xu et al. 2006; Heredia-Arroyo et al. 2010)

respiration of an organic substrate assimilated from the medium has rates varying from 0.01 to 0.6 day^{-1} (Perez-Garcia et al. 2011b). Respiration rates associated with assimilation of organic substrates vary through the cell cycle and complement O_2 consumption rates associated with endogenous respiration, that is, oxidation of an intracellular storage compound, such as starch (Lloyd 1974).

The respiration rate of any organic substrate is intimately geared to growth and cell division. Therefore, dark respiration rates increase with growth rates. Under optimal conditions, respiration rates are about 20–30 % of growth rates (Geider and Osborne 1989). Physiological regulation of respiration is assumed to be controlled by demand for the products of respiration metabolism, such as energy in the form of ATP and NADH and carbon skeletons provided by the organic substrate (Geider and Osborne 1989). With heterotrophic growth conditions, respiration rates equal or exceed the theoretical minimum cost of biomass synthesis. Values for CO_2 generated per carbon incorporated into new biomass (CO_2/C) ranged from 0.4 to 1.4 for several *Chlorella* species and diatoms. This indicates that biomass synthesis during heterotrophic growth conditions can proceed at nearly maximum theoretical efficiency. This happens since CO_2/C ratios for autotrophic growth are much lower than in heterotrophic growth (Raven 1976). Cellular assimilation of organic compounds is influenced by the amount of CO_2 dissolved in the culture's growth medium; a decline in organic matter per cell weight produced was greater when the supply of CO_2 was low (Pipes and Gotaas 1960).

3.3 Biomass Composition

The growth regime (photo-autotrophic, heterotrophic, or mixotrophic), together with the microalga strain and supplied organic substrate, greatly influences the molecular composition of microalgae biomass. The composition of biomass is usually reported as the percentage of biomass per dry weight of protein, lipids, or carbohydrates or any other specific molecule. Table 3 presents percentage ranges of dry weight biomass for the main cellular components reported under photo-autotrophic, heterotrophic, or mixotrophic cultivation of various species (Cheirsilp and Torpee 2012; Cheng et al. 2009a; Lu et al. 2010; Park et al. 2012; Xu et al. 2006). In general, energy storage molecules, such as lipids and carbohydrates (starch and glycogen) are accumulated under heterotrophic and mixotrophic conditions; therefore, the content in biomass of these compounds is higher than under photo-autotrophic conditions (Choix et al. 2012a, 2014). For example, lipid content in *Chlorella protothecoides* cultured under heterotrophic conditions with different carbon sources always reached from 50 to 60 % of cells dry weight in comparison to 15 % reached under photo-autotrophic conditions (Table 3) (Cheng et al. 2009a; Xiong et al. 2008; Lu et al. 2010; Xu et al. 2006; Yan et al. 2011; Leyva et al. 2014, 2015).

Mixotrophic cultivation of various microalgae species promotes a higher percentage of lipids in biomass in comparison to photo-autotrophic cultivation. Several examples show this phenomenon in various species of microalgae. Park et al. (2012) isolated 14 Chlorophyta strains from a cold northern Canadian province and compared their biomass composition. All newly isolated strains produced higher lipid content under mixotrophic cultivation than under photo-autotrophic cultivation. Similarly, lipid production in mixotrophic cultures of *Chlorella* sp. and *Nannochloropsis* sp. was notably higher than under photo-autotrophic and heterotrophic cultivation (Cheirsilp and Torpee 2012). *C. sorokiniana* was found to be well suited for lipid production. This was based on its high biomass production rate and lipid content, reaching 51 % under mixotrophic conditions. In this study, expression levels of *acc*D (heteromeric acetyl-CoA carboxylase beta subunit), *acc*1 (homomeric acetyl-CoA carboxylase), and *rbc*L (ribulose 1, 5-bisphosphate carboxylase/oxygenase large subunit) genes were studied with real-time PCR. Increased expression levels of *acc*D reflect the increased lipid content in the stationary phase of mixotrophic growth, but expression of the *acc*1 gene remained low. This result suggests that this gene may not be critical for accumulating lipids. Additionally, reduction of expression of the *rbc*L gene during mixotrophic conditions on glucose indicated that using glucose reduces the importance of this gene during photosynthesis (Wan et al. 2011). Compared with photo-autotrophic cultivation, lipid productivity of *Nannochloropsis* sp. under mixotrophic cultivation with glycerol was improved by 40 to 100 % (Das et al. 2011). Supplemented with glucose, lipid productivity of *Nannochloropsis oculata*, *Dunalliela salina*, and *Chlorella sorokiniana* under mixotrophic cultivation were 1.1–1.6 times, 1.8–2.4 times, and 4.1–8.0 times more productive than under photo-autotrophic cultivation, respectively (Chojnacka and Noworyta 2004). Production of lipids by *Scenedesmus obliquus*, under mixotrophic cultivation with a supply of glucose, could be as high as 270 mg L^{-1} day^{-1}, which is about 50 times greater than production under photo-autotrophic cultivation (Wang et al. 2014). Cheng et al. (2009a) compared lipid production of *Chlorella vulgaris* under photo-autotrophic, mixotrophic, and heterotrophic cultivation. Experimental results found that, by adding 1 % (w/v) glucose, production of lipids by *C. vulgaris* under mixotrophic cultivation was, respectively, 1.5 times and 13.5 times greater than under heterotrophic and photo-autotrophic cultivation.

Even though biomass production and lipid contents in biomass are affected by the microalgae's growth regime, the composition of lipids is not greatly affected.

Fatty acid (reported as fatty acids methyl esters, FAMEs) composition profiles in *Chlorella* spp. in studies of photo-autotrophic, heterotrophic, and mixotrophic cultivation is dominated by three fatty acid methyl esters: palmitic acid methyl ester ($C_{17}H_{34}O_2$), linoleic acid methyl ester ($C_{19}H_{34}O_2$), and oleic acid methyl ester ($C_{19}H_{36}O_2$). These fatty acids make up 70–90 % of the FAMEs under all tested conditions (Cheirsilp and Torpee 2012; de-Bashan et al. 2002a; Gao et al. 2010; Leyva et al. 2015; Lu et al. 2010; Rosenberg et al. 2014; Xu et al. 2006; Zhang et al. 2014). Not all fatty acids in the lipids can be transferred to FAMEs (biodiesel), since the percentage of lipids suitable for biodiesel production, namely saponifiable lipids, varies with strains and cultivating conditions. Thus, the productivity of saponifiable lipids should be regarded as a more precise measurement of microalgal biodiesel productivity potential (Wang et al. 2014).

Nitrogen depletion generally leads to lipid and carbohydrates accumulation (Brányiková et al. 2011; Choix et al. 2012a, b, 2014; Dragone et al. 2011; Illman et al. 2000; Khozin-Goldberg and Cohen 2011; Leyva et al. 2015; Li et al. 2008). For example, the lack of a nitrogen source (urea) increases the lipid content in *Chlorella* spp. (Hsieh and Wu 2009). Lack of a nitrogen source is the main reason why *C. sorokiniana* cells shifted from growth phase to stationary phase with high lipid content in mixotrophic cultures (Wan et al. 2011). When nitrogen is not present, accumulation of lipids is attributed to mobilization of lipids from chloroplast membranes by the enzyme 1,5-biphosphate carboxylase/oxygenase (EC 4.1.1.39, Rubisco) (García-Ferris et al. 1996). This is supported by the fact that development of chloroplasts and their light harvesting chlorophyll-protein complexes is dependent on nitrogen. Breakdown of chloroplasts for internal supply of nitrogen under nutrient reduction and dark conditions leads to cell survival and growth in the face of prolonged nutrient shortage if an external carbon source is not supplied. Batch cultivation experiments with *C. vulgaris* showed that acetyl-CoA carboxylase (EC 6.4.1.2) activity is directly related to fatty acid accumulation, especially in the final days of cultivation, when nitrogen levels in the medium are depleted (Leyva et al. 2014, 2015). Limiting N or P also influences the accumulation of carbohydrates in microalgae cells (Dragone et al. 2011). Brányiková et al. (2011) states that to produce biomass with high starch content, it is necessary to suppress cell division by limiting N or P, but not disturbing synthesis of starch in chloroplasts. These examples refer to work with *Chlorella* spp. Nutrient starvation has similar effects on lipid and carbohydrate accumulation in other microalgae, such as *Monodus subterraneus*, *Nannochloropsis* sp., and *Parietochloris incisa* (Khozin-Goldberg and Cohen 2006; Merzlyak et al. 2007; Pal et al. 2011; Solovchenko et al. 2008). Carbohydrate composition in *Chlorella*, specifically starch content, is modified when it is immobilized with the plant growth-promoting bacteria *Azospirillum brasilense* (Choix et al. 2012a, b, 2014). Heterotrophic cultivation with glucose or acetate of this co-immobilized cell system had higher total carbohydrate and starch content per culture and per cell than cultures with *Chlorella* spp. immobilized alone. The presence of *A. brasilense* cells in the immobilized system significantly prolonged production of starch in comparison to cultures with *Chlorella* spp. immobilized alone. These results demonstrate that microalgae immobilized with specific beneficial bacteria is a promising strategy to enhance productivity of desirable compounds in microalgae cultures.

3.4 Response to Light in Mixotrophic Metabolism

During mixotrophic cultivation, microalgae can grow both photo-autotrophically and heterotrophically and can use inorganic and organic carbon sources. Inorganic carbon is fixed through photosynthesis, which is influenced by illumination conditions, while organic compounds are assimilated through aerobic respiration, which is affected by the availability of organic carbon and oxygen (Wang et al. 2014). Since organic compounds can be assimilated under mixotrophic cultivation, growth of microalgae does not strictly depend on photosynthesis; hence light is not an absolute limiting factor for growth. Growth limitation by low illumination and growth photo-inhibition by high intensity of light can be significantly reduced in mixotrophic cultures (Wang et al. 2014). Photo-oxidative damages (especially in closed photobioreactors, where oxygen accumulation occurs) may also be reduced because of oxygen consumption during the heterotrophic metabolism. This compensates for increased oxygen production of the photo-autotrophic metabolism; thereby, oxygen cannot reach toxic levels (Chojnacka and Marquez-Rocha 2004; Vonshak et al. 2000).

Mixotrophic cultures have a 20–40 % higher growth rate at any given light intensity in comparison with photo-autotrophic cultures (Vonshak et al. 2000). Cells of the cyanobacteria *Spirulina platensis* grown under mixotrophic conditions exhibit a modified metabolic response to light in comparison to the one exhibited by cells growing under photo-autotrophic conditions (Vonshak et al. 2000). For example, its maximum photosynthetic rate and light saturation value in mixotrophic cultures were higher than those of the photo-autotrophic cultures. Dark respiration and light compensation point were also significantly higher in these cells. As expected, the mixotrophic cultures grew faster and achieved a higher concentration of biomass than the photo-autotrophic cultures. In contrast, the growth rate of the photo-autotrophic cultures was more sensitive to light. The differences between the two cultures were also apparent in their responses to exposure to very high photon flux density of 3000 μmol m^{-2} s^{-1}. The light-dependent O_2 production rate and the maximal efficiency of photosystem II declined more rapidly in photo-autotrophically grown cells than in mixotrophically grown cells, as a result of exposure to high photon flux density. Although both cultures recovered from the high photon flux density stress, the mixotrophic culture recovered faster and reached a higher concentration of biomass. Based on these results, growth of *S. platensis* with a fixed carbon source has a significant effect on photosynthetic activity (Vonshak et al. 2000).

Another interesting aspect is the reduction, or even complete stopping, of photo-inhibitory effects while microalgae are growing mixotrophically. This is explained either by the protective influence of the organic nutrient on cell metabolism or the increase of light intensity required to photo-inhibit cell growth (Chojnacka and Noworyta 2004). Presence of light, however, can also photo-inhibit uptake of organic carbon by affecting the balance between reduced and oxidized energy-carrying molecules (ATP and NADH), as a consequence of photosynthetic activity. For example, in *C. kessleri* cultures, high light intensities (above 120 μmol m^{-2} s^{-1}) inhibited consumption of organic soluble compounds in a growth medium based on wastewater. However, consumption of soluble compounds was independent of light intensity in *C. protothecoides* (Li et al. 2012). If large-scale cultivation is the ultimate goal, only those microalgae strains which are not sensitive to photo-inhibition of organic carbon uptake are suitable for mixotrophic cultivation (Li et al. 2012).

The effect of light intensity on accumulating biomass, removing wastewater nutrients and producing biodiesel through microalgae mixotrophic cultivation, was investigated using *C. kessleri* and *C. protothecoides* (Li et al. 2012). The experimental light intensities were 0, 15, 30, 60, 120, and 200 μmol m^{-2} s^{-1}. The results show that light intensity had a profound impact on responses in both species. The dependence of these responses on light intensity varied. For *C. kessleri*, optimum light intensity was 120 μmol m^{-2} s^{-1} for biomass productivity and removal of NH_3-N and PO_4-P, except for removal of organic carbon measured as chemical oxygen demand (COD). For *C. protothecoides*, optimum light intensity was 30 μmol m^{-2} s^{-1}. Light intensity also greatly affects the composition and quantity of biodiesel derived from algae cultivated under different lighting conditions. The major components of biodiesel produced from microalgae biomass were 16-C and 18-C fatty acids methyl esters (FAME). The highest biodiesel contents were 24.19 % and 19.48 % of dried biomass for *C. kessleri* and *C. protothecoide*, respectively. Both species were capable of removing wastewater nutrients with high removal efficiencies under all lighting conditions (Li et al. 2012).

Mixotrophic cultivation is also associated with lower emission of CO_2 than heterotrophic cultivation on the basis of per unit biomass/lipid production. This happens because part of the CO_2 release can be compensated by photosynthesis (Chen et al. 2011; Xiong et al. 2010). Compared with heterotrophic cultivation, mixotrophic cultivation of *C. protothecoides* released 61.5 % less CO_2 with production of the same yield of lipid (Xiong et al. 2010). Nevertheless, the enzyme Rubisco, which is responsible for CO_2 fixation, remains functional in cells at the heterotrophic phase (Xiong et al. 2010). During mixotrophy, growth is influenced by the medium supplement of glucose during both the light and dark phases; hence, there is less loss of biomass during the dark phase characterized by CO_2 emission (Wang et al. 2014).

3.5 *Advances of Metabolic and Physiological Characteristics of a Strain*

One aspect often mentioned to overcome bottlenecks in microalgal cultivation is to produce biomass using a high-performance strain. Ideally, microalgae strains used for bio-refining of valuable compounds should present the following metabolic and physiological characteristic (Brennan and Owende 2010; Wijffels et al. 2010):

- High production of biomass (fast growth)
- High production of a valuable compound (high percentage of compound weight per unit of biomass)
- Able to growth heterotrophically, using cheap carbon substrates
- High biomass or valuable compound yield per unit of light and/or organic substrate

- Robust growth (able to maintain productivity under changing conditions)
- Good ecological dominance (able to maintain high productivity under non axenic conditions)
- Growth at a wide range of pH
- High CO_2 sinking capacity
- Insensitivity to oxygen (resist photo-oxidative stress)
- Self-flocculation characteristics
- Excretion from the cells of valuable compounds (to facilitate product extraction)
- Large cells with thin cell walls (to facilitate product extraction)
- Resistant to possible inhibitory effect of organic compounds

Currently, none of the known microalgal strains is capable of meeting all these requirements with high efficiency. Optimization of these characteristics can lead to a robust and efficient process of scaling-up and eventually to successful commercialization of products. In general, there are two possible approaches to improve the metabolic and physiological characteristics of biorefinery microalgae cultures: isolation of strains with desirable characteristics (bio-prospecting) and, designing improved strains by genetic/metabolic engineering.

3.6 Bio-prospecting of Species for Valuable Compound Production

Each microalgal species and strain has different capacities to assimilate different organic compounds, and these characteristics must be considered when determining its use in producing desirable compounds (Kröger and Müller-Langer 2011). Only a few microalgal strains are cultivated commercially for various high-value products. These include: *Chlorella prothotecoides, Crypthecodinium cohnii, Dunaliella salina, Haematococcus pluvialis,* and *Nannochloropsis gaditana.* These are not necessarily the best strains for the production of cheap chemicals, such as biodiesel. Consequently, it is necessary to search for new strains or modify the strains so that production of valuable compounds becomes economically viable. In this context, efforts at bio-prospecting of heterotrophic microalgae should focus on: testing strains under several carbon sources at different concentrations (like those tests depicted on Fig. 3), including testing under mixotrophic conditions and testing for growth on cheap organic substrates, such as food waste, industrial by-products, or industrial waste.

Barclay and Apt (2013) and Charles et al. (2002) described approaches and protocols for bio-prospecting and isolating microalgae from natural habitats. These include bio-prospecting strategies focused on: (a) broad collections of algae; (b) collections on specific habitat types; (c) isolating extremophilic microalgae; (d) a target chemical composition or product; (e) heterotrophic growth potential; (f) wastewater tolerance; (g) pesticide tolerance, and (h) targeting various strain capa-

Fig. 3 Bio-prospecting heterotrophic capabilities of *Chlorella vulgaris* UTEX 2714 under completely dark conditions: (**a**) 50 mL tube-cultures after 4 days of cultivation on different organic substrates, initial concentration of substrate was constant in all cultures and (**b**) 250 mL flask cultures

bilities. Serial plating (Park et al. 2012) and serial dilution (Bhatnagar et al. 2011) are two common isolation methods; however, more innovative approaches have been successfully implemented. For example, selection of "xylose-enhanced" strains was carried out by cyclic exposure to UV radiation and fluorescent light of microplates containing *Chlorella* spp. (Hawkins 1999).

Bio-prospection can lead to the discovery of microalgae with novel heterotrophic growth capabilities. Girard et al. (2014) tested the capability of *Scenedesmus acutus*, *Scenedesmus obliquus*, *Chlorella vulgaris*, and *Chlorella protothecoides* to grow on lactose and cheese whey as carbon sources under heterotrophic and mixotrophic conditions. They found that only *Scenedesmus obliquus* can use these substrates for growth. Wang et al. (2010) isolated a *Chlorella* sp. strain tolerant of high nitrogen loads (81–178 NH_4-N mg L^{-1}) in dairy manure. *Chlorella* sp. was able to remove all the nitrogen under heterotrophic cultivation within 21 days. Three mixotrophic microalgae strains of *Chlamydomonas globosa*, *Chlorella minutissima*, and *Scenedesmus bijuga* were isolated after long-term enrichment of cultures containing industrial wastewater (Bhatnagar et al. 2011). In another example, native microalgal strains were isolated from carpet industry wastewater. Some of the isolated strains had heterotrophic growth capability and also reduced the nutrient content on medium (Chinnasamy et al. 2010). Park et al. (2012) analyzed biomass and lipid production of 14 microalgae isolates (in the Chlorophyta family). Each isolated strain has a distinctive linear correlation between biomass and lipid productivity under photoautotrophic and mixotrophic conditions, but only one strain (*Chlamydomonas debaryana*-AMLs1B) held promise for biorefinery purposes. So far, bio-prospecting for strains for heterotrophic and mixotrophic growth has potential, but few strains were isolated to show the usefulness of this approach.

3.7 Genetic and Metabolic Engineering for Improving Existing Strains

Genetic and metabolic engineering can be used to expand heterotrophic and mixotrophic capabilities of strains by enabling a strain to growth on a specific organic substrate or developing a metabolic pathway for a novel product synthesis by inserting the genes for these pathways. Genetic and metabolic engineering are likely to improve the metabolic performance of microalgal strains producing valuable compounds during the coming decades (Brennan and Owende 2010; León-Bañares et al. 2004; Wang et al. 2009). There is an increasing interest in use transgenic microalgae as green cell factories capable of producing biofuels and valuable proteins and carbohydrates (León-Bañares et al. 2004). Nowadays few microalgae, including the green algae *Chlamydomonas reinhardtii*, *Nannochloropsis gaditana*, and *Ostreococcus tauri* or the diatom *Phaeodactylum tricornutum* are currently established as platforms for genetic and metabolic engineering (Wijffels et al. 2013). Cellular transformation techniques have been much better developed in cyanobacteria and transformation systems need to be further developed for microalgae (Wijffels et al. 2013). A promising strategy is transformation of heterologous (prokaryotic and eukaryotic) genes in established eukaryotic hosts, such as *Chlamydomonas reinhardtii* (Wijffels et al. 2013).

Successful metabolic engineering of strains involves the sequencing of a wild type strain, a systematic description of the organism genotype and phenotype using 'omics' techniques, then the application of molecular and genetic techniques for metabolic engineering. Metabolic engineering of a specific strain has one of the following goals: (a) expand metabolic capabilities by adding new genes to an organism; (b) add genes encoding for transporters of products and intracellular toxic compounds so cells can excrete them faster; (c) increase formation of a precursor of a valuable compound; (d) remove precursor catabolism; (e) promote synthesis and storage of compounds in intracellular compartments; and (f) promote efficient product-secretion signaling to avoid product feedback inhibition (Lee et al. 2008; Wang et al. 2009; Wijffels et al. 2013). Metabolic engineering tools to achieve these goals include: insertion of heterologous genes in the genome; insertion of heterologous genes extra-chromosomally via conjugation with replicable plasmids; resistance markers; and random mutagenesis procedures (León-Bañares et al. 2004). Using these methods, various *de novo* metabolic pathways have been introduced into cyanobacteria species, such as *Synechocystis* spp. and *Synechococcus* spp. (Ducat et al. 2012; Gao et al. 2012) and the microalgae *Nannochloropsis gaditana* (Radakovits et al. 2012). To gain the best possible insight into metabolic pathways leading to the product of interest, systems biology, a bioengineering research field based on computational and mathematical modeling of complex biological systems, is used to integrate data from culture's performance measurements and diverse state-of-the-art technologies. These include genome sequencing, transcriptomics, metabolomics, proteomics, metabolic modeling (fluxomics), and bioinformatics.

Two key aspects for successful application of metabolic engineering are the availability of well-annotated genomes and the availability of quantitative tools for metabolic modeling that permit understanding and manipulation of the genome. Metabolic models, but most specifically stoichiometric metabolic network (SMN) models, are at the core of the systems biology approach by providing a computational platform for omics data analysis and phenotype prediction. Figure 4 presents the metabolic network formed by a model of *Chlamydomonas reinhardtii* metabolism developed by Chang et al. (2011). This network accounts for 1080 genes associated with 2190 reactions and 1068 unique metabolites. SMN models have mainly focused on the derivation of flux-balance analysis or FBA (Orth et al. 2010), which can be used either for designing the bioreactor's operation conditions or targeting metabolic reactions and genes for metabolic engineering (Oberhardt et al. 2009). For example, Perez-Garcia et al. (2014) used a metabolic network model of the ammonia-oxidizing bacteria *Nitrosomonas europaea* to find bioreactor operation conditions to avoid the emission of greenhouse gases from biological nitrogen-removal processes.

Legal regulations for large-scale outdoor production of genetically modified microalgae are not yet in place, but efforts to establish regulations are in progress. As a starting point, it is necessary to provide risk assessment studies before outdoor industrial applications of genetically modified cyanobacteria and microalgae (Wijffels et al. 2013). An argument against genetic engineering of strains is that many microorganisms suitable for outdoor mass culture are yet to be discovered; therefore, development of new strains is unnecessary (Rodolfi et al. 2009). Nevertheless, it may be prudent to limit projections to what can be achieved with natural strains. For example, the Aquatic Species Program of the US Department of Energy collected over 3000 strains of oil-producing photosynthetic microorganisms, which after screening, isolation, and characterization efforts, narrowed the collection of promising organisms to 300 species, mostly green algae and diatoms (Sheehan et al. 1998).

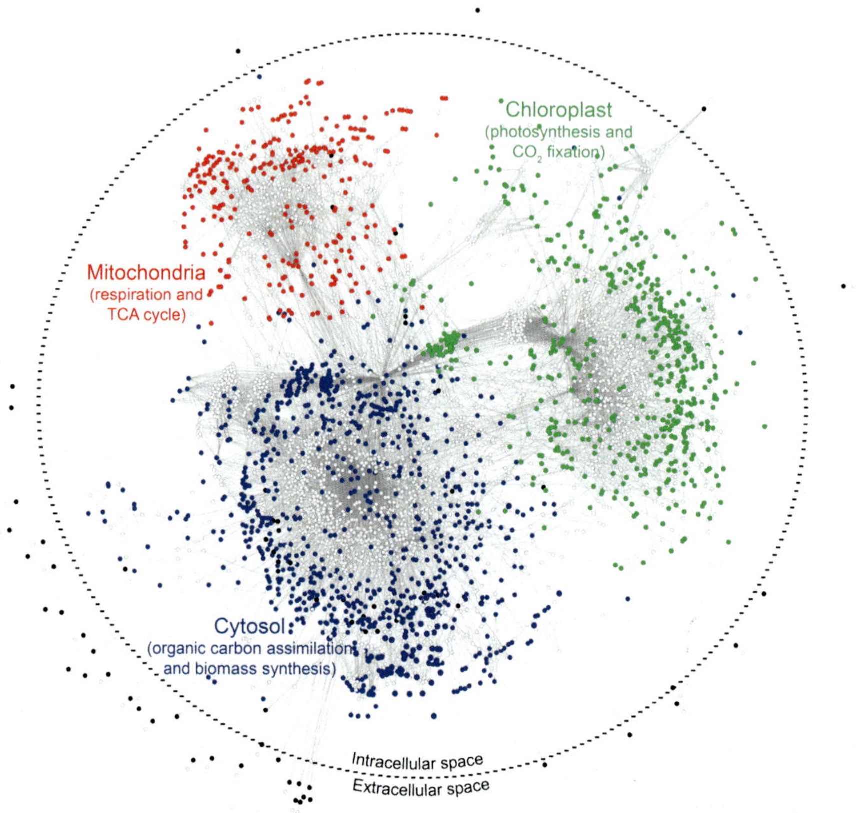

Fig. 4 Genome scale metabolic network of *Clamydomonas reinhardtii* used for systems biology and metabolic engineering studies. The *big circle* formed by a *slashed line* represent the boundary of cellular metabolism. Intracellular *blue* nodes represent metabolites occurring in cytoplasm, Golgi apparatus, nucleus and glyoxysome; *green* nodes represent metabolites occurring in chloroplast and thylakoid lumen; *red* nodes represent metabolites occulting in mitochondria; *black* nodes on the "extracellular space" represent metabolites able to be assimilated and/or secreted from the growth medium, white small nodes represent metabolic reactions, *grey edges* connect metabolic compounds with reactions; the metabolic network was created for the *C. reinhardtii* metabolic model published by Chang et al. (2011) using the software Cytoscape 3.0.1 (Cytoscape consortium, San Diego, USA)

4 Organic Carbon Substrates

4.1 Bulk Organic Compounds

Although heterotrophic and mixotrophic microalgae can use a wide range of carbon and nitrogen sources, from a commercial perspective, the most economic organic substrates for heterotrophic cultivation are glucose, glycerol, and acetate (Barclay et al. 2013; Perez-Garcia et al. 2011b). However many carbon sources, such as arabinose, citrate, fructose, malate, lactic acid, lactose, peptone, urea, fulvic acids, ethanol, methanol, and sucrose, have been tested for heterotrophic cultivation of microalgae. *Chlorella vulgaris* cultures on the above mentioned carbon sources reached significantly lower biomass concentration in comparison to cultures on acetate or glucose (Perez-Garcia et al. 2011a). Microalgal cells growing on "less preferred" substrates require a lag time (acclimation period) to develop the specific transport systems necessary for compound uptake. For example, consumption of "less preferred" substrates is aborted because the enzymes that catalyze their uptake and assimilation cannot be synthesized in the presence of the preferred substrate in *Crypthecodinium cohnii* (Ratledge and Wynn 2002). Lag time is also dependent on the strain, bioreactor configuration, and environmental growth conditions.

Although using glucose, acetate, or glycerol as organic substrate is pertinent for research, these compounds are too expensive for an economically viable bio-refinery production process. For example, glucose is about 80 % of the costs of the medium, when cultivating heterotrophic *Chlorella* spp. for biodiesel production (Kröger and Müller-Langer 2011). Alternative, low-cost substitutions to glucose, such as starch and cellulose-hydrolyzed solutions are ideal carbon sources for heterotrophic microalgae cultivation because similar biomass production can be achieved as using glucose, but at a lower cost (Li et al. 2007). According to two rigorous techno-economic analyses (Liang 2013; Tabernero et al. 2012), production of biofuel and valuable chemicals using heterotrophic and mixotrophic microalgae is not viable if the carbon source is a pure chemical or is it based on land crops, such as sugarcane, established to supply microalgae media. Production of valuable compounds by heterotrophic and mixotrophic microalgae are only viable if organic substrates are supplied from wastes or by-products of other processes. Consequently, studies have investigated microalgae growth on several alternative organic substrates, such as wastewater, industrial by-products, and lignocellulosic materials. The following sections discuss the main findings of those studies.

4.2 Wastewater

Most domestic and industrial wastewaters contain organic carbon, nitrogen, phosphorus, and other minor compounds. This composition makes wastewater suitable for growing microalgae. Besides growing microalgal biomass for valuable compound production, the wastewater is simultaneously treated (Christenson and Sims 2011; Rawat et al. 2011). This double benefit has attracted extensive attention in recent years. Yet, this approach has its drawbacks: (a) some wastewater may be too toxic to support microalgal growth; (b) with outdoor treatment, microalgal growth and wastewater treatment efficiency is significantly affected by seasonal changes; and (c) competition among the microbial community in the wastewater may lead to very slow algal growth. Hence, using microalgae to treat wastewater is problematic for high biomass productivity (Liang 2013; Valderrama et al. 2002).

Soluble organic compounds in wastewater can be assimilated by heterotrophic and mixotrophic microalgae. Organic carbon in wastewater is usually reported as chemical oxygen demand (COD). Heterotrophic and mixotrophic activity in wastewater is usually inferred from the decrease of COD concentration during experimental cultivation (Hu et al. 2012; Li et al. 2011). Hu et al. (2012) observed a decrease in COD concentration from 2.5 to 0.5 g L^{-1} in mixotrophic cultures of *Auxenochlorella protothecoides* in municipal wastewater. This cultivation presented a COD removal efficiency of 78.9 %. Probably the remaining 0.5 g L^{-1} of COD belonged to the non-biodegradable COD fraction. *Chlorella pyrenoidosa*, growing in sterilized sewage, was able to use some of the organic matter, as indicated by a decrease in soluble biological oxygen demand (BOD) and dissolved volatile solids (VSS). However, not all organic chemical substances in the polluted or unpolluted environments can be used as a carbon source for microalgae since some are known to inhibit microalgal growth (Chen 1996; Perez-Garcia et al. 2011a). Various studies show that supplementing readily-biodegradable carbon sources, such as glucose or acetate, is necessary to achieve the dual goal of high nutrient removal efficiency and high biomass production (Perez-Garcia et al. 2011a; Rawat et al. 2011). Naturally, adding extra nutrients for the purpose of increasing biomass production can lead to a wastewater even more polluted than the original one, even after cultivation of the microalgae. Nitrogen removal from wastewater can be achieved by heterotrophic and mixotrophic cultivation of microalgae (Hu et al. 2012; Perez-Garcia et al. 2010, 2011a). Hu et al. (2012, 2013) observed ammonium removal efficiencies from 52 to 72 % in mixotrophic cultures of *Auxenochlorella protothecoides* and *Chlorella* sp. Similarly, total nitrogen removal efficiencies ranged from 64 to 74 %. Removing phosphorus from wastewater can also be achieved with heterotrophic and mixotrophic cultivation of algae. For example, mixotrophic cultures of *A. protothecoides* removed 75 % of the initial phosphorus concentration of 125 mg-PO_4 L^{-1} in municipal wastewater (Hu et al. 2012). In their study, removing phosphorus from mixotrophic cultures was a consequence of microalgae uptake, together with chemical precipitation of dissolved phosphorus (orthophosphate).

Under realistic conditions of cultivation, to increase the algae competitiveness against fast-growing native bacteria in the wastewater and, at the same time, produce high quantities of biomass and lipids is still a major research challenge. The use of a microalgae consortium may increase the robustness of a microalgae population in wastewater under any cultivation regime. Photo-autotrophic mixed cultures, dominated by *C. vulgaris* populations, showed higher biomass and lipid production than photo-autotrophic pure *C. vulgaris* cultures. In addition, lipid production in photo-autotrophic mixed cultures was higher than that in heterotrophic culture (Table 4, Zhang et al. 2014). A microalgal consortium, mostly formed by *Chlorella*, *Chlamydomonas*, *Scenedesmus*, and *Gloeocystis* species cultivated in wastewater from the carpet industry, using raceway ponds, showed higher biomass production and removal of nitrogen and phosphorus than 13 pure strains (Chinnasamy et al. 2010).

Using competitive local strains from wastewater environments is another option. For example, native microalgae isolated from carpet industry wastewaters, such as

Table 4 Summary of studies of microalgae cultivated under heterotrophic or mixotrophic conditions for mass production purposes

Species	Organic carbon source	Algae growth mode	Culture operating mode	Target product	Maximum specific growth rate (day^{-1})	Maximum biomass concentration ($g\ L^{-1}$)	Maximum biomass productivity ($g\ L^{-1}\ day^{-1}$)	Maximum product productivity ($mg\ L^{-1}\ day^{-1}$)	Reference
Algae consortium	Treated carpet industry wastewater	Mixotrophic	Batch	Lipids	–	0.41	0.041	4.9	Chinnasamy et al. (2010)
	Untreated carpet industry wastewater	Mixotrophic	Batch	Lipids	–	0.39	0.039	4.7	Chinnasamy et al. (2010)
	Organic carbon in municipal wastewater	Heterotrophic	Batch	Lipids		0.20	0.026	13	Zhou et al. (2013)
	Organic carbon in municipal wastewater	Mixotrophic	Batch	Lipids	–	2.50	0.333	170	Zhou et al. (2013)
Auxenochlorella protothecoides	Organic carbon in municipal wastewater	Mixotrophic	Batch	Lipids	–	2.51	0.209	43	Hu et al. (2012)
Botryococcus braunii	Treated carpet industry wastewater	Mixotrophic	Batch	Lipids	–	0.37	0.037	3.5	Chinnasamy et al. (2010)
	Untreated carpet industry wastewater	Mixotrophic	Batch	Lipids	–	0.34	0.034	4.4	Chinnasamy et al. (2010)

(continued)

Table 4 (continued)

Species	Organic carbon source	Algae growth mode	Culture operating mode	Target product	Maximum specific growth rate (day^{-1})	Maximum biomass concentration ($g\ L^{-1}$)	Maximum biomass productivity ($g\ L^{-1}\ day^{-1}$)	Maximum product productivity ($mg\ L^{-1}\ day^{-1}$)	Reference
Chlamydomonas globosa	Glucose	Mixotrophic	Batch	Biomass	–	0.22	0.031	–	Bhatnagar et al. (2011)
	Poultry litter aqueous extract	Mixotrophic	Batch	Biomass	–	0.22	0.031	–	Bhatnagar et al. (2011)
	Poultry litter aqueous extract + glucose and nitrate	Mixotrophic	Batch	Biomass	–	1.28	0.181	–	Bhatnagar et al. (2011)
	Treated carpet industry wastewater	Mixotrophic	Batch	Biomass	–	0.16	0.024	–	Bhatnagar et al. (2011)
	Untreated carpet industry wastewater	Mixotrophic	Batch	Biomass	–	0.01	0.002	–	Bhatnagar et al. (2011)
Chlamydomonas reinhardtii	CO_2	Photo-autotrophic	Batch	Biomass	0.20	0.18	0.014	–	Blifernez-Klassen et al. (2012)
	Cellulose (carboxymethyl cellulose)	Mixotrophic	Batch	Biomass	0.27	0.36	0.028	–	Blifernez-Klassen et al. (2012)
	Cellulose (filter paper)	Mixotrophic	Batch	Biomass	0.31	0.42	0.032	–	Blifernez-Klassen et al. (2012)

Chlorella kessleri	Organic carbon in municipal wastewater	Heterotrophic	Batch	Lipids	–	0.40	0.080	40	Li et al. (2011)
	Organic carbon in municipal wastewater	Mixotrophic	Batch	Lipids	–	0.80	0.150	80	Li et al. (2011)
	Organic carbon in municipal wastewater	Mixotrophic	Batch	Lipids	0.80	1.40	0.312	68	Li et al. (2012)
Chlorella minutissima	Glucose	Mixotrophic	Batch	Biomass	–	0.22	0.031	–	Bhatnagar et al. (2011)
	Treated carpet industry wastewater	Mixotrophic	Batch	Biomass	–	0.21	0.030	–	Bhatnagar et al. (2011)
	Untreated carpet industry wastewater	Mixotrophic	Batch	Biomass	–	0.32	0.046	–	Bhatnagar et al. (2011)
	Poultry litter aqueous extract	Mixotrophic	Batch	Biomass	–	0.29	0.042	–	Bhatnagar et al. (2011)
	Poultry litter aqueous extract + glucose and nitrate	Mixotrophic	Batch	Biomass	–	2.60	0.371	–	Bhatnagar et al. (2011)

(continued)

Table 4 (continued)

Species	Organic carbon source	Algae growth mode	Culture operating mode	Target product	Maximum specific growth rate (day^{-1})	Maximum biomass concentration ($g\ L^{-1}$)	Maximum biomass productivity ($g\ L^{-1}\ day^{-1}$)	Maximum product productivity ($mg\ L^{-1}\ day^{-1}$)	Reference
Chlorella protothecoides	Acid hydrolysate of sweet sorghum juice	Heterotrophic	Batch	Lipids	–	3.30	0.660	343	Gao et al. (2010)
	Cassava powder hydrolysate	Heterotrophic	Batch	Lipids	–	4.67	0.900	360	Lu et al. (2010)
	Cassava powder hydrolysate	Heterotrophic	Fed-batch	Lipids	–	53.60	7.660	3790	Lu et al. (2010)
	Corn powder hydrolysate	Heterotrophic	Batch	Lipids	–	3.92	0.653	359	Xu et al. (2006)
	Corn powder hydrolysate	Heterotrophic	Fed-batch	Lipids	–	15.50	2.039	1121	Xu et al. (2006)
	Organic carbon in municipal wastewater	Heterotrophic	Batch	Lipids	–	0.40	0.080	56	Li et al. (2011)
	Organic carbon in municipal wastewater	Mixotrophic	Batch	Lipids	–	0.75	0.150	105	Li et al. (2011)
	Organic carbon in municipal wastewater	Mixotrophic	Batch	Lipids	0.25	0.78	0.195	33	(Li et al. 2012)
	Fructose	Heterotrophic	Batch	Lipids	–	3.90	0.780	374	Gao et al. (2010)
	Glucose	Heterotrophic	Fed-batch, N-sufficient	Lutein	1.06	48.0	4.800	20	(Shi et al. 2002)
	Glucose	Heterotrophic	Fed-batch, N-limited	Lutein	1.04	46.90	4.300	22	Shi et al. (2002)
	Glucose	Heterotrophic	Batch	Lipids	–	3.74	0.623	342	Xu et al. (2006)

	Glucose	Heterotrophic	Fed-batch (3 L)	Lipids	–	15.50	2.066	936	Li et al. (2007)
	Glucose	Heterotrophic	Fed-batch (750 L)	Lipids	–	12.80	1.706	816	Li et al. (2007)
	Glucose	Heterotrophic	Fed-batch (8000 L)	Lipids	–	14.20	1.893	832	Li et al. (2007)
	Glucose	Heterotrophic preliminary	Fed-batch	Lipids	–	16.80	2.191	1209	Xiong et al. (2008)
	Glucose	Heterotrophic optimized	Fed-batch	Lipids	–	51.20	7.358	3700	Xiong et al. (2008)
	Glucose	Heterotrophic	Batch	Lipids	–	12.80	2.560	1160	Yan et al. (2011)
	Glucose	Heterotrophic	Batch	Lipids	–	3.39	0.660	310	Lu et al. (2010)
	Glucose	Heterotrophic	Batch	Lipids	–	16.20	3.200	1560	Xiong et al. (2010)
	Glucose	Sequential photo-heterotrophic	Batch	Lipids	–	32.40	6.400	2900	Xiong et al. (2010)
	Glucose	Sequential photo-heterotrophic	Fed-batch	Lipids	–	55.10	23.900	11800	Xiong et al. (2010)

(continued)

Table 4 (continued)

Species	Organic carbon source	Algae growth mode	Culture operating mode	Target product	Maximum specific growth rate (day^{-1})	Maximum biomass concentration ($g L^{-1}$)	Maximum biomass productivity ($g L^{-1} day^{-1}$)	Maximum product productivity ($mg L^{-1} day^{-1}$)	Reference
	Invertase hydrolysate of sweet sorghum juice	Heterotrophic	Batch	Lipids	–	5.10	1.020	535	Gao et al. (2010)
	Jerusalem artichoke hydrolysate, as medium supplement	Heterotrophic	Batch	Lipids	–	16.50	4.100	1700	Cheng et al. (2009b)
	Jerusalem artichoke hydrolysate, direct use	Heterotrophic	Batch	Lipids	–	15.70	3.600	1600	(Cheng et al. 2009b)
	Molasses	Heterotrophic	Batch	Lipids	–	17.90	3.580	1440	Yan et al. (2011)
	Molasses	Heterotrophic	Fed-batch	Lipids	–	97.10	13.092	7469	Yan et al. (2011)
	Sucrose	Heterotrophic	Batch	Lipids	–	1.20	0.240	33	Gao et al. (2010)
	Sucrose hydrolysate	Heterotrophic	Batch	Lipids	–	9.10	2.180	970	(Cheng et al. 2009a)
	Sugar cane juice hydrolysate	Heterotrophic	Batch	Lipids	–	9.30	2.230	962	Cheng et al. (2009a)
	Sugar cane juice hydrolysate	Heterotrophic	Fed-Batch	Lipids	–	24.20	3.420	1472	Cheng et al. (2009a)
	Whey	Heterotrophic	Batch	Lipids		9.10	1.300	550	Espinosa-Gonzalez et al. (2014)

	Whey	Heterotrophic	Fed-batch	Lipids		17.20	2.450	791	Espinosa-Gonzalez et al. (2014)
Chlorella pyrenoidosa	Glucose	Heterotrophic	Fed-batch	Biomass	–	104.90	0.613	–	Wu and Shi (2007)
	Glucose	Heterotrophic	Fed-batch	Biomass	–	116.20	1.020	–	Wu and Shi (2007)
	Glucose	Heterotrophic	Batch	Fatty acids	2.20	6.50	1.083	25	Pleissner et al. (2013)
	Food waste hydrolysate	Heterotrophic	Batch	Fatty acids	1.10	19.00	3.450	35	Pleissner et al. (2013)
Chlorella saccharophila	Treated carpet industry wastewater	Mixotrophic	Batch	Lipids	–	0.16	0.016	2.7	Chinnasamy et al. (2010)
	Untreated carpet industry wastewater	Mixotrophic	Batch	Lipids	–	0.23	0.023	4.1	Chinnasamy et al. (2010)
Chlorella sorokiniana	Acetate	Heterotrophic	Batch	Biomass		1.50	0.500		Ogbonna et al. (2000)
	Acetate	Mixotrophic	Batch	Biomass		1.10	0.370		Ogbonna et al. (2000)
	Acetate	Heterotrophic	Batch[b]	Total carbohydrates	0.72	0.11	0.037	55	Choix et al. (2012)
	Glucose	Mixotrophic	Batch	Lipids	0.44	1.32	0.580	56	Wan et al. (2011)
	Glucose	Heterotrophic	Batch[b]	Total carbohydrates	0.67	0.09	0.033	66	Choix et al., (2012)

(continued)

Table 4 (continued)

Species	Organic carbon source	Algae growth mode	Culture operating mode	Target product	Maximum specific growth rate (day^{-1})	Maximum biomass concentration (g L^{-1})	Maximum biomass productivity (g L^{-1} day^{-1})	Maximum product productivity (mg L^{-1} day^{-1})	Reference
Chlorella vulgaris	Acetate	Heterotrophic	Batch	Lipids	–	0.99	0.087	27	Liang et al. (2009)
	Acetate	Heterotrophic	Batch	Biomass	0.49	0.24	0.047	–	Perez-Garcia et al. 2011)
	Acetate	Heterotrophic	Batch[b]	Total carbohydrates	0.73	0.15	0.057	52	Choix et al. (2012)
	Fulvic acid	Heterotrophic	Batch	Biomass	0.34	0.11	0.022	–	Perez-Garcia et al. (2011)
	Glucose	Heterotrophic	Batch	Lipids	–	1.21	0.151	35	Liang et al. (2009)
	Glucose	Mixotrophic	Batch	Lipids	–	1.70	0.254	54	Liang et al. (2009)
	Glucose	Heterotrophic	Batch[b]	Biomass	0.15	0.02	0.007	–	Perez-Garcia et al. (2010)
	Glucose	Mixotrophic	Batch[b]	Biomass	0.72	0.11	0.027	–	Perez-Garcia et al. (2010)
	Glucose	Heterotrophic	Batch	Biomass	0.42	0.16	0.031	–	Perez-Garcia et al. (2011)
	Glucose	Heterotrophic	Batch[b]	Total carbohydrates	0.70	0.10	0.033	72	Choix et al. (2012)
	Glycerol	Heterotrophic	Batch	Lipids	–	0.72	0.102	22	Liang et al. (2009)
	Urea	Heterotrophic	Batch	Biomass	0.08	0.10	0.002	–	Perez-Garcia et al. (2011)

Chlorella sp.	Xylose	Mixotrophic	Batch	Biomass	–	–	–	–	Hawkins (1999)
	Dairy manure, anaerobically digested, X10 dilution	Mixotrophic	Batch	Lipids	0.28	1.57	0.074	6	Wang et al. (2010)
	Dairy manure, anaerobically digested, X20 dilution	Mixotrophic	Batch	Lipids	0.40	1.71	0.081	11	Wang et al. (2010)
	Glucose	Photo-autotrophic	Batch	Lipids	–	0.60	0.085	12	Cheirsilp and Torpee (2012)
	Glucose	Heterotrophic	Batch	Lipids	–	0.70	0.100	13	Cheirsilp and Torpee (2012)
	Glucose	Mixotrophic	Batch	Lipids	–	1.35	0.192	30	Cheirsilp and Torpee (2012)
	Glucose	Mixotrophic	Fed-batch	Lipids	–	4.50	0.450	112	Cheirsilp and Torpee (2012)
	Acidogenically digested swine wastewater	Mixotrophic	Semi-continuous	Lipids		0.78	0.28	72	Hu et al. (2013)

(continued)

Table 4 (continued)

Species	Organic carbon source	Algae growth mode	Culture operating mode	Target product	Maximum specific growth rate (day^{-1})	Maximum biomass concentration ($g\ L^{-1}$)	Maximum biomass productivity ($g\ L^{-1}\ day^{-1}$)	Maximum product productivity ($mg\ L^{-1}\ day^{-1}$)	Reference
Crypthecodinium cohnii	Acetic acid	Heterotrophic	Fed-batch	Docosahexaenoic acid	–	109.00	6.540	46	De Swaaf et al. (2003a)
	Glucose	Heterotrophic	Fed-batch	Docosahexaenoic acid	–	26.00	5.200	14	De Swaaf et al. (2003a)
	Ethanol	Heterotrophic	Fed-batch	Docosahexaenoic acid	–	83.00	9.054	53	De Swaaf et al. (2003b)
Dunaliella salina	Glucose	Mixotrophic	Batch	Lipids	0.34	0.52	0.180	12	Wan et al. (2011)
Dunaliella tertiolecta	Treated carpet industry wastewater	Mixotrophic	Batch	Lipids	–	0.38	0.038	4.6	Chinnasamy et al. (2010)
	Untreated carpet industry wastewater	Mixotrophic	Batch	Lipids	–	0.28	0.028	4.2	Chinnasamy et al. (2010)
Haematococcus pluvialis	Acetate	Mixotrophic	Batch	Carotenoids	0.23–0.07	1.20	0.075	0.3	Göksan et al. (2010)
	Acetate	Sequential photo-mixotrophic	Batch	Carotenoids	0.24–0.10	1.51	0.094	0.5	Göksan et al. (2010)
Galdieria sulphuraria	Glucose	Heterotrophic	Fed-batch	Phycocyanin	–	109.00	7.266	473	Graverholt and Eriksen (2007)
	Glucose	Heterotrophic	Continuous	Phycocyanin	–	83.30	4.580	100	Graverholt and Eriksen (2007)

Nannochloropsis oculata	Glucose	Mixotrophic	Batch	Lipids	0.32	0.61	0.052	14	Wan et al. (2011)
Nannochloropsis sp.	Carbon dioxide	Photo-autotrophic	Batch	Lipids	–	0.40	0.057	156	Cheirsilp and Torpee (2012)
	Glucose	Heterotrophic	Batch	Lipids	–	0.45	0.064	13	(Cheirsilp and Torpee 2012)
	Glucose	Mixotrophic	Batch	Lipids	–	1.25	0.178	43	Cheirsilp and Torpee (2012)
	Glucose	Mixotrophic	Fed-batch	Lipids	–	5.87	0.587	148	Cheirsilp and Torpee (2012)
	Carbon dioxide	Photo-autotrophic in blue light	Batch	Lipids	0.64	0.37	0.046	55	Das et al. (2011)
	Glycerol	Mixotrophic in blue light	Batch	Lipids	0.66	0.58	0.072	111	Das et al. (2011)
	Carbon dioxide	Photo-autotrophic in white light	Batch	Lipids	0.58	0.35	0.044	50	Das et al. (2011)
	Glycerol	Mixotrophic in white light	Batch	Lipids	0.61	0.48	0.060	88	Das et al. (2011)
Pleurochrysis carterae	Treated carpet industry wastewater	Mixotrophic	Batch	Lipids	–	0.38	0.038	4.6	Chinnasamy et al. (2010)
	Untreated carpet industry wastewater	Mixotrophic	Batch	Lipids	–	0.28	0.028	4.2	Chinnasamy et al. (2010)

(continued)

Table 4 (continued)

Species	Organic carbon source	Algae growth mode	Culture operating mode	Target product	Maximum specific growth rate (day^{-1})	Maximum biomass concentration ($g\ L^{-1}$)	Maximum biomass productivity ($g\ L^{-1}\ day^{-1}$)	Maximum product productivity ($mg\ L^{-1}\ day^{-1}$)	Reference
Scenedesmus bijuga	Glucose	Mixotrophic	Batch	Biomass	–	0.21	0.030	–	Bhatnagar et al. (2011)
	Treated carpet industry ww	Mixotrophic	Batch	Biomass	–	0.49	0.070	–	Bhatnagar et al. (2011)
	Untreated carpet industry ww	Mixotrophic	Batch	Biomass	–	0.44	0.064	–	Bhatnagar et al. (2011)
	Poultry litter aqueous extract	Mixotrophic	Batch	Biomass	–	0.34	0.049	–	Bhatnagar et al. (2011)
	Poultry litter aqueous extract + glucose and nitrate	Mixotrophic	Batch	Biomass	–	2.19	0.313	–	Bhatnagar et al. (2011)
Scenedesmus obliquus	Carbon dioxide	Photoautotrophic	Batch	Lipids	0.27	0.80	0.061	37	Girard et al. (2014)
	Whey	Heterotrophic	Batch	Lipids	0.70	2.70	0.207	28	Girard et al. (2014)
	Whey	Mixotrophic	Batch	Lipids	1.08	3.60	0.277	38	Girard et al. (2014)
Scenedesmus sp.	Acetate	Mixotrophic	Batch	Lipids	–	–	0.080	30	Park et al. (2012)

Schizochytrium limacinum	Glucose	Heterotrophic	Batch	Docosahexaenoic acid	0.57	18.47	3.08	510	Chi et al. (2007)
	Glycerol (pure)	Heterotrophic	Batch	Docosahexaenoic acid	0.56	14.43	2.41	420	Chi et al. (2007)
	Glycerol (crude from refinery)	Heterotrophic	Batch	Docosahexaenoic acid	0.68	18.04	3.06	510	Chi et al. (2007)
Schizochytrium mangrovei	Glucose	Heterotrophic	Batch	Lipids	2.10	7.00	1.170	21	Pleissner et al. (2013)
	Food waste hydrolysate	Heterotrophic	Batch	Lipids	4.70	14.00	2.333	27	Pleissner et al. (2013)
Synechococcus sp.[a]	Glucose	Mixotrophic	Batch	Biomass	–	–	0.300	–	Kang et al. (2004)
Spirulina platensis[a]	Acetate	Mixotrophic	Batch	Phycocyanin	0.52	1.81	–	246	Chen et al. (1996)
	Molasses	Mixotrophic	Batch	Biomass	0.15	2.90	0.320	–	Andrade and Costa (2007)
	Glucose	Mixotrophic	Batch	Phycocyanin	0.62	2.66	–	322	Chen et al. (1996)
	Glucose	Mixotrophic	Batch	Biomass	1.05	–	–	–	Vonshak et al. (2000)
Spirulina sp.[a]	Glucose	Heterotrophic	Batch	Biomass	0.04	0.67	0.320	–	Chojnacka and Noworyta (2004)
	Glucose	Mixotrophic	Batch	Biomass	0.05	0.54	0.360	–	Chojnacka and Noworyta (2004)

Data from some photoautotrophic cultures is included for comparison. The table only presents average values reported in the corresponding reference

[a]Cyanobacteria species

[b]Cells immobilized in alginate beads together with plant growth promoting bacteria *Azospirillum brasilense*

Chlamydomonas globosa, *Scenedesmus bijuga*, and *Chlorella minutissima* were able to grow mixotrophically and removed nitrogen and phosphorus far better in dark, colored, and opaque wastewaters rich in organic and inorganic nutrients than in the standard laboratory medium (Bhatnagar et al. 2011). Three strains of microalgae (*Scenedesmus* sp. ZTY2, *Scenedesmus* sp. ZTY3, and *Chlorella* sp. ZTY4) isolated from a domestic wastewater treatment plant were heterotrophically cultivated in real domestic wastewater with no illumination. The isolated strains contained good lipid content, from 55 to 80 % of cell dry weight. Efficiencies of removal of dissolved organic carbon ranged from 55 to 65 % (Zhang et al. 2013a). It is important to emphasize that under mixotrophic cultivations, microalgal consortia achieved higher biomass production than heterotrophic cultivations (Table 4).

New kinds of wastewater are being tested as potential sources of organic substrate for heterotrophic microalgal growth. For example, dairy manure contains levels of acetic and propionic acid that is readily available for microalgae cultivation (Wang et al. 2010). Also food waste hydrolysate has been used as culture medium and nutrient source for heterotrophic microalgae cultivation (Pleissner et al. 2013). Some wastewaters need a pre-treatment to make them available as carbon sources for microalgae cultivation. The pre-treatment eliminates side effects of solid particles and indigenous bacteria that may compete and outgrow the microalgae population. Large amounts of glucose, free amino nitrogen, and phosphate were recovered from food waste by fungal hydrolysis using *Aspergillus awamori* and *A. oryzae. Schizochytrium mangrovei* and *Chlorella pyrenoidosa* grew well on the recovered nutrients. Growth rate, biomass, and lipid production were higher in cultures cultivated on food waste hydrolysate than in control cultures growing on conventional medium with glucose. Impressively, at the end of the cultivation, 10–20 g of biomass, out of the original 100 g food waste (as dry weight) were produced and were rich in carbohydrates, lipids, proteins, and saturated and polyunsaturated fatty acids (Pleissner et al. 2013) (Table 4). In another example, cultivation of microalgae on swine manure proved to be a practical and economical organic substrate for production of algae feedstock (Hu et al. 2013). In their study, acidogenic-anaerobic digestion was used as a pre-treatment of fresh swine manure before it was used as the substrate in heterotrophic cultures of *Chlorella* sp. This yielded nutrient removal rates of 751.33 mg COD L^{-1} day^{-1}, 20.21 mg PO_4-P L^{-1} day^{-1} and 60.39 mg NH_3-N L^{-1} day^{-1}. At the same time, the experiment achieved lipid productivity of 3.63 g m^2 day^{-1} or 10×10^3 L ha^{-1} $year^{-1}$ (Hu et al. 2013). In another experiment, wastewater produced from hydrothermal liquefaction of biomass was mixotrophically grown in mixed cultures of microalgae and bacteria. This wastewater had concentrations of COD from 50 to 130 g L^{-1}, total nitrogen from 5 to 20 g L^{-1}, ammonia from 3 to 12 g L^{-1} and phosphorus from 0.6 to 2 g L^{-1}. Consequently, this wastewater needed to be diluted (~100 times) to be used for microalgae heterotrophic/mixotrophic cultivation (Zhou et al. 2013).

Microalgae and cyanobacteria respond to many organic pollutants in various ways, such as growth inhibition, bioaccumulation, and biodegradation. Because of their versatile metabolism and their capacity to switch rapidly from one growth mode to another, mixotrophic microalgae can be successfully employed for remediating polluted environments (Subashchandrabose et al. 2013). The ecological advantage of employing mixotrophic cyanobacteria and microalgae for bioremediation is that decreasing concentrations of organic pollutants will have no adverse effect on their growth. In contrast, if a heterotrophic regime is employed, no further biomass production can be expected after degradation of the organic pollutant. The dual-purpose process of chemical production-wastewater treatment, as a general strategy for many microalgal systems, though it looks very attractive, has many challenges, as discussed above. Therefore, with current knowledge, expectations should be cautious. With better strains and better technologies, it may be possible that we can achieve the microalgal potential and make the process efficient and cost effective (Liang 2013).

4.3 *Processed Materials*

To reduce costs for organic carbon sources, various processed materials (mostly wastes) have been examined. These feedstock include: Jerusalem artichoke hydrolyzed by inulinase (Cheng et al. 2009b), sweet sorghum hydrolyzed liquid by acid or invertase (Gao et al. 2010), cassava starch hydrolyzed by amylase and glucoamylase (Lu et al. 2010), molasses hydrolyzed by invertase (Yan et al. 2011), Na-acetate (Perez-Garcia et al. 2011a, b), crude glycerol (Das et al. 2011; Liang et al. 2009), corn powder hydrolysate (Xu et al. 2006), sugar cane juice hydrolysate (Cheng et al. 2009a), and cheese whey (Girard et al. 2014). A summary of these studies is presented in Table 4.

Food industry by-products and wastes are a promising feedstock for production of chemical compounds by heterotrophic microalgae. Molasses, a by-product of sugar refinery, is a cheap carbon source proven to be an excellent alternative to glucose in promoting microalgal lipid accumulation because it contains about 36 % sucrose (Yan et al. 2011). Yet, pre-enzymatic hydrolysis by invertase was required for molasses to be suitable for growing microalgae. Microalgal biomass derived from medium with molasses hydrolysates was higher than a glucose medium. This may happen because molasses contains fructose, which is also a source of carbon. The biomass of *C. protothecoides* was 97.1 g L^{-1} and lipid production was as high as 5.5 g L^{-1} day^{-1} in fed-batch cultures (Table 4). Yan et al. (2011) estimated that the cost of producing lipids from molasses was 50 % of the cost of using a glucose-based medium.

Espinosa-Gonzalez et al. (2014) and Girard et al. (2014) proposed using cheese whey in mixotrophic microalgae cultures. Whey is generated in large quantities as a by-product of cheese production. Its main constituent is lactose (>80 % w/w of the total dissolved solids). Whey has the advantage that it is free of contaminants and can readily be used as a carbon source for microalgae cultivation (Girard et al. 2014). Substituting 40 % (v/v) of microalgae culture medium with whey significantly stimulated growth of *Scenedesmus obliquus*. Consequently, yields of biomass under mixotrophic (3.6 g L^{-1}) and heterotrophic (2.7 g L^{-1}) conditions were significantly greater than under photo-autotrophic conditions (1.2 g L^{-1}). Crude glycerol, a large volume by-product of the biodiesel industry, has been successfully used as an alternative substrate for growing *Schizochytrium limacinum* (Chi et al. 2007) and *Chlorella vulgaris* (Liang et al. 2009). Using glycerol, heterotrophic cultivation of *S. limacinum* reached a biomass concentration of 18 g L^{-1} and production of docosahexaenoic acid of 510 mg L^{-1} day^{-1} (Table 4; Chi et al. 2007).

Microalgae cells do not assimilate complex sugars, such as polysaccharides; therefore, carbon sources, such as vegetable powders rich in polysaccharides, need to be hydrolyzed to break the compounds into di- and mono-saccharides, which can be assimilated by microalgae cells. Xu et al. (2006) optimized microalgae growth on hydrolyzed corn powder, using α-amylase and glucoamylase. This hydrolysate led to high yields of *C. protothecoides* biomass (15.5 g L^{-1}) in batch culture (Table 4). *C. protothecoides* also grew well on medium supplemented with sugar cane juice hydrolysate but not on raw sugar cane juice (Cheng et al. 2009a). The sugar cane juice hydrolysate contained 18.5 % sucrose and 1.7 % other fermentable sugars, such as glucose and fructose, and led to a biomass concentration of 25.4 g L^{-1} (Cheng et al. 2009a). Sweet sorghum hydrolysate liquid also promoted heterotrophic growth of *C. protothecoides,* leading to a biomass concentration of 5.1 g L^{-1} (Gao et al. 2010). Raw sweet sorghum juice contains sucrose, fructose, and glucose (101.7, 33.1, and 25.0 g L^{-1}, respectively). However, sucrose cannot be used by heterotrophic *C. protothecoides*; it must be hydrolyzed to glucose and fructose prior to use as a feedstock. Finally, Lu et al. (2010) cultivated *C. protothecoides* heterotrophically, using cassava hydrolysate powder, which led to an impressive biomass concentration of 53 g L^{-1} in fed-batch cultures (Table 4).

4.4 Lignocellulose Material

Lignocellulose biomass is a carbon-neutral (CO_2 is not emitted during its production), renewable, and sustainable source of organic carbon. Sugars released from this feedstock are inexpensive carbon sources for heterotrophic cultivation of microalgae. While this sounds promising, this carbon source has the following limitations: (1) Pretreatment to break down lignocellulose, such as wood chips or plant fibers, to simple sugars is expensive; (2) By-products from the breakdown are numerous, roughly classified as aliphatic acids, aromatic acids, and aldehydes/ketones. Some of those products may cause inhibition or toxicity to microalgae. Though the options of lignocellulosic material are many, the detoxification processes increase costs; and (3) The major monomeric sugars in cellulosic feedstock are glucose and xylose. Most microalgal species thrive on glucose, but species that can use xylose are few and perhaps limited only to the genus *Chlorella* (Liang 2013). So far, only one *Chlorella* strain used xylose mixotrophycally or when glucose was present in the media (Hawkins 1999). In the dark or with the absence of glucose, this strain did not prosper on xylose. He speculates that a cofactor or other metabolic pathways is necessary to metabolize xylose and that energy or cofactors derived from glucose or light metabolism can initiate the xylose utilization pathway.

Recently, cellulose as a carbon source was used by the microalgae *Chlamydomonas reinhardtii* (Blifernez-Klassen et al. 2012), an ability which had previously been known for degrading cellulose by non-photosynthetic organisms. This opens new prospects for using cellulose waste in microalgae cultivation. In the presence of light, *C. reinhardtii* used cellulose for growth in the absence of other carbon sources by secreting cellulolytic enzymes (endo-β-1,4-glucanases), when the medium was supplemented with carboxymethyl cellulose or filter paper (Table 4). During this mixotrophic regime, it was impossible to discriminate between the contributions of photosynthesis and heterotrophic assimilation of cellulose on cell growth. This cellulose degrading and assimilation phenotype was not observed in the closely related microalga *C. kessleri* (Blifernez-Klassen et al. 2012).

Research in this line should focus on developing cheap, effective, and inexpensive pretreatment techniques, eliminate inhibitory by-products or decrease their concentrations. Considering the many publications and enormous efforts dedicated to biomass pretreatment in recent times, it is reasonable to expect that pretreatment methods will be developed (Liang 2013). For example, removal of degraded by-products from biomass hydrolysates can be achieved by physical (vacuum evaporation), chemical (ion exchange or over-liming) and biological (enzymes or microorganisms that modify the composition of toxic compounds) methods. Finally, future research efforts should also focus on: (a) bio-prospecting of species that grow on xylose; (b) producing strains that use xylose by genetic and metabolic engineering; and (c) couple microalgae growth with other xylose, using microbial species in the same reactor or sequentially,

5 Cultivation Methods

5.1 The Bioreactor

Heterotrophic microalgae are grown in stirred tank bioreactors (STRs), commonly known as fermenters. Because heterotrophic microalgae growth is independent of light energy, much simpler scale-up possibilities are available since a smaller reactor surface-to-volume ratio is used compared with photo-bioreactors (Brennan and Owende 2010). Table 5 summarizes the main technical aspects of cultivation systems for each microalgae growth regime. Fermenters provide a high degree of growth control and also lower harvesting costs because higher cell densities are possible, compared with photo-bioreactors and open ponds. Also, bioreactor set-up costs are lower than for closed photo-bioreactors. Heterotrophic algal growth requires oxygen, mixing, temperature control, and nutrients in a sterile environment. Therefore additional expenses associated to facilities for preparation of growth medium and reactor sterilization must be considered. The

Table 5 Technical aspects for cultivation at different microalgae growth modes

Technical aspect	Photo-autotrophic	Heterotrophic	Mixotrophic
Energy source	Light	Organic carbon	Light and organic carbon
Carbon source	Inorganic carbon	Organic carbon	Inorganic and organic
Use of renewable energy source	Yes	Not in principle	Partially
Biomass/compounds productivity	Low	Medium	High
Light availability requirement	Obligatory	Not required	Not obligatory
Limiting factor for growth	Light	Oxygen	Light and oxygen
CO_2 emission	Negative emission, sink	Positive emission	Neutral, CO_2 produced and consumed
Bioreactor type	Photobioreactor. Open or close in panels or tubes	Fermenters	Photo-bioreactor. Open or close in panels or tubes
Bioreactor's vessels availability	In house crafted	Commercially available	In house crafted
Bioreactor's surface to volume ratio (m^2/m^3)	High	Low	High
Control of operation parameters in bioreactors	High	High	High
Sterility	Usually sanitized	Sterility required	Sterility preferred but not required
Contamination risk	Low	High	Medium
Harvesting difficulty	High due diluted biomass concentration	Low due dense biomass concentration	Low due dense biomass concentration
Bioreactor set up cost	High per unit of volume	Low per unit of volume	High per unit of volume
Bioreactor operation cost[a]	Low per kg of biomass	Medium per kg of biomass	High per kg of biomass

[a]According to Chen et al. (2011)

growth medium and pipelines for inoculation, feeding substrate, and sampling are commonly sterilized by autoclave. With heterotrophic growth, oxygen for respiration is the main limiting factor; therefore, the bioreactor's oxygen supply and good medium mixing are essential.

A significant advantage of heterotrophic cultivation is that STRs are easily scalable up to commercial size (with working volumes of ~10,000 L) and that vessels are available commercially for cultivation of other microorganisms (Barclay et al. 2013; Li et al. 2007). Producing lipids by heterotrophic *C. protothecoides* were successfully scaled up from 250 mL flasks to STRs with net volume of 5, 750, and 11,000 L (Li et al. 2007). In pilot (750 L) and commercial-scale (11,000 L) bioreactors, real-time respiration-rate monitoring used a dissolved oxygen online detector, and the consumption of substrates was monitored by manual sampling. Inoculation

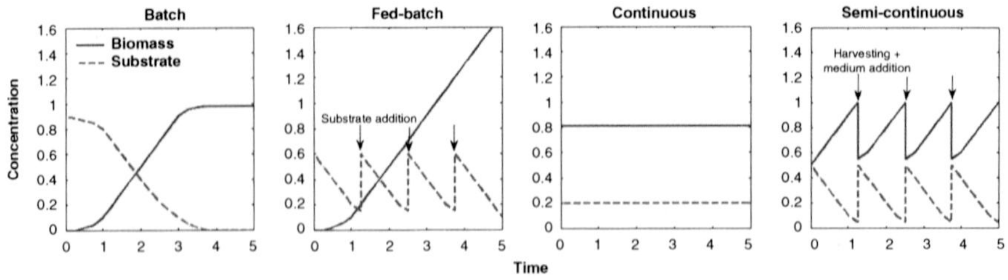

Fig. 5 Typical concentration profiles observed in microalgae cultures under different operation modes. Concentration profiles are similar to those observed in bacteria and fungi fermentations

was performed in three steps: the pre-inoculum *C. protothecoides* cells were first grown in 100 L of medium, then in two 750 L STRs with 400 L of medium each, and finally inoculated in 7,200 L of sterilized medium, with the final work volume reaching about 8,000 L in a 11,000 L STR. While the scaling up did not change the relative fatty acid composition of the biomass as cultivation was scaled up, total lipid content was slightly different (46.1 %, 48.7 %, and 43 % of the cell dry weight) in samples from 5 L, 750 L, and 11,000 L bioreactors, respectively (Li et al. 2007).

The key practical issues required for large-scale heterotrophic cultivation of microalgae are: (a) Good survival of the strain during cultivation; (b) The strains need to be genetically robust and stable under production circumstances and resistant to mutations and infections by other microorganisms; (c) Ideally, strains should be able to grow under extreme conditions, such as high or low pH, high temperatures, or high salinity; (d) Overall low cultivation costs based on the strain to efficiently use inexpensive, common carbon sources, tolerate environmental changes, and generate economic value by generating large quantities of the metabolite(s); and, (e) At the industrial level, the strains must be easy to handle; its cell walls must withstand hydrodynamic and mechanical shear occurring in large bioreactors and it should produce high density biomass, all in minimally modified fermenters used for other microorganisms (Chen and Chen 2006; Day et al. 1991; Wijffels et al. 2013).

5.2 *Mode of Culture Operation*

Heterotrophic and mixotrophic cultures in fermenters or photo-bioreactors can be operated in batch, fed-batch, continuous, and semi-continuous modes (Table 4). Figure 5 illustrates ideal profiles for biomass and substrate concentration observed in fermenters operated in these modes. Batch mode fermentation is commonly adopted for growing heterotrophic microalgae. It achieves biomass concentrations from 0.5 to 20 g L^{-1}. For example, heterotrophic cultures of *Chlorella pyrenoidosa* on food waste hydrolysate reached 19 g L^{-1} (Pleissner et al. 2013). Batch mixotrophic cultures reached biomass concentrations of 3.6 g L^{-1}, using *Scenedesmus obliquus* growing on whey (Girard et al. 2014). Fed-batch operation can reach higher biomass than batch mode. Heterotrophic fed-batch cultures of *Chlorella pyrenoidosa* and *Crypthecodinium cohnii* have reached the remarkable concentrations of 116 g L^{-1} and 109 g L^{-1} biomass growing on glucose and acetate, respectively (De Swaaf et al. 2003a; Wu and Shi 2007). Mixotrophic fed-batch cultures of *Nannochloropsis* sp. reached concentrations of 5.8 g L^{-1} (Cheirsilp and Torpee 2012). The advantage of the fed-batch mode lies in minimal restrictions by maintaining continuous feeding of the nutrient stream. Hence, microbial cells undergo log-phase growth during almost the whole cultivation cycle (Xiong et al. 2008). It appears that continuous (chemostate) and semi-continuous cultivation of heterotrophic and mixotrophic microalgae have been used in only a few studies (Graverholt and Eriksen 2007; Hu et al. 2013). During "chemostat" operations, the fermenter maintains steady chemical conditions. In continuous operation, fresh medium is continuously added to the bioreactor and old culture medium is continuously removed, both at a specific rate (Q; in L s^{-1}). By changing the Q value, the growth rate of microalgae is controlled. Continuous cultures can maintain a constant low substrate concentration and avoid, in practical and easy ways, adverse effects on growth and yield of end products (Graverholt and Eriksen 2007). In semi-continuous mode, a proportion of a batch culture is replaced with fresh medium when the majority of microalgae reach the late logarithmic growth phase. Then, the culture is maintained under batch operation for additional time to increase cell density before the next replacement of medium. The repeated harvesting-regrowing process can be maintained for weeks or several months without apparent decline in growth. In a semi-continuous process, the ratio between the daily-replaced and the total culture volumes defines the hydraulic retention time (HRT), which is a key parameter to control the rate of algae growth and nutrient uptake (Hu et al. 2013).

Sequential regime (two-stage regime) involves first exposing the culture to a photo-autotrophic period, followed by a heterotrophic or mixotrophic period. This strategy increased biomass and product concentrations of microalgal cultures. Xiong et al. (2010) separated the cultivation process into two independent but sequential phases, the "green" stage (nitrogen-sufficient and photosynthetic cultivation) and the "yellow" stage (nitrogen-deficient under heterotrophic growth for accumulating lipids). This regime was more efficient than mixotrophic cultivation alone. *C. protothecoides* cultures under sequential regime reached an impressive biomass concentration above 120 g L^{-1} and a lipid concentration above 40 g L^{-1} (Table 4). The average biomass and lipid productivities during the mixotrophic phase alone were 23.9 and 11.8 g L^{-1} day^{-1}, respectively. In another study, production of carotenoids by *Haematococcus pluvialis* increased ~60 % in cultures using a sequential regime of photo-autotrophic growth period (11 days), followed by a mixotrophic period (5 days), using supplements of acetate. This is in contrast with traditional mixotrophic cultivation, where acetate was added to the medium before inoculation (Göksan et al. 2010; Table 4). Yen and Chang (2013) studied growth of *C. vulgaris* under sequential photo-mixotrophic regimes. The results of the sequential regime reached a biomass of ~64 % over simple photo-autotrophic cultivation. During the mixotrophic stage, the low light intensity limited mixotrophic growth. This indicated that photosynthesis still plays an important role during mixotrophic cultivation. Finally, Zheng et al. (2012) proposed that high density heterotrophic cultivation of *C. sorokiniana* could be used as seed culture for subsequent photo-autotrophic cultivation on a large scale.

5.3 Culture Medium

The composition of the basic culture medium for heterotrophic cultivation is similar to autotrophic cultivation with the sole exception of adding an organic carbon source. The culture medium BG11 has been widely used in photo-autotrophic, heterotrophic, and mixotrophic cultivation of microalgae (Bhatnagar et al. 2011; Cheirsilp and Torpee 2012; Chinnasamy et al. 2010; Göksan et al. 2010; Li et al. 2012; Wan et al. 2011). Also, the medium base supplemented with carbon sources and reduced glycine concentration has been widely used for heterotrophic growth of *C. protothecoides* (Miao and Wu 2006; Shi et al. 2002; Xu et al. 2006). The composition of the commonly used medium base is: (g L^{-1}) 0.7 KH_2PO_4, 0.3 K_2HPO_4, 0.3 $MgSO_4$•7 H_2O, (mg L^{-1}) 3 $FeSO_4$•$7H_2O$, 0.01 vitamin B1, 1 mL L^{-1} A5 trace mineral solution; 1–2 g L^{-1} glycine or 1–7 g L^{-1} yeast extract can be added as a nitrogen source (Cheng et al. 2009a; Hillebrand and Sommer 1999; Xiong et al. 2008). As a general rule, microalgae growing in the dark should have a C:N:P mass ratio of 46.1:7.7:1 (Hillebrand and Sommer 1999).

An optimal concentration of organic carbon in a medium depends on the carbon source, the microalgae strain, and cultivation conditions. An optimal concentration of carbon in the medium for heterotrophic or mixotrophic growth ranges from 10 to 30 g L^{-1}, depending on the type of supplied carbon compound. When glucose is used, highest growth rates of diverse microalgal species occur at concentrations from 7 to 15 g L^{-1} (Bhatnagar et al. 2011; Cheirsilp and Torpee 2012; Liang et al. 2009; Wan et al. 2011). Greater concentrations of biomass can be achieved at glucose concentration above 30 g L^{-1}. However, the biomass produced per unit of sugar consumed significantly decreases (Xiong et al. 2008). In the case of molasses, a concentration of 30 g L^{-1} of reduced sugar in hydrolyzed molasses provided the optimal production of biomass and lipids in cultivation of *C. protothecoides* (Yan et al. 2011). It is important to consider that in batch cultures, when cell density reaches its maximum concentration in the stationary phase, a higher carbon supply does not lead to continuous growth and accumulation of product (Cheng et al. 2009a). Inhibition of growth occurs in some microalgae even when the carbon source was applied at a low concentration, as occurred in *C. reinhardtii*. Inhibition occurred when acetate was >0.4 g L^{-1} (Cheirsilp and Torpee 2012). Under mixotrophic growth conditions, 10 and 20 g L^{-1} of glucose improved cell growth significantly compared with medium containing 50 and 100 g L^{-1} of glucose. At 100 g L^{-1} (glucose solution at 10 % w/v), the glucose inhibited the microalga. Similarly, in *Crypthecodinium cohnii* cultures, glucose concentrations above 25 g L^{-1} inhibited growth (De Swaaf et al. 2003a). Hence, inhibition is strain-dependent and needs to be determined for each strain and set of culturing conditions (Liang et al. 2009).

According to Xiong et al. (2008), among three inorganic nitrogen sources (urea, potassium nitrate, and ammonium nitrate) and two organic nitrogen sources (glycine and yeast extract), yeast extract at a concentration of 1–4 g L^{-1} is the most suitable to provide high biomass and lipid yield in heterotrophic *Chlorella* spp. However, this nitrogen source is more expensive than the inorganic sources. In general, accumulation of lipids and carotenoids (astaxanthin and lutein) in microorganisms is stimulated by an excess of carbon and a limitation in one or more of the other nutrients, especially nitrogen (De Swaaf et al. 2003a; Leyva et al. 2014; Shi et al. 2002). For example, accumulation of microbial lipids is often a two-phase process. In the first phase, cells undergo exponential division. In the second phase, the growth rate decreases from a nutritional limitation and lipids accumulate. Cheng et al. (2009a) found the maximum lipid contents, conversion ratio of sugar-to-biomass, and conversion ratio of sugar-to-lipids are achieved at different carbon-to-nitrogen ratios (C/N). This indicates the critical importance of the C/N ratio and nitrogen concentration for microalga growth and accumulation of lipids. Additionally, high concentrations of nitrogen (>1 g L^{-1}) leads to low efficiency of sugar consumption. Typically, C/N in growth medium of 10 is used for lipid production in cultures of *Chlorella* spp. However, optimal values have been reported to be as high as a C/N of 26 (Cheng et al. 2009a). Nitrogen (NO_3-N) concentrations in the growth medium, by itself, influences the amount of lipid contents and variety of fatty acids in *Chlorella vulgaris* and *C. sorokiniana* at a concentration from 2.5 to 5.6 mg L^{-1}, as optimal to achieve maximum lipids production (Cha et al. 2011). Shen et al. (2010) tested the influence of various concentrations of urea, yeast extract, and nitrate for growing *C. protothecoides*. Among the tested conditions, the highest yield of lipids and biomass were obtained at the low concentration of 0.33 g L^{-1} of NO_3-N medium. Further, urea has been used as a nitrogen source in heterotrophic cultures of *C. protothecoides*, where optimal production of lipids occurred at a concentration of 3.6 g L^{-1} (Shi et al. 2002).

It is important to consider that some organic substrates can change the pH of the medium and other properties such as viscosity and gas-liquid transfer coefficient. When sodium or potassium salt of acetate is used as a substrate, pH increases. This happens because the remaining Na^+ or K^+ couples with hydroxyl ions (OH^-) or other anions to form alkalis. This phenomenon also occurs if reactors are neutralized with alkali. Since metallic hydroxides are stronger than organic acids, the media must be neutralized or at least brought to a non-inhibitory pH by adding an acid, for example, acetic acid into the cycle (Ratledge et al. 2001).

5.4 *Oxygen Supply*

Oxygen is a key factor in heterotrophic cultivation of microalgae. Independent of the organic substrate or the microalgae species, growth rates are enhanced by higher levels of aeration (Griffiths et al. 1960). Limited oxygen in a culture reduces the specific growth rate of microalgae cells, lowering biomass productivity when cell density is high (Wu and Shi 2007). For example, De Swaaf et al. (2003a) showed that high cell density of *Crypthecodinium cohnii* cultivated on acetic acid was a result of high oxygen demand. To maintain aerobic conditions, a high stirring speed had to be maintained during a large part of the process, hence, oxygen transfer is likely to be a limiting factor during a commercial scale cultivation of algae at high density.

5.5 *Mixing, Viscosity, and Foam*

Mixing is one of the most important operations when cultivating microalgae. Mixing within the fermenter is necessary for uniformly distributing nutrients and for gas exchange. Viscosity of the medium affects circulation, increasing with cell concentration or production of viscous cellular material. Many fermenters achieve adequate mixing with impellers and baffles (Behrens 2005; Dorian 1995). An alternative approach is based on air lifting. Fermenters with an internal plenum or draft tube have true airlifting capacity. Air is supplied on one side of the plenum or draft tube, creating a density difference between the side that is aerated and the side that is not. To be effective, airlifting usually requires a substantial airflow rate because sparging of air must supply all the energy for mixing and mass movement within the fermenter. High viscosity in cultures requires higher impeller speed or airflow, which increases power consumption and operational costs (Behrens 2005; Dorian 1995). Adding commercial polysaccharide-hydrolyzing enzymes decreases viscosity, which leads to an immediate increase in the concentration of dissolved oxygen, allowing a reduction in the speed of stirrers (De Swaaf et al. 2003a). Foam control is implemented as in yeast or bacterial fermentations. Usually, this involves monitoring foam production in the culture with sensors and using antifoam agents (De Swaaf et al. 2003a).

5.6 *Temperature*

The less efficient heat transfer in large-scale reactors may result in temperature problems from the combination of biological heat production and power input for mixing (De Swaaf et al. 2003a). Further, slightly differences in temperature can affect the metabolism of a given strain. For instance, maintenance of *C. protothecoides* culture in 30 L fed-batch cultures at 32 °C for 84 h resulted in a 19.9 % increase in lutein content but a 13.6 % decrease in cell dry weight concentration, compared with cultivation at 28 °C. Greater lutein production results from a combination of limited nitrogen and high temperature stress (Shi et al. 2002).

5.7 Light in Mixotrophic Cultivation

Mixotrophic cultivation is better when executed in closed photo-bioreactors (Bassi et al. 2014; Chen et al. 2011). These cultivation systems have transparent walls and vessel shapes (tubular and flat plate photo-bioreactors) that maximize the exposure of cells to light (Bassi et al. 2014). During mixotrophic growth, inorganic carbon is fixed through photosynthesis that is influenced by illumination conditions. Organic compounds are assimilated through aerobic respiration that is affected by availability. In mixotrophic cultivation, among all the environmental factors that affect the regime, the amount and quality of light are most important. Light not only affects microalgae photosynthesis, productivity, cell composition, and metabolic pathways, but also determines the economic efficiency of the cultivation process (Chen et al. 1996; Chojnacka and Noworyta 2004; Li et al. 2012). Light intensity profoundly affects mixotrophic growth of microalgae, independent of the carbon source. It is possible to control light intensity in photo-bioreactors by varying the number of fluorescent lamps, irradiance, and distance between the lamps and bioreactor's surface. The optimum light intensity for microalgae biomass and desired product production is not only strain-dependent, but also relies on the suitability of other factors of the environment, especially temperature and nutrient supply (Li et al. 2012).

Assimilation of carbon also modifies the cell's response to light. For example, as mentioned on Sect. 3 of this review, the maximum photosynthetic rate, light saturation constant, dark respiration rate, and light compensation point of mixotrophic cultures of the cyanobacteria *Spirulina platensis* were significantly higher than these rates in photo-autotrophic cultivation. As expected, mixotrophic cultures grew faster and reached higher biomass than photo-autotrophic cultivation. The growth rate of photo-autotrophic cultures was more sensitive to light. The differences between the two cultures were also apparent in their responses to exposure to extreme high intensities of light of 3000 μmol photon m^2 s^{-1}. Although both cultures recovered from the stress of high photon flux density, the mixotrophic culture recovered faster and reached higher biomass concentration (Vonshak et al. 2000). Most species that are capable of mixotrophic growth need light. Usually the lower limit is ~50 μmol photon m^2 s^{-1}. These photons are not used for the photosynthetic cycles, but rather for enzyme activation for substrate assimilation (Chojnacka and Marquez-Rocha 2004; Chojnacka and Noworyta 2004). It is important to know, in advance, if a desired growth process is truly heterotrophic, mixotrophic, or photo-heterotrophic. If even a marginal amount of light is needed, this has to be considered within the reactor design, since light sources have to be evenly integrated, leading to rising costs or declining yields (Kröger and Müller-Langer 2011).

Light wavelength also affects microalgal productivity and product profile. For example, *Nannochloropsis* sp., was grown in phototrophic and mixotrophic conditions, with glycerol as the carbon source, under three primary monochromatic light wavelengths (red, green, and blue) and white light as the control. Lamps with light-emitting diodes (LEDs) were used to control light chromatic properties. The maximum specific growth rate was higher on blue > white > green > red lights. The intracellular fatty acid composition of these cultures varied with exposure to the different wavelengths and the absolute fatty acid content were significantly in favor of blue light at a wave length of 470 nm (Das et al. 2011).

Turbidity that reduces light penetration in the culture is an important parameter determining the suitability of a culture medium for mixotrophic (and photo-autotrophic) microalgae growth (Wang et al. 2010). Turbidity is specifically important in growth medium based on wastewater. Turbidity is caused by suspended matter or impurities that interfere with the clarity of the water, leading to light scattering and absorption rather than transmitted in straight lines through a water body (Wang et al. 2010). The impurities responsible for turbidity include clay, silt, suspended inorganic and organic matter, soluble colored organic compounds, plankton, and other microorganisms. Diluting wastewater in growth media affects turbidity and nutrient concentrations in the microalgae cultures and is a common solution for cultivation of microalgae (Hu et al. 2013).

Mixotrophic growth is also controlled by the interaction between organic carbon and nitrogen. Similar to heterotrophic cultures, controlling the C/N ratio is fundamental for optimizing reactor operations (Pagnanelli et al. 2014; Silaban et al. 2014). As in all photosynthetic processes, the supply of CO_2 is essential for high productivity of algae. Ideally, CO_2 comes from the air supplied to the culture, but the concentration is too low to make high productivity possible. Alternatively, the CO_2 is a residual gas that bubbles through the water column. As gas bubbling already requires energy, it is important that the mass transfer of CO_2 is efficient (Wijffels et al. 2010).

5.8 *Advances in Bioreactor Designs*

Fermenters are adapted for most heterotrophic cultivation at the laboratory and pilot scale. Also, fermenters are suitable for commercial production of high-value products, such as food supplements and pharmaceuticals, including EPA, DHA, lutein, phycocyanin, and astaxanthin (Barclay et al. 2013; Chen and Chen 2006). Fermenters are not ideal for mixotrophic cultivation. However, fermenters made of transparent materials are used in laboratories to expose the cultures to light. Conventional fermenters and photo-bioreactors are suitable at the laboratory or pilot scale, but are very expensive at the industrial scale. Therefore, hampering production of cheap products, such as biofuels at industrial scale (Molina Grima et al. 1999; Tabernero et al. 2012). According to the economic study by Tabernero et al. (2012), the dearth of suitable bioreactors is one of the reasons preventing a biofuel facility from profitability. To make biofuel and bulk chemicals production from heterotrophic microalgae, a significant innovation is required. The conventional fermentation technologies that work very well for producing pharmaceutical compounds or high-value food supplements are just too expensive. Innovative reactor design must consider the gigantic volume of culture that is needed. At this scale, prevention of contamination from unwanted microbial species and maintenance of good mixing and oxygen supply must be resolved at the same time. With our current knowledge, we are not there.

A good example of innovative reactor designs is the OMEGA (Offshore Membrane Enclosures for Growing Algae) system developed by NASA for cultivating microalgae from wastewater contained in floating bag photo-bioreactors deployed in marine environments. This design eliminates competition with agriculture for water, fertilizer, and land. OMEGA bioreactors use the mechanical power of waves to mix the floating photo-bioreactors and use waste CO_2-rich flue gas. The reactor's wall is in contact with seawater, which is an effective way to dissipate heat (Carney et al. 2014; Wiley et al. 2013).

For evaluating bioreactors designs, it is important to compare performance of the different systems under the same operational conditions for extended periods (Wijffels et al. 2010). To facilitate quick development of the technology, research at the laboratory scale, pilot scale, and demonstration scale should run in parallel with a good interchange of information. This implies that a technology developed in the laboratory can be tested under realistic conditions in less time and research at the laboratory scale can resolve problems that would be encountered at larger scales (Wijffels et al. 2010).

5.9 *Model-Based Bioprocess Optimization*

Several mathematical modeling approaches can be used to optimize productivity of microalgae cultivation. In general many of these model-based optimization approaches belong to chemical and bioprocess engineering disciplines. They involve the practices of experimental design, model development, process design, and optimization by analyzing physical and metabolic data. Model-based process optimization is an iteractive practice where experiments in the laboratory are used to improve models and vice versa (Koide et al. 2009; Mandenius and Brundin 2008). Examples of modeling approaches for microalgae include metabolic control analysis, neural network modelling, and the response surface method (Mandenius and Brundin 2008; Nwabueze 2010; Wang et al. 2009).

Model based process optimization has been used to increase productivity of heterotrophic and mixotrophic microalgae cultivation. For instance, the Box–Wilson central composite design (CCD) is a useful computational approach that is widely used in optimizing cultivation processes. CCD was used by Hu et al. (2013) to develop a quadratic mathematical model for predicting the optimum dilution rate and hydraulic retention time for mass production of and nutrient removal by *Chlorella* sp. from swine wastewater nutrients. The development of the CCD mathematical model is based on values of the response variables, such as biomass production and nutrient removal rates in real experiments. In another example, Wu and Shi (2007) developed a hybrid neural network model for heterotrophic growth of *Chlorella* sp. and used it to optimize operational parameters of carbon source concentration and feeding rate of fed-batch cultures. After optimizing the operational parameter, culture productivity increased from 0.613 to 1.020 g L^{-1} h^{-1}. Metabolic network analysis discovered that in the two-stage culturing mode, CO_2 released during heterotrophic growth was fixed by the still active Rubisco enzyme that catalyzes CO_2 fixation. Thus, the two-stage culturing scheme provided higher carbon consumption efficiency and lower CO_2 emission, compared with heterotrophic culturing. Considering the double benefits of increased lipid yield and higher CO_2 fixation, the two-stage cultivation model shows promise in the microalgae-to-biofuel field. It will be very interesting to see how the same two-stage system affects the outcomes from non-sugar carbon sources (Xiong et al. 2010). Process performance values of the above examples are presented in Table 4.

5.10 Use of Mixed Microbial Cultures and Microalgae Consortia

Mixed microbial cultures has been proposed as an alternative to pure cultures to reduce reactor operation costs associated with sterilization and rigorous containment under heterotrophic conditions. The use of mixed microbial cultures has been investigated mainly for microalgae growth in wastewater or in open ponds.

The highly variable composition of wastewater can limit the growth potential of specific microalgae. Therefore, it is essential to select strains capable of growing in a variety of wastewater under different climatic conditions over year-long operations. Bhatnagar et al. (2011) develop a robust microalgal consortium (strains of *Chlamydomonas* sp., *Chlorella* sp., and *Scenedesmus* sp.) capable of mixotrophic growth on a variety of extracts of poultry manure and carpet industry wastewater. Higher biomass and lipid production were obtained in cultures with the consortium than with single strain cultures under the same conditions. In another example, a microalgae consortium cultivated in carpet industry wastewater in raceway ponds provided higher biomass productivity and removal of nitrogen and phosphorus than single strain cultures. The consortium was developed by growing pure cultures of wastewater microalgae isolates and mixing them together in equal quantities This consortium was used for wastewater treatment and biomass production and was robust enough to withstand competition and predation in open cultivation systems (Chinnasamy et al. 2010).

An innovative approach is to use mixed cultures of microalgae and bacteria to treat wastewater (de-Bashan et al. 2002b, 2004; Hernandez et al. 2006; Covarrubias et al. 2012; Cruz et al. 2013). This approach is used to increase production of biomass and metabolites and remove nutrients from wastewaters by immobilizing microalgae strains with plant growth-promoting bacteria that also promotes and positively affect the metabolisms of the microalgae. When *C. vulgaris* was immobilized in alginate beads with *Azospirillum brasilense* (a microalgae growth-promoting bacteria; MGPB) and grown autotrophically on synthetic wastewater growth medium, *A. brasilense* mitigated environmental stress for the microalgae and removed significantly more nitrogen and phosphorus from the wastewater (de-Bashan et al. 2002b; de-Bashan and Bashan 2010; Hernandez et al. 2006). In another study, major cell growth occurred at pH 8 when *A. brasilense* was immobilized with

Chlorella, compared to restricted growth when *Chlorella* was grown alone under autotrophic conditions (de-Bashan et al. 2005). In a similar study by Perez-Garcia et al. (2010), the growth of *C. vulgaris*, under mixotrophic regimes, was higher when *A. brasilense* was present in the alginate beads. These studies were repeated under heterotrophic conditions and all, regardless of the cultivation regime, showed greater production of useful metabolites, such as carbohydrates (mainly starch), fatty acids, lipids, and pigments (de-Bashan et al. 2002a; Choix et al. 2012a, b, 2014; Leyva et al. 2014, 2015). In a different approach, Zhou et al. (2013) used mixed cultures of low-lipid, fast-growing mixotrophic algae (*Chlorella* spp. *Scenedesmus obliquus*, and seven other less abundant microalgae) to treat wastewater and residues from hydrothermal liquefaction of biosolids. This mix was then used to produce biomass to feed the hydrothermal liquefaction reactor to produce biofuels. The oxygen provided during algal photosynthesis reduced the energy input for aerobic breakdown of wastewater contaminants and the CO_2-rich gas and residues that were produced during liquefaction were recycled back to the algal–bacterial cultivation system for reuse. Itineration of the cycle leverages the nutrient content of wastewater into high bioenergy quantities, which can be many times the original wastewater energy content (Zhou et al. 2013).

6 Harvesting and Downstream Processes

Economical and feasible production of biofuels from heterotrophic and mixotrophic microalgae will only be possible if biofuel production is combined with production of bulk chemicals and animal feed ingredients. It is expected that the biofuel industry and the bulk chemical industry will operate in parallel on the same biomass. Therefore, research and development of downstream processes to separate the biomass into its different molecular fractions is of utmost importance. Downstream processes involve technologies related to biomass harvesting, cell disruption, and extraction and separation technologies all working on algal biomass as raw material (Fig. 1). Industrially useful compounds such as ω-3-fatty acids, carbohydrates, pigments, vitamins, and proteins should maintain their functionality after downstream processing. At the same time, scalability, low energy costs, and ease of use also need to be taken into account.

Some successful techniques commonly used in downstream processes are briefly summarized. Biomass recovery is a key step, accounting from 20 to 30 % of the total operation cost of biomass production. A comprehensive description of the different harvesting methods is reviewed by Molina Grima et al. (2003). They show that for large scale production, one of the best methods is centrifugation after flocculation of the cells. Once the biomass cake is recovered, a common and widely employed method to extract oil from biomass is supercritical fluid extraction (Tabernero et al. 2012). This method recovers lipids for food supplements, such as PUFAs, or for biodiesel. In the case of biodiesel production, transesterification of extracted lipids is the major method to produce biodiesel. Alkali-catalyzed transesterification is efficient, cheap, and scalable to commercial levels (Liang 2013). An innovative approach to improve lipid extraction from biomass is to use nanomaterials, such as CaO and MgO, to extract lipids without cell disruption and as biocatalyst carriers or heterogeneous catalysts in lipid transesterification to biodiesel (Zhang et al. 2013b). Another suggested approach is to break the cells by combining a low-power electromagnetic field, a process known as quantum fracturing, and adding CO_2 for reducing the increasing pH in this process. These conditions extract the lipids with high efficiency in a single-step without chemicals, no initial dewatering, and no heavy machinery (Tabernero et al. 2012).

Whole biomass transformation methods can be employed as an alternative to avoid extraction of specific compounds from biomass. Thermochemical conversion processes convert all compounds in biomass (lipids, proteins and carbohydrates) into fuel oil or gases in a single step. Thus, this kind of transformation maximizes utilization of microalgae biomass. Thermochemical processes typically include: gasification, combustion, pyrolysis, and liquefaction. While the first two processes transform algal biomass to various gas fuels, the latter two processes produce liquid fuels (Liang 2013). Among the different thermochemical conversion processes, hydrothermal liquefaction is particularly attractive. This process converts lipids, proteins and carbohydrates into oil fuel, so the oil yield is much higher than the lipid content of the algal feedstock, as opposite to transesterification that transformed only lipids (Zhou et al. 2013).

Some limitations in downstream processes can be handled during the upstream part of the process. One approach is to avoid energy and time-consuming product extraction from biomass by developing and designing direct catalytic "sun-to-product pathways", in which photon energy and atmospheric CO_2 is directly converted into compounds that do not need further transformation after biomass production. Examples of these products would be ethanol and hydrogen. Advanced synthetic biology strategies are already being used to re-design microalgae metabolism for direct synthesis of these kinds of "ready-to-use" compounds (Robertson et al. 2011; Yu et al. 2011).

7 Market Products and Techno-economics

The economics of industrial production can be potentially transformed by industrial biotechnology and integrated biorefineries that produce multiple streams of valuable products (Erickson et al. 2012). To create this potential, integrated biorefineries must efficiently and simultaneously convert a broad range of industrial biomass feedstocks into affordable biofuels, energy, and a wide range of chemicals and materials. These goals are met by integrating fuel and chemical production within a single operation. High value products become an economic driver that provides higher margins of revenue to support production of low-value products (fuels), leading to a profitable biorefinery operation that also has a positive energy balance (OECD 2011). The biorefinery scheme presented in Fig. 1 illustrates this concept.

The following streams of compounds and products are derived from refined microalgae biomass (Bassi et al. 2014; Hudek et al. 2014; Wijffels et al. 2010).

- Lipids (triacylglycerides and isoprenoids) for biodiesel;
- Lipids (ω-3 fatty acids, DHA and EPA) for nutritional supplements;
- Lipids, hydrocarbons in general, as feedstock to produce bulk chemicals and fuels;
- Proteins (soluble proteins) for nutritional supplements and personal-care products;
- Proteins (insoluble proteins) for animal feeds;
- Carbohydrates (starch, glycogen, and cellulose) as feedstock to produce bulk chemicals and fuels (i.e. bioethanol);
- Pigments (chlorophyll, β-carotene, lutein, astaxanthin) for nutrient supplements and drugs; and
- Oxygen (from photosynthesis) for general use, such as aquaculture.
- Hydrogen (from photosynthesis) used as biofuel

Nowadays, commercially successful heterotrophic microalgae production is only done for specific niche markets of high-value products, such as pigments (astaxanthin and lutein) and ω-3-fatty acids (DHA and EPA) using strains of *Chlorella* spp., *Crypthecodinium* spp., *Haematococcus* spp., *Nitzschia* spp., and *Schizochytrium* spp. (Barclay et al. 2013; Chen and Chen 2006). As best of authors' knowledge, only *Chlorella* spp. and *Haematococcus pluvialis* strains are cultured mixotrophycally for commercial purposes (Hudek et al. 2014). Nevertheless, successful commercialization of biofuels from heterotrophic and mixotrophic microalgae relies on an integrated biorefinery. A number of biofuels and added-value chemicals can be refined from heterotrophic and mixotrophic microalgae (Bassi et al. 2014). Specifically, heterotrophic microalgae is attractive for obtaining products from cellular storage compounds, such as lipids and starch, while mixotrophic microalgae is of interest for obtaining pigments, lipids, proteins and alkanes. Heterotrophic microalgae can potentially produce large amounts of hydrocarbons and polysaccharides that can be converted into organic building blocks (ethylene, propylene, adipic acid, and furanics) for polymers and plastics (Wijffels et al. 2013). Considering the average biomass composition obtained from microalgae under different growth modes (Table 3) and the streams of compounds obtained from biorefined biomass, it is possible to anticipate the potential of each growth mode to obtain certain products, as shown in Table 6.

Biorefinery techniques should initially focus on isolating proteins and lipids because these are the largest fractions of microalgae. Other fractions, such as carbohydrates and pigments, will add significant value to the total production process when separated (Vanthoor-Koopmans et al. 2013). The main bottleneck is to separate the different fractions without damaging one or more of the product fractions. Technologies to overcome these bottlenecks need to be developed, and they should be applicable for a variety of end products of sufficient quality at large quantities. To that end, developed techniques should be gentle, inexpensive, and consume less energy. Further, locating microalgae production facilities near waste treatment

Table 6 Potential compounds and products obtained from different microalgae growth modes

Fraction/stream	Compounds	Products	Autotrophic growth	Heterotrophic growth	Mixotrophic growth
Lipids	Triacylglycerides	Biodiesel	Medium	High	High
	Isoprenoids	Biodiesel	Medium	High	Medium
	ω-3 fatty acids	Nutritional supplements	High	Medium	High
	Hydrocarbons	Biofuels, polymers, and plastics	Medium	High	Medium
Carbohydrates	Starch and glycogen	Biofuels, polymers, and plastics	Low	High	Low
Proteins	Soluble proteins	Nutritional supplements and drugs	High	Low	High
	Insoluble proteins	Animal feed	High	Low	Medium
Pigments	Chlorophylls	Nutritional supplements and drugs	High	Low	Medium
	β-carotene	Nutritious supplements and drugs	High	Low	Medium
	Lutein	Nutritional supplements and drugs	Medium	High	High
	Astaxanthin	Nutritional supplements and drugs	Medium	High	High

facilities, such as dairy and swine manure, food waste, or industrial organic waste could be environmentally effective and economically viable in the near future. However, heterotrophic growth is highly specific to the organic feed and microalgal species. Therefore, these waste streams have to be homogeneous and continuous if the product is biomass from a specific microalga with defined biomass composition (Kröger and Müller-Langer 2011).

Heterotrophic and mixotrophic microalgae cultivation open opportunities for the energy industry. For instance, biodiesel with good characteristics has been obtained from heterotrophic cultivation of *C. protothecoides*. For these reasons, this specific microalga is very attractive to the biodiesel production industry (Xu et al. 2006). In an experimental series by Xu et al. (2006) and Miao and Wu (2006), oil extracted from heterotrophically grown microalgae was transesterified to biodiesel. The fuel was comparable to fossil diesel in density, viscosity, and heating value, and complied with the US standard for biodiesel (ASTM 6751). However, the quality of heterotrophic microalgal lipids regarding biodiesel is inferior to vegetable oil because it contains excessive free fatty acids. These fatty acids complicated the alkaline-transesterification process (Miao and Wu 2006).

Another strategy to produce biofuels from microalgae is to digest or ferment the whole biomass into methane or ethanol, respectively or process it into fuel gas (syngas, a mixture of H_2, CO, and CO_2), hydrogen, or crude oil by thermo-chemical processes, such as gasification, liquefaction, and pyrolysis (Tabernero et al. 2012). Hydrogen is another compound that can be obtained by biological or thermo-chemical processes; however, this route is in the early stages of development (Wijffels et al. 2013). The disadvantage of using the whole biomass only for energy is that the value of more expensive compounds, such as proteins and pigments is lost (Tabernero et al. 2012).

7.1 Techno-economic Analysis

With current technologies, commercial success of either photo-autotrophic, heterotrophic, or mixotrophic microalgae biorefineries is unlikely. However, enormous opportunities exist to improve the current upstream and downstream technologies that can either reduce the production cost of valuable compounds or add more value. In the techno-economic analyses by Tabernero et al. (2012), a conservative and an optimized production cost of one kilogram of heterotrophic *C. protothecoides* biomass was US\$1.4 kg^{-1} and US\$1.19 kg^{-1}, respectively (Table 7). These values are determined by dividing the yearly total production cost (US\$12.4 million $year^{-1}$) by the biomass production capacity (22.7 million kg $year^{-1}$) and then adding an investment cost of US\$0.93 kg^{-1} $year^{-1}$ for a conservative estimate or US\$0.68 kg^{-1} $year^{-1}$ for an optimized estimate. These values were estimated for a biorefinery producing biomass in 465 continuously stirred bioreactors each of 150,000 L and producing 10 million L $year^{-1}$ of biodiesel. Optimized estimates considers a higher percent of lipid recovered from biomass (from 55 to 100 %), a

Table 7 Cost of biomass production, prices in United State dollars (USD) at a rate of Euro to dollar of 1.26 (October 2014)

	Photo-autotrophic[a]	Heterotrophic[b]	Mixotrophic
Conservative cost	US\$5 kg^{-1}	US\$1.4 kg^{-1}	Unknown
Optimized cost	US\$0.5 kg^{-1}	US\$1.19 kg^{-1}	Unknown
Reference	Wijffels et al. 2010	Tabernero et al. 2012	

[a]In 100 ha of flat panel photobioreactors
[b]In 465 continuous stirrer tank bioreactors (CSTR or fermenters) of 150,000 L, assuming oil content in biomass of 44 %

reduced number of bioreactors (from 465 to 257), and no oil extraction costs, based on a novel cell disruption technology using magnetism. In contrast, using a conservative approach, Wijffels et al. (2010) estimated that industrial photo-autotrophic cultivation in flat panel reactors can produce microalgal biomass at a cost of US\$11.3 kg^{-1} per hectare (all estimates are based on a conversion of 1.26 USD/EUR of October 2014) The main costs of this feasibility analysis were power and labor. If the system is scaled up, labor cost is significantly reduced. A conservative cost for biomass production on 100 h of flat panel photo-bioreactors is about US\$5 kg^{-1}. This cost is acceptable for production of biomass for high-value compounds but unacceptable for production of biofuels. At this scale, more than 24 % of the cost would be for energy, such as pumping water and sparging of air and CO_2 in the system. However, production costs can be reduced by appropriate technologies. Wijffels et al. (2010) estimate an optimized cost of US\$0.5 kg^{-1}, considering the following technical aspects: free cultivation of CO_2 and nutrients from wastes, reduction of energy input by 10 %, an increase of photosynthetic efficiency of the microalgae strain from 5 to 7 %, photo-bioreactors placed in a location with high levels of sunshine. The conservative and optimized costs of biomass production are summarized in Table 7. Estimates of production cost at the commercial scale of mixotrophic biomass were not found in the literature.

Once the cost of producing a kilogram of biomass is established, it is possible to analyze the total value of that kilogram of biomass after being refined into various products. Following the Wijffels et al. (2010) analysis, we assumed that microalgae biomass is refined into products for bulk chemical markets; these products and their market prices are similarly listed in Tables 8 and 9. Wijffels et al. (2010) and Xu et al. (2006) assumed different biomass compositions, depending on the growth mode of microalgae; a composition of 40 % lipids, 50 % proteins, and 10 % carbohydrates was assumed for photo-autotrophic growth and a composition of 55 % lipids, 10 % protein, 15 % carbohydrates, and 20 % unusable compounds, such as ash and nucleic acids, was assumed for heterotrophic growth. In both analyses, 25 % of the lipids fraction is used to produce bulk chemicals with a value of US\$2.5 kg^{-1} and the remaining 75 % was for biodiesel production with a value of US\$0.72 kg^{-1} (= US\$0.63 L^{-1}). The protein fraction was subdivided into a water-soluble fraction of 20 % with a food value of US\$6.3 kg^{-1} and a water-insoluble fraction of 80 % with feed value of US\$0.95 kg^{-1}. Finally, 100 % of the carbohydrate

Table 8 Value of biomass and profit obtained from photo-autotrophic microalgae

Product	Percentage of biomass dry weight composition (%)	Price per kg of product (US\$ kg^{-1})	Product value in 1000 kg of biomass[a] (US\$)
Biofuels from lipids	30	0.72	\$214
Bulk chemicals from lipids	10	2.52	\$252
Nutritious supplements from soluble protein	10	6.30	\$630
Feeds from insoluble protein	40	0.95	\$378
Bulk chemicals from carbohydrates	10	1.26	\$126
Oxygen	–	0.20	\$322
N removal	–	2.58	\$170
Total	100		\$2,099
Conservative production cost[b]			\$5,000
Optimized production cost[b]			\$500
NET PROFIT Conservative[b]			**-\$2,901**
NET PROFIT Optimized[c]			+\$1,599

[a]Product value in a 1000 kg of biomass = % of composition (in decimals) multiplied by the price of kg of product multiplied by a thousand
[b]From Table 7 multiplied by a thousand
[c]Net profit = Total value of a 1000 kg of biomass minus production cost

fraction has a market value of US\$1.26 kg^{-1} and consist of cellular storage products, such as starch, glucans, and glycerol that can be used as chemical building blocks or production of energy. Besides these main products, there are additional by-products or valuable environmental services, such as reducing nutrients in waste streams and production of oxygen. In wastewater treatment, removal of nitrogen compounds by nitrification and denitrification is an expensive process. The cost of nitrogen removal is US\$2.52 kg^{-1}. Microalgae contain 70 kg of nitrogen per 1000 kg of microalgae. If microalgae production is combined with wastewater treatment, we can save US\$176 for nitrification and denitrification per ton of microalgae produced. Photo-autotrophic algae produces oxygen. Per ton of microalgae, 1600 kg of oxygen-rich gas is produced with an approximate value of US\$0.2 kg^{-1}, resulting in US\$ 322 ton^{-1}. The percentage of biomass composition and the market price of those components or fractions are listed in Tables 8 and 9.

Analyses presented in Tables 8 and 9 show that, if microalgal biorefining is used, the total value of the biomass produced using photo-autotrophic or heterotrophic cultivation of microalgae (US\$2,099 and US\$1,249 per 1000 kg, respectively) is higher than the optimized costs for microalgae production, but lower than the conservative costs. Considering that conservative costs were estimated with current technology and operating standards, a general analysis shows that heterotrophic cultivation is currently too expensive to be implemented in a sustainable and commercially feasible microalgae biorefinery. This high cost results mainly from the many bioreactors required by the plant and the yields of oil (Tabernero et al. 2012).

Table 9 Value of biomass and profit obtained from heterotrophic microalgae

Product	Percentage of biomass dry weight composition (%)	Price per kg of product (US$/kg)	Product value in 1000 kg of biomass[a] (US$)
Biofuels from lipids	41	0.72	$296
Bulk chemicals from lipids	14	2.52	$348
Nutritious supplements from soluble protein	3	6.30	$162
Feeds from insoluble protein	8	0.95	$73
Bulk chemicals from carbohydrates	15	1.26	$194
Oxygen			$0
N removal			$176
Total	100		$1,249
Conservative production cost[b]			$1,478
Optimized production cost[b]			$1,196
NET PROFIT Conservative[b]			**–$229**
NET PROFIT Optimized[c]			+$54

[a]Product value in a 1000 kg of biomass = % of composition (in decimals) multiplied by the price of kg of product multiplied by a thousand
[b]From Table 7 multiplied by a thousand
[c]Net profit = Total value of a 1000 kg of biomass minus production cost

Further, the analysis shows that heterotrophic biomass has a lower value than photo-autotrophic biomass because it contains less soluble proteins, which is the biomass fraction with highest market value.

This analysis also illustrates the necessity to refine microalgae biomass into various products. If we assume an optimized heterotrophic biomass production cost of US$1.18 kg^{-1}, the algae contains 55 % lipids and that is possible to transform 100 % of these lipids into biodiesel, the cost to produce 1 kg of biodiesel from heterotrophic microalgae would be US$1.81 kg^{-1}. This is more than double the US$0.715 kg^{-1} the average market value of petroleum diesel in October 2014 in the United States (source: U.S. Energy Information Administration. ~US$0.715 kg^{-1} = US$0.63 L^{-1}, considering a biodiesel density of 0.88 kg L^{-1}). For this reason, refining microalgal biomass into different products is essential to increase the total value of the biomass. If microalgal biomass is refined, its total value approaches US$1.25 kg^{-1} (Table 9), slightly higher than the total cost for microalgae production (US$1.18 kg^{-1}). These results reiterate that the economics of microalgal biofuel production would not be competitive with traditional fossil fuels if a large scale facility were to be built today (Davis et al. 2011). The crucial question is whether microalgal biomass can be produced at a cost below US$1.18 or US$0.5 kg^{-1}, using heterotrophic or photo-autotrophic cultivation (Wijffels et al. 2010).

Our analysis, combined with studies by Brentner et al. (2011), Davis et al. (2011), Kröger and Müller-Langer (2011), Tabernero et al. (2012), and Wijffels et al. (2010) demonstrate that economically feasible co-production of heterotrophic microalgae biofuels and other chemicals is possible if the current technology develops and *significantly* reduces production and refining cost. In every analysis, the two highest costs are attributed to bioreactor set up and harvesting costs. Therefore, new technological developments should focus mainly on these aspects of the bioprocess. Sections 3, 4, 5, and 6 of this chapter discuss in detail technological advances currently under development. A sensitivity analysis of a mathematical model of microalgae biomass production process pointed out that the content of valuable compounds in biomass is the most sensitive factor affecting costs of bioreactors operations. This means that, by increasing the content of valuable compounds in biomass, it is possible to effectively reduce operation costs. This emphasizes the importance of strain selection and development of the appropriate processes of cultivation and harvesting (Davis et al. 2011). Additionally, the performance and economic feasibility of a microalgal biorefinery will vary greatly with location, depending on the physical and technological environment in which it is located and the cost of labor at the site. For mixotrophic microalgae, parameters, such as ambient temperature and solar intensity can clearly affect the potential growth rate of microalgae and costs of bioreactors (Brentner et al. 2011).

Mixotrophic growth offers several features, which should be examined more closely. An increase in growth rate and production of valuable compounds by adding an organic substrate to a photo-autotrophic culture is possible. Therefore, the use of a mixotrophic regime is a logical step to stabilize and enhance the production process. High-production refineries will be built in the coming years and mixotrophy is a good strategy to minimize their default risk (Kröger and Müller-Langer 2011). It remains to be investigated if, and for which species, the addition of organic substrates provide a more robust and efficient operation. The possibilities of mixotrophy should also be considered when designing a microalgae production facility, since they may have an impact on the production rate and reliability and, therefore, economics (Kröger and Müller-Langer 2011).

8 Final Remarks

Heterotrophic and mixotrophic cultivation of microalgae offers an alternative to photo-autotrophic cultivation to improve the economic feasibility of microalgae biorefined products. Although microalgae are not yet cultivated at large scales for bulk applications, there are opportunities to develop this process in a sustainable way in the foreseeable future. With the current technologies in place, it remains unlikely, however, that the process will be developed for biodiesel as the sole end-product from microalgae biomass because no biofuel production from microalgae comes close to current market prices of fossil fuels. Consequently, to develop a more sustainable and economically feasible process, all biomass components, such as proteins, lipids, and carbohydrates should be used to obtain compounds with high market value. Therefore, in biorefining of microalgae, the selective separation and use of the functional biomass components is of utmost importance. If, in addition, production of microalgae biomass is done on wastewater or industrial wastes as the primary substrate, without special production of growth medium ingredients, and the cultivation is done on a large scale at low costs, production of bulk chemicals and fuels from microalgae may become economically feasible (Wijffels et al. 2010).

For producing biofuels and bulk chemicals from lipids and carbohydrates, cultivating microalgae under heterotrophic and mixotrophic regimes offers great promise because high production of biomass and lipids or carbohydrates (but not proteins) can be expected, specially using fed-batch cultivation modes. Livestock manures, molasses, food wastes, and whey from the dairy industry were shown to be promising organic substrates for heterotrophic and mixotrophic cultivation. Further research on these substrates should analyze, in detail, yields, volumetric productivity, and product concentration values of different microalgal strains on these potential substrates.

To make heterotrophic and mixotrophic cultivation commercially viable, research should focus on:

- Lowering the cost of carbon substrates;
- Bio-prospecting or metabolically engineering microalgae strains able to assimilate multiple substrates, resist changing environmental conditions, and technical condition prevailing during industrial production and have fast growth;
- Determining the potential of consortia of microalgae or microalgae with bacteria in natural co-cultivation (microorganisms from similar habitat) or in synthetic biology assemblage (microorganisms not from the same origin);
- Design suitable bioreactors at industrial scales that are made of cheap materials adjusted to the needs of microalgae cultivation; and
- Optimizing downstream processes of transforming microalgal biomass to various produces, including fuels.

With intensive research and development currently underway in these fields, it is reasonable to anticipate that large-scale biofuel production combined with an industry of high-value products from heterotrophic and mixotrophic microalgae, probably fueled by waste products from other industries, will be a reality in the foreseeable future.

Acknowledgement We thank Ira Fogel of CIBNOR Mexico for editing the manuscript. Time for writing was provided by The Bashan Foundation, USA.

Acronyms List

μ^{max}	Maximum growth rate
BOD	Biological oxygen demand
C/N	Carbon to nitrogen ratio of growth media
CCD	Central composite design

COD	Chemical oxygen demand
DHA	Docosahexaenoic acid
EC	Enzyme commission numer
EM pathway	Embden-Meyerhof pathway
EPA	Eicosapentaenoic acid
FBA	Flux balance analysis
HRT	Hydraulic retention time
LEDs	Light-emitting diodes
mRNA	Messenger RNA
OMEGA	Offshore Membrane Enclosures for Growing Algae
PCR	Polymerase chain reaction
PP pathway	Pentose phosphate pathway
PUFAs	Polyunsaturated fatty acids
SMN	Stoichiometric metabolic network
STR	Stirred tank bioreactor
TCA cycle	Tricarboxylic acid cycle
VSS	Dissolved volatile solids

References

Andrade MR, Costa JAV (2007) Mixotrophic cultivation of microalga *Spirulina platensis* using molasses as organic substrate. Aquaculture 264:130–134

Aubert S, Gout E, Bligny R, Douce R (1994) Multiple effects of glycerol on plant cell metabolism. Phosphorus-31 nuclear magnetic resonance studies. J Biol Chem 269:21420–21427

Barclay W, Apt K (2013) Strategies for bioprospecting microalgae for potential commercial applications. In: Richmond A, Hu Q (eds) Handbook of microalgal culture: applied phycology and biotechnology, 2nd edn. Wiley-Blackwell, Chichester, pp 69–79

Barclay W, Apt K, Dong XD (2013) Commercial production of microalgae via fermentation. In: Richmond A, Hu Q (eds) Handbook of microalgal culture: applied phycology and biotechnology, 2nd edn. Wiley-Blackwell, Chichester, pp 134–145

Bassi A, Saxena P, Aguirre AM (2014) Mixotrophic algae cultivation for energy production and other applications. In: Bajpai R, Prokop A, Zappi M (eds) Algal biorefineries, Vol. 1. Cultivation of cells and products. Springer, New York, pp 177–202

Behrens PW (2005) Photobioreactor and fermentors: the light and the dark sides of the growing algae. In: Andersen RA (ed) Algal culturing techniques. Elsevier Academic Press, New York, pp 189–203

Bhatnagar A, Chinnasamy S, Singh M, Das KC (2011) Renewable biomass production by mixotrophic algae in the presence of various carbon sources and wastewaters. Appl Energy 88:3425–3431

Blifernez-Klassen O, Klassen V, Doebbe A, Kersting K, Grimm P, Wobbe L, Kruse O (2012) Cellulose degradation and assimilation by the unicellular phototrophic eukaryote *Chlamydomonas reinhardtii*. Nat Commun 3:1214

Bollman RC, Robinson GGC (1977) The kinetics of organic acid uptake by three Chlorophyta in axenic culture. J Phycol 13:1–5

Borowitzka MA (1999) Commercial production of microalgae: ponds, tanks, tubes and fermenters. J Biotechnol 70:313–321

Boyle NR, Morgan JA (2009) Flux balance analysis of primary metabolism in *Chlamydomonas reinhardtii*. BMC Syst Biol 3:4. doi:10.1186/1752-0509-3-4

Brányiková I, Marsálková B, Doucha J, Brányik T, Bisová K, Zachleder V, Vitova M (2011) Microalgae – novel highly efficient starch producers. Biotechnol Bioeng 108:766–776

Brennan L, Owende P (2010) Biofuels from microalgae – a review of technologies for production, processing, and extractions of biofuels and co-products. Renew Sustain Energy Rev 14:557–577

Brentner LB, Eckelman MJ, Zimmerman JB (2011) Combinatorial life cycle assessment to inform process design of industrial production of algal biodiesel. Environ Sci Technol 45:7060–7067

Carney LT, Reinsch SS, Lane PD, Solberg OD, Jansen LS, Williams KP, Trent JD, Lane TW (2014) Microbiome analysis of a microalgal mass culture growing in municipal wastewater in a prototype OMEGA photobioreactor. Algal Res 4:52–61

Cha TS, Chen JW, Goh EG, Aziz A, Loh SH (2011) Differential regulation of fatty acid biosynthesis in two Chlorella species in response to nitrate treatments and the potential of binary blending microalgae oils for biodiesel application. Bioresour Technol 102:10633–10640

Chang RL, Ghamsari L, Manichaikul A, Hom EFY, Balaji S, Fu W, Shen Y, Hao T, Palsson BØ, Salehi-Ashtiani K, Papin JA (2011) Metabolic network reconstruction of *Chlamydomonas* offers insight into light-driven algal metabolism. Mol Syst Biol 7:518. doi:10.1038/msb.2011.52

Charles DF, Knowles D, Davis RS (2002) Protocols for the analysis of algal samples collected as part of the U.S. Geological Survey National Water-Quality Assessment Program. Report No. 02–06, Philadelphia

Cheirsilp B, Torpee S (2012) Enhanced growth and lipid production of microalgae under mixotrophic culture condition: effect of light intensity, glucose concentration and fed-batch cultivation. Bioresour Technol 110:510–516

Chen F (1996) High cell density culture of microalgae in heterotrophic growth. Trends Biotechnol 14:421–426

Chen GQ, Chen F (2006) Growing phototrophic cells without light. Biotechnol Lett 28:607–616

Chen F, Zhang Y, Guo S (1996) Growth and phycocyanin formation of *Spirulina platensis* in photoheterotrophic culture. Biotechnol Lett 18:603–608

Chen CY, Yeh KL, Aisyah R, Lee DJ, Chang JS (2011) Cultivation, photobioreactor design and harvesting of microalgae for biodiesel production: a critical review. Bioresour Technol 102:71–81

Cheng Y, Lu Y, Gao C, Wu Q (2009a) Alga-based biodiesel production and optimization using sugar cane as the feedstock. Energy Fuel 23:4166–4173

Cheng Y, Zhou W, Gao C, Lan K, Gao Y, Wu Q (2009b) Biodiesel production from Jerusalem artichoke (*Helianthus Tuberosus* L.) tuber by heterotrophic microalgae *Chlorella protothecoides*. J Chem Technol Biotechnol 84:777–781

Chi Z, Pyle D, Wen Z, Frear C, Chen S (2007) A laboratory study of producing docosahexaenoic acid from biodiesel-waste glycerol by microalgal fermentation. Process Biochem 42:1537–1545

Chinnasamy S, Bhatnagar A, Hunt RW, Das KC (2010) Microalgae cultivation in a wastewater dominated by carpet mill effluents for biofuel applications. Bioresour Technol 101:3097–3105

Choix FJ, de-Bashan LE, Bashan Y (2012a) Enhanced accumulation of starch and total carbohydrates in alginate-immobilized *Chlorella* spp. induced by *Azospirillum brasilense*: II. Heterotrophic conditions. Enzyme Microb Technol 51:300–309

Choix FJ, de-Bashan LE, Bashan Y (2012b) Enhanced accumulation of starch and total carbohydrates in alginate-immobilized *Chlorella* spp. induced by *Azospirillum brasilense*: I. Autotrophic conditions. Enzyme Microb Technol 51:294–299

Choix FJ, Bashan Y, Mendoza A, de-Bashan LE (2014) Enhanced activity of ADP glucose pyrophosphorylase and formation of starch induced by *Azospirillum brasilense* in *Chlorella vulgaris*. J Biotechnol 177:22–34

Chojnacka K, Marquez-Rocha F (2004) Kinetic and stoichiometric relationships of the energy and carbon metabolism in the culture of microalgae. Biotechnology 4:21–34

Chojnacka K, Noworyta A (2004) Evaluation of *Spirulina* sp. growth in photoautotrophic, heterotrophic and mixotrophic cultures. Enzyme Microb Technol 34:461–465

Christenson L, Sims R (2011) Production and harvesting of microalgae for wastewater treatment, biofuels, and bioproducts. Biotechnol Adv 29:686–702

Covarrubias SA, de-Bashan LE, Moreno M, Bashan Y (2012) Alginate beads provide a beneficial physical barrier against native microorganisms in wastewater treated with immobilized bacteria and microalgae. Appl Microbiol Biotechnol 93:2669–2680

Cruz I, Bashan Y, Hernandez-Carmona G, de-Bashan LE (2013) Biological deterioration of alginate beads containing immobilized microalgae and bacteria during tertiary wastewater treatment. Appl Microbiol Biotechnol 97:9847–9858

Das P, Lei W, Aziz SS, Obbard JP (2011) Enhanced algae growth in both phototrophic and mixotrophic culture under blue light. Bioresour Technol 102:3883–3887

Davis R, Aden A, Pienkos PT (2011) Techno-economic analysis of autotrophic microalgae for fuel production. Appl Energy 88:3524–3531

Day JD, Edwards AP, Rodgers GA (1991) Development of an industrial-scale process for the heterotrophic production of a micro-algal mollusc feed. Bioresour Technol 38:245–249

De Swaaf ME, Sijtsma L, Pronk JT (2003a) High-cell-density fed-batch cultivation of the docosahexaenoic acid producing marine alga *Crypthecodinium cohnii*. Biotechnol Bioeng 81:666–672

De Swaaf ME, Pronk JT, Sijtsma L (2003b) Fed-batch cultivation of the docosahexaenoic-acid-producing marine alga *Crypthecodinium cohnii* on ethanol. Appl Microbiol Biotechnol 61:40–43

de-Bashan LE, Bashan Y (2010) Immobilized microalgae for removing pollutants: review of practical aspects. Bioresour Technol 101:1611–1627

de-Bashan LE, Bashan Y, Moreno M, Lebsky VK, Bustillos JJ (2002a) Increased pigment and lipid content, lipid variety, and cell and population size of the microalgae *Chlorella* spp. when co-immobilized in alginate beads with the microalgae-growth-promoting bacterium *Azospirillum brasilense*. Can J Microbiol 48:514–521

de-Bashan LE, Moreno M, Hernandez J-P, Bashan Y (2002b) Removal of ammonium and phosphorus ions from synthetic wastewater by the microalgae *Chlorella vulgaris* coimmobilized in alginate beads with the microalgae growth-promoting bacterium *Azospirillum brasilense*. Water Res 36:2941–2948

de-Bashan LE, Hernandez J-P, Morey T, Bashan Y (2004) Microalgae growth-promoting bacteria as "helpers" for microalgae: a novel approach for removing ammonium and phosphorus from municipal wastewater. Water Res 38:466–474

de-Bashan LE, Antoun H, Bashan Y (2005) Cultivation factors and population size control the uptake of nitrogen by the microalgae *Chlorella vulgaris* when interacting with the microalgae growth-promoting bacterium *Azospirillum brasilense*. FEMS Microbiol Ecol 54:197–203

Dorian PM (1995) Bioprocess engineering principles. Academic, London

Dragone G, Fernandes BD, Abreu AP, Vicente AA, Teixeira JA (2011) Nutrient limitation as a strategy for increasing starch accumulation in microalgae. Appl Energy 88:3331–3335

Droop MR (1974) Heterotrophy of carbon. In: Stewart WDP (ed) Algal physiology and biochemistry. Blackwell Scientific, Oxford, pp 530–559

Ducat DC, Avelar-Rivas JA, Way JC, Silvera PA (2012) Rerouting carbon flux to enhance photosynthetic productivity. Appl Environ Microbiol 78:2660–2668

Erickson B, Nelson WP (2012) Perspective on opportunities in industrial biotechnology in renewable chemicals. Biotechnol J 7:176–185

Espinosa-Gonzalez I, Parashar A, Bressler DC (2014) Heterotrophic growth and lipid accumulation of *Chlorella protothecoides* in whey permeate, a dairy by-product stream, for biofuel production. Bioresour Technol 155:170–176

Gao C, Zhai Y, Ding Y, Wu Q (2010) Application of sweet sorghum for biodiesel production by heterotrophic microalga *Chlorella protothecoides*. Appl Energy 87:756–761

Gao Z, Zhao H, Li Z, Tan X, Lu X (2012) Photosynthetic production of ethanol from carbon dioxide in genetically engineered cyanobacteria. Energy Environ Sci 5:9857–9865

García-Ferris C, De Los RA, Ascaso C, Moreno J (1996) Correlated biochemical and ultrastructural changes in nitrogen-starved *Euglena gracilis*. J Phycol 32:953–963

Geider RJ, Osborne BA (1989) Respiration and microalgal growth: a review of the quantitative relationship between dark respiration and growth. New Phytol 112:327–341

Girard J, Roy M, Hafsa MB, Gagnon J, Faucheux N, Heitz M, Tremblay R, Deschênes J (2014) Mixotrophic cultivation of green microalgae *Scenedesmus obliquus* on cheese whey permeate for biodiesel production. Algal Res 5:241–248

Göksan T, Ak I, Gökpinar S (2010) An alternative approach to the traditional mixotrophic cultures of *Haematococcus pluvialis* Flotow (Chlorophyceae). J Microbiol Biotechnol 20:1276–1282

Graverholt OS, Eriksen NT (2007) Heterotrophic high-cell-density fed-batch and continuous-flow cultures of *Galdieria sulphuraria* and production of phycocyanin. Appl Microbiol Biotechnol 77:69–75

Griffiths DJ, Thresher CL, Street HE (1960) The heterotrophic nutrition of *Chlorella vulgaris* (brannon no. 1 strain): with two figures in the text. Ann Bot 24:1–11

Hawkins RL (1999) Utilization of xylose for growth by the eukaryotic alga, *Chlorella*. Curr Microbiol 38:360–363

Heredia-Arroyo T, Wei W, Hu B (2010) Oil accumulation via heterotrophic/mixotrophic *chlorella protothecoides*. Appl Biochem Biotechnol 162:1978–1995

Hernandez JP, de-Bashan LE, Bashan Y (2006) Starvation enhances phosphorus removal from wastewater by the microalgae *Chlorella* spp. co-immobilized with *Azospirillum brasilense*. Enzyme Microb Technol 38:190–198

Hillebrand H, Sommer U (1999) The nutrient stoichiometry of benthic microalgal growth: redfield proportions are optimal. Limnol Oceanogr 44:440–446

Hirano A, Ueda R, Hirayama S, Ogushi Y (1997) CO_2 fixation and ethanol production with microalgal photosynthesis and intracellular anaerobic fermentation. Energy 22:137–142

Hong SJ, Lee CG (2007) Evaluation of central metabolism based on a genomic database of *Synechocystis* PCC6803. Biotechnol Bioprocess Eng 12:165–173

Hsieh CH, Wu WT (2009) Cultivation of microalgae for oil production with a cultivation strategy of urea limitation. Bioresour Technol 100:3921–3926

Hu B, Min M, Zhou W, Li Y, Mohr M, Cheng Y, Lei H, Liu Y, Lin X, Chen P, Ruan R (2012) Influence of exogenous CO_2 on biomass and lipid accumulation of microalgae *Auxenochlorella protothecoides* cultivated in concentrated municipal wastewater. Appl Biochem Biotechnol 166:1661–1673

Hu B, Zhou W, Min M, Du Z, Chen P, Ma X, Liu Y, Lei H, Shi J, Ruan R (2013) Development of an effective acidogenically digested swine manure-based algal system for improved wastewater treatment and biofuel and feed production. Appl Energy 107:255–263

Hudek KL, Davis LC, Ibbini J, Erickson LE (2014) Commercial products from algae. In: Bajpai R, Prokop A, Zappi M (eds) Algal biorefineries, Vol. 1. Cultivation of cells and products. Springer, New York, pp 275–295

Illman AM, Scragg AH, Shales SW (2000) Increase in *Chlorella* strains calorific values when grown in low nitrogen medium. Enzyme Microb Technol 27:631–635

John RP, Anisha GS, Nampoothiri KM, Pandey A (2011) Micro and macroalgal biomass: a renewable source for bioethanol. Bioresour Technol 102:186–193

Kanehisa M, Goto S (2000) KEGG: Kyoto encyclopedia of genes and genomes. Nucleic Acids Res 28:27–30. Database website: http://www.genome.jp/kegg/

Kang R, Wang J, Shi D, Cong W, Cai Z, Ouyang F (2004) Interactions between organic and inorganic carbon sources during mixotrophic cultivation of *Synechococcus* sp. Biotechnol Lett 26:1429–1432

Khozin-Goldberg I, Cohen Z (2006) Effect of phosphate starvation on lipid and fatty acid composition of the eustigmatophyte *Monodus subterraneus*. Phytochemistry 67:696–701

Khozin-Goldberg I, Cohen Z (2011) Unraveling algal lipid metabolism: recent advances in gene identification. Biochimie 93:91–100

Koide T, Lee Pang W, Baliga NS (2009) The role of predictive modelling in rationally re-engineering biological systems. Nat Rev Microbiol 7:297–305

Kröger M, Müller-Langer F (2011) Impact of heterotrophic and mixotrophic growth of microalgae on the production of future biofuels. Biofuels 2:145–151

Lebsky VK, Gonzalez-Bashan LE, Bashan Y (2001) Ultrastructure of interaction in alginate beads between the microalga *Chlorella vulgaris* with its natural associative bacterium *Phyllobacterium myrsinacearum* and with the plant growth-promoting bacterium *Azospirillum brasilense*. Can J Microbiol 47:1–8

Lee SK, Chou H, Ham TS, Lee TS, Keasling JD (2008) Metabolic engineering of microorganisms for biofuels production: from bugs to synthetic biology to fuels. Curr Opin Biotechnol 19:556–563

León-Bañares R, González-Ballester D, Galván A, Fernández E (2004) Transgenic microalgae as green cell-factories. Trends Biotechnol 22:45–52

Leyva LA, Bashan Y, Mendoza A, de-Bashan LE (2014) Accumulation of fatty acids in *Chlorella vulgaris* under heterotrophic conditions in relation to activity of acetyl-CoA carboxylase, temperature, and co-immobilization with *Azospirillum brasilense*. Naturwissenschaften 101:819–830

Leyva LA, Bashan Y, de-Bashan LE (2015) Activity of acetyl-CoA carboxylase is not directly linked to accumulation of lipids when *Chlorella vulgaris* is co-immobilized with *Azospirillum brasilense* in alginate under autotrophic and heterotrophic conditions. Ann Microbiol 65:339–349.

Li X, Xu H, Wu Q (2007) Large-scale biodiesel production from microalga *Chlorella protothecoides* through heterotrophic cultivation in bioreactors. Biotechnol Bioeng 98:764–771

Li Y, Horsman M, Wang B, Wu N, Lan CQ (2008) Effects of nitrogen sources on cell growth and lipid accumulation of green alga *Neochloris oleoabundans*. Appl Microbiol Biotechnol 81:629–636

Li Y, Zhou W, Hu B, Min M, Chen P, Ruan RR (2011) Integration of algae cultivation as biodiesel production feedstock with municipal wastewater treatment: strains screening and significance evaluation of environmental factors. Bioresour Technol 102:10861–10867

Li Y, Zhou W, Hu B, Min M, Chen P, Ruan RR (2012) Effect of light intensity on algal biomass accumulation and biodiesel production for mixotrophic strains *Chlorella kessleri* and *Chlorella protothecoides* cultivated in highly concentrated municipal wastewater. Biotechnol Bioeng 109:2222–2229

Liang Y (2013) Producing liquid transportation fuels from heterotrophic microalgae. Appl Energy 104:860–868

Liang Y, Sarkany N, Cui Y (2009) Biomass and lipid productivities of *Chlorella vulgaris* under autotrophic, heterotrophic and mixotrophic growth conditions. Biotechnol Lett 31:1043–1049

Lloyd D (1974) Dark respiration. In: Stewart WDP (ed) Algal physiology and biochemistry. Blackwell Scientific Publications, Oxford, pp 505–529

Lu Y, Zhai Y, Liu M, Wu Q (2010) Biodiesel production from algal oil using cassava (*Manihot esculenta* Crantz) as feedstock. J Appl Phycol 22:573–578

Mandenius CF, Brundin A (2008) Bioprocess optimization using design-of-experiments methodology. Biotechnol Prog 24:1191–1203

Marquez FJ, Sasaki K, Kakizono T, Nishio N, Nagai S (1993) Growth characteristics of *Spirulina platensis* in mixotrophic and heterotrophic conditions. J Ferment Bioeng 76:408–410

Merzlyak MN, Chivkunova OB, Gorelova OA, Reshetnikova IV, Solovchenko AE, Khozin-Goldberg I, Cohen Z (2007) Effect of nitrogen starvation on optical properties, pigments and arachidonic acid content of the unicellular green alga *Parietochloris incisa* (trebouxiophyceae) chlorophyta. J Phycol 43:833–843

Miao X, Wu Q (2006) Biodiesel production from heterotrophic microalgal oil. Bioresour Technol 97:841–846

Molina Grima E, Fernández FGA, García Camacho F, Chisti Y (1999) Photobioreactors: light regime, mass transfer, and scaleup. J Biotechnol 70:231–247

Molina Grima E, Belarbi E, Fernández FGA, Robles Medina A, Chisti Y (2003) Recovery of microalgal biomass and metabolites: process options and economics. Biotechnol Adv 20:491–515

Neilson AH, Lewin RA (1974) The uptake and utilization of organic carbon by algae: an essay in comparative biochemistry. Phycologia 13:227–264

Nwabueze TU (2010) Basic steps in adapting response surface methodology as mathematical modelling for bioprocess optimisation in the food systems. Int J Food Sci Technol 45:1768–1776

Oberhardt MA, Chavali AK, Papin JA (2009) Flux balance analysis: interrogating genome-scale metabolic networks. Method Mol Biol 500:61–80

OECD (2011) Future prospects for industrial biotechnology. OECD Publishing, Paris, http://dx.doi.org/10.1787/9789264126633-en

Ogbonna JC, Yoshizawa H, Tanaka H (2000) Treatment of high strength organic wastewater by a mixed culture of photosynthetic microorganisms. J Appl Phycol 12:277–284

Ogbonna JC, Ichige E, Tanaka H (2002) Interactions between photoautotrophic and heterotrophic metabolism in photoheterotrophic cultures of *Euglena gracilis*. Appl Microbiol Biotechnol 58:532–538

Orth JD, Thiele I, Palsson BO (2010) What is flux balance analysis? Nat Biotechnol 28:245–248

Pagnanelli F, Altimari P, Trabucco F, Toro L (2014) Mixotrophic growth of *Chlorella vulgaris* and *Nannochloropsis oculata*: interaction between glucose and nitrate. J Chem Technol Biotechnol 89:652–661

Pal D, Khozin-Goldberg I, Cohen Z, Boussiba S (2011) The effect of light, salinity, and nitrogen availability on lipid production by *Nannochloropsis* sp. Appl Microbiol Biotechnol 90:1429–1441

Park KC, Whitney C, McNichol JC, Dickinson KE, MacQuarrie S, Skrupski BP, Zou J, Wilson KE, O'Leary SJB, McGinn PJ (2012) Mixotrophic and photoautotrophic cultivation of 14 microalgae isolates from Saskatchewan, Canada: potential applications for wastewater remediation for biofuel production. J Appl Phycol 24:339–348

Perez-Garcia O, de-Bashan LE, Hernandez J, Bashan Y (2010) Efficiency of growth and nutrient uptake from wastewater by heterotrophic, autotrophic, and mixotrophic cultivation of *Chlorella vulgaris* immobilized with *Azospirillum brasilense*. J Phycol 46:800–812

Perez-Garcia O, Bashan Y, Puente ME (2011a) Organic carbon supplementation of sterilized municipal wastewater is essential for heterotrophic growth and removing ammonium by the microalga *Chlorella vulgaris*. J Phycol 47:190–199

Perez-Garcia O, Escalante FME, de-Bashan LE, Bashan Y (2011b) Heterotrophic cultures of microalgae: metabolism and potential products. Water Res 45:11–36

Perez-Garcia O, Villas-Boas SG, Swift S, Chandran K, Singhal N (2014) Clarifying the regulation of NO/N_2O production in *Nitrosomonas europaea* during anoxic–oxic transition via flux balance analysis of a metabolic network model. Water Res 60:267–277

Pipes WO, Gotaas HB (1960) Utilization of organic matter by Chlorella grown in sewage. Appl Microbiol 8:163–169

Pleissner D, Lam WC, Sun Z, Lin CSK (2013) Food waste as nutrient source in heterotrophic microalgae cultivation. Bioresour Technol 137:139–146

Pulz O (2001) Photobioreactors: production systems for phototrophic microorganisms. Appl Microbiol Biotechnol 57:287–293

Radakovits R, Jinkerson RE, Fuerstenberg SI, Tae H, Settlage RE, Boore JL, Posewitz MC (2012) Draft genome sequence and genetic transformation of the oleaginous alga *Nannochloropsis gaditana*. Nat Commun 3:686. doi:10.1038/ncomms1688

Ratledge C, Wynn JP (2002) The biochemistry and molecular biology of lipid accumulation in oleaginous microorganisms. Adv Appl Microbiol 51:1–51

Ratledge C, Kanagachandran K, Anderson AJ, Grantham DJ, Stephenson JC (2001) Production of docosahexaenoic acid by *Crypthecodinium cohnii* grown in a pH-auxostat culture with acetic acid as principal carbon source. Lipids 36:1241–1246

Raven JA (1976) The quantitative role of 'dark' respiratory processes in heterotrophic and photolithotrophic plant growth. Ann Bot 40:587–602

Rawat I, Ranjith Kumar R, Mutanda T, Bux F (2011) Dual role of microalgae: phycoremediation of domestic wastewater and biomass production for sustainable biofuels production. Appl Energy 88:3411–3424

Richmond A (1986) Cell response to environmental factors. In: Richmond A (ed) Handbook for microalgal mass culture. CRC Press, Boca Raton, pp 69–99

Robertson DE, Jacobson SA, Morgan F, Berry D, Church GM, Afeyan NB (2011) A new dawn for industrial photosynthesis. Photosynth Res 107:269–277
Rodolfi L, Zittelli GC, Bassi N, Padovani G, Biondi N, Bonini G, Tredici MR (2009) Microalgae for oil: strain selection, induction of lipid synthesis and outdoor mass cultivation in a low-cost photobioreactor. Biotechnol Bioeng 102:100–112
Rosenberg JN, Kobayashi N, Barnes A, Noel EA, Betenbaugh MJ, Oyler GA (2014) Comparative analyses of three *Chlorella* species in response to light and sugar reveal distinctive lipid accumulation patterns in the microalga *C. sorokiniana*. PLoS One 9:e92460. doi:10.1371/journal.pone.009246
Sheehan J, Dunahay T, Benemann JR, Roessler P (1998) A look back a the U.S. Department of Energy's Aquatic Species Program-Biodiesel from algae. U.S. Department of Energy, Golden, CO. Report No. NREL/TP-580-24190
Shen Y, Yuan W, Pei Z, Mao E (2010) Heterotrophic culture of *Chlorella protothecoides* in various nitrogen sources for lipid production. Appl Biochem Biotechnol 160:1674–1684
Shi X, Jiang Y, Chen F (2002) High-yield production of lutein by the green microalga *Chlorella protothecoides* in heterotrophic fed-batch culture. Biotechnol Prog 18:723–727
Silaban A, Bai R, Gutierrez-Wing MT, Negulescu II, Rusch KA (2014) Effect of organic carbon, C:N ratio and light on the growth and lipid productivity of microalgae/cyanobacteria coculture. Eng Life Sci 14:47–56
Solovchenko AE, Khozin-Goldberg I, Didi-Cohen S, Cohen Z, Merzlyak MN (2008) Effects of light intensity and nitrogen starvation on growth, total fatty acids and arachidonic acid in the green microalga *Parietochloris incisa*. J Appl Phycol 20:245–251
Subashchandrabose SR, Ramakrishnan B, Megharaj M, Venkateswarlu K, Naidu R (2013) Mixotrophic cyanobacteria and microalgae as distinctive biological agents for organic pollutant degradation. Environ Int 51:59–72
Tabernero A, Martín del Valle EM, Galán MA (2012) Evaluating the industrial potential of biodiesel from a microalgae heterotrophic culture: scale-up and economics. Biochem Eng J 63:104–115
Ueda R, Hirayama S, Sugata K, Nakayama H (1996) Process for the production of ethanol from microalgae. US Patent 5,578,472
Ueno Y, Kurano N, Miyachi S (1998) Ethanol production by dark fermentation in the marine green alga, Chlorococcum littorale. J Ferment Bioeng 86:38–43
Valderrama LT, Del Campo CM, Rodriguez CM, de-Bashan LE, Bashan Y (2002) Treatment of recalcitrant wastewater from ethanol and citric acid production using the microalga *Chlorella vulgaris* and the macrophyte *Lemna minuscule*. Water Res 36:4185–4192
Vanthoor-Koopmans M, Wijffels RH, Barbosa MJ, Eppink MHM (2013) Biorefinery of microalgae for food and fuel. Bioresour Technol 135:142–149
Vonshak A, Cheung SM, Chen F (2000) Mixotrophic growth modifies the response of *Spirulina* (Arthrospira) *platensis* (Cyanobacteria) cells to light. J Phycol 36:675–679
Wan M, Liu P, Xia J, Rosenberg JN, Oyler GA, Betenbaugh MJ, Nie Z, Qiu G (2011) The effect of mixotrophy on microalgal growth, lipid content, and expression levels of three pathway genes in *Chlorella sorokiniana*. Appl Microbiol Biotechnol 91:835–844
Wang Y, Chu J, Zhuang Y, Wang Y, Xia J, Zhang S (2009) Industrial bioprocess control and optimization in the context of systems biotechnology. Biotechnol Adv 27:989–995
Wang L, Li Y, Chen P, Min M, Chen Y, Zhu J, Ruan RR (2010) Anaerobic digested dairy manure as a nutrient supplement for cultivation of oil-rich green microalgae *Chlorella* sp. Bioresour Technol 101:2623–2628
Wang J, Yang H, Wang F (2014) Mixotrophic cultivation of microalgae for biodiesel production: status and prospects. Appl Biochem Biotechnol 172:3307–3329
Wijffels RH, Barbosa MJ, Eppink MHM (2010) Microalgae for the production of bulk chemicals and biofuels. Biofuels Bioprod Biorefin 4:287–295
Wijffels RH, Kruse O, Hellingwerf KJ (2013) Potential of industrial biotechnology with cyanobacteria and eukaryotic microalgae. Curr Opin Biotechnol 24:405–413

Wiley PE, Harris L, Reinsch SS, Tozzi S (2013) Microalgae cultivation using offshore membrane enclosures for growing algae. J Sustain Bioenergy Syst 3:18–32

Wu Z, Shi X (2007) Optimization for high-density cultivation of heterotrophic *Chlorella* based on a hybrid neural network model. Lett Appl Microbiol 44:13–18

Xiong W, Li X, Xiang J, Wu Q (2008) High-density fermentation of microalga *Chlorella protothecoides* in bioreactor for microbio-diesel production. Appl Microbiol Biotechnol 78:29–36

Xiong W, Gao C, Yan D, Wu C, Wu Q (2010) Double CO_2 fixation in photosynthesis-fermentation model enhances algal lipid synthesis for biodiesel production. Bioresour Technol 101:2287–2293

Xu H, Miao X, Wu Q (2006) High quality biodiesel production from a microalga *Chlorella protothecoides* by heterotrophic growth in fermenters. J Biotechnol 126:499–507

Yan D, Lu Y, Chen Y, Wu Q (2011) Waste molasses alone displaces glucose-based medium for microalgal fermentation towards cost-saving biodiesel production. Bioresour Technol 102:6487–6493

Yang C, Hua Q, Shimizu K (2000) Energetics and carbon metabolism during growth of microalgal cells under photoautotrophic, mixotrophic and cyclic light-autotrophic/dark-heterotrophic conditions. Biochem Eng J 6:87–102

Yang C, Hua Q, Shimizu K (2002) Integration of the information from gene expression and metabolic fluxes for the analysis of the regulatory mechanisms in *Synechocystis*. Appl Microbiol Biotechnol 58:813–822

Yen HW, Chang JT (2013) A two-stage cultivation process for the growth enhancement of *Chlorella vulgaris*. Bioprocess Biosyst Eng 36:1797–1801

Yu WL, Ansari W, Schoepp NG, Hannon MJ, Mayfield SP, Burkart MD (2011) Modifications of the metabolic pathways of lipid and triacylglycerol production in microalgae. Microb Cell Fact 10:91. doi:10.1186/1475-2859-10-91

Zhang X, Chen F, Johns MR (1999) Kinetic models for heterotrophic growth of *Chlamydomonas reinhardtii* in batch and fed-batch cultures. Process Biochem 35:385–389

Zhang TY, Wu YH, Zhu SF, Li FM, Hu HY (2013a) Isolation and heterotrophic cultivation of mixotrophic microalgae strains for domestic wastewater treatment and lipid production under dark condition. Bioresour Technol 149:586–589

Zhang XL, Yan S, Tyagi RD, Surampalli RY (2013b) Biodiesel production from heterotrophic microalgae through transesterification and nanotechnology application in the production. Renew Sustain Energy Rev 26:216–223

Zhang K, Sun B, She X, Zhao F, Cao Y, Ren D, Lu J (2014) Lipid production and composition of fatty acids in *Chlorella vulgaris* cultured using different methods: photoautotrophic, heterotrophic, and pure and mixed conditions. Ann Microbiol 64:1239–1246

Zheng Y, Chi Z, Lucker B, Chen S (2012) Two-stage heterotrophic and phototrophic culture strategy for algal biomass and lipid production. Bioresour Technol 103:484–488

Zhou Y, Schideman L, Yu G, Zhang Y (2013) A synergistic combination of algal wastewater treatment and hydrothermal biofuel production maximized by nutrient and carbon recycling. Energy Environ Sci 6:3765–3779

Design of Closed Photobioreactors for Algal Cultivation

Martin Koller

Abstract Apart from their indispensable role as solar-driven oxygen factories, microalgae act as powerful microbial cell factories for production of various intra- or extracellular bio-products like proteins, lipids, pigments, well-known and exotic carbohydrates, biopolyesters, antibiotics or bio-hydrogen. These products can serve the demands of various markets such as the fuel- and energy sector, cosmetic industry, pharmaceutical industry, convenience- and functional food, and agriculture, or even constitute novel raw-materials for manufacturing of biodegradable plastic materials.

Efficient output of these products by using selected microalgal species requires the adaptation of the cultivation system to the special requirements of different microalgae. Factors like protection against microbial contamination, optimized nutrient supply, tailored illumination, sufficient outgassing of the produced oxygen, and maintaining pH-value and temperature in the optimum range have to be taken into account when designing an algae-based production platform for bio-products.

Simple, well-known open cultivation systems are operating at typical natural environmental conditions which are far below the real biosynthetic potential of these microbial cell factories. As a common consequence, such systems only produce modest cell densities at low volumetric productivity. Closed systems allow for the adaptation of process conditions to the optimum values inherent in the different species, provide the possibility to implement more effective illumination systems, prevent water loss by evaporation, avoid the entrance of competing microbes into the system, and circumvent the release of the algal cells into the environment. Hence, high output for desired algal bio-products requires the development of sophisticated closed photobioreactor (PBR) systems; they are designed based both on deep understanding for microbial processes and on process engineering know-how. Such optimized design, mimicking nature's strategies for light harvest, constitutes the pre-requisite for economic success of phototrophic biotechnology that now is already announced since decades. The chapter at hands offers a detailed overview of different used types of photobioreactors for cultivation of microalgae,

M. Koller (✉)
Institute of Chemistry, Department of Physical and Theoretical Chemistry, NAWI Graz, University of Graz, Heinrichstrasse 28/III, 8010 Graz, Austria
e-mail: martin.koller@uni-graz.at

A. Prokop et al. (eds.), *Algal Biorefineries*, DOI 10.1007/978-3-319-20200-6_4

highlighting their opportunities, advantages and constraints, devotes special attention to the scalability of different PBR systems, and provides examples for successful (semi)industrial implementations.

Keywords Air lift reactor • Bioreactor façade • Bubble column • Closed reactor design • Stirred tank reactor (STR) • Cyanobacteria • Flat panel • Illumination • Microalgae • Modular systems • Photobioreactor (PBR) • Productivity • Tubular reactor • Vertical reactor

1 Introduction

Contemporarily, application-oriented farming of microalgae is conceived as a "biorefinery concept" that aims at the complete utilization of the wealth of microalgal major and minor products that originate from the photosynthetic fixation of CO_2 (Bajpai et al. 2014; Hariskos and Posten 2014; Koller et al. 2014; Uggetti et al. 2014). Combining the enormous potential of microalgae for CO_2-sequestration with the plenty of marketable products produced by them, it is evident that efforts to enhance the cultivation strategies to farm these microbes are globally strongly increasing. Since ancient times, these versatile cellular factories are known for their high nutritional value both for humans and for feeding purposes; the systematic application of high-value niche algal products for nutraceutical, pharmaceutical, cosmetic etc. formulations started in parallel with the detailed investigation of the metabolic backgrounds and the development of enhanced cultivation systems (Koller et al. 2014; Spolaore et al. 2006). Since the first oil crisis in the 1970s, the demand for novel engine fuels directed the worldwide attention especially to oleaginous algae that are expected to efficiently convert CO_2 from industrial flue gases into raw materials for 3rd and 4th generation biofuels, hence to provide a solution to energy supply after the ultimate depletion of fossil feedstocks (Chen et al. 2011; Li et al. 2008; Mata et al. 2010; Rodolfi et al. 2009; Scott et al. 2010). Among these energy carriers, biodiesel, jet fuel, diesel, bioethanol, biogas, and bio-hydrogen are described (Patel et al. 2012). All of these product sectors (food and feed, niche products, green energy carriers) require from the algal cells fast growth and high production rates, and, from the engineers and technologists, technological solutions for secure and stable process performance (Pulz 2001; Wang et al. 2012).

Only since the last few decades, microalgal cultivation switches from "wild" techniques of breeding and harvesting, e.g. in crater lakes, lagoons, natural ponds or saline lakes, to controlled farming strategies for production of selected compounds. Figure 1 shows the classical way of farming of the saline β-Carotene producing alga *Dunaliella salina* in a salt lake in California, USA (online resource 1).

Technically advanced systems are found among open racing ponds (or raceway ponds, see Fig. 2 (online resource 2), Fig. 4a, b) which are still subjected to the impact of weather conditions. In most cases, such open cultivation systems are cheaper to set up and easy to handle and maintain. But, an additional drawback

Fig. 1 Cultivation of the saline algal species *Dunalliella salina* in a saline lake; the desired product β-carotene provides for the reddish coloration of the culture. The high salinity of the lake is well visible by the crystallized salt at the water´s edge (online resource 1; with kind permission of R. Malcom Brown Jr., University of Texas)

Fig. 2 Raceway pond with paddle wheels operated in outdoor mode. Different sizes of the ponds are used to prepare the inoculum cultures for the subsequent cultivation stage (online resource 2) (with permission of *Algae Energy*, UK)

arises from the fact that per m^2 of reactor surface 1 m^2 of construction area is needed, hence, open cultivation occupies large land areas! In addition, recent studies indicate additional economic shortcomings in the operation of racing ponds, if high productivities are aspired (Ravikumar 2014). These uncertainties encompass e.g. energy requirements for paddlewheels, the fresh water supply, and energy for water circulation (Rogers et al. 2014). Closed cultivation systems as the focus of the article at hands, by far outperform such simple open systems by their high degree of technological maturity; they often resemble typical bioreactors both of the stirred tank reactor (STR) (Fig. 4f) and the plug flow tubular reactor (PFTR) type (Fig. 4i) (Cardozo et al. 2007; Tredici 2004).

In addition to simple raceway ponds, more sophisticated thin-layer culture can also be used for open outdoor operation. Such systems are applied on industrial scale already since more than three decades; they display a variety of beneficial parameters which cannot be achieved by open raceway technology neither in closed reactor setups. Cell concentrations exceeding 50 g/L were obtained by such thin layer cultures without compromising volumetric productivity (Doucha and Lívanský 2014). Such "open outdoor solar photobioreactor" systems have successfully been applied for high-throughput production of algal oils by *Chlorella sorokiniana* (Li et al. 2013), or starch by *Chlorella vulgaris* Beijerinck (Brányiková et al. 2011). In order to avoid the typical drawbacks of outdoor operated systems such as contamination and weather influence, such thin layer systems can be placed inside glasshouses; this enables high productivities of aspired valued products even under unfavourable climatic conditions, hence, they can be installed in such regions not particularly suited for outdoor algal farming, e.g., in Central Europe (Doucha and Lívanský 2014).

In principal, all technical systems aiming at the production of phototrophic biomass are labelled "photobioreactors" (PBRs) and have in common the major task to bring, beside nutrients, light energy (photons) to the catalytically active cells distributed in the aqueous cultivation medium, and to release the generated O_2 into the headspace. This goes as well for open as for closed set-ups, although, in a narrower sense, PBRs more and more only term closed cultivation systems (Pulz 2001). Figure 3 provides a schematic of the pros and cons of different algal cultivation modes, encompassing natural habitats, open racing ponds and controlled cultivation systems in closed PBRs. As a principle, both open and closed systems can be operated indoor or outdoor. It is well visible from Fig. 3 that the list of "pros" is much longer in the case of closed systems, whereas a high number of "cons" is listed for open systems. Figure 4 displays the basic geometric shapes of the different PBR-devices for algal cultivation.

Similar to the high versatility of microalgae and their nearby omnipresence in all imaginable ecological environments, the number of different attempts to create novel PBR systems for enhanced algal farming seems to be unlimited; almost each and every researcher dealing with the cultivation of algae has his own and special ideas about how to improve such processes (Olivieri et al. 2014; Wang et al. 2012). The design of PBRs currently in operation is mainly based on "trial and error" more than on comprehensive modelling on the photoautotrophic processes as the

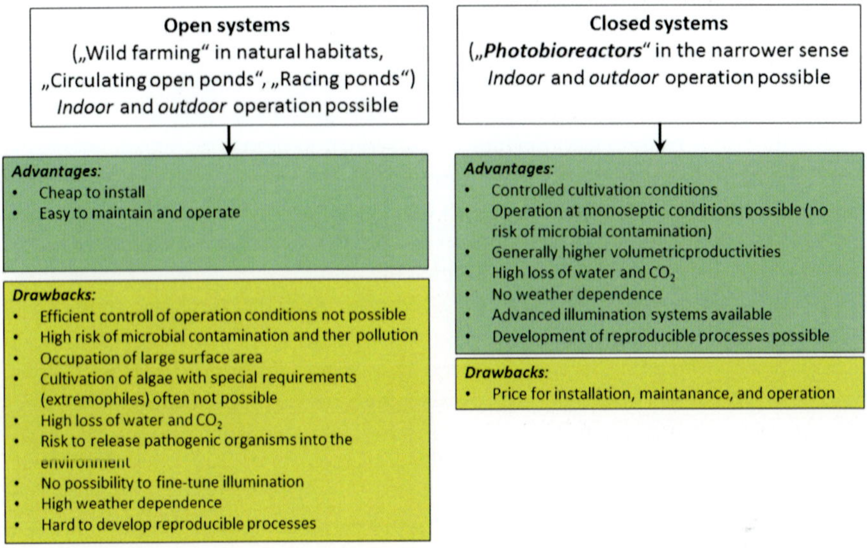

Fig. 3 Comparison of pros and cons of open and closed systems for algal farming

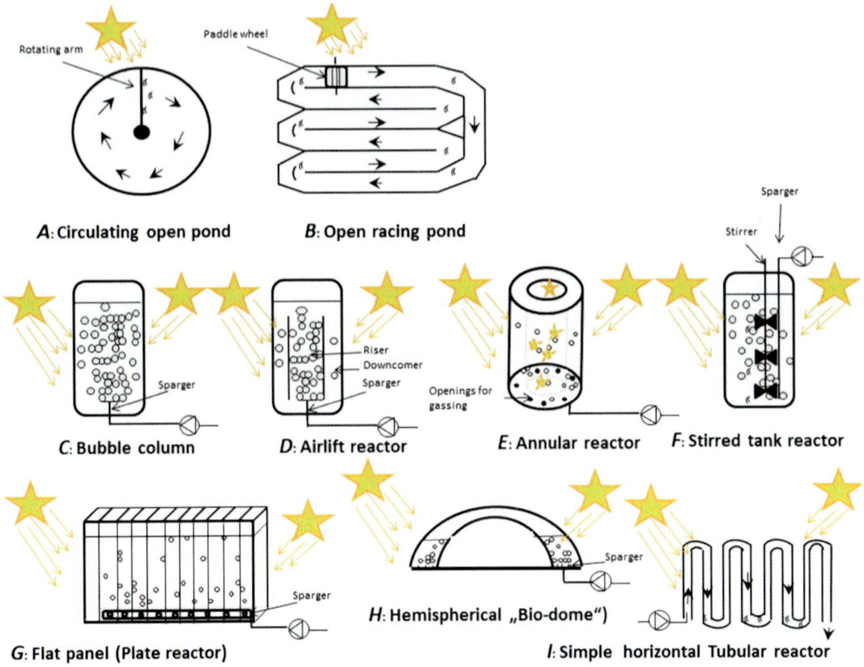

Fig. 4 Different closed PBR devices for algal cultivation. The stars indicate the location of the light source, arrows indicate the flow direction of the cultivation broth. *A* (circulating pond) and *B* (raceway pond) constitute open cultivation systems; *C* (bubble column), *D* (airlift reactor), *E* (annular reactor), and *F* (Stirred tank PBR) represent PBRs of the cylindrical tank type; *G* (typical flat panel) and *H* (dome-shaped PBR) are varieties of flat plate reactors; *I* shows a simple variant of a horizontal tubular PBR

fundamental of all algal cultivations and the simulation of three phase fluid dynamics comprising bubbles, cells, and liquids (Merchuk et al. 2007). Therefore, established algal cultivation techniques, especially using open systems, are characterized by low cell densities and moderate growth rates; this necessarily results in productivities of algal biomass and subsequently of algal products that are insufficient for any industrial implementation of such processes. This can easily be understood by the fact that such systems very closely resemble the natural environment of algae, characterized by fluctuating environmental conditions, instable and often limited and unbalanced nutrient flow, and, from the microbiologist's point of view, heterosepsis by exposure of the production strain to a consortium of different microbial species. Such set-ups "close to nature" have to come along without major possibilities towards technological optimization, and are characterized by a considerable evaporation of water due to high surface-to-volume-ratios.

As a pre-requisite to obtain the above discussed increase in biomass and product formation, strict process control and protection of the algal culture against microbial competitors and other pollutants is required. In the microalgal case, process control not only encompasses the maintenance of optimum ranges for substrate concentration, pH-value, redox potential or temperature, but also of illumination as the ultimate energy supply for phototrophic organisms like algae. Providing optimum values of all these process parameters for the myriad of cells can only be accomplished by taking into account a three-dimensional cultivation system as it is the case in closed bioreactor systems instead of well-known two-dimensional cultivation systems, were optimum conditions of illumination and substrate supply are only approached on the surface of the fermentation broth until a maximum depth hardly surmounting 20 cm. At such depth of the aqueous phase, only the "natural" cell densities in a magnitude of about 0.1 g/L or 1000 cell forming units s (CFUs) per mL, corresponding to a productivity per area of maximum 25 g/m^2 d, can be reached. Such processes are long-lasting with typical duration of 6–8 weeks. What is at least needed are biomass concentrations in the order of dozens of g per liter and short process times in the range of a few days. As a matter of fact, a 13-fold increase in biomass productivity is reported for closed PBR systems if compared to open pond cultivation of the same microalgal species (Chisti 2007). In addition, such higher biomass concentrations also minimize the necessary efforts and consequently costs for cell harvest and downstream processing for recovery of intracellular products, as shown in details for the case of other intracellular bio-products, e.g. poly(hydroxyalkanoates) (PHA) (Koller et al. 2013).

2 Mixing Systems

2.1 General

Desired high biomass concentrations cause self-shading of the cells; this requires technological means to statistically bring the cells in a range of sufficient illumination without endangering them by light inhibition. This is also needed to provide a constant distribution of nutrients and cells in the cultivation medium. Therefore, advanced mixing systems are needed.

2.2 Simple Mixing Systems in Open Ponds

The classical way of growing algae occurs in open ponds, mainly designed as circular flat and racing ponds (see Fig. 4a, b), similar to the design of simple waste water cleaning systems (Phang and Kim-Chong 1988; Sim and Goh 1988). Such basins are simply filled with the aqueous growth medium and inoculated with a pre-culture of the selected algal specie. Agitation is performed by a rotating arm in the most simple case of open spherical circulating ponds, or, in the case of (mainly ellipsoidal) open racing ponds, by so-called paddle wheels that accomplish the circular agitation of the liquid (see also Fig. 1). In the simplest cases, algal farmers even do without any mixing at all; this is the case in natural aquatic habitats (see Fig. 1). Using such "low-tech" techniques, mixing is insufficient and only occurs in the range of the agitators; especially the balanced supply of cells with CO_2 is suboptimal. Due to the lacking mixing and due to the low penetration depth of the light needed for growth and biosynthesis of products, only a low concentration of algal biomass and intracellular accumulation products can be reached. It may even happen that those cells severely lacking light intensity start to metabolize accumulated products such as carbohydrates (starch) *via* dark respiration.

2.3 Mixing in Closed Systems

In contrast to the open systems discussed in Sect. 2.1, closed system have the beneficial advantage of often profiting from the full range of mixing systems applied in the different areas of biotechnology. An exception is provided by the utilization of simple plastic bags used for algal farming; here, no mixing facilities are implemented (see Sect. 8.). Depending on the algal species, high levels of mixing are often necessary to reach a turbulent flow of the culture, in order to optimize the light regime, and to prevent adhesion of cells on the interior PBR surface. The latter refers to the fact that sufficient mixing is also of importance to avoid the settling of algae on the inner surface of the vessel. This microbial layer ("bio-fouling") prevents the sufficient penetration of light to the cells and counteracts efforts towards enhanced productivity (details see later in chapter 3).

Bubble columns and airlift reactors in closed design are suitable engineering tools to provide for enhanced mixing (see 8.3.2 and 8.3.3). Such systems accomplish mixing of the culture broth by bubbling CO_2-enriched air into the culture without implementing additional rotation parts such as stirrers. Here, due to the lack of rotating devices like impellers that generate mechanical stress, the cells are not exposed towards excessive physical strain. Nevertheless, even in the case of bubble columns or airlift reactors, certain cell damage might occur due to the bursting of gas bubbles (Gudin and Chaumont 1991).

A huge range of diverse stirrers (e.g. the propeller stirrer) and impellers can be used to accomplish mixing in STR-type PBRs. Such stirrers are well known from most aerobic and anaerobic bioreactor cultivations of bacteria, yeasts etc. They are highly effective for a homogenous mixing of the cultivation broth and, dependent on the geometry, can create different turbulence regimes, both axial and radial, in the fermentation broth. Nevertheless, their application is restricted to PBRs of the STR type; it is not possible to implement them in the frequent case of tubular PBRs except in the case of small lab-scale devices (see later). In addition, such mixing systems can create considerable shear forces that might, in dependence on the cell wall's robustness, negatively impact the algal culture. Special stirring systems like e.g. the *Intermik* stirrer (produced by the company EKATO©) generally yield less shear to the culture and might be applied in algal cultivation processes; up to date, the implementation of such systems for algae cultivation processes is not described in literature.

Mixing in tubular PBR systems is predominately accomplished by the hydrodynamic flow of the culture, mainly generated by using peristaltic pumps; no rotation devices support such set-ups. The gap between the theoretical biological potential of microalgae and the biomass productivity obtained with algal culture in tubular PBRs is due to a reduced growth rate related to hydrodynamic stress of pumping. The optimal conditions of pumping to produce this significant liquid mixing may produce some cell damage based on the generated shear forces. Hence, compromises have to be taken between the pumping rate and the protection of the cells from mechanic damage (Gudin and Chaumont 1991).

For continuously operated flat panel PBRs, mixing is accomplished by the action of hydrodynamic flow resulting from the action of bubbling (CO_2-enriched) air (Rodolfi et al. 2009; see also flat panel systems implemented on large scale by ECODUNA and PSI in Sect. 8.5.3). Discontinuously and continuously operated flat panel reactors could be supported and enhanced by mixing devices (small stirrers) at the PBRs bottom for a better bubble distribution and better homogenization of the liquid phase. Until today, such attempts are not described for PBR devices on a relevant scale.

3 Illumination and Light Penetration

3.1 General

In order to reach higher productivity, dense algal biomasses and fast kinetics both for algal growth and product formation are required to obtain reasonable output of desired algal products in order to finally guarantee the breakthrough of algal-based biofuel, pigments, lipids and others on the global market; we have to think about volumetric productivities instead of areal productivities, i.e. ideal cultivation conditions have to be guaranteed in the entire volume of a PBR, not only at and near its surface. Light penetration, in most cases, constitutes the rate-determining, committing step for

Fig. 5 Illustration of the limited depth of light penetration in a culture of *Nannochloropsis* sp. in a CSTR PBR from Infors© (own picture Philipp Tuffner)

success of algal cultivations. An example for the low penetration of light especially into dense cultures ("self-shading" of the cells) is well visible in Fig. 5, where a *Nannochloropsis* sp. culture of a concentration of about 3 g/L CDM in the cultivation broth is externally illuminated; the major part of the PBR's interior remains dark!

Light penetration is highly determined and often limited by the depth of the cultivation broth, cell density, the transparency of the photo-bioreactor material such as glass or various plastic materials, and by the turbulence regime in the cultivation system. It has to be emphasized that materials commonly used undergo an aging and alteration process after a long-term use (Doucha and Lívanský 2014); here, plastic walls of the PBR can lose transparency under permanent high illumination; glass walls have to be cleaned manually and maintained carefully in order to avoid glass corrosion by the adhesion of calk lime. As mentioned above, mixing devices have to be selected in order to warrant the statistically well-balanced exposure of the single algal cells to the provided light supply (Koller et al. 2012; Sierra et al. 2008; details see Sect. 2.2).

Many cultivation approaches in closed PBRs, both in discontinuous and continuous mode (Prokop et al. 1967; Prokop and Řičica 1968), often look promising on laboratory scale; nevertheless, upscaling of these processes often turns out to be tricky due to certain special requirements of microalgae in contrast to other well-known whole-cell biocatalysts like bacteria, fungi and yeast that are cultivated under controlled conditions in optimized submersed fermentation processes since many years; this is especially true, beside mixing and gassing systems, regarding the implementation of adequate illumination. The first generation of simple and inexpensive closed systems like hanging plastic bags (see also Fig. 11) soon underlined the limitation of the scalability of such systems; light penetration became limiting already at reaction volumes exceeding 50 L, impeding their further scale-up and successful industrial implementation (Trotta 1981).

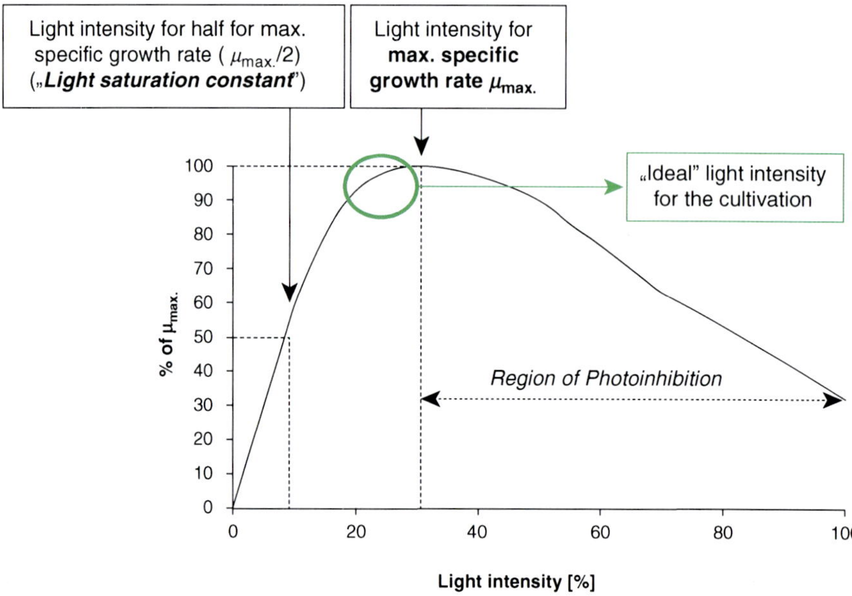

Fig. 6 Specific growth rate μ at different light regimes

Photosynthesis rate of microalgae is directly proportional to light intensity, until at high illumination intensity, photo-inhibition takes place that causes damage of the photosynthetic receptor system (Fig. 6). This fact again underlines to necessity of a complete mathematical modelling of the photosynthetic characteristics of each algal species before a scale up a process can successfully be performed (Merchuk et al. 2007). As a common feature of most microalgal species, the intracellular photosynthesis apparatus is saturated at about 30 % of the total terrestrial solar radiation, hence at about 2000 $\mu E/m^2s$ (Pulz 2001). Figure 6 provides a graphic showing the effects of different light regimes on algal cultivation. Here, the "light saturation constant" is indicated as the light intensity resulting in a specific growth rate of half the maximum value ($\mu_{max.}/2$); the green circle marks the area of ideal illumination conditions. Technical developments allow for the distribution of excessive illumination to a larger surface in order to come to optimized illumination conditions. This was described by Morita et al. (2000) by providing different inclination angles of the solar light collecting surface.

3.2 Illumination of Indoor and Outdoor Operated Systems

Generally, outdoor-operated PBRs are more frequently subjected towards photo-inhibiting conditions, provoking severe limitation on algal growth kinetics. One solution proposed to reduce the intensity of incident solar radiation and overcome the light saturation effect is "spatial dilution of light" (i.e. distribution of the impinging photon flux on a greater photosynthetic surface area) (Tredici and Chini Zitelli 1998). Classical stirred fermenters with submersed illumination facilities are suitable for a production of about 1 g CDM per day and liter; this can be considered as the upper limit for typical bioreactor designs with surface-to-volume ratios of about 2–8 m^2 per m^3. Such bioreactor and illumination designs are well suited for laboratory research, but hard to scale-up. This indicates that the surface-to-volume ratio has to be increased significantly if constructing large-scale PBR facilities. A tenfold surface-to-volume ration (20–80 m^2 per m^3), together with a thickness of the layer not exceeding 5 mm and radiations of 1150 $\mu E/m^2 s$ is reported to allow the production of up to 5 g CDM $L^{-1} day^{-1}$ (Chini Zitelli et al. 2000). Attempts are described for using different portion of the solar light spectrum in divers ways; the U.S. Pat. No. 6,603,069 describes a full spectrum solar energy system based on a hybrid solar concentrator that collects, separates, and distributes the visible portion of sunlight, while simultaneously generating electricity from the infrared portion of the spectrum.

Analyzing outdoor operated solar thin layer photobioreactor systems, Doucha and Lívanský (2014) report that a layer thickness of 0.7 typical for such systems enables a maximal utilization of sunlight, which cannot be reached by any other open or closed systems. Operating such systems in glass houses provides a further possibility of extra illumination by applying e.g., fluorescent lamps.

For indoor operated systems, both external and internal illumination is possible. The simplest case of external illumination is the utilization of solar light that penetrates transparent buildings (glasshouses) where the PBR is located and operated. The prime-example, using a glasshouse of an area exceeding 10,000 m^2, is detailed in Sect. 8.5.3. Artificial external illumination provides the possibility to trigger light intensities and to direct the illumination angle. For small PBRs of the STR-type (see Sect. 8.3), well-known tubular halogen lamps, fluorescent lamps, or high pressure sodium lamps as used for aquaria can be placed in optimized positions next to the PBR. Generally, in PBRs, such artificial light sources are problematic for most algal species. Often, such lamps must be placed in a safe distance from the algae to avoid light inhibition by local over-lighting and temperature increase. Additionally, although high pressure sodium lamps provide high light output, the lighting level is not uniform along the tube. Most of all, both sodium and fluorescent lamps produce light of wavelengths which the algal cells cannot use as energy source for photosynthesis. More sophisticated approaches use light emitting diode (LED) lamps irradiating light of exactly the wave length required by the algal species to be cultivated. Frequent combinations are LEDs emitting in the red plus blue range, skipping the green range that is not absorbed by microalgae. The exact wavelength required by a new algal species can be determined by the well-known experiment of Engelmann, where the photosynthetic activity of the algal culture is determined by measuring the oxygen evolution at different applied wavelengths. As an example, a new LED illumination device for cylindrical PBRs was developed by Jacobi and colleagues (2012). Here, an ideally illuminated volume is achieved by focusing the light toward the center of the PBR, thereby compensating the self-shading effect of the cells. It was shown by the authors that, during cultivations in batch mode, this illumination device can be successfully applied for determination of growth rates and photo conversion efficiencies. A wide variety of different parameters can be examined like the effect of different illumination conditions (light intensity, frequency of day/night cycles, flashing light, wave length, etc.) and thereby, for each single application, specific parameters can be examined (Jacobi et al. 2012). Another example of the use of LED-illumination on larger scale is provided by the company Photon Systems Instruments (PSI; Czech Republic); their PBR systems, consisting of flat panel PBR modules, bear a panel of LED lamps at the top of each module; the "standard panels" are bicolor (white/red or blue/red) with light intensities of 400 μE/m^{2}s per diode (online resource 3).

Fresnel lenses and light guides to focus, transport and distribute light were used by Zijffers and colleagues (2008) and Ogbonna et al. (1999) with promising results on laboratory scale (see next paragraph). Enhanced light transfer in bubble sparged PBR for hydrogen production and CO_2 mitigation was recently described by Berberoglu and co-workers (2007), using the filamentous cyanobacterium *Anabaena variabilis*.

Theoretical considerations by Ogbonna and colleagues (1996) resulted in a new illumination concept for PBRs. Here, the authors regarded PBRs as systems consisting of a myriad of light sources and their environments. The authors designed a model PBR consisting of four "cells" connected to each other in an alveolar manner like honeycombs. Each cell was equipped with fluorescence- or halogen lamps of adjustable light intensities. The algal strain used for the investigations was *Chlorella pyrenoidosa*. This concept enables a constant illumination in the entire multi-cell PBR systems and a high flexibility towards different phototrophic species with different light requirements. Of course, such complex systems are highly complicated and expensive to be scaled up, but certainly provide a precious tool for laboratory research under constant illumination conditions. The system was further enhanced by integrating multiple optical fibers per PBR cell. The sun light was collected by Fresnel lenses. The new system was additionally equipped with a light tracking sensor so that the lenses rotate with the position of the sun (Ogbonna et al. 1999).

A two-stage experimental PBR with a total volume of 450 L, solely based on linear Fresnel lenses of the SOLARGLAS™ type as solar concentrators is reported by Masojidek and colleagues (2009). The daily courses of irradiance and its distribution inside the cultivation broth were studied in two unit types. The supra-high irradiance units in the 'roof' achieved a maximum summer value exceeding 60,000 μE/m^{2}s (!), while irradiance in the vertical-facade units was lower than in the 'ambient'. In model cultivations of *Arthrospira platensis* at much higher solar irradiances than those usually available outdoors in summer it turned out that this organism is astonishingly robust to photo-inhibition. A two-stage cultivation process of *Haematococcus pluvialis* was investigated with respect to correlations between photochemical activities and astaxanthin production. The culture was first grown in low-irradiance units, and then exposed to supra-high irradiance when the rate of astaxanthin production was 30–50 % higher than in the culture exposed to ambient irradiance. The authors report an increase in the astaxanthin content of 25 % at supra-high irradiance if compared to ambient irradiance conditions (Masojídek et al. 2009).

The application of optical fibres to place illumination into a PBR was already suggested in 1989 by Mignot and colleagues. The optical fibre device consisted of a bundle of hundreds of optical fibres that are placed in a hollow steel barrel. Such systems work well on a laboratory scale; as a drawback, optical fibres are cost demanding; in addition, the scale up of such systems is rather difficult to realize.

Light Diffused PBR El-Shishtawy et al. (1997) introduced the concept of induced and diffused PBR for distributing the light homogenously inside the bioreactor. It consists of two parts, first a diffusion plate made of two transparent poly(methylmetacrylate) (PMMA, "Plexiglas") sheets. One of the sheets is treated with dot printing on one side for diffusion of light. A reflection sheet, consisting of poly(ethyleneterephtalate) (PET) is placed on the printed surface and fastened tightly between the two plates to create the diffusion plate. Using this PBR setup, the authors succeeded in production of bio-hydrogen with the phototrophic organism *Rhodobacter sphaeroides* RV, a highly light requiring process. This PBR system could also be operated efficiently for farming other phototrophic cells, especially microalgae. It has to be emphasized that the device needs further engineering efforts before it can be produced in a relevant size; the device described by the authors is thermostated just be being placed in a water bath. A schematic is provided in Fig. 7.

4 Bio-Fouling at the PBR Interior Surface and How to Prevent It

As the major drawback of closed PBRs, adhesion of algal cells at the inner wall of the PBR, the so called "bio-fouling", results in decreased penetration of light to the reactor's interior and consequently restricted availability of the light energy for the cells. For this effect, a high dependence on the material as well as on the PBR geometry is reported; flat panel PBRs are generally more accessible towards bio-fouling than e.g. tubular reactors. As one of the few advantages of the STR-type PBR, it can be stated that, based on the flow characteristics, high mixing, and low surface-to-volume-ratio, this PBR type displays lowest problems with bio-fouling. Continuous and random stirring avoids bio-fouling by generating turbulences, thus washing away the cells. This effect can be enhanced by switching to stirrer types generating radial flow characteristic in addition to axial flow. Further, stirrers that operate close to the interior wall of the PBR are the devices of choice to minimize the adhesion of algal cells at the PBR's inner wall.

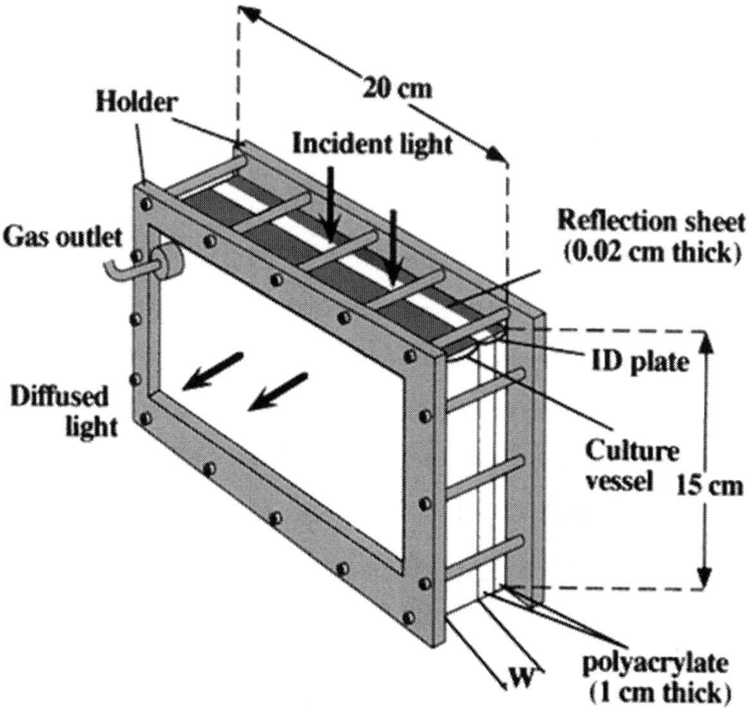

Fig. 7 Schematic of the induced and diffused PBR according to El-Shishtawy et al. (1997) (with permission of Springer)

As mentioned before, efficient stirrers can only be implemented in the case of PBRs of the cylindrical vessel type. In the tubular PBR case, one has to rely on the washing effect of the hydrodynamic streams generated by pumping. Especially for horizontal tubular PBRs, bio-fouling constitutes a severe problem to be solved, mainly in bending parts of the tubes. Innovative strategies were developed to prevent bio-fouling in tubular PBRs. For example, foam balls made of poly(urethane) were used to prevent the deposition of the culture in the inner walls of a helical tubular PBR device (Tsygankov et al. 1998). A totally different system is implemented in the horizontal BIOCOIL PBRs (details see Sect. 8.5.4), where the "cleaning system" consists of a scouring pad that is held motionless during operation but travels through the tube, scraping the sides when the flow is reversed. After the scouring pads have completed the cycle, the pump is then turned off and the flow returned to the original direction. The algae then settle out in a settling tank where they concentrate and can easily be harvested or recycled.

Dome-shaped PBRs ("bio-domes", see 8.5.4) can be equipped with a special mechanical requisite for removing attached cells; here the aeration tube is connected to a moving device and thus continuously scratches the inner surface of the PBR (Sato et al. 2006).

5 Avoiding Contamination of the Algal Culture and Release of Algal Cells to the Environment

Open systems are not protected from the surrounding; hence, they are easily susceptible for contaminations by unwanted microorganisms like other eukaryotic microbes (ciliates, fungi, other algae), prokaryotes (bacteria, archaea), and viruses, and are exposed to dust and spoilage by excretion of birds etc. Such contamination can destroy whole cultivation batches. One could argue that this problem might be overcome by selecting specialized algal production strains that can be cultivated under extreme environmental conditions (reviewed by Koller et al. 2012) such as high salinity (described for *Dunaliella*; Ginzburg 1988; Fisher et al. 1994), high temperatures as known for certain thermophile cyanobacteria like *Cyanidium caldarium* or *Synechococcus* (Castenholz 1969a, b; Yamaoka et al. 1978), extremely low temperatures like preferred by the kryophilic "snow- or ice algae" *Raphidonema nivale*, *Chlamydomonas nivalis*, *Chloromonas pichinchae*, *Cylindrocystis brébissonii*, *Koliella antarctica*, and *Chloromonas rubroleosa*, *Chlamydomonas* sp. (Hoham 1975; Hoham and Blinn 1979; Leya et al. 2000; Ling and Seppelt 1993; Kol 1969; Remias et al. 2005; Stein and Bisalputra 1969; Vona et al. 2004) or extreme pH-values (*Spirulina*) (Ling and Seppelt 1993; Ogbonna et al. 2007; Tatsuzawa et al. 1996; Yamaoka et al. 1978). Such extreme conditions provide the microalgal production strain with advantages during cultivation against competing microbial species. Nevertheless, also such systems do not absolutely guarantee a contamination-free running of the process, especially during long-term operations. In addition, it has to be emphasized that pond systems classically occupy large areas and, in many cases, it is not possible to cultivate a selected algal species with such special requirements to the culture conditions at sufficient cell densities in such simple systems (Mata et al. 2010).

Application of transgenic algae might be a research area of increasing activity in the next years (Radakovits et al. 2010). This is due to the attempts to increase the production of valued intracellular products, or to increase the productivity for molecular hydrogen as a new "green" energy carrier. Algae are, in comparison e.g. to higher plants, rather easy to undergo genetic engineering due to the lacking cell differentiation. Successful genetic modifications are reported using the genera *Phaeodactylum*, *Chlamydomonas*, and the biopolyesters-accumulating cyanobacteria *Synechococcus* and *Synechocystis*. It is important to emphasize that, similar to higher transgenic organisms, transgenic algae potentially pose a considerable risk if exposed to ecosystems. Therefore, they have to be banned from cultivation in open systems, especially outdoors, and should only be farmed in closed indoor systems under strict control and regulation (Pulz and Gross 2004).

In addition, also some algal wild-types are known to bear risks by toxic compounds produced by them; an example is provided by the production of neurotoxins by several Dinoflagellata (Wang 2008). Such neurotoxins can easily enter the nutritional chain if the algae are consumed by molluscs or fishes. If cultivated at high concentrations, these algae might also bear a health risks if released into the environment; closed PBR systems again provide a possibility to prevent the cells from being released into the environment. The same is valid for the alga *Cephaleuros virescens*, a well-known parasite in tea-, coffee-, and pepper plantations (Betula et al. 1986).

In the case of open thin layer systems, a certain degree of protection against contamination can be provided by putting these systems into glasshouses; in addition, this strategy contributes to prevent the unwanted release of algal cells into the environment (Doucha and Lívanský 2014).

6 CO_2 Import, Prevention of CO_2 Loss, and O_2 Export

It is a well-known feature that O_2 at higher concentrations display product-inhibition in the photosynthetic apparatus of algae (Doucha and Lívanský 2014). The CO_2-O2 – balance has to be adjusted in such a way that Ribulose-1,5-bisphosphate-carboxylase/-oxygenase (*RuBisCo*, EC 4.1.1.39), the enzyme mainly responsible for CO_2 fixation, is sufficiently supplied with CO_2, but does not utilize O_2 for photorespiration, i.e. the light-dependent oxidative formation of CO_2 that results in loss of algal cell mass (Birmingham et al. 1982). Apart from photorespiration, O_2 at high concentration can create free radicals by the light used for illumination. These radicals can severely damage the membranes of the algal cells. Therefore, O_2 must outgas before reaching inhibiting levels. High concentrations of O_2 are a serious problem especially at high area-to-surface ratios and high concentration of active algal biomass, hence, especially in large PBR systems. O_2 removal is a more crucial task in closed PBRs than in open systems, where O_2 is easily released through the large liquid surface into the atmosphere. Special technical requisites are possible and successfully tested to enhance O_2 release from large, closed PBR systems. Most conveniently, O_2 can be removed in non-illuminated bubble columns that are connected to the light harvesting flat or tubular modules as indicated in the horizontal tubular PBR-setups in Figs. 21 and 22.

The optimal values for CO_2 supply are classically within a narrow range for most phototrophic beings; hence, the minimum levels are not too far away from already inhibiting concentrations. Nevertheless, values for the optimum CO_2 concentration ranges vary considerably between different algal species. CO_2-import is classically accomplished by mixtures of air and CO_2 that are used for gassing the cultivation broth; hence, substrate supply is often coupled with mixing of the culture broth. Regarding the cultivation conditions, CO_2 solubility is strongly determined by the pH-value that is decisive for the CO_2/HCO_3- balance. In addition, solubility of CO_2 is also dependent on temperature and salinity of the cultivation medium. Generally spoken, direct CO_2 fixation occurs under low pH conditions and high concentrations of CO_2, hence under conditions detrimental for ideal algal growth due to their suppression of the photosynthetic system. It is a non-trivial question how exactly HCO_3 contributes to photosynthetic carbon fixation that therefore is investigated intensively since decades, showing that different mechanisms where developed to increase the concentration of free CO_2 for photosynthesis (Azov 1982; Raven and Beardall 2003).

Closed systems are important to protect against CO_2 loss. In this context, it has to be considered that the high surface-to-volume ratio in open systems favors the loss of significant shares of the imported CO_2. Therefore, for high-productive cultivation of most microalgal species, large scale closed photo-bioreactor systems, adapted to the microbial requirements, have to be developed and designed.

In order to close material cycles, flue gases from industry are more and more implemented to supply algal cultures with CO_2 in an inexpensive way (Koller et al. 2012); such attempts are reported for cultivations in different PBR facilities, also for outdoor operated thin layer systems (Doucha et al. 2005). It has to be emphasized that also the application of biogenic CO_2 from biogas plants and other sources is contemporarily gaining increasing importance; residual algal biomass is one of the conceivable substrates for biogas production (Collet et al. 2011). One the one hand, this provides a cheap method of CO_2-supply for the algal cultivation and, on the other hand, serves for refining of biogas. In contrast to contemporarily applied chemical or physical biogas conditioning techniques, the "algal-driven" approach should overall be less cost intensive, especially in the case of smaller biogas plants (Mann et al. 2009). This strategy should go in parallel to the utilization of diverse inexpensive natural sources of nitrogen, phosphate and mineral-rich fertilizers for algal farming. Such "green fertilizers" can be found in divers (agro)industrial waste water bodies, or even from the hydrolysis of residual algal biomass remaining after product isolation, and from the digestate of residual algal biomass in biogas plants (reviewed by Koller et al. 2012).

Table 1 collects the major differences of open- and closed operated PBR systems, summarizing the facts discussed in the prior paragraphs.

7 Adaptation of Process- and Reactor Design to the Requirements of the Algal Production Strain

The huge variety of microalgal species that are reported to produce diverse marketable products inherently implies the need of adaptation of cultivation systems to the special needs of the different species. The same is valid for the fundamental kinetics of formation of envisaged products; here, products that are formed in parallel to algal biomass growth (products of the primary metabolism) need different adaptation of the bioreactor system than secondary metabolites that are produced only after the formation of catalytically active biomass. In many cases, microalgal growth and product formation do not occur simultaneously; often, the cells have to be exposed to nutritional or environmental stress conditions (Rodolfi et al. 2009). In this case, novel process engineering approaches have to be developed, allowing the independent optimization of both microalgal growth and product formation. This can be realized in two-stage chemostat systems, where

Table 1 Comparison of the advantages and drawbacks of open and closed PBR systems

Criterion	Open systems	Closed systems
Space occupation	High	Very low (dependence on geometry of the PBR; horizontal vs. vertical set-ups)
Risk of microbial contamination	Very high Can be reduced by operating open thin layer systems in glasshouses	No (if obeying good microbiological working practices and problem-free operation of the PBR device)
Risk of release of transgenic or pathogenic algae	Very high Can be reduced by operating open thin layer systems in glasshouses	No (if obeying good microbiological working practices and problem-free operation of the PBR device)
Water evaporation	Very high	No
Investment costs	Very low Medium in the case of open thin layer systems	Medium to High (depending on PBR type)
CO_2-supply	Mainly restricted to CO_2-exchange with air	Highly effective methods for CO_2-supply by technical requisites (sparger etc.)
Risk of CO_2-loss	High	Low
O_2-removal	Restricted to CO_2-exchange with air	Degassing sectors available in several PBR types
Flexibility towards various production processes (strains, cultivation conditions, products)	Not possible	Easily possible
Reproducibility of the cultivation set-ups and standardization	Hardly possible in the case of raceway ponds Possible in the case of open thin layer systems	Possible (establishment of standardized production protocols)
Possibility for process control	Not possible in the case of raceway ponds Possible in the case of open thin layer systems	Possible
Lag-phases and "dead time"	High	Low (after adequate preparation of inoculum culture)
Dependence on weather conditions	Absolute! Can be overcome in the case of open thin layer systems in glasshouses	Not at all (providing artificial illumination)
Volume of cultivation broth	Very high	Low
Typical biomass concentrations to be reached	Very low (typically in range below 1 g/L)	Medium to High; needs still to be optimized for most processes (typically in range of some g/L)
Surface-to-volume-ratio	High (depending on the depth of the pond)	High (depending on the PBR type)

(continued)

Table 1 (continued)

Criterion	Open systems	Closed systems
Risk of light inhibition	High	No (at suitable technological illumination solutions like "light dilution" etc.)
Technically advanced illumination systems	Restricted to solar irradiation	Variety of advanced illumination systems reported
Selection of defined spectral ranges for illumination and possibility for "light dilution"	Not possible Can be overcome in the case of open thin layer systems in glasshouses (external illumination)	Possible
Suitability for mixotrophic or heterotrophic cultivation	No	Yes

microalgal growth is followed by product formation under nutritional stress conditions (Koller et al. 2012; Rodolfi et al. 2009). Especially in the still new field of cyanobacterial ("blue algae") bio-polyester production, a two-stage process appears to be the method of choice. Here, generation of catalytically active biomass should occur under nutritionally balanced conditions in a first vessel, whereas bio-polyester formation by accumulation of poly(hydroxyalkanoates) (PHA) occurs under conditions of nitrogen- or phosphate limitation in a second vessel (Koller and Muhr 2014). This is similar to the strategy described for formation of polyunsaturated fatty acids accumulated by *Nannochloropsis* sp. only after the termination of multiplication of algal cells.

It is a common feature of microalgae to readily adapt to strongly fluctuating process conditions during biosynthesis, such as salinity, temperature, pH-value and illumination (light intensity, dark-light cycles and spectral range). The extents of production of biomass, lipids, pigments, biopolyesters and carbohydrates can vary considerably depending on the conditions the organisms are exposed to (Rao et al. 2007). This is especially valid regarding cyanobacterial biopolyester production (Abed et al. 2009). Further, stress provoked by excessive illumination, salinity and temperature is decisive for the change of the pigment pattern during cultivation of an algal species, typically characterized by an increased carotenoid-to-chlorophyll ratio. This is extensively studied in the case of *Dunaliella* sp. (Ben-Amotz and Avron 1983; Borowitzka and Brown 1974; Borowitzka et al. 1984) and for *Nannochloropsis* sp. (Lubián et al. 2000). Therefore, before selection of the appropriate production parameters, the decision has to be made as the case arises to which final product the nutrient flux should predominantly be directed. These facts are decisive for the design of an adequate photobioreactor system facilitating high-performance cultivation of microalgae. Such systems should be flexible concerning different microalgal species to be cultivated therein, and concerning different final products.

8 Different Types of Closed PBRs

8.1 General and Basic Geometric Types

All the facts discussed above raise the challenges as well for the PBR type (tubular, tank, flat plate), as for the process-engineering design, and for the operation mode (discontinuous, fed-batch, semi-continuous, continuous). Comprehensive adaptation of the PBR set-up to the metabolic and kinetic ongoings offers the route for novel nutrient supply regimes as needed for enhanced efficiency in microalgal cultivation (Posten 2009). As a common principal, one has to consider that continuous illumination of the culture is desired to sufficiently provide light energy to the photosynthetic pigments. Therefore, most PBRs nowadays are designed in such a way that the reactor surface directed towards the light source is enlarged, e.g. by lamination.

Starting from observing nature, where leafs of higher plants take over the role of light captures, engineers started to develop ideas how an ideal closed PBR might look like. These leafs can be mimicked by flat thin layer plates or a series of parallel transparent tubes of thin diameters. These considerations resulted in construction of thin PBR lumina exposed to light; this can be accomplished by the design of the tubes of reduced diameter. Based on the lessons learnt from nature, laminar, flat type configurations should even be more favorable in scavenging solar light. Nevertheless, the tubular (or pipe) design principle is the most common strategy for contemporarily described closed PBR systems, both vertically packed sets of straight horizontal tubes (Molina et al. 2001) and coiled arrangements (Concas et al. 2010; Morita et al. 2000).

All modern PBRs try to balance between a thin layer of culture suspension, optimized light application, low pumping energy consumption, investment costs (CAPEX), and microbial purity. Many different systems have been tested, but only a few approaches were able to perform on an industrial scale. One has to consider that all described types of PBRs display advantages and drawbacks (see Table 2); none of the systems absolutely fulfils all of the criteria of a "perfect" PBR in terms of light distribution, resistance time, gas transfer, shear forces, and accessibility towards upscaling.

It is not easy to give a detailed compilation of the myriad of different PBRs described in literature during the last two or even three decades. The subsequent sections are dedicated to the description of the most important types of PBRs and the elucidation of their technological principles, benefits and restrictions. Although this number of different closed PBR types is vast, it is possible to trace them back to three ancestral geometric types:

- The **vertical column (cylindrical) type** encompasses all tank reactors like the bubble column (Fig. 4c), diverse airlift rectors (Fig. 4d), annular reactors (Fig. 4e) and STR-type PBRs (Fig. 4f). Such PBRs can conveniently be operated both in continuous, semi- and discontinuous mode, and can be equipped with means both for internal and external illumination.

Table 2 Characteristics, advantages and drawbacks of different closed PBR systems

Criterion	Bubble column	Airlift PBR	STR-type PBR	Flat panels	Horizontal tubular PBR
Surface-to-volume-ratio	Low (2–8 m^2 per m^3)	Low (2–8 m^2 per m^3)	Low (2–8 m^2 per m^3)	High (20–80 m^2 per m^3)	High (up to 100 m^2 per m^3)
Mixing efficiency	High	Very high	Very high	Medium	Medium
Mixing systems	*Via* gassing (bubbling of CO_2-enriched air)	*Via* gassing (bubbling of CO_2-enriched air)	Different stirrer types (impellers etc.); magnetic bars for small laboratory PBR devices	By hydrodynamic flow, peristaltic pumps	By hydrodynamic flow, peristaltic pumps
k_La	High	High	High	Low	Low
Operation mode	Continuous, semi-continuous and discontinuous	Continuous, semi-continuous and discontinuous	Continuous, semi-continuous and discontinuous	Continuous, semi-continuous and discontinuous	Continuous, semi-continuous and discontinuous
Risk of photo-inhibition	Low	Low	Low	Medium	High
Risk of self-shading of cells	Medium – high	Medium – high	Medium – high	Low (at thin panel thickness)	Low (at thin tube diameter)
Risk of bio-fouling	Low	Low	Medium	High	High
Possibility to work contamination-free	Yes	Yes	Yes	Yes	Yes
Investment costs	Low	Medium	Medium	Medium – High	Medium – High
Space occupation	Low	Low	Low	Medium	Medium
Ideal for products:	Protein, bioactive compounds, pigments	Protein, bioactive compounds, pigments	Protein, Lipids[a], Pigments, bioactive compounds, pigments, PUFAs, Poly(hydroxyalkanoates)[a]	Storage lipids, Pigments	Storage lipids, Pigments
O_2-release	Easy	Easy	Easy	More difficult	Very difficult

(continued)

Table 2 (continued)

Criterion	Bubble column	Airlift PBR	STR-type PBR	Flat panels	Horizontal tubular PBR
Scalability	Difficult due to illumination requirements	Difficult due to illumination requirements	Difficult due to illumination requirements	Most conveniently by combining single modules to larger units	Technical realization of large tubular systems is very easy
	Mixing efficiency decreases with increasing scale	Mixing efficiency decreases with increasing scale	Ideal for large-scale mixotrophic and heterotrophic cultivations at low light requirements	Design of huge single modules is expensive and not reasonable	*But*: severe restrictions by difficult outgassing of O_2 and difficult control of the process parameters over the entire length of the tubular system
	Upscaling of improved modifications ("Annular reactor") expensive			Scalability of special modifications like the "Bio-dome" appears highly doubtful	Risk of bio-fouling rises with increasing length of the system
					Outgassing might be facilitated in specially inclined modifications ("Christmas tree PBR")
Shear forces	Low	Low	High	Low	High

[a]Preferably in multistage PBR-set ups (cascade of STRs)

- **Flat panels** include prismatic (Fig. 4g), cuboid, and hemispherical ("bio-domes", Fig. 4h) varieties mimicking the light harvesting strategy of plants by leaves; this reactor type can also be operated in continuous, semi- and discontinuous mode
- **Tubular PBRs** include all modifications of the classical PFTR (Fig. 4i) in straight, conical, and helical arrangement of the tubes; they can also be operated in continuous, semi-continuous and discontinuous mode

Table 2 compares the characteristics, potentials and limitations of different types of closed PBRs. In addition, a tentative estimation of the most preferable PBR for different final products is provided. This estimation is based on the reaction kinetics of product formation and the light requirement of the diverse product formation processes. Based in theoretical considerations (Braunegg et al. 1995), autocatalytic processes (production of algal biomass as protein source, compounds associated to biomass growth like light-harvesting pigments and membrane-lipids [PUFAs], bacteriocins etc.) should most preferably be accomplished in reactors of the STR-type or other vertical tubes (e.g. bubble column, airlift reactor), whereas such products that are produced as secondary metabolites (PHA, storage lipids etc.) should rather be produced in such process-engineering devices that most closely resemble the classical plug flow reactor (i.e. horizontal tubular reactors, flat panels with long resistance time, multi-stage cascade of CSTR-type PBRs). Heterotrophic cultivations with restricted need for illumination should in any case be accomplished in vertical tubes like STRs due to the better possibility to control these systems regarding the nutrient supply; such attempts are explicitly described for production of PUFAs by *Crypthecodinium* sp. (Pulz 2001) or *Nannochloropsis* sp. (Koller et al. 2012).

8.2 *Foil Bioreactors ("Plastic Bag Type" PBR)*

The pressure of market prices has led the development of foil-based PBR types. Cheap poly(vinyl chloride) (PVC) or poly(ethylene) (PE) foils can be mounted to form bags or vessels (pillow-shaped "plastic sleeves") which cover the algae suspension and expose it to the light. This can be accomplished both in- and outdoors. Both advanced open systems in the form of plastic sleeves, as well as closed plastic pillars and containers are used, in many cases as disposable one-way systems. Often, such bags and sleeves directly float on the surface of fish farming water at the coastal area. Such systems, both a "hanging" variant for indoor use, as a "floating" variant to be used in indoor basins or directly on the sea, are shown in Fig. 8 (online resource 4). It is well visible that such simple systems lack any mixing pre-requisites, and are very limited in its possibilities for gas exchange.

The pricing ranges of these PBR types have been enlarged with the foil systems. It has to be kept in mind that these systems have a limited sustainability as the foils have to be replaced from time to time. For full price-performance balances, the investment for required support systems has to be calculated as well. In principal, these systems are characterized by low productivities, extremely low cell densities, and completely lacking possibilities to control and adapt the cultivation conditions.

Fig. 8 Plastic sleeves (**a**) and plastic bags (**b**) as most simple systems for closed algal cultivation (online resource 4; with kind permission of *Oilseed Crops*)

For cultivation of *Tetraselmis* sp., frequently applied methods mainly use PE bags and simple transparent glass-fibre cylinders (up to 500 L; reviewed by Chini Zittelli et al. 2006); these devices are usually kept indoors and illuminated by artificial light sources; as major drawback, these systems operate at low productivity (Fulks and Main 1991).

A sustainable solution can be identified in the fabrication of plastic foils consisting of such materials that are directly produced by the phototrophic microbes cultivated therein. This is the case for production of PHA biopolyesters by cyanobacteria (Abed et al. 2009). These PHA are known to be degraded also in marine environment within a defined period, dependent on the thickness of the sheets, and the exact PHA composition on the monomeric level (Imam et al. 1999; Reddy et al. 2003). For realization, some items have to be raised: the PHA-based foil has to display sufficient transparency, certain stability against degradation for cultivation during several months, and flexibility sufficient for process them towards sheets by film blowing. A solution might be the formation of flexible blends with other biologically benign plastics, such as poly(lactate) (PLA) or poly(ε-caprolactone) (PCL). Nevertheless, there is still a lot of R&D-work to be done before PHA-production processes by cyanobacteria in such simple reactor facilities are optimized to provide the needed volumetric productivities.

8.3 Vertical, Cylindric PBR Systems (Tanks)

8.3.1 General

Generally, vertical, cylindrical PBR systems (tank reactors) show the advantage of high mass transfer (excellent import of CO_2 and export of O_2), good mixing at low shear forces, high potential to scale-up in case of low required illumination levels, low accessibility towards bio-fouling, facile techniques of sterilization, and reduced photo-inhibition. On the other hand, they display a limited surface for illumination (suboptimal high-to-volume ratio, see Fig. 5) especially at up-scaled devices in case of algal species with high demands for illumination, feature a high energy demand, high fragility of the PBR material (advanced transparent materials are needed!), complicated gas transfer at the top regions, and temperature control.

Made of glass or plastic (mainly PE or PMMA) tubes, this PBR type has succeeded on production scale. The tubes are predominantly oriented vertically, and are supplied from a central utilities installation with pumps, sensors, nutrients and CO_2.

Vertical tank PBRs without stirring facilities are divided into simple bubble columns and technologically more advanced airlift PBRs.

8.3.2 Bubble Columns

Bubble columns are simply agitated by bubbling CO_2 or CO_2-enriched air from a sparger at the PBR's bottom without any special internal constructions and completely lack any moving parts (Fig. 4c). At larger devices of this type, perforated plates are used instead of sparger in order to break up and redistribute coalesced bubbles. The most important benefit of bubble column PBRs is the low investment costs of such devices. Mixing of the culture as well as heat- and mass transfer are sufficient for many requirements in algal cultivation. Illumination is accomplished externally. The photosynthetic efficiency depends to a high extent on the gas flow rate as the liquid more often circulates from the central dark zone to outer (illuminated) zone at high flow rates. As a typical feature, the height of bubble columns amounts to at least the double value of their diameter. Hence, they classically have a more beneficial surface-to-volume-ratio than PBRs of the STR-type which frequently display a height not too different from the diameter.

8.3.3 Airlift Reactors

In airlift reactors (Fig. 4d), mixing properties are enhanced by baffles separating the system into two interconnecting segments. One of the segments ("riser") provides the gas supply, whereas in the second segment ("downcomer"), no gas is supplied to the culture. In the internal loop airlift reactor type, the two segments are separated either by a tube or a split cylinder. Within the internal type, one can distinguish between the internal loop split airlift reactor, and the internal loop concentric reactor (see Fig. 9). The external airlift reactor type is characterized by a constructional separation of "riser" and "downcomer", mostly by a tube (Heijnen et al. 1997). Similar to the bubble column, gassing and mixing is performed by bubbling CO_2-enriched air through a sparger without any stirring; in the case of the airlift reactor type, this bubbling is done at the bottom of the riser. From the technological point of view, the riser acts as a bubble column itself, resulting in a rapid and random upwards movement of the gas bubbles, whereby the upwards movement is enhanced

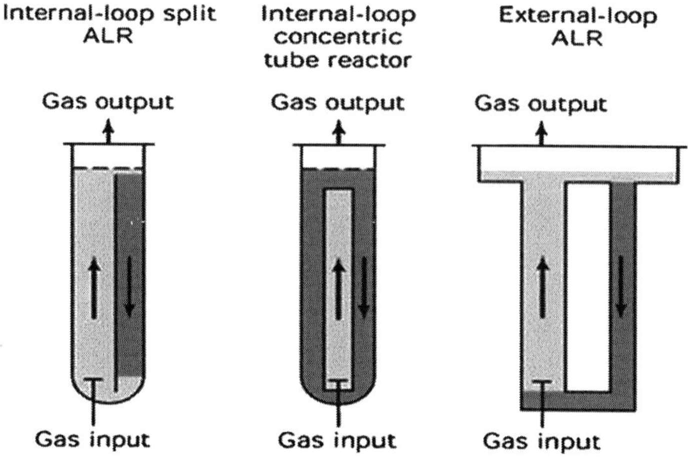

Fig. 9 Different types of airlift bioreactors (Singh and Shaishav Sharma 2012) (with permission of Elsevier)

by the riser's gas holdup. In the subsequent "disengagement zone", the gas has the possibility to remove gas from the culture. Now, the more or less degassed culture passes into the downcomer segment. The retention time in the riser and downcomer segment, and the quantity of degassing in the disengagement zone highly impact the cultivation performance in terms of growth rate and product formation. Also in the case of airlift reactors, illumination is done by external devices.

Some special vertical tubular systems (both bubble columns and airlift reactors) exist that provide additional bubbling by sparging gas also from the side of the reactor instead of merely from the bottom.

As a well-investigated example, Miron et al. (2000) studied the hydrodynamics and mass transfer in a bubble column, split hydrodynamics and mass transfer in bubble column, split cylinder airlift and concentric draft tube sparged airlift reactors for *Phaeodactylum tricornutum*. In all the three bioreactors (each about 60 L of working volume), a biomass concentration of 4 g L^{-1} was achieved with a specific growth rate of 0.022 h^{-1}.

8.3.4 Annular Reactors

In comparison to horizontal tubular PBRs, bubble columns and airlift reactors display a considerably higher dark fraction in the middle of the cylinder. To overcome this problem, PBRs of the bubble column type were designed leaving out the inner reactor space. The so called annular column PBR (Fig. 4e) has been developed at the *Dipartimento di Biotecnologie Agrarie* of the University of Florence and has originally been used to cultivate strains belonging to diverse genera, such as *Nostoc* sp. (Rodolfi et al. 2002), *Nannochloropsis* sp. (Chini Zitelli et al. 2003), *Isochrysis* sp. (Chini Zitelli et al. 2004), and *Tetraselmis* sp. (Chini Zittelli et al. 2006). Such annular jacketed photo-bioreactors until now are used only on laboratory scale; they

Fig. 10 Cultivation of *Tetraselmis* sp. in annular reactors (Chini Zitelli et al. 2006) (with permission of Elsevier)

might cause problems when they are scaled-up to production size. Jacketed reactors consist of three concentric chambers. Fluorescent bulbs are inserted in the inner-most chamber. Surrounding this is the culture chamber agitated with the help of a magnetic stirrer. Temperature is controlled by circulating water in the outermost chamber. The reactor has a high surface-to-volume-ratio. Such systems were designed on larger scale by the Italian company F&M. The constructed annular columns consist of two Plexiglas cylinders of 2 m in high and of 40 and 50 cm in diameter; the cylinders are placed one inside the other, thus forming an annular chamber. The system can even be conceived as an enfolded flat plate reactor. To enhance the contribution of the inner surface to the overall illumination, additional lamps can be fitted there. Compressed air is bubbled at the bottom of the annular chamber through a perforated plastic tube, for mixing and gas exchange. CO_2 from cylinders is injected into the culture through a gas diffuser placed in a non-aerated zone of the annular chamber, as carbon source and for pH-value regulation. Such annular reactors in a size of 120 L were used by Chini Zitelli and colleagues (2006) for cultivation of *Tetraselmis* sp. investigating mass transfer at different aeration rates and the influence of the harvest rate on productivity and biochemical composition. The investigated organism is known for its high potential to produce bioactive compounds and used for feeding bivalve mollusks, shrimp larvae and rotifers (Chini Zittelli et al. 2006). The arrangement of four annular reactors in east-west orientation is provided in Fig. 10.

8.4 Transparent Stirred Tank Reactors (STR-Type)

The simplest approach is the redesign of the well-known glass fermenters, which are state of the art in many biotechnological research and production facilities worldwide. The moss reactors consist of standard glass vessels, which in most cases are externally supplied with light, e.g. by tubular halogen lamps or fluorescence lamps (Fig. 4f). As a major drawback, illumination only occurs in the outer regions of the fermentation broth; the majority of cells remains in the dark section until they reach the outer areas by the action of the stirring system (see also Fig. 1).

Head nozzles are used for sensor (pH-electrode, redox electrode, temperature etc.) installation and for gas exchange. Mixing is most frequently accomplished by impeller stirrers of different sizes and shapes, or, in the case of very small units, even by magnetic stirrers. In the case of larger representatives of this reactor type, baffles are integrated in the interior in order to minimize vortex (visible in Fig. 11). CO_2-supply occurs by bubbling CO_2-enriched air through an aeration tube that ends in a sparger at the bottom of the BR, where the gas is distributed into small bubbles. Well-known manufacturers of transparent bioreactors suitable be used as STR-type PBRs are, *inter alia*, *Infors*©, *Bioengineering*©, or *Biolab*©. Figure 11 shows a Labfors 3 bioreactor (*Infors*, CH) during cultivation of *Nannochloropsis* sp.

Fig. 11 Labfors 3 bioreactor (*Infors*, CH) during cultivation of *Nannochloropsis* sp. (own picture Philipp Tuffner)

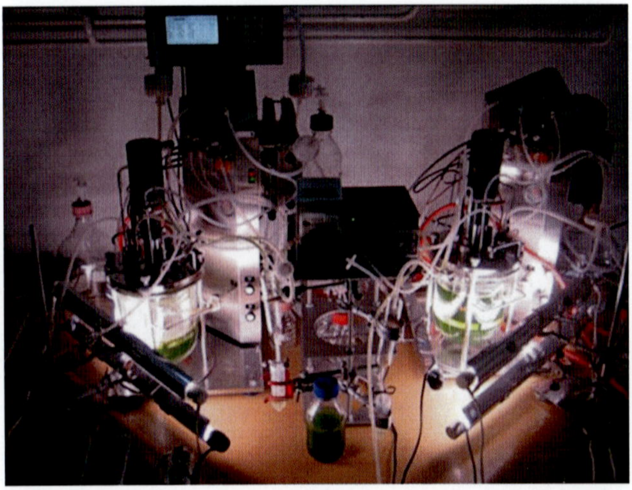

Fig. 12 Two-stage system of continuously operated, externally illuminated STR-type PBRs (own picture Philipp Tuffner)

The STR-type of PBR can conveniently be operated both discontinuously and continuously (single – and multi stage) and is frequently used at many research institutions, although, due to its limited vessel size and unfavourable high-to-volume ratio, hence low surface-to-volume-ratio, it has never been established on an industrially relevant scale. Figure 12 shows a two-stage system of continuously operated, externally illuminated CSTR--type PBRs for production of polyunsaturated fatty acids (PUFAs) by *Nannochloropsis* sp. In this experimental set-up, biomass formation (multiplication of algal cells) was spatially separated from the intracellular accumulation of PUFAs (Koller et al. 2012). Scale-up to relevant volumes is complicated due to the limitation in light supply towards the interior of the cultivation broth. It has to be emphasized that from an industrial scale-up perspective, this PBR type has the lowest potential for success, but is of importance for academic research on laboratory scale to optimize media composition, scrutinize optimum process parameters and stability of algal cultures.

A further development of the STR-type, especially suitable for larger volumes at a first glance, is the internally illuminated stirred tank reactor (STR). Internal illumination can be provided by the use of optical fibres; as a drawback, the fibres constitute "dead sectors" in the cultivation broth, negatively impacting the mixing of the culture. Regarding scalability, such systems are rather complex to realize, because the negative effects of these "dead zones" become more pivotal for the mixing system with increasing scale of the PBR.

In the case of heterotrophic cultivation of algae, excessive illumination is not needed for maintaining the culture. Heterotrophic cultivation is *de facto* completely restricted to closed reactor systems, because a sufficient distribution of heterogeneous nutrients (organic materials) can hardly be accomplished in simple open systems like racing ponds. In this case, "simple" STRs can be used to be operated in continuous or discontinuous mode. This strategy enables an easy control of the process parameters and a triggered feeding regime for supply of heterotrophic carbon sources. Nevertheless, heterotrophic or even mixotrophic cultivation is only possible for a restricted number of algal species, and to an ever more restricted number of algal products to be commercialized (Apt and Behrens 1999). As an example for successful heterotrophic cultivation of alga, one can mention the farming of divers *Chlorella* sp. for production of protein, lipids and pigments (Gao et al. 2010; Liang et al. 2009; Shi et al. 1997, 2000).

8.5 Plate Reactors ("Flat Panels")

8.5.1 General

These PBRs are generally of cuboidal geometry (schematic see Fig. 4g), a considerably higher surface-to-volume-ratio increasing the light harvesting surface for the growth of algae if compared to simple CSTR-type STR systems, and exhibit a short light path of a few to ca. 70 mm (Posten 2009), hence an extremely high percentage of algal cells under permanent illumination (Hu et al. 1998; Zhang et al. 2002). An example designed by Barbosa et al. (2005) displayed a surface-to-volume-ratio of 0.39. Typical transparent materials to be used for constructing flat panel PBRs are poly(carbonate) (e.g. "Lexan"), glass, or PMMA.

Plates of different technical design are mounted to form a small layer of culture suspension, which provides an optimized light supply. In addition, the more simple construction when compared to tubular reactors (see later) allows the application of cheap plastic materials. From the pool of different concepts e.g. meandering flow designs or bottom gassed systems have been realized and shown good output results. Some unsolved issues are material life time stability or the biofilm formation. Applications at industrial scale are bordered by the limited scalability of plate systems, additionally. Most preferably, scale up of flat panel PBRs is accomplished by combining a high number of such plate reactors, resulting in a so called "compact plate PBRs" consisting of parallel plates (Hu et al. 1998; Zhang et al. 2002).

Mixing and transport of the liquid phase normally is achieved by injection of a gas stream (air enriched with CO_2), hence, the simplest flat panel PBRs constitute bubble reactors of special (flat) geometry. A typical example for all modern flat panel PBRs was developed by Hu and colleagues (1998), where flat panel PBRs of narrow light path (optimum growth at light paths of 1 cm), were intensive turbulent flow is generated by streaming compressed air through a perforated sparger tube (see Fig. 13). This reactor type performed well in cultivations of *Chlorococcum littorale* regarding CO_2 sequestration and biomass growth (Hu et al. 1998). Flat plate PBRs are characterized by an open gas transfer area ("open disengagement zone"), thus reducing the need for a dedicated degassing unit. The exact impact of the gassing and CO_2-supply strategy (CO_2-enrichment of air between 5 and 10 vol.%, gassing rates of 0.025–1 vvm) were investigated for a vertical flat plate PBR by Zhang and colleagues (2002). In addition, these authors characterized different flat panel PBRs regarding their volumetric gas transfer performance, hence the oxygen mass transfer coefficient (k_La value). The authors concluded that increasing the high of the vertical flat PBR increases both illumination performance and the k_La.

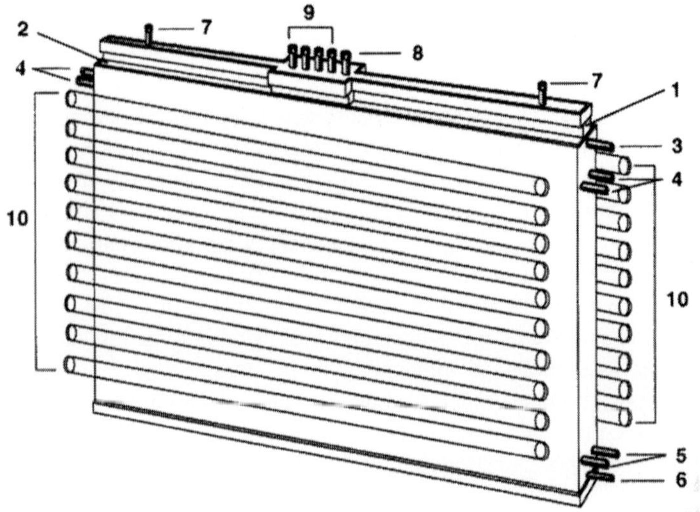

Fig. 13 Prototype vertical flat plate PBR of Hu and colleagues (1998). *1*: inner culture chamber; *2*: outer temperature regulation chamber; *3*: overflow; *4*: inlet cooling water; *5*: outlet cooling water; *6*: perforated sparger tube; *7*: air outlet; *8*: sampling device; *9*: ports for probes and electrodes; *10*: fluorescent lamps (with permission of Springer)

8.5.2 Modifications

As shown later, flat panels can be arranged in horizontal orientation (SOLIX system), or in vertical orientation ("portrait" orientation: see ECODUNA systems "Hanging gardens", "landscape" orientation: see PIS systems).

Plate-airlift reactors are mainly used in discontinuous processes for culturing microalgae. Further developments provide aeration from one side of the panel, baffles to improve agitation, and the possibility to enhance mixing by rotating the panel. Special shapes of the flat panel, e.g. V-shape, can also beneficially impact the mixing behavior; in addition, "dead zones" can be plugged in order to minimize shear forces and bio-fouling (Iqbal et al. 1993). Degen and colleagues (2001) constructed a flat panel PBR that operated as a new type of airlift reactor regarding the mode of circulation, featuring a large riser segment and a small downcomer segment.

The performance of different types of outdoor operated reactors was compared by Tredici and Chini Zittelli (1998) using the cyanobacterial strain *A. platensis* M2. Based on the outcomes (1.93 g $L^{-1}day^{-1}$ CDM in the flat plate PBR, 1.64 g $L^{-1}day^{-1}$ CDM in the tubular PBR system, together with slightly higher output for the target product phycobiliprotein in the flat plate PBR), a better performance of flat plate reactors (in this case so called "vertical alveolar panels") in comparison to coiled tubular PBRs (both rigid and collapsible tubes were investigated) was postulated. The authors hypothesized that this was due to lack of susceptibility to the orthogonal rays during midday that cause light inhibition effects on the culture in tubular systems.

In the case of high biomass concentration in flat panel reactors, additional pumps are needed to assist mixing, causing negative effects on the growing biomass due to shear forces generated by the pumps that might cause damage to the cells. In addition, a significantly higher process energy demand has to be denoted in this case. In addition, static mixers are often implemented to overcome the slow mixing capacities of flat panel systems.

Rodolfi and colleagues (2009) tested a selected algal strain, *Nannochloropsis* sp. that accumulates about 60 % of lipids as secondary metabolites under nitrogen limited conditions, in 20 L alveolar flat panel PBRs. The influence of radiation and nutrient limitation (phosphate, nitrogen) on lipid accumulation was investigated. Both higher lipid content and higher productivity (calculated as the lipid yield per area) where obtained by increased illumination, followed by nutrient deprivation. The experiments were repeated in a 110 L flat panel PBR ("Green wall panel" PBR), consisting of culture chambers made of a 0.3-mm thick flexible PE films, embedded in metal frames, under solar radiation with and without nutrient deprivation. Mixing was accomplished by bubbling compressed air through a sparger tube at the PBR bottom, CO_2 was injected through a gas diffuser in a non-aerated zone. The results showed again the same trend of increased lipid accumulation at low nitrogen levels.

8.5.3 Examples of Successfully Implemented Flat Panels

A scalable photobioreactor system for efficient production has been described by Willson (2009), the so called SOLIX PBR. *Solix Systems Inc.* uses this system to produce refinable algal oil for energy generation purposes. In various embodiments, this system combines increased surface areas to reduce light intensity, an external water basin to provide structural support and thermal regulation at low cost, and membranes for gas exchange. It consists of flexible plastic or composite panels in horizontal orientation joined together to make triangular or other cross-sectional geometries when partially submerged in water. This way, basic designs demands are combined here, including a horizontal low-ceilinged installation without panels which would need frameworks or racks and without the necessity of an expensive external green house. In principal, the system comprises a network of thin, vertical panels maintained in a shallow water bath. The algal cells are cultivated in these panels; the vertical orientation provides "extended surface area" which allows illumination of more surface area at lower intensity per unit area, thus maximizing photosynthetic efficiency. The panels contain sparging tubes which deliver CO_2 as a carbon source and also deliver sparging air to remove the generated oxygen. The SOLIX system is applicable to a wide range of algae species, both freshwater and marine species. Figure 14 shows the SOLIX demonstration plant *Coyote Gulch* in Colorado that harbors a volume of 180,000 L of *Nannochloropsis salina* broth with cell densities of 3–4 g/L on a CDM basis after cell harvest (online resource 5).

Hanging Gardens The company *ECODUNA* (Bruck an der Leitha, Austria) commercializes the so called "Hanging garden" system, consisting of a series of flat panel modules in vertical portrait orientation, providing an extremely large surface area for photosynthesis. The entire biomass transport during the growth process is accomplished at low energy demands hydrostatically at minimum shear forces for stress-free growth of the cells. Hydrostatic transport of the fluid (fermentation broth) allows for uniform flow throughout the entire maturation process without the use of pumps.

Fig. 14 SOLIX demonstration plant (flat panel in horizontal arrangement) in Colorado (online resource 5; with kind permission of Solix Biosystems, Inc.)

In addition, this system follows a "light tracking mechanism" by turning the panels in parallel with the change of the sun's altitude. This technical "trick" is mimicked from the well-known daily movements of sunflower spadices. Therefore, this system is an example how engineering can resorts to strategies that are applied by nature since a very long time! The arrangement of the panels enables low space requirements. The thin light path provides a maximum photosynthetic efficiency, at the same time avoiding photo-inhibition by diluting light from the sun from 80.000 down to 1.500–2.500 lx irradiation, the optimum for algae growth. For this innovation, the company received the "Energy Globe World Award" in 2000. An industrial size production unit has been sold to GMB Vattenfall, the largest thermal energy provider in Europe, to reduce CO_2 from the coal power station Senftenberg, Germany. This plant with an active photosynthetic volume of 50,000 L started its operation in July 2011. Figure 15 (online resource 6) shows an indoor operated "Hanging gardens" system (online resource 3) consisting of a range of basic flat panel modules according to Fig. 4g.

Photon System Instruments (PSI) bioreactors are produced in Brno (Czech Republic; online resource 3). These systems are based on flat panels (see Figs. 16 and 17) and are suitable for large-scale culturing of microalgae and cyanobacteria. They are manufactured as modular systems that are flexible to be adapted to the needs of different cultivation cultivations and different species. PSI systems are characterized by a strict process control that is the pre-requisite for safely and efficiently performing fermentations, especially on larger scale. The core of the

Fig. 15 Indoor operated "Hanging gardens" PBR plant of ECODUNA operated in Vattenfall, Germany (online resource 6; with kind permission of ECODUNA)

Fig. 16 PSI PBR system consisting of 5 25 L models operating at the Washington University, USA (with kind permission by *Photon System Instruments*, Czech Republic)

Fig. 17 120 L module unit of the Photon System Instruments (PSI) PBR (with kind permission by *Photon System Instruments*, spol. s.r.o., Czech Republic)

PSI PBR system is a flat cultivation tank with 25 L or 120 L in volume, consisting of thick glass at the front and back side surrounded by stainless steel and closed at the top by a gas-tight vessel lid that also harbors the electronic probes and ports for gas import and outlet, and an illumination panel with high-intensity LEDs. Figure 17 shows a picture of a single 120 L unit (online resource 3). By combining the respective number of modules, the system can be cascaded to a total volume of up to 1000 L. According to the manufacturers, such systems are well suitable for operation on pilot- and industrial scale; for example, a systems consisting of five 25 L unit modules was recently installed at the Washington University, USA (see Fig. 16; movie clip see online resource 3).

Cultivations can be performed in temperature ranges between 15 and 60 °C; this enables also the cultivation of thermophilic organisms. Similar to the Hanging gardens system, mixing is accomplished simply by the bubbles used for gas import. One of the main differences to similar systems like the "Hanging gardens" (ECODUNA) is the landscape orientation of the modules, and their fixed installation (they do not rotate according to the solar altitude; hence, they are more dependent on the artificial illumination system). As an additional drawback, the systems are not accessible towards sterilization by autoclaving; this is also the case for the ECODUNA system.

The modular PSI PBRs are designed to allow for precise control and optimization of the cultivation conditions by continuous monitoring of fluorescence parameters, photosynthetic efficiency, chlorophyll content, cell density (*via* turbidity measurement), and pH-value of the suspension; based on these parameters, the quantity of CO_2 injection is determined. These PBRs assure the control and monitoring of numerous parameters essential for successful algae cultivation and maximum production of biomass, namely constant temperature, constant illumination and a homogenous light spectral distribution. The performance of the cultivation is additionally controlled by two gas modules: One acts for the control of input gas flow rate and composition (CO_2 concentration), the other measures the CO_2 concentration in the gaseous PBR output stream. This enables the compilation of the complete carbon balance of the process (conversion of CO_2 towards algal biomass and intra- and extracellular products) as the basis to establish mathematical models of the phototrophic bioprocesses.

8.5.4 Dome-Shaped (Hemispherical) PBRs

A special case of flat reactors is the so called "Dome shaped reactor" (Fig. 4h), combining both an excellent mixing behavior and an enlarged light-harvesting surface. Here, the plate is deformed to cylinders, resulting in a kind of "horizontal plate reactor" (Sato et al. 2006). In details, two hemispheric, transparent "domes" are placed one upon the other, resulting in a hemispheric culture chamber. The culture is mixed, degassed, and supplied with CO_2 by bubbling CO_2-enricghed air into the chamber, bubbling around the circular bottom of the bio-dome. The shape of this reactor type, mimicked from earth's geometry, was developed based on computer modelling of flow characteristics for enhanced mixing (computational fluid dynamics, CFD). The air tube is inserted from the top of the apparatus, and connected to a moving device at the bottom. By the movement, this tube scratches the surface and thus performs the cleaning of the transparent interior surfaces. Sato and colleagues investigated the performance of this device in comparison with flat panels and tubular PBRs using the species *Chlorococum littorale*; results obtained amounted to a volumetric productivity for algal biomass of 0.095 g/Lday for the "bio-dome", in comparison to 0.086 g/Lday for the flat plate and 0.146 for the tubular PBR type. The schematic of this "bio-dome" is provided in Fig. 18.

Cooling is provided by spraying water from the top of the dome. Illumination is done by solar light or artificial light at the outer surface, in addition, illumination can be also accomplished by placing lamps at the bottom of the dome. Scale-up of such "bio-domes" is not a trivial task due to highly complex arrangement of all the required single units to create a commercial plant. Nevertheless, in 2001, 1000 so called "bio-domes" of 1.2 m of diameter were installed in Maui (Hawaii) by the company Micro Gaia Inc. for production of astaxanthin (Tredici 2004; online resource 7).

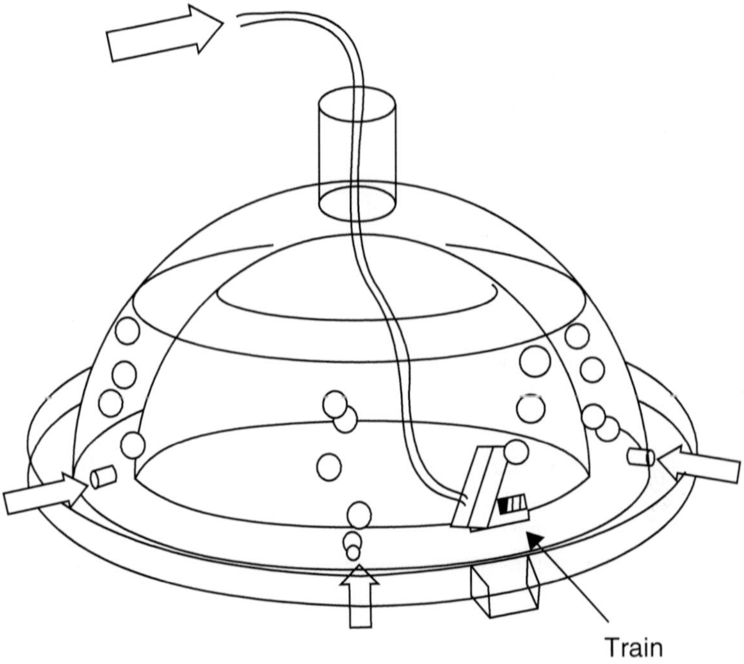

Fig. 18 Dome-shaped PBR according to Sato and colleagues (2006) (with permission of Elsevier)

8.6 Horizontal Tubular Reactors

8.6.1 General

Tubular PBRs are most frequently used as closed PBR systems for mass culture of microalgae, and are often operated outdoors in order to profit from daylight. Among the huge variety of tubular reactors (comprehensively reviewed and compared by Šantek et al. 2006), horizontal tubular reactor systems (Fig. 4i) nowadays are most frequently used for growing algae. They are made of glass or plastic, fermentation broth is circulated by pumps or simple airlift systems, aeration is also accomplished by air pumps or airlifting. The can be constructed as straight, coiled, conical or looped tubes. The exact geometry and arrangement of the tubes determines the culture's contact angle to the illumination source, and the flow characteristics. Similar to flat panel reactors, tubular rectors are characterized by high surface-to-volume-ratios. The tubes have diameters of 10 to maximum 60 mm, and lengths of up to several hundred meters. By adjusting tube diameter, length, and flow rate of the fresh medium, the retention time of the algal cells can be determined. According to the theoretical background, the dilution rate (D), hence flow per volume, determines the maximum specific growth rate of the culture (Braunegg et al. 1995). A schematic of a simple variety of a horizontal tubular PBR is provided in Fig. 19.

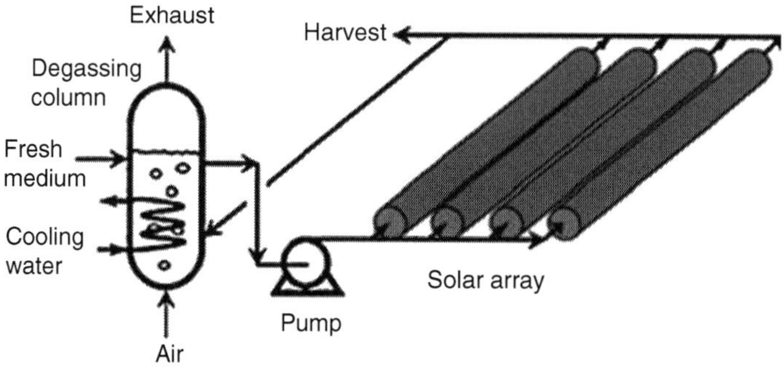

Fig. 19 Schematic of a horizontal tubular PBR (Singh and Shaishav Sharma 2012) (with permission of Elsevier)

Horizontal tube PBRs are characterized by high surface-to-volume-ratios of up to 100/1 (see Table 2). Such systems can consist of glass or plastic; they are suitable to protect against microbial contamination and display large surfaces for enhanced uptake of light due to their better orientation towards the illumination source if compared with vertical tubular systems. Gas is introduced into the tube connection or *via* a dedicated gas exchange unit. The algae containing broth has to be pumped steadily during the growth and production period, consuming a lot of electrical energy. Moreover the shearing stress on the biomass when the cells pass the pumps has negative impacts on the biomass by reducing the growth rates of the algae, and hereby lowering the volumetric productivity of the biomass. Additionally, oxygen produced during the photosynthesis process can hardly be separated, and therefore will increase the dissolved oxygen concentration in the aqueous medium, eventually leading to increasing growth inhibition. Another major problem of horizontal tubular systems is the difficult temperature control. In most cases, such systems have to be operated in thermostated buildings. Fouling on the inside of the reactor is another disadvantage of horizontal tubular PBRs (Singh and Sharma 2012). Horizontal tubular systems are easier to compose to larger systems than vertical tubular systems, mainly by set-ups of parallel tubes; hence, they are more suitable to process high amounts of fermentation broth. As an example, Tredici developed the so called "Near horizontal tubular reactor", where parallel Plexiglas tubes are connected at the top and bottom ends by tubular Plexiglas manifolds and tilted at 5 ° from the surface (Tredici et al. 1998).

8.6.2 Shortcomings

As the main problem of outdoor operated tubular PBRs, the risk of photo-inhibition by excessive solar radiation has to be mentioned; this is especially valid in large, up-scaled systems of high tube diameters. In addition, in long tubular systems it is problematic to maintain temperature and pH-value constant for the entire distance the fermentation broth has to pass through. The same is valid for difficulties in the mass transfer of O_2 and CO_2 in large tubular devices; especially a built-up of O_2 is very likely, resulting in high dissolved oxygen (DO) levels (Molina et al. 2001; Richmond et al. 1993; Torzillo et al. 1986). Additionally, the productivity of tubular PBRs is limited to the intensity of the sun, which itself depends on the time of day, season, and the localization. Attempts have been made to combine concentrated solar energy in horizontal tubular PBRs with photovoltaic cell-battery powered LED lighting to maintain optimum light level for biomass production throughout the day (US 8716010 B2).

8.6.3 Implementation of Horizontal Tubular PBRs

Tubular PBRs are established worldwide from laboratory up to production scale, e.g. for the production of the carotenoid astaxanthin, an important feeding additive for farming of salomonides, from the green algae *Haematococcus pluvialis* (Olaizola 2000), or for the production of food supplement from the green algae *Chlorella vulgaris* (Šantek et al. 2006) or *Phaeodactylum tricornutum* (Molina et al. 2001). These PBRs take advantage from the high purity levels and their efficient outputs. The biomass production can be done at a high quality level and the high biomass concentration at the end of the production allows for an energy-efficient downstream processing. Due to the recent prices of such PBRs, economically feasible concepts today can only be found within high-value markets, e.g. food supplement or cosmetics.

A typical PBR system consisting of narrow arranged horizontal tubes acting as the collector for solar light was developed by Molina and colleagues (2001). Principles of fluid mechanics, gas–liquid mass transfer, and irradiance controlled algal growth were integrated for designing of this PBR in which the culture is circulated by the gas stream provided by an airlift pump. The solar collector unit is connected with an airlift unit in order to remove the O_2 generated in the horizontal part of the PBR (solar collector). The volume amounted to 0.2 m^3; the device was used for continuous outdoor culture of the microalga *Phaeodactylum tricornutum*. The culture performance was assessed under various conditions of irradiance, D-values and liquid velocities through the tubular solar collector. Best results (biomass productivity of 1.90 g L^{-1} day^{-1}) were obtained at a D of 0.04 h^{-1}. Figure 20 shows a schematic of this PBR type.

Olaizola (2000, 2003) reports the production of astaxanthin by *Haematococcus pluvialis* under controlled conditions in a rather simple tubular PBR developed by the company *Aquasearch Inc.*, USA. This PBR consists of tubular serpentines, where the PE tubes, 0.18–0.41 m in diameter, are placed next to each other on the ground. This systems requires quite high flow rate of the culture broth, temperature is kept constant by immersion in a huge water basin. Both temperature and pH value are computer-controlled, which provides for very tight tolerances independent of variability in ambient conditions (Olaizola 2003). In the year 1999, a pilot plant consisting of three modules of 25,000 L each was implemented in Kailuna-Kona,

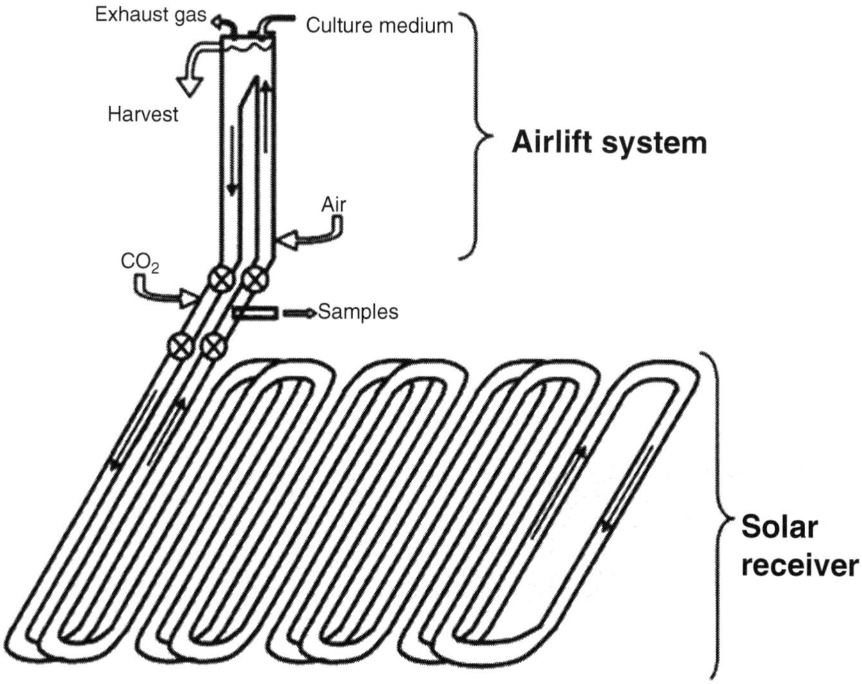

Fig. 20 Schematic of a typical horizontal tubular PBR according to Molina et al. (2001) (with permission of Elsevier)

Hawaii. This system occupies quite large surface areas (about 100 m^2 per 25,000 L module) and displays a rather high total volume, hence resulting in only modest volumetric productivities of max. 0.05 g/Ld.

After an upscaling period of nearly 3 years, a tubular system on an industrial scale was established in the year 2000 in Klötze near Wolfsburg, Germany, by the company *Ökologische Produkte Altmark GmbH* based on the know how developed by the research group of Prof. Pulz (2001). According to the available literature, this system constitutes the largest tubular PBR that ever started commercial production. It consists of compact and vertically arranged horizontal running glass tubes, consisting of 20 modules each with a separate online-control unit, of a total length of 500 km (!) and a total PBR volume of 700 m^3. Mixing is achieved using horizontal tubes. The plant is placed in a glasshouse requiring an area of only about 10,000 m^2 (Šantek et al. 2006); illumination is performed by a horizontal fence of borosilicate glass tubes of 3 m of height. The process uses the CO_2 as unwanted by-product from biogas production of a local composting plant for pine wood chips (Tredici 2004). Based on an produced annual quantity of 130–150 tones dry biomass of the strain *Chlorella* sp. during an 8 month operation period, corresponding to a volumetric productivity of 0.9 g/Lday, this plant was demonstrated to be economically feasible under Central European conditions (Pulz 2001; Šantek et al. 2006). As the major drawback, the shading among the tubes is reported; a possible solution might be a mixotrophic mode of cultivation at lower illumination requirements (Tredici 2004), especially considering the fact that the applied algal species *Chlorella* sp. is known to operate well under mixotrophic conditions (Gao et al. 2010; Liang et al. 2009; Mitra et al. 2012; Shi et al. 1997, 2000).

8.6.4 Helical Tubular PBRs

Many modifications of the helical framework have been proposed to improve the design and light distribution. It was reported that for a given area 60 cone angles of conical helical layout had the maximal photo-receiving area and photosynthetic efficiency of 6.84 % (Morita et al. 2000). Such helical tubular systems have a high surface-to-volume ratio, hence are well suited for optimized illumination. Although they can be operated outdoors using sun light as the only source of illumination, helical PBRs often require artificial illumination that contributes to the production costs of such devices. Hence, such systems are predominately used for high-priced niche products, not for bulk products such as biofuel feedstocks. They are constructed in modules and can theoretically be scaled outdoors up to agricultural scale. A dedicated location is not crucial, similar to other closed systems, and therefore non-arable land is suitable as well. The material choice shall prevent biofouling and ensure high final biomass concentrations. The combination of turbulences and the closed concept are ought to reach a clean operation and a high operational flexibility.

The **BIOCOIL** system, invented by Lee Robinson, is an example for a helical arrangement of helical tubes to create a new type of PBR. Here, transparent PVC tubes of small diameter (2.4 × 5 cm) are wounded around a vertical cylindrical frame; illumination is performed using external light (solar radiation) plus external artificial illumination; it is also possible to provide additional illumination from the interior. This way, a large illuminated surface is provided to ensure a high light-input to the cells (Borowitzka 1996, 1999). Agitation is provided by peristaltic pumps creating compressed air that generates uniform mixing and is also used for preventing bio-fouling by cell adhesion (Šantek et al. 2006). In addition, scouring pads are present in each tubular section to remove attached algal cells (Concas et al. 2010). The main application of the BIOCOIL system is the sequestering of CO_2 and nutrients like phosphate and nitrogen from effluent water. As a further benefit of its vertical arrangement, BIOCOIL PBRs occupy only small areas (Concas et al. 2010). This system was already tested on pilot scale (700 L volume) with different marine algal species (*Tetraselmis* spp., *Isochrysis galbana*, *Phaeodactylum tricornutum*, *Chaetoceros* spp. and *Spirulina* sp.) for extended time periods exceeding 4 months. A 21 L set-up of the BIOCOIL type was operated in semi-continuous mode under steady-state conditions, different D-values. Using *Spirulina* sp. as production strain, 0.4 g/L day of CDM were obtained at the optimum D-value of 0.0078 h^{-1} (Travieso et al. 2001). Figure 21 shows the schematic of the BIOCOIL system.

The "**Christmas tree reactor**" or "fir tree reactor" constitutes a special form of helical tubular PBRs aiming at optimum light exposure of the tubular system (Cotta

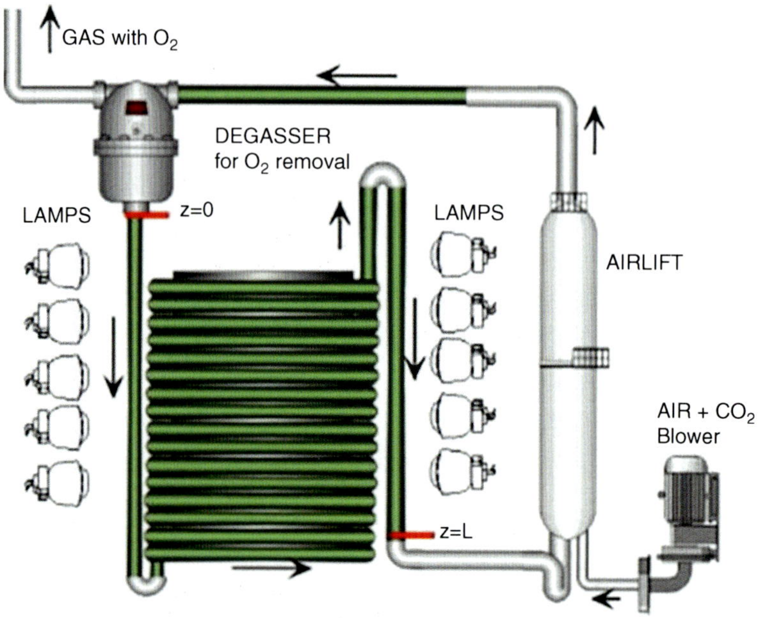

Fig. 21 The BIOCOIL PBR (Concas et al. 2010) (with permission of Elsevier)

2011). They are constructed by coiling straight tubes made of flexible plastic into three dimensional helical frameworks with a desired inclination providing for their special conical shape. They are externally coupled with a gas exchanger and a heat exchanger. A centrifugal pump is used to drive the feed through the system in ascending mode, resulting in a high radial mixing. The Christmas tree reactor is characterized by its compact design and thus a lower space occupation if compared with other tubular PBR systems, and by increased turbulences than e.g. the BIOCOIL PBR. Short light paths are achieved by using an internal tube which also maintains the correct temperature (Jacobi et al. 2012). A pilot plant of a series of Christmas tree reactors is operated in a glass house at *Mitteldeutsches Biosolarzentrum* in Köthen, Germany, and depicted in Fig. 22 (online resource 8).

A typical conical configuration (cylinder with the top at the bottom; "inverse Christmas tree PBR") was investigated by Morita and colleagues (2000) using *Chlorella* sp. on laboratory scale. A total volume of 0.45 L was used; the circulation of medium was just accomplished by the provided gas flow. Temperature was kept constant in a thermosetting unit; additionally, a degasser unit acts to remove the generated oxygen. The experiments were carried out at high irradiation levels of about 980 μE/m^{2}s. This extremely high illumination was re-distributed to the entire large PBR surface at different inclination angles of the tubular system; it turned out that an inclination angle of 60 °C was most beneficial for algal growth if compared to the other investigated set-ups with 180°, 120° and 90°, respectively. This means, the sharper the angle of inclination, the better the light distribution.

Fig. 22 The "Christmas tree PBR", a conical variety of helical tubular PBRs (online resource 8, with kind permission from GICON – Großmann Ingenieur Consult GmbH). These examples are operated indoor in a glass house, using external solar illumination in addition to artificial internal illumination

8.7 Hybrid PBRs, Solid State Systems, and Integration of Closed PBRs in Buildings

Hybrid PBRs combine selected benefits of different PBR types discussed before and might be used for niche applications on smaller scale in the future.

8.7.1 Combinations of Open Tanks and Plat Panels

A new hybrid photobioreactor system was designed by Velea et al. (2014), consisting of an open tank connected to a variable number of transparent flat plate photosynthetic cells aligned in a parallel configuration and two vertical bubble columns in series, interconnected with pipes for algal suspension recirculation with a pump and pipelines for gas bubbling.

The new system was tested using the green alga *C. homosphaera*; it turned out that it enhances irradiance distribution, flow dynamics, and gaseous transfer, and thereby results in increased biomass concentration, dense algal cultures and an extensive sequestration of CO_2.

8.7.2 Combination of Closed Vessels and Plate Panels

The different advantages of tubular photobioreactors, flat plates and vertical vessels can be connected. E.g., a combination of a glass vessel with a thin tube coil allows relevant biomass production rates; until today, such set-ups are restricted to laboratory scale research. Being controlled by a complex process control system for the regulation of the environmental conditions, a high level of biomass production can be reached (Singh and Sharma 2012).

8.7.3 Solar Hybrid PBRs

A photobioreactor system was patented consisting of at least two bioreactor tubes, each having an end and a hollow interior, the ends being connectively joined by one or more connector units having a hollow portion defined by a circumference, a solar concentrator configured to collect and concentrate solar power, at least one light guide (large-core polymer optical fibres to deliver large quantities of visible sunlight into the PBR) associated with the solar concentrator to illuminate the hollow portion of the one or more connector units, and at least one LED illuminating the one or more connector units (US 8716010 B2).

8.7.4 Solid State PBRs

A novel approach for enhanced cultivation of phototrophic cyanobacteria, aiming at the production of valued pigments, was reported by Léonard and colleagues (2010). This new system is based on using solid, non-toxic porous silica matrices, leading to a hybrid PBR where the algae are spatially separated from the liquid cultivation medium. Here, the authors immobilized the active cyanobacterial biomass on acid-exchanged sodium silicate combined with the application of silica nanoparticles for gel-strengthening. Beside its biocompatibility, silica is optically transparent, which is of outstanding importance for photosynthesis to continue after the immobilisation. As a result, a high productivity after immobilisation and a preservation of the photosynthetic pigments of up to 35 weeks was achieved. These are the first studies demonstrating the possibility to encapsulate active phototrophic biomass like cyanobacteria inside porous silica gels whilst maintaining the integrity of their membranes and their photosynthetic activity.

8.7.5 The "Penthouse Roof PBR"

In the Czech Republic, a horizontal tubular PBR was incorporated into the roof of a building with a slope of 40°. The roof harboured a louver consisting of linear glass Fresnel lenses of the SOLARGLAS^{TM} type. The algal suspension flow in horizontal direction in tubes mounted in a mobile metallic frame, located directly beneath the

Fig. 23 Photobioreactor façade of the "BIQ-building" in Hamburg (© Paul Ott, Graz, Austria, with permission by Splitterwerk)

illumination lenses system. This integrated system light collector – PBR is flexible only for temperature conditions between 30 and 40 °C; it serves as a shutter for reducing direct solar irradiation, production of algal biomass, passive cooling of the building, and heated water (Jirka et al. 2002). In addition, O_2 evolved by the algae could be released into the building's interior and positively contribute to the climatic conditions in the rooms. Unfortunately, no information is provided by the authors regarding the actual performance of this integrated system for algal production rates and kinetics.

8.7.6 PBR Façade

A "zero-energy house" was installed in Hamburg (Germany) as the first real-life test for a new façade system that uses living microalgae to provide shade and generate renewable energy and oxygen at the same time (schematic see Fig. 23). The world's first 'bio-adaptive façade' was installed in the "BIQ-building" for the International Building Exhibition (IBA) in Hamburg, which operated through 2013. The concept of the façade, consisting of a high number of interconnected single flat panel modules, was designed in such a way that algae in the bio-reactor façades grow faster in bright sunlight to provide more internal shading.

The PBRs not only produce biomass that can subsequently be harvested, but they also capture solar thermal heat – both energy sources can be used to power the building; in addition, similar to the "Penthouse roof PBR", the generated oxygen could be imported into the building in order to improve the air quality.

9 Scalability and Controllability If Diverse Closed PHR Types

Comparing the characteristics of the diverse PBR types discussed in the prior paragraphs, it is clear that compromises have to be made if choosing the "ideal" PBR type for large scale production; special challenges and difficulties in the upscaling of different PBR prototypes are integrated in the respective chapters and also summarized in Table 2.

To summarize, on the one hand, a large surface-to-volume ratio is beneficial for large scale devices, because all discussed pre-requisites for internal illumination (Fresnel lenses, "honey-comb" reactor type, optical fibres etc.) are expensive to realize on a larger scale. This restricts the application of all column-type PBRs (CSTR, airlift, bubble column) on large scale. Although the annular PBR type was already operated successfully on pilot scale, it's implementation on industrial scale is highly doubtful. The situation changes regarding such special cultivation processes requiring limited illumination, and especially for mixotrophic and heterotrophic set-ups; here, the column type PBR displays a viable tool for large-scale processes. On the other hand, efficient mixing devices (stirrers, impellers), and convenient nutrient supply are restricted to the CSTR-type if designing large scale facilities. Such systems additionally result in the lowest losses of CO_2. Based on the theoretical considerations and on experience obtained by different companies, tubular systems like demonstrated by SOLIX, and especially flat panel systems of modular character as developed by ECODUNA and PSI might be the most promising near-term solutions for large scale algal cultivation. Such systems are also characterized by a low energy demand, if compared to column type PBRs. In the case of flat panels, it will not be reasonable to design large single flat modules, but to combine smaller single modules, achieving higher flexibility of the system; such systems operate well on large scale (see ECODUNA and PSI systems). Concerning the "bio-dome" PBR, a special form of flat panels, upscaling seems to be doubtful anyhow. Regarding the difficulties regarding O_2 outgassing and parameter control over extended tubular lengths that typically show up when tubular PBR systems are scaled up, flexible modular flat panel systems appear most promising for industrial scale photoautotrophic cultivation of microalgae.

10 Concluding Remarks and Lessons Learned

Developing new closed photobioreactors for tailored applications means learning from and mimicking nature. According to our knowledge and experience of today, there is no ideal PBR that constitutes the best solution for each algal cultivation process; requirements of diverse algal strains and special cultivation conditions to produce the final product of choice are to manifold to be satisfied by a restricted number of PBRs flexible for all imaginable algal production processes. This means that the process design needs to be adapted according to the customer's demands (Morweiser et al. 2010). Compromises have to be made regarding light supply techniques, active light harvesting surface, mixing, flow rates, k_La, export of oxygen, import of CO_2, CAPEX *vs.* product prices, and others.

As a non-trivial task, one also has to consider the different kinetics of product formation (primary and secondary metabolites) that severely impacts the choice of the most suitable PBR system. For special products, e.g. cyanobacterial polyesters (PHA), a cascade of PBRs might be a viable process engineering tool in order to provide sufficient retention time for product accumulation, light supply, and possibilities to trigger cultivation conditions in the single vessels of the cascade. Less trivial tasks such as hetero- or mixotrophic production of algal protein will come along with simple process engineering designs at restricted illumination requirements; here, simple bubble columns might be a sufficient solution, also for large-scale production. High-priced products that are produced as a reaction to high illumination, such as pigments, will require novel PBR facilities with enhanced light harvesting pre-requisites, such as PBRs changing their position with the position of sun, and light scavenging devices like e.g. Fresnel lenses, in order to fully profit from the high biological potential of phototrophic microbes.

As a consequence, the development of closed photobioreactor systems remains a challenging task for engineers; this task needs the narrow cooperation with microbiologists, biologists, and experts in the field of mathematic modelling of phototrophic bioprocesses. Until today, especially the lacking synopsis of skills in process design and well-grounded understanding of intracellular activities hampers the broad industrial employment of these powerful phototrophic cell-factories.

Nevertheless, one must not forget that even if a superior PBR system is developed and available, cost efficiency and ecological sustainability of large-scale algal cultivation need additional factors to be considered, such as nitrogen and phosphate supply, enhanced harvesting and downstream processing techniques, and fresh water requirements. Especially the assessment of various waste streams, e.g. industrial gaseous or liquid effluents, need to be assessed regarding their suitability for application as nutrient sources for algal cultivation. Combining the refining of biogas with CO_2-supply of microalgal cells, together with the application of "green fertilizers" for microalgal farming, appears to be a highly promising strategy for future developments.

Designing adequate scalable PBR devices can also result in positive socioeconomic impacts. Developing countries are most vulnerable to the expected and already occurring negative impacts of climate change; most of all, this embodies the proceeding shortage of resources for food, feed, and energy production. Since these countries are often not the powerful decision-makers with adequate possibilities to fight the effects of climate change, their secure supply concerning food, water and energy is severely endangered. Especially for developing and emerging countries, the industrial production of various algal products might contribute to create a broad range of differently qualified jobs, to enter global markets where such prized products can be commercialized, and, if considering green energy carriers accessible from algae, can even provide a certain degree of energetic independence from pressure of the global fossil resource market. The development of efficient PBR systems to be operated on an industrially relevant scale constitutes a pivotal step in this direction.

List of Abbreviations and Symbols

CAPEX	Capital Expenditure (Investment cost)
CDM	Cell dry mass
CFD	Computational fluid dynamics
CFU	Colony forming unit
D	Dilution rate
E	Einstein (1 Mol of photons)
k_La	Oxygen mass transfer coefficient
μ	specific growth rate [1/h]
$\mu_{max.}$	maximum specific growth rate [1/h]
$\mu E/m^2 s$	Mikroeinstein per square meter and second
PE	Poly(ethylene)
PBR	Photobioreactor
PCL	Poly(ε-caprolactone)
PET	Poly(ethyleneterephtalate)
PHA	Poly(hydroxyalkanoate)
PLA	Poly(lactate)
PMMA	Poly(methylmetacrylate)
PUFAs	Polyunsaturated fatty acids
PVC	Poly(vinyl chloride)
STR	Stirred tank reactor
vvm	volume per volume and minutes

References

Abed RMM, Dobretsov S, Sudesh K (2009) Applications of cyanobacteria in biotechnology. J Appl Microbiol 106(1):1–12

Apt KE, Behrens PW (1999) Commercial developments in microalgal biotechnology. J Phycol 35(2):215–226
Azov Y (1982) Effect of pH on inorganic carbon uptake in algal cultures. Appl Environ Microbiol 43(6):1300–1306
Bajpai R, Zappi M, Dufreche S, Subramaniam R, Prokop A (2014) Status of algae as vehicles for commercial production of fuels and chemicals. In: Bajpai R, Prokop A, Zappi M (eds) Algal biorefineries volume 1: cultivation of cells and products. Springer, Dordrecht, pp 3–24
Barbosa MJ, Zijffers JW, Nisworo A, Vaes W, van Schoonhoven J, Wijffels RH (2005) Optimization of biomass, vitamins, and carotenoid yield on light energy in a flat-panel reactor using the A-stat technique. Biotechnol Bioeng 89(2):233–242
Ben-Amotz A, Avron M (1983) On the factors which determine massive β-carotene accumulation in the halotolerant alga *Dunaliella bardawil*. Plant Physiol 72(3):593–597
Berberoglu H, Yin J, Pilon L (2007) Light transfer in bubble sparged photobioreactors for H_2 production and CO_2 mitigation. Int J Hydrog Energy 32(13):2273–2285
Betula C, Campsis L, Sambucus S, Cornus D, Quercus L, Carya L (1986) Hosts of the parasitic alga *Cephaleuros virescens* in Louisiana and new host records for the continental United States. Plant Dis 70(11):1080–1083
Birmingham BC, Coleman JR, Colman B (1982) Measurement of photorespiration in algae. Plant Physiol 69(1):259–262
Borowitzka MA (1996) Closed algal photobioreactors: Design considerations for large-scale systems. J Mar Biotechnol 4(4):185–191
Borowitzka MA (1999) Commercial production of microalgae: ponds, tanks, and fermenters. Prog Ind Microbiol 35:313–321
Borowitzka LJ, Brown AD (1974) The salt relations of marine and halophilic species of the unicellular green alga, *Dunaliella*. Arch Microbiol 96(1):37–52
Borowitzka LJ, Borowitzka MA, Moulton TP (1984) The mass culture of *Dunaliella salina* for fine chemicals: from laboratory to pilot plant. Hydrobiologia 116(117):115–134
Brányiková I, Maršálková B, Doucha J, Brányik T, Bišová K, Zachleder V, Vítová M (2011) Microalgae—novel highly efficient starch producers. Biotechnol Bioeng 108(4):766–776
Braunegg G, Lefebvre G, Renner G, Zeiser A, Haage G, Loidl-Lanthaler K (1995) Kinetics as a tool for polyhydroxyalkanoate production optimization. Can J Microbiol 41(13):239–248
Cardozo KH, Guaratini T, Barros MP, Falcão VR, Tonon A et al (2007) Metabolites from algae with economical impact. Comp Biochem Phys C 146(1):60–78
Castenholz RW (1969a) Thermophilic blue-green algae and the thermal environment. Bacteriol Rev 33(4):476–504
Castenholz RW (1969b) The thermophylic cyanophytes of Iceland and the upper temperature limit. J Phycol 5(4):360–368
Chen CY, Yeh KL, Aisyah R, Lee DJ, Chang JS (2011) Cultivation, photobioreactor design and harvesting of microalgae for biodiesel production: a critical review. Bioresour Technol 102(1):71–81
Chini Zitelli G, Rodolfi L, Tredici MR (2000) Mass cultivation of marine microalgae under natural, mixed and artificial illumination. In: 4th European workshop on biotechnology of microalgae. Bergholz-Rehbrücke, Germany
Chini Zitelli G, Rodolfi L, Tredici MR (2003) Mass cultivation of *Nannochloropsis* sp. in annular reactors. J Appl Phycol 15(2–3):107–114
Chini Zitelli G, Somigli S, Rodolfi L, Tredici MR (2004) Outdoor mass cultivation of *Isochrysis* sp. in annular reactors. In: Abstracts of the first Latinoamerican Congress on Algal Biotechnology (CLABA), Buenos Aires, Argentina, 25–29 Oct 2004, p 45
Chini Zittelli G, Rodolfi L, Biondi N, Tredici MR (2006) Productivity and photosynthetic efficiency of outdoor cultures of *Tetraselmis suecica* in annular columns. Aquaculture 261(3):932–943
Chisti Y (2007) Biodiesel from microalgae. Biotechnol Adv 25(3):294–306
Collet P, Hélias A, Lardon L, Ras M, Goy RA, Steyer JP (2011) Life-cycle assessment of microalgae culture coupled to biogas production. Bioresour Technol 102(1):207–214

Concas A, Pisu M, Cao G (2010) Novel simulation model of the solar collector of BIOCOIL photobioreactors for CO_2 sequestration with microalgae. Chem Eng J 157(2):297–303

Cotta F, Matschke M, Großmann J, Griehl C, Matthes S (2011) Verfahrenstechnische Aspekte eines flexiblen, tubulären Systems zur Algenproduktion (Process-related aspects of a flexible, tubular system for algae production); lecture at DECHEMA 2011

Degen J, Uebele A, Retze A, Schmid-Staiger U, Trösch W (2001) A novel airlift photobioreactor with baffles for improved light utilization through the flashing light effect. J Biotechnol 92(2):89–94

Doucha J, Lívanský K (2014) High density outdoor microalgal culture. In: Bajpai R, Prokop A, Zappi M (eds) Algal biorefineries volume 1: cultivation of cells and products. Springer, Dordrecht, pp 147–173

Doucha J, Straka F, Lívanský K (2005) Utilization of flue gas for cultivation of microalgae (Chlorella sp.) in an outdoor open thin-layer photobioreactor. J Appl Phycol 17(5):403–412

El-Shishtawy RM, Kawasaki S, Morimoto M (1997) Biological H_2 production using a novel light-induced and diffused photoreactor. Biotechnol Tech 11(6):403–407

Fisher M, Pick U, Zamir A (1994) A salt-induced 60-kilodalton plasma membrane protein plays a potential role in the extreme halotolerance of the alga *Dunaliella*. Plant Physiol 106(4):1359–1365

Fulks W, Main KL (1991) Rotifer and microalgae culture systems. In: Fulks W, Main KL (eds) Proceedings of a U.S.-Asia workshop, The Oceanic Institute Honolulu, Hawaii, pp 3–52

Gao C, Zhai Y, Ding Y, Wu Q (2010) Application of sweet sorghum for biodiesel production by heterotrophic microalga *Chlorella protothecoides*. Appl Energy 87(3):756–761

Ginzburg M (1988) *Dunaliella*: a green alga adapted to salt. Adv Bot Res 14:93–183

Gudin C, Chaumont D (1991) Cell fragility—the key problem of microalgae mass production in closed photobioreactors. Bioresour Technol 38(2):145–151

Hariskos I, Posten C (2014) Biorefinery of microalgae–opportunities and constraints for different production scenarios. Biotechnol J 9(6):739–752

Heijnen JJ, Hols J, Van Der Lans RGJM, Van Leeuwen HLJM, Mulder A, Weltevrede R (1997) A simple hydrodynamic model for the liquid circulation velocity in a full-scale two-and three-phase internal airlift reactor operating in the gas recirculation regime. Chem Eng Sci 52(15):2527–2540

Hoham RW (1975) Optimum temperatures and temperature ranges for growth of snow algae. Arct Alp Res 7:13–24

Hoham RW, Blinn DW (1979) Distribution of cryophilic algae in an arid region, the American Southwest. Phycologia 18(2):133–145

Hu Q, Kurano N, Kawachi M, Iwasaki I, Miyachi S (1998) Ultrahigh-cell-density culture of a marine green alga *Chlorococcum littorale* in a flat-plate photobioreactor. Appl Microbiol Biotechnol 49(6):655–662

Imam SH, Gordon SH, Shogren RL, Tosteson TR, Govind NS, Greene RV (1999) Degradation of starch–poly (β-hydroxybutyrate-co-β-hydroxyvalerate) bioplastic in tropical coastal waters. Appl Environ Microbiol 65(2):431–437

Iqbal M, Grey D, Stepan-Sarkissian F, Fowler MW (1993) A flat-sided photobioreactor for culturing microalgae. Aquac Eng 12(3):183–190

Jacobi A, Steinweg C, Sastre RR, Posten C (2012) Advanced photobioreactor LED illumination system: scale-down approach to study microalgal growth kinetics. Eng Life Sci 12(6):621–630

Jirka V, Šourek B, Pokorný J, Štys D, Papácek S, Masojídek J (2002) Microalgal 'penthouse-roof' photobioreactor based on solar concentrators-linear raster lenses. In: Renewable energy, extended abstracts of the World Renewable Energy Congress VII (WREC 2002), Cologne, Germany, 29, pp 359–360.

Kol E (1969) The red snow of Greenland. II. Acta Bot Acad Sci Hung 15(3–4):281–289

Koller M, Muhr A (2014) Continuous production mode as a viable process-engineering tool for efficient poly (hydroxyalkanoate)(PHA) bio-production. Chem Biochem Eng Q 28(1):65–77

Koller M, Salerno A, Tuffner P, Koinigg M, Böchzelt H, Schober S, Pieber S, Schnitzer H, Mittelbach M, Braunegg G (2012) Characteristics and potential of micro algal cultivation strategies: a review. J Clean Prod 37:377–388

Koller M, Niebelschütz H, Braunegg G (2013) Strategies for recovery and purification of poly [(R)-3-hydroxyalkanoates](PHA) biopolyesters from surrounding biomass. Eng Life Sci 13(6):549–562

Koller M, Muhr A, Braunegg G (2014) Microalgae as versatile cellular factories for valued products. Algal Res 6(A):52–63

Léonard A, Rooke JC, Meunier CF, Sarmento H, Descy JP, Su BL (2010) Cyanobacteria immobilised in porous silica gels: exploring biocompatible synthesis routes for the development of photobioreactors. Energy Environ Sci 3(3):370–377

Leya T, Müller T, Ling HU, Fuhr G (2000) Taxonomy and biophysical properties of cryophilic microalgae and their environmental factors in northwest Spitsbergen, Svalbard. In: Proceedings of the 57th Eastern snow conference, Syracuse, New York, pp 199–205

Li Y, Horsman M, Wu N, Lan CQ, Dubois-Calero N (2008) Biofuels from microalgae. Biotechnol Prog 24(4):815–820

Li T, Zheng Y, Yu L, Chen S (2013) High productivity cultivation of a heat-resistant microalga *Chlorella sorokiniana* for biofuel production. Bioresour Technol 131:60–67

Liang Y, Sarkany N, Cui Y (2009) Biomass and lipid productivities of *Chlorella vulgaris* under autotrophic, heterotrophic and mixotrophic growth conditions. Biotechnol Lett 31(7):1043–1049

Ling HU, Seppelt RD (1993) Snow algae of the Windmill Islands, continental Antarctica. 2. Chloromonas rubroleosa sp. nov. (Volvocales, Chlorophyta), an alga of red snow. Eur J Phycol 28(2):77–84

Lubián LM, Montero O, Moreno-Garrido I, Huertas IE, Sobrino C, González-del Valle M, Parés G (2000) *Nannochloropsis* (Eustigmatophyceae) as source of commercially valuable pigments. J Appl Phycol 12(3–5):249–255

Mann G, Schlegel M, Schumann R, Sakalauskas A (2009) Biogas-conditioning with microalgae. Agron Res 7(1):33–38

Masojídek J, Sergejevová M, Rottnerová K, Jirka V, Korečko J, Kopecký J, Zaťková I, Torzillo G, Štys D (2009) A two-stage solar photobioreactor for cultivation of microalgae based on solar concentrators. J Appl Phycol 21(1):55–63

Mata TM, Martins AA, Caetano NS (2010) Microalgae for biodiesel production and other applications: a review. Renew Sust Energ Rev 14(1):217–232

Merchuk JC, Garcia-Camacho F, Molina-Grima E (2007) Photobioreactor design and fluid dynamics. Chem Biochem Eng Q 21(4):345–355

Mignot L, Junter GA, Labbe M (1989) A new type of immobilized-cell photobioreactor with internal illumination by optical fibres. Biotechnol Tech 3(5):299–304

Miron AS, Camacho FG, Gomez AC, Molina GE, Chisti Y (2000) Bubble column and airlift photobioreactors for algal culture. AIChE J 46:1872–1893

Mitra D, van Leeuwen JH, Lamsal B (2012) Heterotrophic/mixotrophic cultivation of oleaginous *Chlorella vulgaris* on industrial co-products. Algal Res 1(1):40–48

Molina E, Fernández J, Acién FG, Chisti Y (2001) Tubular photobioreactor design for algal cultures. J Biotechnol 92(2):113–131

Morita M, Watanable Y, Saiki H (2000) Investigation of photobioreactor design for enhancing the photosynthetic productivity of microalgae. Biotechnol Bioeng 69(6):693–698

Morweiser M, Kruse O, Hankamer B, Posten C (2010) Developments and perspectives of photobioreactors for biofuel production. Appl Microbiol Biotechnol 87(4):1291–1301

Ogbonna JC, Yada H, Masui H, Tanaka H (1996) A novel internally illuminated stirred tank photobioreactor for large-scale cultivation of photosynthetic cells. J Ferment Bioeng 82(1):61–67

Ogbonna JC, Soejima T, Tanaka H (1999) An integrated solar and artificial light system for internal illumination of photobioreactors. Prog Ind Microbiol 35:289–297

Ogbonna KH, Aminigo RE, Abu GO (2007) Influence of temperature and pH on biomass production and protein biosynthesis in a putative *Spirulina* sp. Bioresour Technol 98(11):2207–2211

Olaizola M (2000) Commercial production of astaxanthin from *Haematococcus pluvialis* using 25,000-liter outdoor photobioreactors. J Appl Phycol 12(3–5):499–506
Olaizola M (2003) Commercial development of microalgal biotechnology: from the test tube to the marketplace. Biomol Eng 20(4):459–466
Olivieri G, Salatino P, Marzocchella A (2014) Advances in photobioreactors for intensive microalgal production: configurations, operating strategies and applications. J Chem Technol Biotechnol 89(2):178–195
Online resource 1: http://cfb.unh.edu/phycokey/Choices/Chlorophyceae/unicells/flagellated/DUNALIELLA/Dunaliella_Image_page.html
Online resource 2: http://algae-energy.co.uk/biofuel_production/cultivation/
Online resource 3: http://www.psi.cz/products/photobioreactors/download. 16 Sept 2014
Online resource 4: http://www.oilseedcrops.org/algae/
Online resource 5: http://www.algaeindustrymagazine.com/a-i-m-interview-solix-biosystems-ceo-joel-butler/
Online resource 6: http://www.cleanthinking.de/vattenfall-und-ecoduna-nehmen-algenzuchtanlage-in-betrieb/21608/
Online resource 7: http://www.hawaiibusiness.com/Hawaii-Business/October-2001/Economies-of-Scale/
Online resource 8: http://www.gicon.de/uploads/tx_sbdownloader/Biosolarzentrum_05.jpg
Patel B, Tamburic B, Zemichael FW, Dechatiwongse P, Hellgardt K (2012) Algal biofuels: a credible prospective? ISRN Renew Energy. Article ID 631574, 14 pp
Phang SM, Kim-Chong O (1988) Algal biomass production in digested palm oil mill effluent. Biol Waste 25(3):177–191
Posten C (2009) Design principles of photo-bioreactors for cultivation of microalgae. Eng Life Sci 9(3):165–177
Prokop A, Řičica J (1968) *Chlorella pyrenoidosa* 7-11-05 in batch and in homogeneous continuous culture under autotrophic conditions. Folia Microbiol 13(5):362–372
Prokop A, Řičica J, Málek I, Thomas J (1967) Growth and physiological characteristics of a high temperature strain of *Chlorella pyrenoidosa* in continuous culture. Nature 214:1234–1235
Pulz O (2001) Photobioreactors: production systems for phototrophic microorganisms. Appl Microbiol Biotechnol 57(3):287–293
Pulz O, Gross W (2004) Valuable products from biotechnology of microalgae. Appl Microbiol Biotechnol 65(6):635–648
Radakovits R, Jinkerson RE, Darzins A, Posewitz MC (2010) Genetic engineering of algae for enhanced biofuel production. Eukaryot Cell 9(4):486–501
Rao AR, Dayananda C, Sarada R, Shamala TR, Ravishankar GA (2007) Effect of salinity on growth of green alga Botryococcus braunii and its constituents. Bioresour Technol 98(3):560–564
Raven JA, Beardall J (2003) Carbon acquisition mechanisms of algae: carbon dioxide diffusion and carbon dioxide concentrating mechanisms. In: Anthony WD, Larkum SE, Douglas JA (eds_ Photosynthesis in algae. Springer, Dordrecht, pp 225–244
Ravikumar R (2014) Micro algae in open raceways. In: Bajpai R, Prokop A, Zappi M (eds) Algal biorefineries volume 1: cultivation of cells and products. Springer, Dordrecht, pp 127–146
Reddy CSK, Ghai R, Kalia V (2003) Polyhydroxyalkanoates: an overview. Bioresour Technol 87(2):137–146
Remias D, Lütz-Meindl U, Lütz C (2005) Photosynthesis, pigments and ultrastructure of the alpine snow alga *Chlamydomonas nivalis*. Eur J Phycol 40(3):259–268
Richmond A, Boussiba S, Vonshak A, Kopel R (1993) A new tubular reactor for mass production of microalgae outdoors. J Appl Phycol 5(3):327–332
Rodolfi L, Biondi N, Piccardi R, Ferroni P, Tredici MR (2002) Effect of temperature on growth and bioactivity of two *Nostoc* strains in mass culture. In: Abstracts of the 9th international conference on applied algology, 26–30 May 2002, Almeria, Spain (2002), p 21

Rodolfi L, Chini Zittelli G, Bassi N, Padovani G, Biondi N, Bonini G, Tredici MR (2009) Microalgae for oil: strain selection, induction of lipid synthesis and outdoor mass cultivation in a low-cost photobioreactor. Biotechnol Bioeng 102(1):100–112

Rogers JN, Rosenberg JN, Guzman BJ, Oh VH, Mimbela LE, Ghassemi A, Betenbaugh MJ, Oyler GA, Donohue MD (2014) A critical analysis of paddlewheel-driven raceway ponds for algal biofuel production at commercial scales. Algal Res 4:76–88

Šantek B, Ivančić M, Horvat P, Novak S, Marić V (2006) Horizontal tubular bioreactors in biotechnology. Chem Biochem Eng Q 20(4):389–399

Sato T, Usui S, Tsuchiya Y, Kondo Y (2006) Invention of outdoor closed type photobioreactor for microalgae. Energy Convers Manag 47(6):791–799

Scott SA, Davey MP, Dennis JS, Horst I, Howe CJ, Lea-Smith DJ, Smith AG (2010) Biodiesel from algae: challenges and prospects. Curr Opin Biotechnol 21(3):277–286

Shi XM, Chen F, Yuan JP, Chen H (1997) Heterotrophic production of lutein by selected *Chlorella* strains. J Appl Phycol 9(5):445–450

Shi XM, Zhang XW, Chen F (2000) Heterotrophic production of biomass and lutein by *Chlorella protothecoides* on various nitrogen sources. Enzyme Microb Technol 27(3):312–318

Sierra E, Acien FG, Fernández JM, García JL, González C, Molina E (2008) Characterization of a flat plate photobioreactor for the production of microalgae. Chem Eng J 138(1):136–147

Sim TS, Goh A (1988) Ecology of microalgae in a high rate pond for piggery effluent purification in Singapore. MIRCEN J Appl Microbiol 4(3):285–297

Singh RN, Sharma S (2012) Development of suitable photobioreactor for algae production–a review. Renew Sust Energ Rev 16(4):2347–2353

Spolaore P, Joannis-Cassan C, Duran E, Isambert A (2006) Commercial applications of microalgae. J Biosci Bioeng 101(2):87–96

Stein JR, Bisalputra T (1969) Crystalline bodies in an algal chloroplast. Can J Bot 47(2):233–236

Tatsuzawa H, Takizawa E, Wada M, Yamamoto Y (1996) Fatty acid and lipid composition of the acidophilic green alga *Chlamydomonas* sp. 1. J Phycol 32(4):598–601

Torzillo G, Pushparaj B, Bocci F, Balloni W, Materassi R, Florenzano G (1986) Production of *Spirulina* biomass in closed photobioreactors. Biomass 11(1):61–74

Travieso L, Hall DO, Rao KK, Benıtez F, Sánchez E, Borja R (2001) A helical tubular photobioreactor producing *Spirulina* in a semicontinuous mode. Int Biodeterior Biodegrad 47(3):151–155

Tredici MR (2004) Mass production of microalgae: photobioreactors. In: Richmond A (ed) Handbook of microalgal culture: biotechnology and applied phycology. Blackwell Science Ltd, Oxford/Ames/Carlton, pp 178–214

Tredici MR, Zittelli GC (1998) Efficiency of sunlight utilization: tubular versus flat photobioreactors. Biotechnol Bioeng 57(2):187–197

Tredici MR, Zittelli GC, Benemann JR (1998) A tubular integral gas exchange photobioreactor for biological hydrogen production. In: Zaborsky OR (ed) Biohydrogen. Plenum Press, London, pp 391–401

Trotta P (1981) A simple and inexpensive system for continuous monoxenic mass culture of marine microalgae. Aquaculture 22:383–387

Tsygankov AA, Hall DO, Liu J, Rao KK (1998) An automated helical photobioreactor incorporating cyanobacteria for continuous hydrogen production. In: Zaborsky OR (ed) Biohydrogen. Plenum Press, London, pp 431–440

Uggetti E, Sialve B, Trably E, Steyer JP (2014) Integrating microalgae production with anaerobic digestion: a biorefinery approach. Biofuels Bioprod Biorefin 8(4):516–529

Velea S, Ilie L, Stepan E, Chiurtu R (2014) New photobioreactor design for enhancing the photosynthetic productivity of *Chlorella homosphaera* culture. Rev Chim Bucharest 65(1):56–60

Vona V, Di Martino RV, Lobosco O, Carfagna S, Esposito S, Rigano C (2004) Temperature responses of growth, photosynthesis, respiration and NADH: nitrate reductase in cryophilic and mesophilic algae. New Phytol 163(2):325–331

Wang DZ (2008) Neurotoxins from marine dinoflagellates: a brief review. Mar Drugs 6(2):349–371

Wang B, Lan CQ, Horsman M (2012) Closed photobioreactors for production of microalgal biomasses. Biotechnol Adv 30(4):904–912

Willson B (2009) The Solix AGS system: a low-cost photobioreactor system for production of biofuels from microalgae. IOP Conf Ser Earth Environ Sci 6:192015. doi:10.1088/1755-1307/6/9/192015

Yamaoka T, Satoh K, Katoh S (1978) Photosynthetic activities of a thermophilic blue-green alga. Plant Cell Physiol 19(6):943–954

Zhang K, Kurano N, Miyachi S (2002) Optimized aeration by carbon dioxide gas for microalgal production and mass transfer characterization in a vertical flat-plate photobioreactor. Bioprocess Biosyst Eng 25(2):97–101

Zijffers JWF, Salim S, Janssen M, Tramper J, Wijffels RH (2008) Capturing sunlight into a photobioreactor: ray tracing simulations of the propagation of light from capture to distribution into the reactor. Chem Eng J 145(2):316–327

Tubular Photobioreactors

Giuseppe Torzillo and Graziella Chini Zittelli

Abstract Considerable progress has been made in the past decade in developing the appropriate biotechnology for microalgal mass cultivation aimed at establishing a new agro-industry. However, until today economic constraints currently limit the industrial exploitation of microalgae for feed, food and biofuel production. Large-scale tubular reactors are being operated in Germany and Israel for the production of *Chlorella* and *Haematococcus* respectively. However, because of their high investment costs and energy requirement (particularly for mixing and cooling) their use is limited to the production of high-value products for human nutrition, cosmetics and pharmaceutical applications, and for the preparation of inocula for industrial production of low value commodities (biofuels). Tubular reactors are mandatory for the cultivation of strains that require a strict control of temperature and for the production of biohydrogen and in general volatile compounds. In this chapter, rather than extensively examining the plethora of photobioreactor designs available in the literature, we focus the attention on the main biological and technological constraints affecting their performance, and in the second part of this chapter we briefly describe the tubular reactors that are currently operated at a market size. Finally, principles for guiding optimal photobioreactor design are proposed.

Keywords Areal yield • *Arthrospira* • Culture contamination • Culture density • Fresnel lenses • Growth rate • Helical photobioreactors • Light-dark cycle • Manifold photobioreactors • Microalgae • Mixing • Mixing time • Oxygen build-up • Productivity • Scale-up • Serpentine photobioreactors • Tubular photobioreactor • Turbulence • Tube diameter • Time cycle

---

G. Torzillo (✉) • G. Chini Zittelli
CNR-Istituto per lo Studio degli Ecosistemi – sede di Firenze,
Via Madonna del Piano, 10, 50019 Sesto Fiorentino, Italy
e-mail: torzillo@ise.cnr.it

A. Prokop et al. (eds.), *Algal Biorefineries*, DOI 10.1007/978-3-319-20200-6_5

1 Introduction

Microalgae have mostly been produced for food and feed market, as a source of biomass and bioactive substances. The current market of microalgal biomass is estimated to be around 20,000 t per year; the price ranges between 30 and 300 € kg^{-1}. Mass cultures of microalgae have traditionally been cultivated in open ponds which are much cheaper and easier to operate than photobioreactors (PBRs). Several PBR designs have been proposed in recent years, to grow microalgae as a source of sustainable energy. For the cultivation of most species suitable for biodiesel production and for human consumption, the use of a closed system is advisable. Indeed, most of the species cultivated for oil production require strict control of the temperature between 20 and 30 °C, which is problematical to maintain in open ponds. PBRs are also recommended for the production of high-value compounds for which the strict control of culture variables is necessary in order to satisfy the good manufacturing practice (GMP) requirement for pharmaceutical products. However, it is worth pointing out that although a major advantage of closed PBRs is their ability to prevent contact of the microalgal culture with the atmosphere, the risk of pollution cannot be ruled out. Among the closed systems, tubular PBRs are the most common design developed at an industrial level (Pulz et al. 2013). The advantages and limitations of tubular PBRs have been discussed in several book chapters (e.g. Torzillo 1997; Tredici et al. 2010; Chini Zittelli et al. 2013). Because of the high production cost usually reached with this culture system, the main R&D on PBR design is aimed at achieving high light conversion efficiency, i.e. at pushing productivity well beyond the one currently attained, which – it seems – is the main way to develop cost-effective tubular PBRs.

In this chapter, rather than fully examining the plethora of PBR designs available in the literature, for which readers can consult some excellent recent books (Borowitzka and Moheimani 2013; Richmond and Hu 2013) we would prefer to focus on the main biological and technological constraints affecting the performance of tubular PBRs. However, in the second part of this chapter, we also briefly update and report on some recently proposed tubular PBR designs.

2 Design Criteria for Tubular Photobioreactors

The basic function of a properly designed tubular PBR is to provide a controlled environment in order to achieve optimal growth and/or product formation with the particular microalgal strain employed. A photobioreactor can be considered a four-phase system consisting of a solid (cells), a liquid growth medium, a gaseous phase, and a superimposed light radiation (Posten 2009). Therefore, to achieve a successful PBR design, a background in both chemical engineering and biological science is required. One of the most important aspects is a basic understanding of the morphology and physiology of the organism to be grown. The geometric configuration of the PBR, in particular the characteristics of the circulating device, is important.

The PBR design and its hydrodynamic characteristics must be tailored differently to suit the microorganisms to be cultivated (Acién Fernández et al. 2013). This, together with a correct choice of a location and not only in terms of climatic conditions, but also considering the availability of specialized manpower, can strongly affect the economy of the process. The experience gathered over many years of work performed in the authors' laboratory, as well as in others, indicates that the main factors affecting their performance are (1) the diameter of the tubing (2) the length of the tube, and (3) the mixing.

2.1 *Optimal Tube Diameter and Length*

The choice of the tube diameter represents an important decision for an optimal design of the PBR, since it affects: (1) the surface-to-volume ratio (S/V) and, as a result, the light uptake of the culture; (2) the biomass concentration; (3) the volumetric productivity; (4) the temperature profile of the culture; (5) the concentration of oxygen in the culture; (6) the CO_2 storage capacity of the PBR; (7) the head loss for culture recycling; (8) the length of the tubes.

In Table 1, is shown the variation of productivity obtained in different tube diameters that were experimented with cultures of *Arthrospira platensis* (formerly *Spirulina*) in the authors' laboratory. As can be expected in a photo-limited system, the smaller the diameter, the higher the volumetric productivity. Another advantage that is obtained with a reduction in the diameter is the increase in the biomass concentration of the culture. This aspect is important for culture harvesting particularly with small-size organisms. However, it is worth pointing out that a rise in the volumetric productivity does not necessarily entails a corresponding increase in areal yield (and thus in light conversion efficiency). For example, when two PBR designs made with tubes of 13 and 7.4 cm internal diameter were compared, no significant changes in areal yield were observed (Torzillo et al. 1987). A further reduction in

Table 1 Productivity of *Arthrospira plantensis* cultures grown outdoors in photobioreactors made with tubes of different diameters

Tube internal diameter (cm)	Litre/m (1)	Litre/m^2 (2)	Mean areal density (g/m^2) (2)	S/V (m^{-1}) (2)	Optimal biomass concentration (g/l) (2)	Mean net volumetric productivity (g/l/d)	Mean areal yield ($g/m^2/d$) (2)
14.0	15.4	110	50–70	9	0.4–0.6	0.20	22
13.1	13.3	102	60–80	10	0.6–0.8	0.23	23
7.4	4.3	58	70–80	17	1.2–1.4	0.40	23
5.0	1.96	39	70–90	25	1.4–2.0	0.65	25
2.5	0.49	19.6	70–120	50	3.5–6.0	1.4	27

(1) Volume per linear meter; (2) Surface area refers to the ground occupied by the tubes calculated considering the sum of diameters

the diameter entails a higher areal density, to which corresponds an increase in the areal yield. This is probably due to a better light-dark cycle realized in tubes with shorter optical light path (Table 1). The effect of the tube diameter of a PBR on microalgal growth was investigated by Kobayashi and Fujita (1997). The volumetric productivity increased as the tube diameter decreased from 5-cm to 1.6-cm. However, the biomass yield per m^2 of PBR increased with the diameter demonstrating that the greater volumetric productivity reached with a shorter tube diameter was not high enough to compensate for the greater culture volume attained per unit of surface with a larger diameter PBRs. These authors concluded that tubes narrower than 5 cm are inadequate for mass culture, because their lesser volume reduces the areal yield and the greater drop in pressure (Kobayashi and Fujita 1997). It is important to point out, that PBRs made of small diameter tubes creating long loops can be problematic due to higher head loss. A comparison between different tube diameters are tricky since they are affected by several factors, namely the mixing rate (Reynold number), the O_2 and CO_2 concentrations, the number of passages of the culture through the circulating device, and the temperature profile, which may shade the effect of a better light–dark cycle attainable in a shorter light path. After many years of experiences carried at the author's lab as well as by the group of Prof. Molina Grima at the Department of Chemical Engineering of the University of Almeria (Spain), tube diameter within the range of 5–9 cm has been found to be suitable for promoting adequate light to dark cycle frequencies and a limited energy consumption (Torzillo 1997; Brindley et al. 2004).

The applicable tube length is limited by the oxygen accumulation in the PBR loop. In *Arthrospira* cultures, a concentration of oxygen above 30 mg l^{-1} has a negative effect on both growth and biomass protein content (Torzillo et al. 1986). For example, in PBRs made of tubes with an internal diameter of about 5 cm, in well-growing *Arthrospira* cultures, the oxygen concentration can increase at a rate of 2–3 mg l^{-1} min^{-1}. This can result in an oxygen concentration of up to 70–80 mg l^{-1} at midday even with a gas exchange every 50 s and high turbulence rate (Torzillo et al. 1998). In Fig. 1, daily courses of growth (dry weight increase) are reported as that were measured at the optimal temperature of 35 °C different times of the day in a culture of *Arthrospira* cultured outdoors in a tubular PBR and exposed to low (20 mg l^{-1}) and high (60–80 mg l^{-1}) oxygen concentrations. The reduction in productivity was about 37 % under high-oxygen concentration at the optimal temperature of 35 °C, and increased to 47 % in cultures grown at 25 °C (not shown). The situation became much worse if a suboptimal temperature was superimposed on a high oxygen concentration as the reduction in productivity reached 63 % compared to optimal culture conditions (Fig. 2). The combination of high oxygen and low temperature in outdoor cultures can occur at the beginning of the cold season, when the culture temperature drops greatly below the optimum but the irradiance is high enough to drive photosynthesis. Indeed, it is common that light fluctuations occur within a range of 1–2 h, while the increase in temperature is a slower process and takes about 4–5 h. This kind of de-synchronization between the two most important environmental factors affecting photosynthesis and growth outdoors is found in a unique stress condition under which photoinhibition may indeed be induced at a relatively low light intensity (Vonshak et al. 2001).

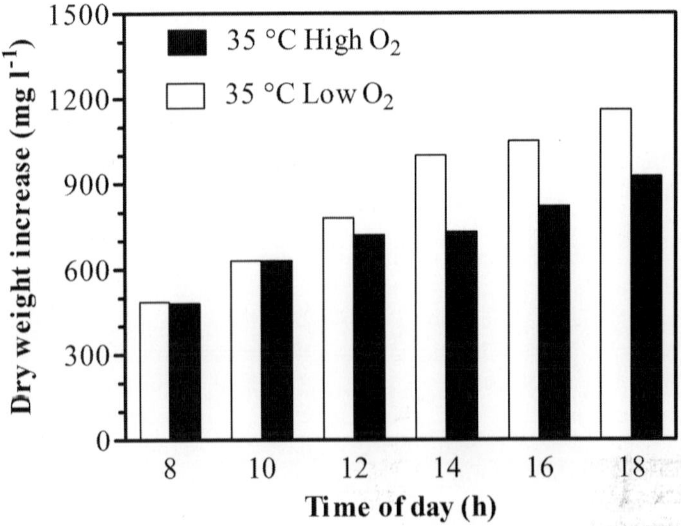

Fig. 1 Effect of oxygen concentration on growth of outdoor *A. platensis* cultures. Cultures were grown in tubular photobioreactors at the optimal temperature of 35 °C and exposed to different oxygen concentrations. In low oxygen cultures, the concentration of O_2 was maintained within 20 mg l^{-1}, by bubbling pure nitrogen in the reactor degasser; while, in high oxygen cultures the O_2 concentration was not controlled therefore it increased up to 70–80 mg l^{-1} in the middle of day as result of the photosynthetic activity

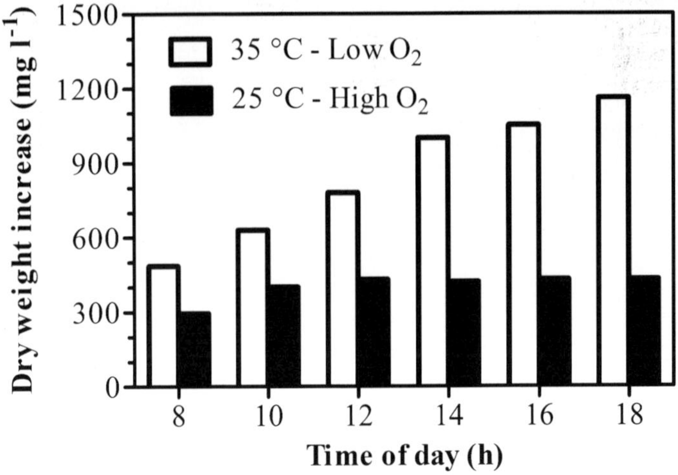

Fig. 2 Effect of combination of high oxygen concentration and suboptimal temperature on growth of cultures of *A. platensis* cultivated outdoors in tubular photobioreactors. Cultures were grown at the optimal temperature of 35 °C and at 25 °C and exposed to high and low oxygen concentrations as described in Fig. 1

2.2 Mixing

Mixing of the culture is necessary (1) to ensure that all the cells are regularly exposed to light, (2) to maintain nutrient supply throughout the PBR, and (3) to diminish the nutritional and gaseous gradients surrounding the cells in actively growing cultures, thus improving the rate of exclusion of cell excretions, including oxygen, on the cell surface Richmond (2013). These parameters are highly interdependent, and in many cases it is difficult to clearly distinguish the effect of any single one of them on culture performance. Indeed, one of the key problems related to illuminating microalgal cultures is that light is not a miscible nutrient, but rather a spatial external physics-dependent and self-distributing element. Light penetration in dense microalgae cultures is subjected to rapid attenuation due to scattering and absorption from the cells (Cornet et al. 1992), thus partitioning off the culture into several compartments with two extremes: (i) a highly illuminated external microalgal layer in which the cultures are very often subjected to light intensity greater than that required for saturation, and (ii) an inner one in which the cells are in the dark. Therefore, mixing the culture represents the most practical way to achieve a more homogeneous exposure of the cells to light.

As pointed out by Grobbelaar (1991) microalgal growth could be influenced by three ranges of fluctuating light/dark cycles, i.e., (1) high frequency fluctuations of 100 ms (10 Hz) and less, (2) medium frequency fluctuation of seconds to minutes, and (3) low frequency cycles of hours to days and years. The first range (>10 Hz) give rise to "flashing-light effect" (Kok 1953; Terry 1986; Nedbal et al. 1986), whereby the efficiency of photosynthesis is increased under specific conditions of illumination. Light fluctuations of medium frequency are particularly prevalent in turbulent mass cultures as well as in surface waters of natural systems. A first attempt to assist microalgal biotechnologists in exploring the relation of turbulent flow to intermittent lighting of *Chlorella* was made by Ippen (Burlew 1953) (Fig. 3). According to his calculations, even with tube diameters as small as 2.54 cm and with a high velocity of 0.7–0.8 m s^{-1}, it is not possible to achieve the required frequencies for flashing light effect. With the most common tube diameter (5–10 cm) and with a velocity of 0.3 m s^{-1}, the frequency of light-dark cycling ranges between 0.083 and 0.042 Hz, which is very far from the minimum frequency required in order to obtain even a partial benefit from a flashing effect with (1 Hz) (Kok 1953; Laws et al. 1983; Terry 1986; Molina et al. 2000). Moreover, it must be pointed out that mixing represents an important item in biomass production costs (about 30 % of the total) (Norsker et al. 2011), and therefore that attempts to increase culture velocity results in a sharp increase in the head loss, since they increase according to the square of the culture velocity. Power consumption for culture mixing ranges from about 100 W m^{-3}, for airlift-driven configuration, to about 500 W m^{-3} for the pumps (Acién Fernández et al. 2013).

The flow pattern within the tube lumen can be regarded as a plug flow with minimal backward and forward mixing. Therefore, considerable spatial gradients of O_2 and CO_2 along the axis may occur and can gain importance if the length of the tubes

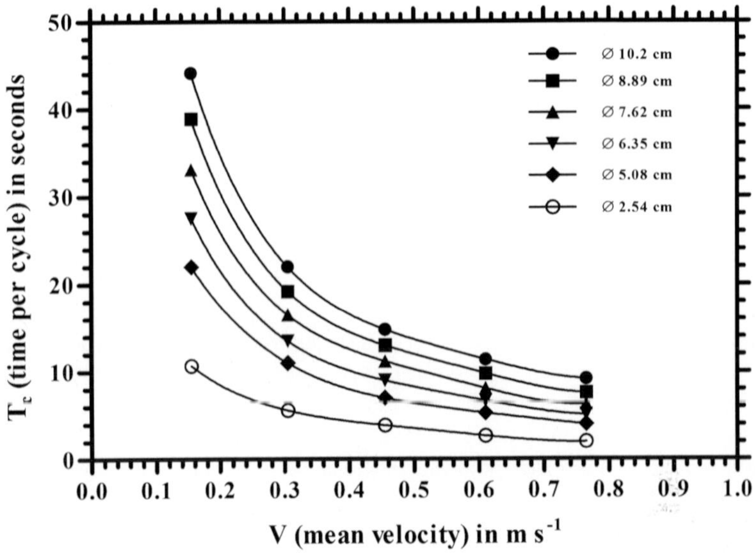

Fig. 3 Time cycles from the top surface to the bottom of the tube and back to the surface of microalgal cells as a function of liquid velocity and tube diameter (Adapted from Burlew 1953)

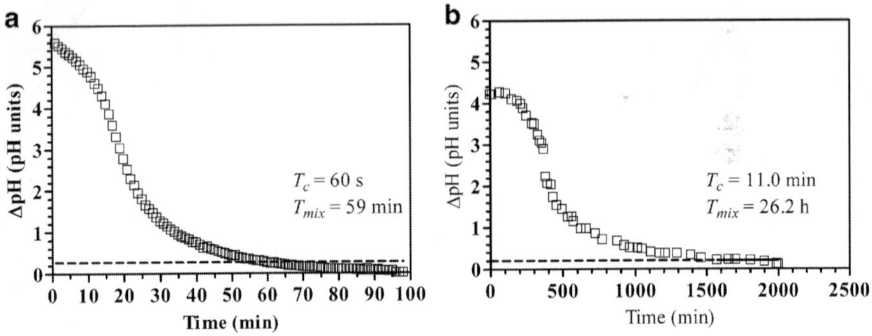

Fig. 4 Time course of pH response to the injection of a concentrated HCl solution at the inlet of the tubular photobioreactor. (**a**) 50 l (5 cm i.d., 23.5 m long circuit), and (**b**) 110 l (3 cm i.d., 133 m long circuit). *Dashed line* indicates 0.05 $\Delta pH(t_0)$ used for calculation of mixing time

is increased, and may affect the microalgal growth (Dillschneider and Posten 2013). Measuring the mixing time can provide useful information about nutrient homogeneity within the culture. Mixing time can be defined as the time necessary to achieve a homogeneous mixture after the injection of a tracer solution. It is usually measured by a signal-response method that uses a tracer and pH electrode (Camacho Rubio et al. 1999; Preuvost et al. 2006; Torzillo et al. 2014). In Fig. 4 are shown two examples of mixing time measurements carried out in a 50-l tubular PBR (Fig. 4a) and a 110-l tubular PBR (Fig. 4b). Because of the reduced tube diameter (i.e., from

5-cm and 3-cm, respectively), just a 2-fold increase in the PBR volume, results in a 26-fold increase in the mixing time. The increase in mixing time is clearly correlated to the increase in the length of the tubular PBR, a factor that must be taken into account in the scale-up of the PBR in order to guarantee a sufficient nutrient homogeneity distribution within the PBR.

2.3 Scale-Up Criteria

In biotechnological processes, scale-up criteria are usually based on a determination of the relationship between the culture parameters and fluid patterns inside the PBRs, which, of course, do not scale linearly with the PBR size. For example, while culture mixing in laboratory-scale PBR may be able to ensure that cells move along the light gradient within the same time scale as certain biochemical processes (i.e. the PQ pool turnover, ~10 ms), this does not hold for large scale PBRs used to cultivate photosynthetic microorganisms. Therefore, it seems conceivable that kinetic models obtained in small-scale PBRs might be very far from the reality of what occurs in larger scale PBRs, where longer mixing times are involved and culture homogeneity is difficult to achieve. Mixing time is certainly a key parameter to be taken into account when scaling up stirred tank PBRs. However, as discussed in the previous section, the increased mixing time is clearly correlated to an increase in the length of the tubular PBR, a factor that must be considered if this parameter is used for PBR scale up. Moreover, as previously stated, there is no guarantee that keeping the mixing time constant will not affect the performance of a scaled-up PBR (Oncel and Sabankay 2012).

Molina Grima et al. (2000), proposed a scale-up criteria of tubular PBRs based on the increase of tube diameter, provided that the light/dark frequency between light and dark zones of the culture in the tube lumen is kept within 1–1.5 Hz. Their findings confirmed that the benefit on yield by applying flashing light is higher in cultures exposed to higher light intensities. However, above those frequencies that were achieved by increasing the speed, productivity was not affected. This indicates that turbulence is a poor mechanism for creating an intermittent effect in cultures growing in lighted tubes, since the vertical motion of cells is a result of a random mixing pattern induced by the culture speed. The authors reached another important conclusion, namely that the maximum scalable tube diameter is about 10 cm. Larger diameters would require unrealistically high culture speeds (>2 m s^{-1}), which could result in damage to the cells, and would require that the PBR be constructed with very resistant material, one able to resist the highest mechanical stress. Therefore, the use of plastic materials, which are normally cheaper compared to glass tubes, would have to be excluded. Sensitivity of the organism to shear stress represents one of the biggest constraints for the scaled-up PBR, since it influences the choice of the circulating device and poses a limit on the circulation speed of the culture, which in turn affects the design of the PBR (e.g. the length of the tubes), as well as some operational factors such as: (1) turbulence degree, (2) oxygen build-up, and (3) CO_2 supply to the culture (Torzillo et al. 2003; Acién Fernández et al. 2013).

According to Amanullah et al. (2004), an important and non-negligible correlation between turbulence intensity and the microorganisms' shape can be expected, especially when said microorganisms grow to sizes near that of the turbulent Kolmogorov eddy scale (η) expected for the system. The size of the smallest eddies inside the system for highly turbulent flows can be calculated according to Kresta and Brodkey (2004).

$$\eta = (\nu^3/\varepsilon)^{1/4} \quad (1)$$

where ν ($m^2 s^{-1}$) is the kinematic viscosity and ε ($m^2 s^{-3}$) is the rate of energy dissipation. Being η (m) the smallest eddy size for a given culture vessel, its calculation is useful in biotechnology to asses potential cell damage deriving from the interaction between such eddies and the cells. When the Kolmogorov scale (η) falls within the same scale of magnitude as the cell size, the energy entrained in the eddies can be entirely transferred as shear stress on cell walls. Culture grown in tubular PBRs may be subjected to shear stress mainly during the passage of cells through the pump rather than the tubes themselves. High shear stress may occur at the turbine tips, therefore keeping the distance between the blades and casing relatively wide (0.5 cm) is recommended. Experiments carried out on a *Chlamydomonas reinhardtii* culture grown in a microcosm stirred with a spinning plate (Leupold et al. 2013) have shown no negative effects on photosynthetic activity with a tip speed up to 49 cm s^{-1}, but a further increase in the tip speed to 203 cm s^{-1} caused a continuous decline in photosynthetic activity by 15 % compared to unstirred cultures. However, quantitative studies are necessary to better characterize the shear sensitivity of *C. reinhardtii*, particularly during the sulfur starvation process when the organism's ability to repair the damage caused by the pump may be significantly reduced. The effects of hydrodynamic stress due to centrifugal and air-lift pumps and nozzles on two model microalgae, *Chlorella vulgaris* and *Scenedesmus dimorphus* was studied by Scarsella et al. (2012). They concluded that *C. vulgaris* appear to be less resistant to shear than *S. dimorphus*. The use of centrifugal pumps for culture recycling affected the growth of *C. vulgaris,* while nozzles effects were limited to aggregate breakage of coenobes of *S. dimorphus*.

Oxygen accumulation represents a major obstacle in the scale-up of tubular PBRs by increasing the tube length. The oxygen production of a microalgal culture can be calculated according to the following equation (Pirt et al. 1983).

$$O_2\left(g\,l^{-1}\right) = \left(\mu X\,Tc\right)/\,Y_0 \quad (2)$$

Where μ is the growth rate of the culture (h^{-1}), X is the dry weight ($g\ l^{-1}$), Tc is the time cycle of the PBR (h), that is, the time interval between two degassing stations, Y_0 (dimensionless) is the ratio between the biomass synthesized and the oxygen released. For *Arthrospira* grown with nitrate as the nitrogen source, Y_0=0.507 (i.e.,

Table 2 Guidelines for an optimal tubular photobioreactor design

Challenges	Possible workarounds	References
Uniform illumination of the culture	In general, illuminating cultures from both sides of the PBR is preferable to illuminating them from just one side	Giannelli and Torzillo (2012); Scoma et al. (2012)
Low dark-to-illuminated culture volume ratio	It is advisable that the ratio of dark to total culture volume should be kept as small as possible ($\leq$0.05)	Perner-Nochta and Posten (2007)
Low O_2 partial pressure	It is advisable to reduce the O_2 in the reactor as much as possible since it inhibits growth	Torzillo et al. (1986); Torzillo et al. (1998)
High light surface to volume ratio (S/V)	Shorter light paths are preferable since cells need less time to go back and forth between the photic and dark volume of the reactor thus improving the light-dark regime	Richmond (2013); Hu et al. (1996)
High ratio between illuminated area and ground area occupied by the reactor (A_R/A_G)	This ratio gives an indication of the "*dilution factor*" that can be sensed by the reactor. In principle, the mean light irradiance recorded on a horizontal surface should be reduced by a factor corresponding to the (A_R/A_G) ratio, so that the average light seen by the cells should fall in the linear part of the photosynthesis to light irradiance curve (P/I) of the strain	Posten (2009); Wijffels and Barbosa (2010)
Orientation of vertical PBRs	It depends on latitude. East/West oriented flat plate and vertical arranged (fenced) tubular reactors intercept more light than North/South for latitudes above 35 °N, while at lower latitudes the result is the opposite	Cuaresma et al. (2011); Hu et al. (1996)
Reduced mixing time	Long mixing times may also cause excessive build-up of O_2 in the reactor and inhibit growth. They may occur in long tubular reactors. Shorter mixing times are more easily attainable in flat PBRs	Giannelli and Torzillo (2012); Oncel and Sabankay (2012)
Turbulent mixing	In order to reach turbulent flow, the Reynolds number should be over 4000 (e.g. tubular reactors). Application of CFD to optimize the mixing has great potential for the design of a well-mixed PBR	Pruvost et al. (2006); Gang et al. (2009)

1.97 g of O_2 are released per g of biomass synthesized). From Eq. 2 it is possible to calculate the maximum permissible tube length for a given culture velocity, which allows to maintain the level of oxygen below the inhibition threshold for the culture. The experience gained by the author and well as by others (Molina et al. 1999) indicated that the optimal length for tube diameters within 5 and 10 cm ranges between 80 and 120 m. This distance seems also adequate to minimize the CO_2 loss (Molina et al. 1999). Table 2 summarizes the main guidelines for an optimal tubular PBR design.

2.4 Culture Contamination in Tubular Photobioreactors

Large scale microalgae cultures like terrestrial crops can be attacked by pests and weeds with devastatating effects on culture. Since the start of microalgal biotechnology at the Carnegie Institution of Washington (Burlew 1953), the choice of closed system was based on the conviction that it would allow easy cleaning and prevent contamination. However, a number of recent evidences demonstrated that this conviction is only partially true (Hoffman et al. 2008; Forehead and O'Kelly 2013; Carney and Lane 2014). In fact, cultures grown in closed systems are usually affected by contaminant in spite of their protection from outside atmosphere. Indeed, it has been found that in many cases the vehicle of the contamination is represented by the water used for medium preparation (Wang et al. 2013). Surprisingly, in some cases the contamination is higher in tubular system than in open ponds. For example the cyanobacterium *Leptolyngbya*, one of the most common contaminants found in *Arthrospira* cultures in tubular systems, is highly adhesive, resulting in rapid fouling formation on the internal tube surface, which prevents light penetration, and causing frequent shout-down of the PBR for mechanical cleaning and sterilization (Torzillo 1980). This problem can become particularly serious with aged plastic tubes which reduced smoothness. For PBRs designed as a loop removal of the biofilm can be achieved by introducing a spongy of the same diameter in the tubes which is pushed by the culture flow. This technique has been demonstrated very effective to remove biofilm of *Leptolyngbya* which is then collected at the exit of the circuit (Torzillo 1980).

Because microalgae biomass production is usually carried out in monocultures and selected few microalgal species are in wide scale application, there is an increased potential for parasites to have a devastating effect on monoculture. As commercial scale production continues to expand with a widening variety of applications, including biofuel, food and pharmaceuticals, the parasites accompanying microalgae will become of greater interest and potential economic impact. A number of important microalgal parasites have been identified in mass culture systems in the last few years and this number is to grow as the number of commercial microalgae ventures increases (Carney and Lane 2014). For example, some species of fungi belonging to class of *Chytrids* can attack cultures of microalgae such as *Chlorella* and *Haematococcus* (Hoffman et al. 2008) and cause the loss of the culture. Some parasites are difficult to eliminate even with aggressive pesticide since in their cycle include the formation of spores. Recently, the use of pulsed electric fields has been proposed for the control of predators in industrial scale microalgae (Rego et al. 2015). Routine monitoring and early detection of pest species is the most important requirement for large-scale cultivation. Knowledge gained from long-term experience operations may allow for the rapid identification of common pest species and the environmental conditions, which promote their presence. When possible molecular-based techniques are the most reliable and sensitive for the purpose of identifying which parasite may be present using Sanger, shotgun or next generation sequencing and then monitoring for these specifically using qPCR or phylochip technology (Carney and Lane 2014). FlowCAM, that is, a flow-cytometer that employ continuous digital imaging to measure the number, size and shape of microscopic particles in a fluid medium to provide warning signal of ciliate contamination in *Nannochloropsis* cultures have been proposed (Day et al. 2012). To accomplish mitigation for any culture system it is required: (1) information on the means by which a contaminant enters in the PBR (vector); (2) the amount (rate) of contaminant entry through the vector (i.e., inoculum size), the time (lag phase) it takes for a contaminant to be detected under light microscopic. Mathematical models that can predict the behavior of the contaminant in a culture have been proposed by Forehead and O'Kelly (2013). Despite a paucity of information on the economic impact of parasitism on the nascent algae biomass industry, there is a general consensus that parasites constitute a serious threat to microalgae industrial development. At this stage of knowledge's, in many cases, the easiest solution and the least technologically demanding and least satisfying response to parasite contamination is the salvage harvest, which of course, must be done upon detection of a parasite species and prior to serious loss of biomass. The success of salvage harvest is dependent on both the early detection and quantification of the contaminating parasite and the operator experience necessary to find the optimal compromise between maximizing of biomass yield against the potential catastrophic loss.

3 Types of Tubular Photobioreactors

This paragraph is subdivided according to tubular PBR design and for each design we describe those PBRs that are widely used at research level or pilot scale, however some examples of the world's largest commercial plants in operation are also presented.

3.1 Serpentine Photobioreactors

Serpentine PBRs are the oldest closed system tested (Burlew 1953). They consist of straight tubes connected by U-bends to form a flat loop (the photostage) that may be arranged either vertically or horizontally (Chini Zittelli et al. 2013). Gas exchange and nutrient addition normally take place in a separate vessel and culture circulation (at flow rates between 20 and 30 cm s^{-1}) is achieved by a pump or an airlift (Tredici et al. 2010). Torzillo et al. (1986), developed a tubular photobioreactor for the production of *Arthrospira*. It consisted of flexible polyethylene sleeves 14 cm in diameter and 0.3 mm think connected each other by PVC (polyvinylchloride) bends to form a loop, each bend incorporating a narrow tube for oxygen release. The pilot plant was composed of two units (500 m long, volume 8 m^3, surface area of 80 m^2). A similar design of tubular photobioreactor for mass culture of *Portphyridium* was also developed by Gudin and Chaumont (1983). The loop was made of polyethylene tubes of 6.4 cm diameter and 1500 m length. The temperature control of the culture was achieved by either floating or submerging the upper layer on or in a pool filled with water by adjusting the amount of air in the tubes of the lower layer.

In the recent years, PBRs of this type are being operated by Molina Grima and co-workers in Spain (Molina Grima 1999). The systems consist of a tubular photostage, made of Plexiglas tubes (from 2.6 to 5 cm in diameter) joined to form about 100-m long horizontal loop, which is connected to a 3–3.5-m high airlift where dissolved oxygen, pH and temperature probes are inserted. The Spanish group (Almeria) has carried out extensive research activity with serpentine PBRs on the influence of the principal variables that regulate growth of photosynthetic cells, among which average irradiance, gas–liquid mass transfer, temperature control, fluid dynamics and mixing (Acién Fernández et al. 1997, 1998; Sánchez Mirón et al. 1999; Molina Grima et al. 1999).

Following the design first developed by Torzillo et al. (1993), a two-layer 4,000-l horizontal tubular PBR, made of 10-cm diameter Plexiglas tubes connected by U-joints to form a single 400-m long loop, has been tested at the University of Almeria and set up by Cajamar in a greenhouse near Almeria. The pilot scale PBR has been used for production of lutein-rich biomass of *Scenedesmus almeriensis* achieving an annual average lutein productivity of 2.9 kg ha^{-1} d^{-1} (Fernández Sevilla et al. 2010). The system has been recently redesigned by the Department of Chemical Engineering of the University of Almeria and an industrial size plant is now composed of ten 2.8 m^3 vertical serpentine units (Fig. 5). Each unit occupies a surface area of about 50 m^2 and consists of 400-m long tube, 0.09-m in diameter Plexiglas tubes running in a fence-like structure. The loop outlet of each solar receiver is connected to a 3.5-m high and 0.4 m in diameter bubble column connected to the inlet of the loop, for degassing and heat exchange (Acién Fernández et al. 2013; Fernández et al. 2014). The excess of oxygen is removed in the column by constant air bubbling and the culture temperature controlled by passing tap water from a close farmer reservoir through a stainless steel heat exchanger inserted within the bubble column (Molina Grima personal communication). The microalgal culture is continuously re-circulated between the loop and the column by a centrifugal pump located at the bottom of the column and the pH is controlled by on-demand injection of pure CO_2. The culture is harvested as an overflow at the top of the column when fresh medium is poured into the bubble column. A mean volumetric productivity of 0.43 g l^{-1} d^{-1} (corresponding to a total annual production of 4.6 t) was attained with *Nannochloropsis gaditana* (Molina Grima et al. 2013). The system is fully automatized and can be operated by a single worker. This technology consumes significant energy for liquid circulation through the tubes, required for maintaining the proper fluid dynamics conditions and to overcome the pressure drop due to frictional losses within the tubes. To a lesser extent, energy is spent in aeration, harvesting, and water recycling. The total amount of energy consumed for biomass production was estimated about 600 MJ d^{-1} (Molina Grima et al. 2013).

Important innovations in terms of mixing and mass transfer were made in the "windy, wavy and wiped" tubular serpentine PBR (www PBR), a two-phase flow

Fig. 5 Vertical tubular serpentine photobioreactor developed by the Department of Chemical Engineering of the University of Almeria (Spain) and operated in a 30 m^3 plant installed at the Estacíon Esperimental de Cajamar "Las Palmerillas" (Almeria) (With permission of Fundacíon CAJAMAR, Spain)

vertical serpentine especially designed by Microphyt SAS (Baillargues, France) to cultivate fragile and/or slow growing microalgal species (Muller-Feuga et al. 2012; Chini Zittelli et al. 2013). The www PBR design forms a 50 m long tubular fence, consisting of 12 double layers of tubes with a total loop of 1,200 m made by connecting flanged glass tubes, 8 m long and 8 cm in external diameter, and U-bends by means of silicone seals and clamps. The loop outlet is connected to a 3 m high downcomer tube connected to the inlet of the loop. The whole PBR has a vertical height of 3 m and a width of 0.3 m, its total volume is 5.4 m^3 and its footprint area 48 m^2 (Muller-Feuga et al. 2012).

A low pressure pumping device circulates the culture suspension through the loop at a velocity of about 0.3 m s^{-1}. Air is injected continuously at the lowermost point of the loop, just downstream of the circulation pump, to achieve a mean gas velocity of 1.0 m s^{-1} with the result that the two fluids circulate co-currently with a differential velocity (wind velocity) of 0.7 m s^{-1}. According to the concept of this two-phase flow PBR, the whole length of the piping between inlet and outlet was the site of mass transfer between the circulating gas and liquid. Compared to similar two-phase flow PBRs (i.e., Near-horizontal manifold type PBR) (Tredici 2004), the www TPBR shows a significantly higher differential velocity between gas and liquid, a key factor for high mass transfer (Babcock et al. 2002; Muller-Feuga et al. 2012). The flow conditions of the www PBR proved satisfactory as neither harmful speed variation nor small bubble generation was observed. Oxygen stripping proved to be efficient as the concentration of dissolved oxygen never exceeded twice the saturation with air (200 %) at peak sunlight (Muller-Feuga et al. 2012). The www TPBR are being operated in Baillargues (south of France), inside a greenhouse, to produce *Neochloris oleoabundans*, *Porphyridium cruentum*, and other fragile species (Muller-Feuga et al. 2012). The www PBR technology can be particularly addressed to the commercial production of microalgae biomass for markets with the highest income such as aquaculture, cosmetics, and nutraceuticals.

Cellana LLC (Kailua-Kona, Hawaii, USA) continuing the activity of Aquasearch, Inc., and Mera Pharmaceuticals, Inc., utilizes a 25,000 l horizontal serpentine PBR made of large (38 cm in diameter), low-density polyethylene tubing for culturing the green stage of *Haematococcus. pluvialis* (Olaizola 2003; Huntley and Redalje 2007). Temperature is controlled by immersion of the PBR in a water pond. The culture grown in the PBR is used to inoculate raceway ponds in which astaxanthin and oil accumulation is induced (Huntley and Redalje 2007).

A horizontal serpentine PBR cooled by immersion in a water pool is used to cultivate marine microalgae by Fitoplancton Marino SL (Cadiz, Spain). The company supplies lyophilized biomass and slurries of several microalgae for aquaculture use (http://www.fitoplanctonmarino.com/ accessed 13 March 2015).

3.2 *Manifold Photobioreactors*

In manifold PBRs, a series of parallel tubes is connected at the ends by two manifolds, one for distribution and the other for collection of the culture suspension. The main advantages of these systems over serpentine loop PBRs are the reduction of head losses and lower oxygen concentrations (due to the distance of suspension flow between the tube inlet and degasser), two factors that facilitate scale-up to industrial size (Tredici et al. 2010).

Richmond and co-workers (1993), for the first time, devised a system made of parallel sets of 20-m-long tubes connected by manifolds in which the culture was circulated by an airlift. Productivities of 0.55 g l^{-1} d^{-1} were attained with *Arthospira*. Tredici and co-workers developed the Near-Horizontal Tubular Reactor (NHTR), with tubes inclined from 5 to 20° to the horizontal so that mixing could be achieved by air bubbling (Tredici and Chini Zittelli 1998; Tredici et al. 2010). Besides simplicity of operation and low maintenance cost, this internal gas exchange PBR has other advantages among which low shear stress and reduced wall fouling due to the scouring effect of the gas bubbles. The length of tubes is, however, limited to about 40 m due to reduced mass transfer (Babcock et al. 2002).

A vertical manifold PBR known as BioFence was developed by Applied Photosynthetics Limited (APL) (Manchester, UK) in the late 1990s. The BioFence consists of an array of rigid or flexible transparent tubes racked together in banks and connected by manifolds in a fence-like structure. The culture suspension is circulated between the photostage and a holding tank by a centrifugal pump or by an airlift (Tredici et al. 2010). BioFence systems are currently commercialized by Varicon Aqua Solutions Ltd. (UK) which offers several sizes of modular PBRs that are used as small cultivation units for research, and microalgae production for aquaculture, cosmetic and nutraceutical sectors (http://www.variconaqua.com; accessed 9 March 2015),

Industrial-scale plants based on vertically arranged manifold PBRs are operated by Roquette Klötze GmbH & Co. KG since 2000 (Germany) (www.algomed.de) for the production of a dozen of microalgae like *Chlorella, Spirulina, Nannochloropsis,* sold in the markets for dietary supplements, food and cosmetic, by Salata GmbH (Ritschenhausen, Germany), which produces various microalgae for the cosmetic, food and pharmaceutical markets, and by Algatechnologies Ltd (Ketura, Israel) (www.algatech.com) for astaxanthin production from *Haematococcus pluvialis.* In the first two facilities, the tubular PBR modules were designed and installed by IGV GmbH (Pulz et al. 2013). At Roquette Klötze the microalgae are grown in a greenhouse covering 1.2 ha with a total cultivation volume of around 600 m^3, distributed in 500 km of tubes (Fig. 6). It is one of the largest microalgae production plants in Europe with an annual autotrophic production in Klötze of 30–50 t y^{-1} (Ullmann, personal communication).

A similar PBR has been put in operation at Ritschenhausen (Germany) by Salata GmbH. In modules of sizes ranging from 15,000 to 42,000 l (Waldeck 2012), different microalgae and cyanobacteria are cultured with productivities in the range of 0.2–0.8 g l^{-1} d^{-1} (Pulz et al. 2013). The largest PBR developed by IGV-GmbH has been built in 2011 in Jerez (southern Spain) and is operated by Biotechnologia des microalgas (BTM). This is a single unit PBR of 85 m^3 volume composed of 35 km glass tubes and occupies a footprint area of 1,000 m^2. The photostage of this type of PBR demands an investment of approximately € 220 per m^2 footprint (Pulz et al. 2013).

Algatechnologies Ltd carries out both the green and the red stage of *H. pluvialis* culture in vertical and horizontal manifold PBRs with 300 km long tubes in a 1.2-ha plant. Compared to the Roquette Klötze and Salata plants in Germany, productivity in Algatechnologies plant is much favored by the high solar radiation available in the Arava desert (Southern Israel) (Tredici et al. 2010).

Among commercial PBRs currently in operation, special attention is paid to the plant recently built at the Secil Cement Company in Pataias (Portugal) by AlgaFuel, S.A., a spin-off company from Necton S.A. The prototype unit is already operational and uses microalgae to capture the flue gases evolved from the Cement plant. The plant, one of the largest ever realized, consists of 19 modular PBRs, has a total volume of about 1,300 m^3, and occupies a 1-ha ground area. The photostage is made from acrylic tubes (330 km total length) placed horizontally in a fence-like structure to form vertical array walls (Verdelho 2012). The plant is now producing food grade *Chlorella vulgaris.* The downstream processing includes harvesting by ultrafiltration that concentrates the biomass to 5-10 % dry weight, pasteurization and spray drying (http://www.algaeindustrymagazine.com/secil-a4f-form-algafarm-jv/ accessed 14 March 2015). Investment cost for the whole plant (PBRs, control cabinet, biomass harvesting and storage, etc.) was of about 1800 € m^{-2} (Verdelho 2012).

Fig. 6 Tubular manifold photobioreactors installed at Roquette Klötze & Co. KG (Germany) (Photo ©: Jörg Ullmann (Courtesy of J. Ullmann))

3.3 Helical Photobioreactors

Helical PBRs consist of small-diameter flexible tubes wound around an upright supporting structure. This design was used in the 1950s to grow *Chlorella* and later adopted by Šetlík and collaborators (1967) and Jüttner (1982). A 120-l helical bubble PBR was used to grow *Anabaena siamensis* and *A. platensis* outdoors. With *A. platensis*, a mean volumetric productivity of 0.9 g l^{-1} d^{-1} and a photosynthetic efficiency (PE) of 6.6 % (PAR) were achieved (Tredici and Chini Zittelli 1998).

A *Biocoil* consisting of a photostage of polyethylene or PVC tubing (between 2.5 and 5 cm diameter) wound helically around a cylindrical support (8 m in height with a core diameter of 2 m) was devised and patented by Robinson and Morrison (1987). Several parallel bands of tubes were connected via manifolds to the pumping system, allowing more even flow and shorter tube length thus minimizing oxygen build-up (Tredici et al. 2010). A heat exchange or evaporative cooling provided temperature control. Several marine species and the cyanobacterium *Arthrospira* have been cultivated in 1000-l *Biocoil* in Perth, Western Australia (Borowitzka 1999). However, limited discussion of the *Biocoil* during recent years suggests that it no longer has substantial potential of use.

A conical, instead of cylindrical, framework has also been suggested, as it improves the spatial distribution of tubes for sunlight capture (Morita et al. 2000). Small conical biocoil units have been and are being operated mainly for experimental and teaching purposes (Morita et al. 2000; Travieso et al. 2001; Briassoulis et al. 2010; Raes et al. 2014). A 'Christmas Tree' shaped pilot PBR designed by GICON (www.gicon.de) is tested at the Anhalt University in Köthen (Germany) as recent implementation of this technology of microalgae cultivation.

Helical type systems have the great advantage to allow deployment of relatively long tubes on a small land area as compared to the other PBR categories. Still the cleaning problems are not easy to solve while the hydrodynamic stress vary from low to high depending on microalgae species. At present, no commercial application of this design is known.

3.4 Tubular PBRs for Special Applications

A tubular serpentine PBR that makes use of condensed solar radiation has been set-up at the Academic and University Centre of Nové Hrady (Czech Republic) (Masojídek et al. 2003). The PBR is based on solar concentrators (linear Fresnel lenses), mounted in a climate-controlled greenhouse, that focus the direct component of incident sunlight on the cultivation tubes. The cultivation unit (a 24-m long loop made of six parallel horizontal glass tubes connected by U-bends) is placed on a movable frame, which enables automatic focusing of direct solar irradiance. The PBR has been used to study acclimation of microalgae to supra-high solar irradiance, since irradiance values as much as 3.5 times the naturally available are

Fig. 7 Tubular serpentine photobioreactor based on solar concentrators – linear Fresnel lenses developed at the Center for Biotechnology, University of South Bohemia, Nové Hrady (Czech Republic) (Courtesy of Prof. J. Masojídek)

achieved (Masojídek et al. 2003). Based on this first design, an improved, two-stage PBR consisting of two vertical and two inclined units, was constructed and tested (Fig. 7) (Masojídek et al. 2009). With the novel 450-l PBR it is possible to carry out a two-stage process of biomass production: (i) growth of the selected microalga under lower irradiances in the vertical units, and (ii) exposure of the culture to supra-high irradiance in the inclined units. This process has been tested for the production of *H. pluvialis*: the green phase in vertical and the red phase in the inclined units (Masojídek et al. 2009). The system allows to cultivate microalgae at much higher irradiance than that naturally available, which may be useful at high latitudes, and exposes the culture at ultrahigh irradiance necessary to induce stress and achieve high-value product accumulation (pigments, superoxide-dismutase, etc.). The main limitation of the technology is the high cost of the PBR, about €1,000 m^{-2}, comprising glass tubes, Fresnel lenses, light tracking system, degasser, heat exchanger, pump, sensors, and control box (Masojídek et al. 2003).

To circumvent the problem of light saturation effect, which is actually a great drawback in microalgal biotechnology, Torzillo and co-workers have recently proposed a new tubular PBR design in which the culture tubes are arranged on an 8×8 square pitch cell connected by U-bends and immersed in a light scattering silica nanoparticle suspension (Giannelli and Torzillo 2012) (Fig. 8). Culture is circulated

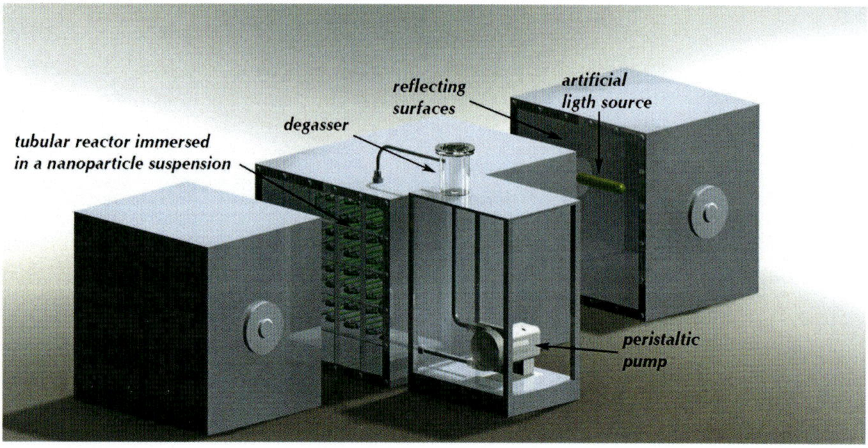

Fig. 8 General view of the 110 l tubular photobioreactor placed in a container filled with a light scattering nanoparticle suspension, which also used for culture thermoregulation. The reactor is made up of 8 tube layers connected by each other by U-bends to form a loop of 133 m long (Modified from Giannelli and Torzillo 2012)

with a peristaltic pump and illuminated with artificial light supplied by the light bulbs (2×2000 W) from opposite square sides of the PBR. A motorized Venetian mirror allows for the reduction of the light intensity when necessary. The system's main advance was the increase in the light dilution ratio, namely the ratio between the total surface of the tubes (i.e., PBR cross section) and that of the opposite transparent faces of the PBR whereby light was supplied to the culture. This ratio was about 4, meaning that the incident light can be reduced by an analogous factor. However, this dilution factor would mainly indicate a mere geometric reduction of incident light, if no scattering nanoparticles were suspended in the water bath, allowing a much more uniform distribution of light on the tube culture, and supplied at lower intensity. The latter is a condition for reducing light dissipation via fluorescence and non-photochemical quenching (heat). The PBR was tested for hydrogen production by means sulfur-deprived *C. reinhardtii* cultures. With this PBR design it was possible to obtain more than 3.12 l of pure H_2 per culture cycle, which was almost twice higher than that obtained per unit of volume with the 50-l horizontal PBR (Torzillo et al. 2014).

Among tubular PBRs experimented with at small-scale level special attention is recently devoted on floating systems, i.e., those PBRs that are deployed in water bodies and are anchored to piers or moorings to prevent them from drifting away with waves and currents (Chini Zittelli et al. 2013). The Offshore Membrane Enclosures for Growing Algae (OMEGA) PBR has been designed by NASA. It represents a system for culturing microalgae using wastewater in floating PBRs deployed in marine environments and thereby eliminating competition with agriculture for water, fertilizers, and land. A 110-l prototype of OMEGA system made of two LDPE floating tubular elements (3-m long and 0.11 m in diameter) connected to an external gas exchanging column (Wiley et al. 2013), was tested with *Chlorella vulgaris* for 23 days during April and May in California, reaching biomass yields of

4–21 g m^{-2} d^{-1} with an average of 14.1 g m^{-2} d^{-1} (Trent et al. 2012). No data on operative problems and PBR resistance are available. The influence of PBR material [polyurethane vs. low-density polyethylene (LLDPE)] and shape (rectangular vs. tubular) on biofouling and the influence of biofouling on microalgae productivity were investigated (Harris et al. 2013). LLDPE tubes had less biofouling and were easier to clean than all the other plastics tested. Biofouling accumulated primarily on wetted sides of the tubes and correlations between biomass accumulation, surface coverage, and light transmittance revealed that both thick and thin biofouling layers affect light transmittance, as does a harsh cleaning method. Twelve weeks of biofouling on LLDPE decreased microalgae productivity, suggesting the need for a cleaning cycle (Harris et al. 2013).

4 Concluding Remarks

The main drawbacks of tubular PBRs (and of PBR in general) are the limited possibility of being scaled-up, the high capital and operating costs and the negative energy balance (Lardon et al. 2009; Clarens et al. 2010; Borowitzka and Moheimani 2013; Tredici et al. 2015). However, large-scale microalgae cultivation in tubular systems is a well-established and reliable technology, able to allow a production process in highly controlled and automatic conditions compared to open systems like raceways. In recent years, large scale tubular PBRs have been constructed and operated in Germany and Israel for the production of *Chlorella* biomass and *Haematococcus,* respectively. However, to date, due to their high investment costs and energy requirements, these PBRs are suitable for production of high-value products for human nutrition, cosmetics and pharmaceutical applications, fine chemicals and industrial-scale inocula production, but not for low-value biocommodities, unless the investment cost and energy consumption for mixing is consistently reduced. Acién Fernández et al. (2013) estimated the production costs on facilities of ten 3-m^3 tubular PBRs operated in continuous mode for 2 years. They concluded that, although the yield of the facility was close to maximum expected for the location of Almería (Spain), the production cost of the *Scenedesmus almeriensis* biomass was not lower than 69 € kg^{-1}. Labor and depreciation costs were the major factors contributing to this high cost. The scale-up to a production capacity of 200 t y^{-1} should allow to reduce the production cost up to 12.6 € kg^{-1}.

Therefore, the future of microalgal biotechnology using closed systems (PBRs) is dependent, to a large extent, on three factors: (1) the ability to reduce the production costs and thus make microalgal biomass a commodity traded in large amounts, not limited to the health food market; (2) the development of cost-effective PBRs; (3) to attain a positive energy balance of the whole cultivation process.

Acknowledgements The research leading to these results has received funding from the European Union Seventh Framework Program (FP7/2007-2013) under grand agreement number 308518 (Cyanofactory).

List of Acronyms

A_R/A_G	ratio between illuminated area and ground area of the reactor
CFD	computational fluid dynamics
NHTR	near-horizontal tubular reactor
LLDPE	low-density polyethylene
OMEGA	offshore membrane enclosed growing algae
PBR	photobioreactor
PE	photosynthetic efficiency
PFD	photon flux density
PQ	plastoquinone
S/V	surface to volume ratio
SOD	superoxide-dismutase
wwwPBR	wind, wavy, wiped tubular photobioreactor

References

Acién Fernández FG, García Camacho F, Sánchez Pérez JA, Fernández Sevilla JM, Molina Grima E (1997) A model for light distribution and average solar irradiance inside outdoor tubular photobioreactors for the microalgal mass culture. Biotechnol Bioeng 55:701–714

Acién Fernández FG, García Camacho F, Sánchez Pérez JA, Fernández Sevilla JM, Molina Grima E (1998) Modelling of biomass productivity in tubular photobioreactors for microalgal cultures. Effects of dilution rate, tube diameter and solar irradiance. Biotechnol Bioeng 58:605–611

Acién Fernández FG, Fernández Sevilla JM, Molina Grima E (2013) Photobioreactors for the production of microalgae. Rev Environ Sci Biotechnol 12:131–151

Amanullah A, Buckland BC, Nienow AW (2004) Mixing in the fermentation and cell culture industries. In: Paul EL, Atiemo-Obeng VA, Kresta SM (eds) Handbook of industrial mixing. Wiley-Interscience, New York, pp 345–390

Babcock RW, Malda J, Radway JC (2002) Hydrodynamics and mass transfer in a tubular airlift photobioreactor. J Appl Phycol 14:169–184

Borowitzka MA (1999) Commercial production of microalgae: ponds, tanks, tubes and fermenters. J Biotechnol 70:313–321

Borowitzka MA, Moheimani NR (2013) Algae for biofuels and energy. Developments in applied phycology 5, Springer+Business Media Dordrecht, Dordrecht, p 285

Briassoulis D, Panagakis P, Chionidis M, Tzenos M, Lalos D, Tsinos A, Berberidis K, Jacobsen A (2010) An experimental helical–tubular photobioreactor for continuous production of *Nannochloropsis* sp. Bioresour Technol 101:6768–6777

Brindley C, Garcia-Malea MC, Aciér-Fernández FG, Fernández Sevilla JM, García Sánchez JL, Molina Grima E (2004) Influence of power supply in the feasibility of *Phaeodactlylum tricornutum* cultures. Biotechnol Bioeng 87:723–733

Burlew JS (1953) Algal culture from laboratory to pilot plant. Carnegie Institution of Washington Publication N. 600, Carnegie Institution, Washington, DC, pp 235–281

Camacho Rubio F, Acién Fernández FG, Sánchez Pérez JA, Garcia Camacho F, Molina Grima E (1999) Prediction of dissolved oxygen and carbon dioxide concentration profiles in tubular photobioreactors for microalgal culture. Biotechnol Bioeng 62:71–86

Carney LT, Lane TW (2014) Parasites in algal mass culture. Front Microbiol 5(278):1–8

Chini Zittelli G, Biondi N, Rodolfi L, Tredici MR (2013) Photobioreactors for mass production of microalgae. In: Richmond A, Hu Q (eds) Handbook of microalgal culture: applied phycology and biotechnology, 2nd edn. Wiley, Oxford, pp 225–266

Clarens AF, Resurreccion EP, White MA, Colosi LM (2010) Environmental life cycle comparison of algae to other bioenergy feedstocks. Environ Sci Technol 44:1813–1819

Cornet J, Dussap C, Duberteret G (1992) A structural model for simulation of cultures of the cyanobacterium *Spirulina platensis* in photobioreactors: I. Coupling between light transfer and growth kinetics. Biotechnol Bioeng 40:817–825

Cuaresma M, Janssen M, Vílchez C, Wijffels RH (2011) Horizontal or vertical photobioreactors? How to improve microalgae photosynthetic efficiency. Biores Technol 102:5129–5137

Day JG, Thomas NJ, Achilles-Day UEM, Leakey RJG (2012) Early detection of protozoan grazers in algal biofuel cultures. Biores Technol 114:715–719

Dillschneider R, Posten C (2013) Closed bioreactors as tools for microalgae production. In: Lee JW (ed) Advanced biofuels and bioproducts. Springer Science+Business Media, New York, pp 629–649

Fernández Sevilla JM, Acién Fernández FG, Molina Grima E (2010) Biotechnological production of lutein and its applications. Appl Microbiol Biotechnol 86:27–40

Fernández I, Acién Fernández FG, Berenguel M, Guzmán JL (2014) First principles model of a tubular photobioreactor for microalgal production. Ind Eng Chem Res 53:11121–11136

Forehead H, O'Kelly C (2013) Small doses, big troubles: modelling growth dynamics of organisms affecting microalgal production cultures in closed photobioreactors. Biores Technol 129:329–334

Gang Y, Li Y, Shen G, Wang W, Lin C, Wu H, Chen Z (2009) A novel method using CFD to optimize the inner structure parameters of flat photobioreactors. J Appl Phycol 21:719–727

Giannelli L, Torzillo G (2012) Hydrogen production with the microalga *Chlamydomonas reinhardtii* grown in a compact tubular photobioreactor immersed in a scattering nanoparticle suspension. Int J Hyd Energy 37:16951–16961

Grobbelaar JU (1991) The influence of light/dark cycle in mixed algal cultures on their productivity. Bioresour Technol 38:189–194

Gudin C, Chaumont D (1983) Solar biotechnology study and development of tubular solar receptors for controlled production of photosynthetic cellular biomass for methane production and specific exocellular biomass. In: Palz W, Pirrwitz D (eds) Energy from biomass. D. Reidel, Dordrecht, pp 184–193

Harris L, Tozzi S, Wiley P, Young C, Richardson Tra-My J, Clark K, Trent JD (2013) Potential impact of biofouling on the photobioreactors of the Offshore Membrane Enclosures for Growing Algae (OMEGA) system. Bioresour Technol 144:420–428

Hoffman Y, Aflalo C, Zarka A, Gutman J, James T, Boussiba S (2008) Isolation and characterization of a novel chytrid species (phylum Blastocladiomycota), parasitic on the green alga *Haematococcus*. Mycol Res 112:70–81

Hu Q, Guterman H, Richmond A (1996) A flat inclined modular photobioreactor for outdoor mass cultivation of photoautotrophs. Biotechnol Bioeng 51:51–60

Huntley ME, Redalje DG (2007) CO_2 mitigation and renewable oil from photosynthetic microbes: a new appraisal. Mitig Adapt Strat Glob Chang 12:73–608

Jüttner F (1982) Mass cultivation of microalgae and photosynthetic bacteria under sterile conditions. Proc Biochem 7:2–7

Kobayashi K, Fujita K (1997) Tube diameter on tubular photobioreactor for microalgal culture and its biomass. J Chem Eng Japan 30:339–341

Kok B (1953) Experiments on photosynthesis by *Chlorella* in flashing light. In: Burlew JS (ed) Algal culture from laboratory to pilot plant. Carnegie Institution of Washington Publication N. 600, Carnegie Institution, Washington, DC, pp 63–75

Kresta SM, Brodkey RS (2004) Turbulence in mixing applications. In: Paul EL, Atiemo-Obeng VA, Kresta SM (eds) Handbook of industrial mixing. Wiley-Interscience, New York, pp 345–390

Lardon L, Hélias A, Sialve B, Steyer JP, Bernard O (2009) Lyfe-cycle assessment of biodiesel production from microalgae. Environ Sci Technol 43:6475–6481

Laws EA, Terry KL, Wickman J, Chalup MS (1983) A simple algal production system designed to utilize the flashing light effect. Biotechnol Bioeng 25:2319–2335

Leupold M, Hindersin S, Gust G, Kerner M, Hanelt D (2013) Influence of mixing and shear stress on *Chlorella vulgaris*, *Scenedesmus obliquus*, and *Chlamydomonas reinhardtii*. J Appl Phycol 2:485–495

Masojídek J, Papáček S, Sergejevová M, Jirka V, Červený C, Kunc J, Korečko J, Verbovikova O, Kopecký J, Štys D, Torzillo G (2003) A closed solar photobioreactor for cultivation of microalgae under supra-high irradiance: basic design and performance. J Appl Phycol 15:239–248

Masojídek J, Sergejevová M, Rottnerová K, Jirka V, Korečko J, Kopecký J, Zaťková I, Torzillo G, Štys D (2009) A two-stage solar photobioreactor for cultivation of microalgae based on solar concentrators. J Appl Phycol 21:55–63

Molina Grima E, Acién Fernández FG, Camacho G, Chisti Y (1999) Photobioreactors: light regime, mass transfer, and scale-up. J Biotechnol 70:231–247

Molina Grima E, Acién Fernández FG, Camacho G, Camacho Rubio F, Chisti Y (2000) Scale-up of tubular photobioreactors. J Appl Phycol 12:355–368

Molina Grima E, Acién Fernández FG, Medina AR (2013) Downstream processing of cell mass and products. In: Richmond A, Hu Q (eds) Handbook of microalgal culture: applied phycology and biotechnology, 2nd edn. Wiley, Oxford, pp 267–309

Morita M, Watanabe Y, Saiki H (2000) Investigation of photobioreactor design for enhancing the photosynthetic productivity of microalgae. Biotechnol Bioeng 69:63–668

Muller-Feuga A, Lemar M, Vermel E, Pradelles R, Rimbaud L, Valiorgue P (2012) Appraisal of a horizontal two-phase flow photobioreactor for continuous production of delicate microalgae species. J Appl Phycol 24:349–355

Nedbal L, Tichý V, Xiong VF, Grobbelaar JU (1986) Microscopic green algae and cyanobacteria in high-frequency intermittent light. J Appl Phycol 8:325–333

Norsker NH, Barbosa MJ, Vermuë MH, Wijffels RH (2011) Microalgal production-A closer look at the economics. Biotechnol Adv 29:24–27

Olaizola M (2003) Commercial development of microalgal biotechnology: from the test tube to marketplace. Biomol Eng 20:459–466

Oncel S, Sabankay M (2012) Microalgal biohydrogen production considering light energy and mixing time as the two key features for scale-up. Bioresour Technol 121:228–234

Perner-Nochta I, Posten C (2007) Simulations of light intensity variation in photobioreactors. J Biotechnol 131:276–285

Pirt SJ, Kee YK, Walach MR, Pirt MW, Balyuzi HHM, Bazin ML (1983) A tubular bioreactor for photosynthetic production of biomass from carbon dioxide: design and performance. J Chem Tech Biotechnol 33B:35–58

Posten C (2009) Design principles of photo-bioreactors for cultivation of microalgae. Eng Life Sci 3:165–177

Pruvost J, Pottier L, Legrand J (2006) Numerical investigation of hydrodynamic and mixing conditions in a torus photobioreactor. Chem Eng Sci 61:4476–4489

Pulz O, Broneske J, Waldeck P (2013) IGV GmbH experience report, industrial production of microalgae under controlled conditions: innovative prospects. In: Richmond A, Hu Q (eds) Handbook of microalgal culture: applied phycology and biotechnology, 2nd edn. Wiley, Oxford, pp 445–460

Raes EJ, Isdepsky A, Muylaert K, Borowitzka MA, Moheimani NR (2014) Comparison of growth of *Tetraselmis* in a tubular photobioreactor (Biocoil) and a raceway pond. Appl Phycol 26:247–255

Rego A, Redondo LM, Geraldes V, Costa L, Navalho J, Pereira MT (2015) Control of predators in industrial scale microalgae cultures with pulsed electric fields. Bioelectrochemistry 103:60–64

Richmond A (2013) Biological principles of mass cultivation of phototrophic microalgae. In: Richmond A, Hu Q (eds) Handbook of microalgal culture, applied phycology and biotechnology. Wiley & Sons, Ltd. Published 2013 by Blackwell Publishing Ltd., Oxford, UK, pp 171–204

Richmond A, Hu Q (2013) Handbook of microalgal culture applied phycology and biotechnology, 2nd edn. Wiley Blackwell, Chichester, p 705

Richmond A, Boussiba S, Vonshak V, Kopel R (1993) A new tubular reactor for mass production of microalgae outdoors. J Appl Phycol 5:327–332

Robinson LF, Morrison AW (1987) Improvements relating to biomass production. EU Patent 0239272 A2

Sánchez Mirón A, Contreras Gómez A, García Camacho F, Molina Grima E, Chisti Y (1999) Comparative evaluation of compact photobioreactors for large-scale monoculture of microalgae. J Biotechnol 70:249–270

Scarsella M, Torzillo G, Cicci A, Belotti G, De Filippis P, Bravi M (2012) Mechanical stress tolerance of two microalgae. Proc Biochem 47:1603–1611

Scoma A, Giannelli L, Faraloni C, Torzillo G (2012) Outdoor H_2 production in a 50-L tubular photobioreactor by means of a sulfur-deprived culture of the microalga *Chlamydomonas reinhardtii*. J Biotechnol 157:620–627

Šetlík I, Komárek J, Prokeš B (1967) Short account of the activities from 1960 to 1965. In: Nečas J, Lhotský O (eds) Annual report of the laboratory of experimental algology and department of applied algology for the year 1966. Knihtisk, Prague, pp 5–36

Terry KL (1986) Photosynthesis in modulated light: quantitative dependence of photosynthetic enhancement on flashing rate. Biotechnol Bioeng 28:988–995

Torzillo G (1980) Two years of mass culture of *Spirulina maxima* in tubular system (in Italian). In: Materassi R (ed) Proceedings of conference "Perspectives of mass cultivation of *Spirulina* in Italy". Florence, Accademia dei Georgofili, 20–21 November 1980

Torzillo G (1997) Tubular reactors. In: Vonshak A (ed) *Spirulina platensis (Arthrospira)*, physiology, cell biology and biotechnology. Taylor & Francis, London, pp 101–115

Torzillo G, Pushparaj B, Bocci F, Balloni W, Materassi R, Florenzano G (1986) Production of *Spirulina* biomass in closed photobioreactors. Biomass 11:61–74

Torzillo G, Bocci F, Pushparaj B, Materassi R (1987) Studies on the production of *Spirulina* biomass through outdoor culture in tubular photobioreactors. In: Grasso G, Delmon B, Molle JF, Zibetta H (eds) Biomass for energy and industry. Proceedings of the international conference on biomass for energy and industry, Orléans, France, 11–15 May 1987. Elsevier, London/New York, pp 608–614

Torzillo G, Carlozzi P, Pushparaj B, Montaini E, Materassi R (1993) A two-plane tubular photobioreactor for outdoor culture of *Spirulina*. Biotechnol Bioeng 42:891–898

Torzillo G, Bernardini P, Masojidek J (1998) On-line monitoring of chlorophyll fluorescence to assess the extent of photoinhibition induced by high oxygen concentration and low temperature and its effect on the productivity of outdoor cultures of *Spirulina platensis* (cyanobacteria). J Phycol 34:504–510

Torzillo G, Pushparaj B, Masojídek J, Vonshak A (2003) Biological constraints in algal biotechnology. Biotechnol Bioprocess Eng 8:338–348

Torzillo G, Scoma A, Faraloni C, Giannelli L (2014) Advances in the biotechnology of hydrogen production with the microalga *Chlamydomonas reinhardtii*. Crit Rev Biotechnol. doi:10.3109/07388551.2014.900734

Travieso L, Hall DO, Rao KK, Benitez F, Sanchez E, Borja R (2001) A helical tubular photobioreactor producing *Spirulina* in a semicontinuous mode. Int Biodeter Biodegr 47:151–155

Tredici MR (2004) Mass production of microalgae: photobioreactors. In: Richmond A (ed) Handbook of microalgal cultures, biotechnology and applied phycology. Blackwell Science, Oxford, pp 178–214

Tredici MR, Chini Zittelli G (1998) Efficiency of sunlight utilization: tubular versus flat photobioreactors. Biotechnol Bioeng 57:187–197

Tredici MR, Chini Zittelli G, Rodolfi L (2010) Photobioreactors. In: Flickinger MC, Anderson S (eds) Encyclopedia of industrial biotechnology: bioprocess, bioseparation, and cell technology, vol 6. Wiley, Hoboken, pp 3821–3838

Tredici M, Bassi N, Prussi M, Biondi N, Rodolfi L, Chini Zittelli G, Sampietro G (2015) Energy balance of algae biomass production in a 1-ha "Green Wall Panel" plant: how to produce algae biomass in a closed reactor achieving a high Net Energy Ratio. Appl Energy. http://dx.doi.org/10.1016/J.apenergy.2015.01.086

Trent J, Wiley P, Tozzi S, NcKuin B, Reinsch S (2012) The future of biofuels: is it in the bag? Biofuels 3:521–524
Ugwu CU, Aoyagi H, Uchiyama H (2008) Photobioreactors for mass cultivation of algae. Biores Technol 99:4021–4028
Verdelho Vieira V (2012) (Emerging) Industrial production of *Nannochloropsis* microalgae biomass around the world. http://www.acpnonfood.com/WS8.2-20130429-(Vitor%20Verdelho).pdf. Accessed 12 Mar 2015
Vonshak A, Torzillo G, Masojidek J, Boussiba S (2001) Sub-optimal temperature induces photoinhibition in dense outdoor cultures of the alga Monodus subterraneus (Eustigmatophyta). Plant Cell Environ 24:113–118
Waldeck P (2012) Closed photobioreactor designs: from lab to industrial scale production of microalgal biomass. Oral presentation at ABO Algae Biomass Summit, Denver, USA, 24–27 September
Wang H, Zhang W, Chen L, Wang J, Liu T (2013) The contamination and control of biological pollutants in mass cultivation of microalgae. Biores Technol 128:745–750
Wijffels R, Barbosa MJ (2010) An outlook on microalgal biofuels. Science 329:796–799
Wiley P, Harris L, Reinsch S, Tozzi S, Embaye T, Clark K, McKuin B, Kolber Z, Adams R, Kagawa H, Richardson TMJ, Malinowski J, Beal C, Claxton MA, Geiger E, Rask J, Campbell JE, Trent JD (2013) Microalgae cultivation using offshore membrane enclosures for growing algae (OMEGA). J Sustain Bioengy Syst 3:18–32

Photobioreactors with Internal Illumination

Magda Sergejevová, José Romel Malapascua, Jiří Kopecký, and Jiří Masojídek

Abstract There has been an increasing commercial interest in phototrophic culturing of microalgae for all-year biomass production for food and feed supplements, as a source of valuable and bioactive compounds, and most recently biofuels. Indoor photobioreactors (PBRs) with artificial illumination make it possible to grow microalgae strains under well-controlled physico-chemical conditions aimed to directed production of desirable compounds. By far, light is one of the most important factors for growth and it significantly influences the yield and composition of microalgae biomass.

PBRs with interior illumination make it possible to use light (and energy) efficiently as the light sources are placed inside the microalgae culture. In our laboratory, various types of PBRs with artificial illumination – tubular, panel as well as column – have been constructed and used. Recently, 10- and 100-L pilot PBRs with internal lighting have been tested for microalgae growth. Tubular light sources based on white, high-intensity light-emitting diodes were submerged in microalgae culture. This set-up of PBRs makes it possible to adjust a wide range of culture conditions – temperature, light intensity, mixing and nutrient supply – required by individual strains.

The tested PBRs with internal illumination can be used for the production of microalgae strains (containing various bioactive compounds) that require delicate and well-controlled culture conditions, or for the preparation of seed cultures to inoculate large cultivation systems. The principle of these PBRs with internal illumination can be used for construction of large-scale commercial plants of thousands of litres for mass microalgae production.

M. Sergejevová
Laboratory of Algal Biotechnology, Centre Algatech, Institute of Microbiology CAS, v.v.i., Opatovický mlýn, CZ-379 81 Třeboň, Czech Republic

J.R. Malapascua • J. Kopecký • J. Masojídek (✉)
Laboratory of Algal Biotechnology, Centre Algatech, Institute of Microbiology CAS, v.v.i., Opatovický mlýn, CZ-379 81 Třeboň, Czech Republic

Faculty of Science, University of South Bohemia, CZ-370 05 České Budějovice, Czech Republic
e-mail: masojidekj@seznam.cz

A. Prokop et al. (eds.), *Algal Biorefineries*, DOI 10.1007/978-3-319-20200-6_6

Keywords Biomass • *Chlorella* • Growth rate • Mass culture • Microalgae • Internal illumination • Light-emitting diode • Light-dark cycle • Photobioreactor • Productivity • Phototrophic regime • Scale-up

1 Introduction

In microalgal biotechnology, suitable species can be grown as productive strains in aquacultures facilitating efficient manipulation of the cultivation process. Dense, well-mixed *mass culture of microalgae*[1] (>0.5 g biomass per litre) with sufficient nutrition and gas exchange represents an artificial system, which is completely different from optically-thin natural phytoplankton populations. In their biomass microalgae produce various bioactive and valuable substances such as pigments, antioxidants, lipids, polyunsaturated fatty acids (PUFA), polysaccharides, or immunologically effective compounds (for recent review see e.g. Koller et al. 2014a). Due to metabolic flexibility of microalgae, novel pathway discoveries and genetic manipulation have to be considered in a biosynthetic perspective. Hence, microalgae are an ideal platform for the large-scale production of high-value compounds because they represent fast-growing, solar-powered 'biofactories' with low nutrient requirements.

In the past six decades, there have been numerous attempts by researchers and companies to commercialize microalgal production. They have been cultivated for biomass as food and feed supplements, due to its potential to enhance the nutritional value of conventional food, as dietary supplements, a means of waste water treatment, and as a source of bioactive and novel compounds for pharmacology, cosmetics and the chemical industry (for a recent review e.g. Koller et al. 2014a). Most recently, microalgae have been explored as an important commodity for novel food (e.g. Plaza et al. 2008), a source of polyhydroxyalkanoates as a base of biodegradable plastics (e.g. Koller et al. 2014b) and also a potential source of biofuels – hydrogen, biogas, bio-ethanol and bio-diesel (for a recent review e.g. Wijffels and Barbosa 2010) to supplement current transportation fossil fuels.

1.1 Cultivation Systems

A number of cultivation systems and technologies have been developed to grow microalgal mass cultures *phototrophically*, using both natural and artificial light. The choice of a suitable cultivation system and the adjustment of the cultivation

[1]The term has no taxonomic significance; it is used in a broad sense for oxygenic phototrophic microorganisms, which include prokaryotic cyanobacteria and various eukaryotic algae and diatoms.

regime must be worked out for each individual strain. In every cultivation system, several basic features must be considered: illumination, mixing of culture, temperature control, nutrition and gas exchange (supply of CO_2 and O_2 degassing).

Simplifying again, two basic approaches are used in microalgal mass production: the first applies to cultivation in open reservoirs that are relatively large in area, while the second represents closed vessels – *photobioreactors*.[2] An overview of culture systems used for the mass cultivation of microalgae has recently been presented (Tredici 2004; Zittelli et al. 2013).

A seed culture (inoculum) is usually prepared in laboratory cultivations. The transfer of laboratory to the outdoor culture is scaled up in stepwise fashion, starting with the culture in a dilution ratio of approximately 1 to 5–10. It is advisable not to expose diluted laboratory cultures outdoors to full sunlight during the first few days, in order to avoid the risk of photoinhibition. However, a minimum biomass concentration corresponding to about 5–10 g/m^2 (~100–200 mg chlorophyll m^{-2}) is recommended.

1.2 Culture Principles

In the photoautotrophic cultivation of microalgae, the light may be supplied by the sun or an artificial light source. Sunlight is free and abundant and its use decreases the necessity for investment or electricity costs. Natural sunlight is widely used in the large-scale outdoor cultivation of microalgae. However, sunlight is primarily restricted to the surface and its availability for outdoor cultivation varies with the diurnal cycle, changing weather conditions, and with the season and the choice of location. This affects the growth rate and metabolism of a microalgae culture.

Apart from light, the growth of microalgae biomass is further influenced by a number of physico-chemical conditions: a suitable temperature and pH, a sufficient gas exchange (supply of carbon and removal of oxygen) and a supply of nutrients in the growth medium. Since microalgal mass cultures grow in dense suspensions (as compared with natural populations of phytoplankton), efficient mixing (turbulence) is critical to expose cells and supply them continuously with light and to allow for the efficient transfer of mass (heat, nutrient and gases). Microscopic observation of the culture is also essential in order to detect morphological changes of cells and contamination by other microorganisms (microalgae, bacteria, fungi and protozoa). In comparison with open systems where contamination by other microorganisms, especially other strains of microalgae and pollution from the environment are a constant problem, PBRs can be effectively sterilized. The cultivation process is thus similar to the production of microorganisms in fermenters with the only difference that an artificial source of light has to be used.

[2] The term *photobioreactor* is used for closed or semi-closed vessels with no direct contact between the culture and the atmosphere that can illuminated naturally sun irradiance or using artificial light sources.

Inorganic media are usually used for phototrophic microalgae cultivation which not only contain macro- and microelements, but also trace elements. In phototrophic cultivation, sufficient CO_2 (and/or bicarbonate) supply is a crucial point as the ambient CO_2 concentration of 350–400 ppm (0.03–0.04 vol.-%) is very low, with a one-order higher level being required in microalgal mass cultures. In the case of mixotrophic cultivation regimes, organic compounds (e.g. acetate, sugars) are added as a carbon source. In practice, semi-batch or semi-continuous culture regimes are usually adopted in outdoor PBRs, that is, a part of the culture is harvested continuously at regular intervals.

Photobiochemical monitoring methods reflect the general status of the cells' photosynthetic apparatus and are thus often used to adjust the appropriate cultivation conditions. Oxygen production and electron transport monitoring (by chlorophyll fluorescence) have been widely used as reliable and sensitive techniques to monitor the photosynthetic activity of various photosynthetic organisms including microalgae (Masojídek and Torzillo 2008; Masojídek et al. 2011a). From these correlations, the growth and productivity of microalgal cultures can be estimated (Masojídek et al. 2009, 2011b; Malapascua et al. 2014).

A more detailed description of culture maintenance and photosynthesis monitoring is described (e.g. Richmond 2013; Grobbelaar 2013) and elsewhere in this book (Masojídek et al.).

2 Open Outdoor Systems

Natural or artificial ponds, raceways (shallow race-tracks mixed by paddle wheels) and sloping cascades represent open cultivation systems for microalgae with the culture having direct contact with the environment. Numerous variations of open ponds are used: according to local requirements, climate conditions, and materials available (concrete, PVC, fiberglass). The culture depth may vary between 10 and 30 cm. To improve growth, tanks, raceways and ponds are mixed by impellers, rotating arms, paddle wheels, or by a stream of CO_2-enriched air supplied into the culture (see Masojídek et al. in this book). Large-scale commercial cultivation of microalgae is mostly carried out in outdoor open reservoirs, utilizing solar light energy. However, the productivities of these outdoor open cultivation units are very low (~1 g DM m^{-2} day^{-1}) due to the lack of mixing and insufficient CO_2 supply. The cultures are usually grown at a biomass concentration ranging between 0.5 and 1 g DM L^{-1} depending on the culture depth. These cultures are considered 'photo-limited' as they are operated at high culture depth.

In contrast, in another type of open-system – sloping cascades – some improvements have been made. The microalgae suspension flows over inclined-surface platforms in such a way that the layer thickness remains below 1 cm and the turbulent flow created by the arrangement prevents self-shading (for an overview see Masojídek et al. in this book). A high exposed surface-to-total volume ratio of up to 130 m^{-1} can be operated in these systems and give high areal or volumetric productivities. At times, such high productivities as over 40 g dry matter m^{-2} day^{-1} can be achieved in cascade cultivation units, even in temperate climate zones. Due to the very short optical path, high biomass densities between 15 and 35 g DM L^{-1} can be reached in these units, which is advantageous for harvesting and processing. Thin-layer sloping cascades combine the advantages of open systems (direct sun irradiance, easy heat derivation, simple cleaning and maintenance, lower biomass costs and efficient degassing) with positive features of closed systems (operation at high biomass densities achieving high volumetric productivity).

Although open systems cost less to build and operate, are more durable and have a larger production capacity, compared to more sophisticated PBRs, open systems have serious intrinsic disadvantages, such as: a susceptibility to contamination by other microalgae; difficulty to maintain a suitable culture regime; limited irradiance per cell; massive water loss due to evaporation; and a low cell concentration and biomass productivity.

Generally, the cultivation in open reservoirs is mostly suitable for resistant microalgal strains that grow rapidly (*Chlorella*), or under very selective conditions, for example at high alkalinity or salinity (*Arthrospira, Dunaliella*).

3 Closed and Semi-closed Systems – Photobioreactors

As compared to open systems, closed PBRs are more flexible and can be better maintained and optimized according to the physiological characteristics of the microalgal species involved. Recently, closed PBRs using natural or artificial illumination have attracted an increasing interest for the production of many pharmaceutically important compounds from microalgae, especially those which are not obtainable by chemical synthesis. Important points for the design and construction of PBRs have to be considered: light source, mixing, temperature control, gas exchange, sterility, reliability of operation, friction of cells (surface, tube diameter or panel shape, type of pumps), prevention of fouling, cleaning and construction material (glass, plastic).

In contrast to open systems, PBRs have a number of advantages: reproducible culture conditions with regard to environmental influences; a possibility to use artificial light effectively; low risk of contamination; low CO_2 losses; and smaller area requirements. On the downside, PBRs are: more difficult to clean; the construction material partially decreases sunlight penetration; and the system must be cooled and degassed effectively since over-optimal temperature or any excessive oxygen produced by the growing cultures can reduce growth.

A variety of PBRs – using either natural or artificial illumination – have been designed in the 1980–1990s consisting of glass or transparent plastic tubes, columns or panels, positioned horizontally or vertically, arranged as serpentine loops, fences, flexible coils, or as a series of panels or column gardens; these act as a photostage in which the microalgal suspension is continuously mixed. The most widely adopted outdoor, closed PBRs are tubular, column and flat plate types. An overview of closed or semi-closed culture systems and their use for mass cultivation of microalgae outdoors has been recently published (e.g. Pulz et al. 2013; Zittelli et al. 2013).

Commonly, transparent materials, glass or plastics (Plexiglas, polyethylene, polycarbonate, etc.) are used for the construction of the PBR photostage. In comparison with plastics, glass is better for cleaning and disinfection as it is less prone to the formation of biofilms. From physical point of view, light transmittance of glass is constant (a 3-mm thick glass transmits 90–92 % light spectrum) while plastic surfaces can be scratched or etched by mechanical or chemical cleaning and also suffer from continuous decrease in light penetration due to 'blinding'.

In PBRs, the culture suspension is circulated by a pump – or more preferably by air-bubbling or air-lifting (injecting a stream of compressed air in an upward-pointing tube). Peristaltic and membrane pumps are physically more 'friendly' to cells than centrifugal pumps, which can cause higher sheer stress. Cooling is maintained by submerging the units in a pool of water (e.g. Torzillo et al. 1996, 1998), using heat exchangers (e.g. Masojídek et al. 2003, 2009), or by spraying water onto the PBR surface (e.g. Vonshak et al. 2001). In the 2000s, indoor PBRs based uniquely on solar concentrators – linear Fresnel lenses – were tested at the Institute of Physical Biology in Nové Hrady, Czech Republic (Fig. 1) when the tubular loop of the PBR was placed in the focus of the raster lenses (Masojídek et al. 2003, 2009).

In model cultivations, cultures of the cyanobacterium *Arthrospira platensis* were grown at much higher solar irradiances than those usually recorded outdoors in summer, indicating that this organism is resilient to high-irradiance. The system was used to study the effect of microalgal acclimation (the cyanobacterium *Arthrospira platensis*) to supra-high solar irradiance with values up to 6000 μE m^{-2} s^{-1} (or *μmol photons* $m^{-2} s^{-1}$, *alternatively μmol quanta m-2* s^{-1}), making the approach rather unique.

Slow-growing strains that are sensitive to contamination and culture conditions are grown in PBRs, e.g. *Nannochloropsis, Haematococcus, Tetraselmis, Phaeodactylum, Skeletonema, Pavlova, Thalassiosira, Nostoc, Navicula, Isochrysis, Chaetoceros*, etc.

3.1 Photobioreactors with Artificial Illumination

In long-term periods of bad weather (low irradiance, suboptimal temperature, heavy rainfall, etc.), it is impossible to grow microalgae outdoors; in some cases it is even necessary to stop the cultivation. For year-round controlled biomass production PBRs are often placed in greenhouses. The fluctuations in sunlight intensity can be subsidised by the application of artificial illumination and supply light continuously. The artificial light sources are stable and controllable, and they can be easily integrated with the PBR design. Commercial large-scale PBRs for mass microalgae production furnished with artificial light sources or their combination with natural

Fig. 1 An indoor PBR with a total volume of 120 L based uniquely on solar concentrators – linear Fresnel lenses. It was mounted in a greenhouse in the biotechnology hall of the Institute of Physical Biology, University of South Bohemia in Nové Hrady, Czech Republic. The tubular loop of the PBR was automatically kept in the focus of the solar concentrators which allowed for a maximum value of up to 6000 μE m^{-2} s^{-1} on the tube surface. This dual purpose system was designed for microalgae cultivation in temperate climate zone and with surplus heat being used for warming service water

light have been developed in the 1980–1990s (for an overview see Pulz et al. 2013). The main design criterion is to supply light efficiently.

One disadvantage of using artificial light sources is high capital investment and operation costs (Blanken et al. 2013). Their efficiency can be insufficient when located outside and at a distance from the cultivation vessel; thus, a large part of the photon flow may be lost due to scattering or absorption in the material. Moreover, light sources produce heat and the efficient temperature control may be required to avoid overheating the microalgae suspension.

Artificial light sources – filament bulbs, fluorescent tubes or high-intensity discharge lamps (pressure sodium lamps) – have been used in indoor PBRs since the early days of microalgae biotechnology development. Already in 1964 a pilot system, 160-L indoor tubular PBR was constructed in Třeboň which consisted of 200 m of glass tubes and U-bends (inner diameter of 32 mm) connected in helical loops with high-pressure mercury lamps placed inside (Fig. 2) (Šetlík et al. 1967). The productivity of these indoor PBRs was up to 3 g DM L^{-1} day^{-1}. Starting in the 1990s, novel light sources – light-emitting diodes (LED) – have been employed in

Fig. 2 Indoor tubular PBR with a volume of 160 L. It was constructed at the Laboratory of Algal Production Technology, Institute of Microbiology in Třeboň, Czech Republic in 1965. The unit consisted of 200 m of glass tubes with a diameter of 32 mm arranged through connecting to helical loops with artificial illumination by high-pressure sodium lamps placed inside the loop

laboratory panel or column PBRs for microalgae growth (e.g. Lee and Palsson 1994, 1995; Nedbal et al. 1996, 2012; Cuaresma et al. 2009; Jacobi et al. 2012). Light-emitting diodes (LEDs) are the most advanced, efficient, and long-life (~50,000 h) light sources with high energy-to-biomass conversion efficiency. Since, these advanced light sources have been continuously improved over the last decade, their PAR efficiency has rapidly increased and the costs have significantly decreased.

In comparison to other lighting systems, LEDs have a relatively low construction cost, are design flexible and can operate with low electrical voltage (<4 V) and current (<1 A). Besides, LEDs have high luminous intensity and high energy conversion efficiency (~85 %) with low heat production. One other advantage of LED light sources is the possibility to produce intermittent light of medium frequency (10–1000 Hz). Light intermittency, i.e. short L/D cycles of tens to hundreds of milliseconds corresponding to the time scale of the rate-limiting dark reactions of photosynthesis enhances light use efficiency and subsequently growth rate as light energy is utilized with maximum efficiency (Richmond 2013; Zarmi et al. 2013).

Some designs are worth mentioning. In a 2-L cylindrical PBR for microalgae cultivation, the LED light source was mounted as an external jacket focused on the cultivation chamber (Jacobi et al. 2012). Another model described by Markou (2014) was a half-litre column PBR with internal illumination, provided by a rod covered with LEDs that was placed inside a glass tube mounted in the centre of the growth chamber. In principle, this design is similar to our models of PBRs described further as concerns all-side light supply to microalgae culture.

Recently, some sophisticated indoor systems have been designed which feature a unique combination of cultivator and monitoring devices to measure the physiological status of the microalgae culture. For example, Photon Systems Instruments Inc. (Drasov, Czech Republic; http://www.psi.cz) have produced a modular indoor PBR panel for microalgae cultivation using multicolour external LED panels (for illustration see Fig. 3). The core of the PBR is a flat panel (of 120 L) which may be cascaded to increase the total volume up to 1000 L. The system has precise control of the culture conditions (optical density, the culture pH, and CO_2 supply as well as photosynthetic activity). A unique feature of the illumination system is the possibility of intermittent illumination that may be used to modulate metabolic processes in cells.

3.2 *Photobioreactors with Internal Illumination*

The utilization of light energy has been thought to require the use of PBRs with large exposed surfaces, such as panel PBRs. However, with advances in illumination technology, efficient utilization of light from artificial sources has become possible, even with PBRs that have a relatively low exposed surface-to-volume ratio. To reduce the loss of light from artificial lighting, PBRs with internal illumination were designed in order to minimise light losses in microalgae culturing. If artificial light sources are setup correctly, the light conversion efficiency and biomass productivity of microalgae cultures can be even higher than that of sunlight. Moreover, continuous illumination which can be provided by artificial light sources is advantageous since biomass losses due to respiration in diurnal dark periods can be avoided. One important point is to optimise illumination intensity, light path and effective mixing to supply cells with light and nutrients evenly and facilitate efficient gas exchange. Thus, in the PBR with internal illumination various means of efficient mixing are essential: air-lift or bubble-column systems as well as mechanical impellers or magnetic and mechanical stirrers, or combinations thereof can be used.

Most internally-radiated PBRs have adopted the basic configuration of a cylindrical vessel that eases culture maintenance and system cleaning. One possible solution is to construct a PBR with one or more light sources evenly placed inside the cultivation vessel (e.g. Radmer 1989; Ogbonna et al. 1996; Suh and Lee 2001; Chiang et al. 2011; Choi et al. 2011; Wang et al. 2014), usually fluorescent lamps shielded by glass tubes. Here, the construction complexity, cooling (due to heat produced by light sources) and fouling of the system may cause some maintenance problems.

Fig. 3 An indoor flat-panel PBR for cultivation of microalgae with external illumination by light-emitting diodes (Institute of Microbiology, Třeboň, Czech Republic). The PBR consists of a stainless steel frame holding glass plates; it has a volume of 80 L and has a short light path of 6 cm. The system is mixed by air bubbling (+1–2 % CO_2) and temperature is controlled by Peltier (thermoelectric) cells. The illumination panel consists of 2200 high-intensity white LEDs with adjustable light intensity. This PBR can generate the L/D cycles modulated in the microsecond scale (produced by Photon Systems Instruments Inc., Drasov, Czech Republic; http://www.psi.cz)

Another design option of PBRs with internal illumination represent an annular arrangement of two concentric cylinders (one placed inside another) where the suspension of microalgae is kept in the interspace and the lighting system is placed in the inner cylinder (e.g. Ogbonna et al. 1996; Chini Zittelli et al. 2000; Suh and Lee 2003; Choi et al. 2011; Pegallapati and Nirmalakhandan 2011; Pegallapati et al. 2012a, b, 2013, 2014). Compared to the previous example, this system has some advantages as concerns cleaning, light source cooling and maintenance.

A complex solution to illuminating PBRs is based on collecting natural or artificial light and transferring it inside the cultivation vessel through light guides. A variety of light-harvesting systems have been worked out. These can be based on a combination of Fresnel lenses and optical fibres guiding solar irradiance to light radiators – vertically etched transparent glass, acrylic or quartz rods (e.g. Ogbonna et al. 1999), or a combination of Fresnel lenses and optical fibres guiding natural or artificial light to energy-excited polymethylmethacrylate optical fibres (Hirata et al. 1996; An and Kim 2000; Csögör et al. 2001; Chen et al. 2006a, b, c, 2008). A unique but rather complex design has been a PBR employing solar concentrators (Fresnel lenses in combination with solar-energy-excited optical fibres) for internal illumination and tungsten filament lamps for direct external illumination of the glass cultivation vessel (Chen et al. 2008). Here again, the construction complexity, and fouling of the system may cause some maintenance problems. In these systems cooling is usually much less problematic since the light is transferred by guides which mostly avoid heat transfer.

Until recently PBRs with interior lighting have mostly been operated for research purposes as laboratory systems with the volume up to tens of litres. Scaling up to 100's or 10,000's of litres is technically complicated (Rorrer and Mullikin 1999), but feasible and becoming more economical for mass production of some valuable compounds since the demand for space, light use efficiency and the control of culture conditions have been more favourable as compared to PBRs with external illumination.

At present, some internally-illuminated PBRs are produced and used commercially, for example the TreeLife PBR that is designed as a 250-L column with nine fluorescent tubes embedded evenly in microalgae culture (Fig. 4) (Microlife Cosmetics Italy; http://www.microlife.biz/it/prodotti/treelife). A large-scale commercial PBR are indoor tanks with internal illumination and a volume of thousands of litres, for example for the cultivation of the green microalga *Haematococcus* to produce valuable carotenoid astaxanthin tanks (AstaReal AB, Gustavsberg, Sweden; http://www.bioreal.se); however, not much information has been available about technical design of these PBRs.

3.3 Photobioreactors with Internal LED Illumination

In our laboratory, two PBR models with internal LED illumination were designed and constructed for the cultivation of microalgae. Both – 10- and 100-L PBRs were designed as vertical columns made of glass with LED light sources placed in cultivation chamber, i.e. submerged in the microalgae culture. The requirements respected in the design have been: easy maintenance and cleaning, sterilisability, adjustment of culture regime (irradiance, temperature, etc.) in a wide range of conditions, intensive mixing, effective gas exchange, temperature regulation, and the control and contamination. The important criteria were to have an efficient, high intensity source of illumination, good light distribution and low power consumption. One of the important presumptions was a possibility to scale-up to thousands

Fig. 4 An indoor internally-illuminated PBR TreeLife produced by Microlife Cosmetics Italy (http://www.microlife.biz/it/prodotti/treelife). The 250-L PBR (height of 1560 mm with; diameter of 430 mm) is designed as a column with nine fluorescent tubes embedded evenly in microalgae culture (Centro Interdipartimentale di ricerche per la gestione delle risorse idrobiologiche e per l'aquacoltura, Universita degli Studi di Napoli, "Federico II", Portici, Italy)

of litres for mass microalgae production under well-controlled conditions. A transparent vessel made of glass allowed the use of a combination of natural light and artificial illumination for microalgae cultivation to compensate for respiration losses at night, or low-irradiance periods.

The first PBR – model LED-PBR-10 – consists of a vertical glass column with a working volume of 8 L and four internal light sources evenly submerged in a microalgae culture (Fig. 5). Each of the internal LED light sources was made of a rectangular aluminium rod (holder) with LED strips placed along on all four sides; the holder was then placed into a glass tube with a closed bottom end. In this arrangement all light was evenly dispersed in the microalgae suspension and used for photosynthesis. The tubular light sources were mounted vertically in a lid of the PBR. This set-up maintained adequate heat transfer from the light source into the microalgae culture and temperation prevented the system from overheating. The temperature (in the range between 20 and 40 °C) was controlled via a water circuit with a temperature controller connected to the double-jacket of the cultivation vessel. A stream of filtered air (containing 1–2 % CO_2) was bubbled from a loop placed at the vessel bottom maintaining sufficient mixing of microalgae culture to supply cells evenly with light, nutrients and remove oxygen produced in photosynthesis. In this way, a slight overpressure inside the cultivation vessel was maintained (~0.1 bar) that prevented microalgae culture from external contamination. All inputs for medium, air supply and temperature, and pH-value sensors were inserted through the lid of the PBR. In this type of PBR the thickness of photic microalgae layer (light path) varied between 3.5 and 5.5 cm.

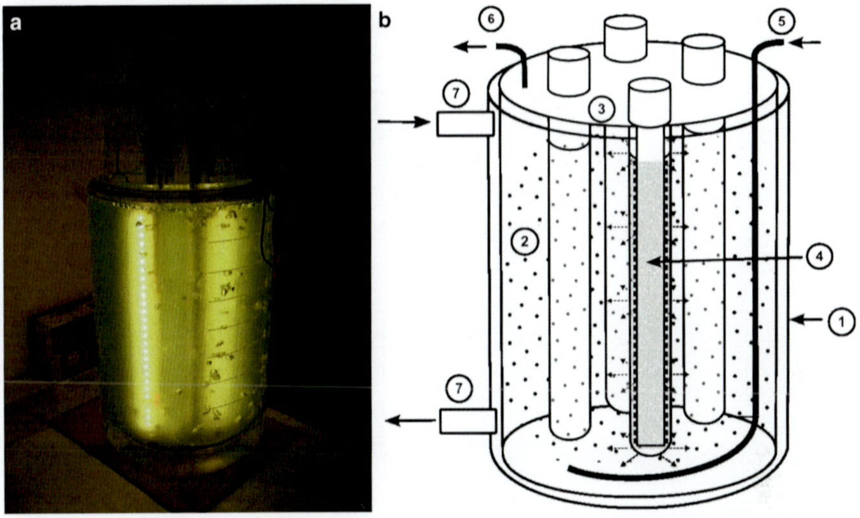

Fig. 5 A column PBR, LED-PBR-10 of 8-L working volume with internal illumination. (**a**) Demonstration photo of the PBR used for cultivation trial placed on laboratory bench (height ~ 400 mm, inner diameter ~190 mm). (**b**) Schematic diagram of the PBR consisting of a glass column (*1*) filled microalgae culture (*2*) in which four light sources based on white, high intensity LEDs are submerged (*4*). The tubular light sources are mounted to the lid of the PBR (*3*). The temperature is controlled via the water circuit of a temperature controller connected to a double-jacket (*7*) of the cultivation vessel. The stream of filtered air (containing 1–2 % CO_2) is supplied by a loop (*5*) to mix and supply CO_2 to cells and remove oxygen produced in photosynthesis (*6*)

The second model, LED-PBR-100 – was a larger annular PBR that consisted of a vertical glass column (inner diameter of 40 cm) with a working volume of 90 L in which another glass column with a LED light source was inserted; microalgae culture was grown in the interspace (Fig. 6a). The light source was a metallic holder densely covered with LED strips (neutral-white and warm-white light) that was inserted into a glass column with a closed bottom (inner diameter of 12.5 cm) (Fig. 6b). The holder – an aluminium tube of 6 cm in diameter – served as a heat exchanger that was connected to the water circuit of a temperature controller for cooling of the light source. An additional cooling loop was also placed in the cultivation vessel if the cultivation temperature was below 30 °C. At higher cultivation temperatures this loop was connected to a circulation thermostat to utilise the heat produced by LEDs for culture warming. The stream of filtered air (containing 1–2 % CO_2) was from a perforated loop placed at the vessel bottom maintaining the mixing of microalgae culture and supplied cells with CO_2, light, and nutrients while removing oxygen produced in photosynthesis. Additional mixing was produced by a stream of air blown inside the vessel through a bottom outlet valve. In the LED-PBR-100 model the thickness of photic layer (light path) was about 14 cm. Before culture inoculation the cultivation vessels can be sterilised in an autoclave or illuminated by an UV lamp placed into the growth medium.

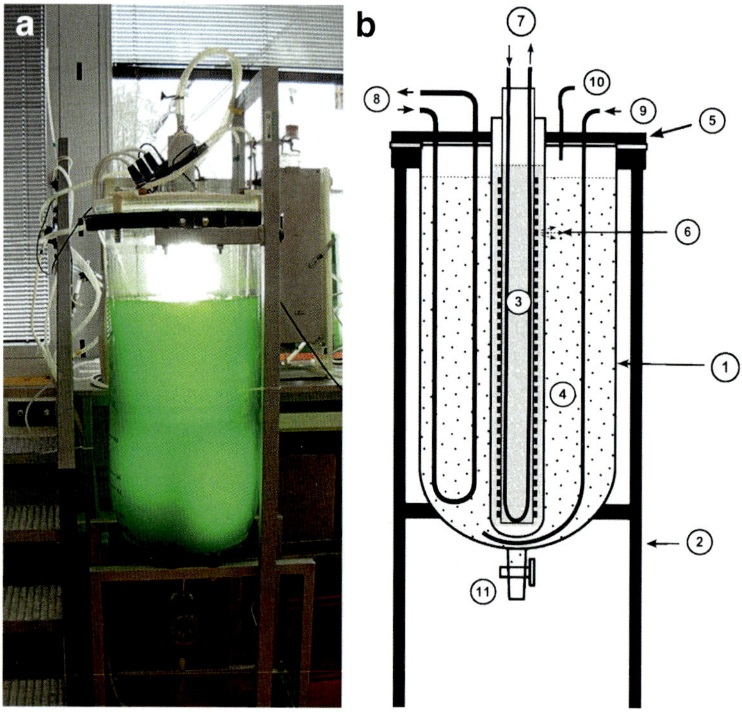

Fig. 6 An annular PBR, model LED-PBR-100 of 90-L working volume with internal illumination. (**a**) Demonstration photo of the PBR used for cultivation trial (height ~ 900 mm, internal diameter ~400 mm). (**b**) Schematic diagram of the PBR consisting of a glass column (*1*) supported by a metal stand (*2*) in which one, central light source (*3*) is submerged in microalgae culture (*4*). The light source mounted to the lid of the PBR (*5*) is a metal holder densely covered by LED strips (*6*; neutral-white and warm-white light) that is inserted into a glass column with a closed bottom. The holder, an aluminium tube, serves as a heat exchanger (*7*) that is connected to the water circuit of a temperature controller for cooling the light source. An additional cooling loop (*8*) was also placed in the cultivation vessel. The stream of filtered air (containing 1–2 % CO_2) is supplied *via* a perforated loop (*9*) to mix and supply CO_2 to cells and remove oxygen produced in photosynthesis (*10*). Additional mixing is maintained by a stream of air blown inside the vessel through a bottom outlet valve (*11*)

The LED intensity in both LED-PBR-10 and LED-PBR-100 can be regulated continuously up to the intensity between 2000 and 3000 μE m^{-2} s^{-1} of photosynthetically active irradiance at the surface of the light source to allow for suitable light intensity according to biomass density and strain requirements in order to maintain good growth rate, but avoid photoinhibition in the diluted culture.

In an exemplary series of experiments the cultures of the freshwater microalga *Choricystis* sp. *strain 1983 (Trebouxiophyceae) [cf. C. parasitica (BRANDT) PRÖSCHOLD et DARIENKO]* was grown in both LED-PBR-10 and LED-PBR-100 in a mineral medium at 30 °C. Some characteristics of these systems are shown in Table 1.

Table 1 Characteristics of the LED-PBR-10 and LED-PBR-100 units tested in these trials

PBR unit	Max. light intensity at output (μE m^{-2} s^{-1})	Light path (mm)	S/V (m^{-1})	Biomass productivity (g L^{-1} day^{-1})	Power input to light source (W)	Power input to suspension (W L^{-1})	Light use[a] (W g^{-1})
LED-PBR-10 HL	2500	35–55	16	0.10	280	28	16.1
LED-PBR-10 LL	3500	35–55	16	0.23	400	40	11.7
LED-PBR-100	2200	140	4	0.09	1500	15	11.6

The 10 L PBR was operated in a low light (*LL*) and high light (*HL*) regime, the 100 L unit in low light. The light was measured close to the light sources using the LI-190SA quantum sensor (cosine-corrected up to 80° angle of incidence) coupled to the LI-250 light meter (Li-Cor, USA). The ratio between the exposed area to the total volume S/V (m^{-1}), biomass productivity (g L^{-1} day^{-1}) power input of light sources (under continuous illumination) per volume of culture (W L^{-1}) and light use as power input per biomass are shown

[a]Calculated as total irradiance of PBR in Watts per total biomass after 2 weeks of trial

In the smaller LED-PBR-10 (with four light sources), two light regimes were used differing in the maximum irradiance intensity produced by light sources which was 2500 and 3500 μE m^{-2} s^{-1}, respectively (Table 1). The distance between the light source and the inner wall of the vessel was about 4.5 cm. In the larger LED-PBR-100, the maximum irradiance intensity produced by one central light source was 2200 μE m^{-2} s^{-1} which was comparable to the low-light regime of the smaller LED-PBR-10, but the distance between the light source and the inner wall of the cultivation vessel was about 14 cm, much longer than in the LED-PBR-10. The ratio between the exposed area to the total volume (S/V) in low and high-irradiated LED-PBR-10 and LED-PBR-100 was 16, 16 and 4 m^{-1}, respectively (Table 1).

In the LED-PBR-10 the growth of the *Choricystis* cultures was directly dependent on light intensity (mean cell irradiance). At higher irradiance the biomass density increased from a diluted culture (0.1 g L^{-1}) to about 3.4 g L^{-1} in 2 weeks which was twice higher than that in the culture grown at a lower irradiance (1.9 g L^{-1}) (Fig. 7). The sigmoidal shape of the growth curve in the culture exposed to higher irradiance showed the lag phase which was probably cause by photo-stress in the diluted culture. In the larger LED-PBR-100, the culture growth was slower as the biomass density reached only 1.4 g L^{-1} in 2 weeks, less than that in the low light culture and the culture was still growing. The slower growth was apparently caused by a long light path.

The biomass productivities of the cultures in lower and higher-irradiated LED-PBR-10 and LED-PBR-100 were 0.10, 0.23 and 0.09 g (DM) L^{-1} day^{-1}, respectively. Interestingly, when we considered the ratio between the power input of light sources (under continuous illumination) and the volume of culture, about 40, 28 and 15 W

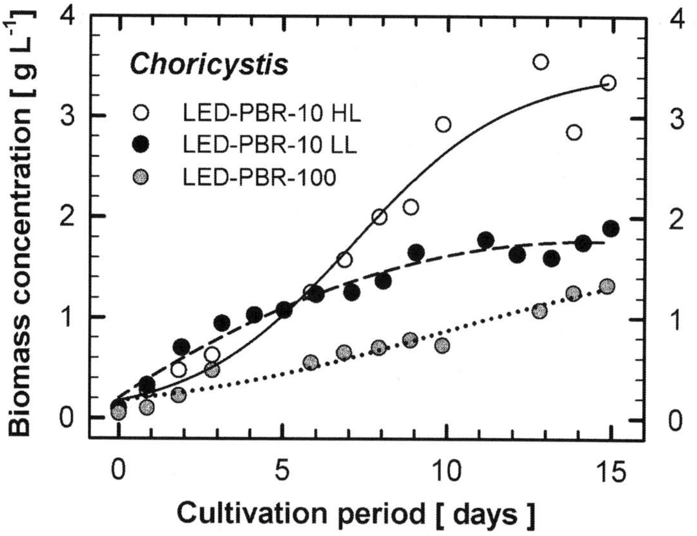

Fig. 7 Growth curves of *Choricystis* cultures cultivated in two column PBRs LED-PBR-10 under low light (LL) and high light (HL) regimes and in LED-PBR-100 during a 14-day experimental trial

per 1 L of microalgae culture was supplied to lower and higher-irradiated LED-PBR-10 and LED-PBR-100, respectively (Table 1). If we correlated power input with the biomass productivity, it corresponded to 16.1, 11.7 and 11.6 W per g, respectively. This shows that low-irradiated LED-PBR-10 and LED-PBR-100 were comparable, even when the culture grew in the large PBR with only one central light source and long light path meaning that the light energy was well utilised.

Photosynthetic activity of the *Choricystis* culture was measured daily at fixed daytime (09:00 h) by saturation pulse analysis of chlorophyll fluorescence quenching using a fluorimeter (pulse-amplitude-modulation fluorimeter PAM-2500, H. Walz, Germany) (for more details of technique see Masojídek et al. in this book). The maximum photochemical yield of PSII, F_v/F_m and relative electron transport rate, rETR were calculated from rapid light-response curves (using saturation-pulse analysis of fluorescence quenching) as described in Malapascua et al. (2014). The maximum quantum yield of PSII, F_V/F_M of dark adapted microalgae culture can be estimated by normalizing the variable fluorescence, F_V to the maximum PSII quantum efficiency of photosystem II, F_M. The actual photochemical yield of photosystem II (PSII), Y_{PSII} which estimates the efficiency at which certain light intensity absorbed is used for photochemistry (Genty et al. 1989). A widely used variable in photosynthetic studies is the relative PSII electron transport rate, rETR calculated as Y_{II} multiplied by irradiance E. At a given photosynthetically active photon flux density (PPFD), rETR provides an estimate of the quantum yield of linear electron flux through PS II, which might be correlated with the overall photosynthetic performance of the microalgal culture (Juneau et al. 2005; Baker 2008).

In green microalgae when cultures are healthy, the values of the F_V/F_M ratio range from 0.7 to 0.8 (Masojídek et al. 2013). In our cultivation trials all cultures were in relatively good physiological state as the maximum photochemical efficiency of PSII, F_v/F_m was between 0.65 and 0.75 (Fig. 8a). In the culture grown in the LED-PBR-10 at higher irradiance the subsequent decrease of the F_v/F_m value was found during the trial which was probably caused by shade adaptation of the dense culture and its decreased maximum photosynthetic performance compared to the start of the trial. The highest F_v/F_m values were found in the culture grown in the LED-PBR-100 as this was in a good physiological state, growing at lower irradiance with maximum photosynthetic efficiency even after 2 weeks of growth. The courses of the relative electron transport rate rETR coincided with previous results. The culture grown in the low-irradiated LED-PBR-10 was photosynthetically very active due to lower biomass density allowing more light penetration to the photic zone. rETR was continuously decreasing during the trial as the culture became dense (Fig. 8b). The culture grown in the LED-PBR-100 showed an increase of rETR after one week of experiment, but due to a long light path the supply of light was not sufficient. The trends of rETR measurements correlated with lower F_v/F_m values (Fig. 8).

The so-called rapid light-response curves (RLC), i.e. the dependency of photosynthetic electron transport (rETR) on irradiance intensity (E) reflects photosynthetic activity and the physiological status of the culture (Kromkamp et al. 1998; White and Critchley 1999; Ralph and Gademann 2005). The measurement of RLC clearly indicated the photosynthetic performance of cultures (Fig. 9). On Day 2,

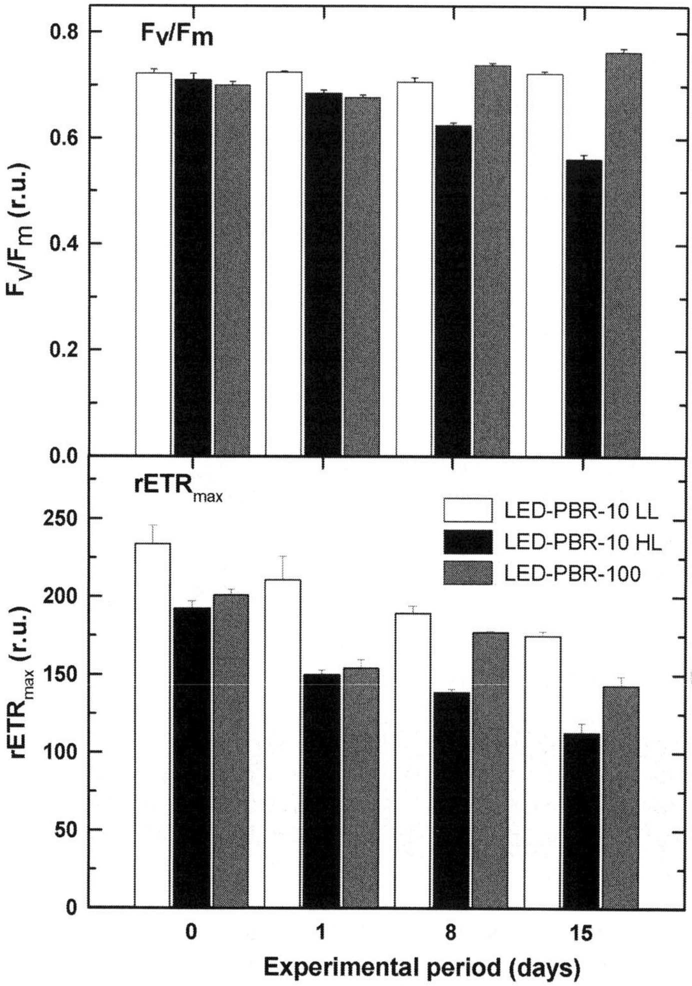

Fig. 8 Changes of the maximum quantum yield of PSII, F_V/F_M and relative electron transport rate, rETR of *Choricystis* cultures grown in two column PBRs LED-PBR-10 under low light (LL) and high light (HL) regimes and in LED-PBR-100 during a 14-day experimental trial. Values are presented as a mean (n = 3) with SE indicated by error bars

when the cultures were already acclimated to growth conditions, the highest photosynthetic activity was found in low-light culture in the LED-PBR-10 as it showed higher initial slope (photochemical efficiency) of RLC and higher rETR of 210 while the high-light exposed culture had lower photochemical efficiency and by about 30 % lower rETR (=150). In comparison to the latter, the culture grown in the LED-PBR-100 revealed a similar $rETR_{max}$ value, but higher photochemical efficiency (slope) and the curve was 'bending' earlier (~800 μE m^{-2} s^{-1}) due to down-regulation as it was acclimated to lower irradiance.

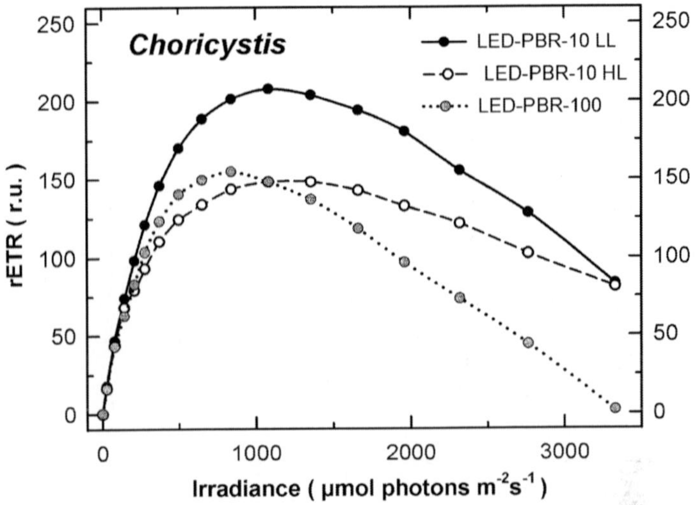

Fig. 9 Rapid light-response curves of rETR (=$Y_{PSII} \times E$) of *Choricystis* cultures grown in two column PBRs LED-PBR-10 under low light (LL) and high light (HL) regimes and in LED-PBR-100 measured on Day 2 of a 14-day experimental trial

In LED-PBR-10 and LED-PBR-100, the width-to-height ratio was 0.48 and 0.44, respectively. The smaller LED-PBR-10 had a shorter light path; thus logically a higher growth rate and biomass density can be achieved, but the handling of multiple light sources and their cleaning was more complicated. In this respect the larger LED-PBR-100 with one compact light source was better – with easy cleaning and maintenance and a higher volume, although the light path is longer which determines a slower growth rate and subsequently a lower biomass density.

When we consider the ratio between the height and diameter (width) of a PBR, we have to find some compromise between the light path and volume of the microalgae culture. The larger volume of the culture can be used for production of desired amount of microalgae culture of some delicate strains, or as a seed culture to inoculate larger cultivation systems.

4 Conclusions

Internally-illuminated PBRs are mostly designed as closed reservoirs with light sources embedded in microalgae culture in order to minimise light losses and maximise light-use efficiency for photosynthesis. One possible solution is to construct a PBR with light sources placed inside the cultivation vessels. Another design option of internally-illuminated PBRs represents an annular arrangement of two concentric cylinders where the suspension of microalgae is kept in the interspace and the lighting system is placed in the inner cylinder. In internally-illuminated PBRs the light is supplied in several ways: either by solely artificial sources filament bulbs, fluorescent tubes or high-intensity discharge lamps, or using light radiators (glass, acrylic or quartz rods) to which natural or artificial light is transferred through light guides. Lately, novel light sources, LEDs – most advanced, efficient and long-lived with high energy-to-biomass conversion efficiency – have been used in PBRs.

Thus, we have constructed and successfully tested two models of glass-column PBRs for microalgae cultivation in which LED light sources are placed in cultivation chamber, i.e. submerged in the microalgae culture. The features respected in the design are: easy maintenance and cleaning, sterilisability, light intensity regulation, adjustment of culture regime in a wide range of conditions, thorough mixing, effective gas exchange, efficient temperature regulation and the control of contamination.

The tested PBRs with internal illumination can be used for the growth of delicate microalgae strains (containing bioactive compounds) that require preciously controlled culture conditions. Another application of these PBRs can be the preparation of seed monocultures to inoculate large cultivation systems.

Finally, one of the outcomes of this study has been to test the PBRs with LED interior illumination in a pilot scale which can be scaled up to large-scale commercial systems of thousands of litres for mass microalgae production under well-controlled conditions. In our opinion it is feasible although an open question remains whether individual PBR modules are to be linked in a series, or huge vessels (like fermenters) with internal illumination should be constructed. Here, we have to consider and test costs and technical feasibility of large PBR construction.

Acknowledgements The authors thank Mr Pavel Souček, Mr. Viktor Korbel and Ms. Karolina Ranglová for technical assistance and Mr Jason Dean for language corrections. This study was supported by the Czech Academy of Sciences, by the Ministry of Education, Youth and Sports, projects Algatech CZ.1.05/2.1.00/03.0110 'Centre for Algal Biotechnology', Algain CZ.1.07/2.3.00/30.0059 and Algaman CZ.1.07/2.3.00/20.0203 and by National Programme of Sustainability I, project Algatech Plus LO1416.

Abbreviations

Chl	chlorophyll
DM	dry mass
E	irradiance
F_v, F_m	variable and maximum fluorescence in dark-adapted state
F', F_m'	steady-state and maximum fluorescence in light-adapted state
F_v/F_m, $\Delta F'/F_m'$	maximum, resp. actual photochemical yield of PSII
L/D cycle	light-dark cycle
LED	light-emitting diode

LED-PBR-10, LED-PBR-100	LED-illuminated PBRs with a volume of 10 and 100 L, respectively
OD	optical density
PAM	pulse-amplitude-modulation
PAR	photosynthetically active radiation
PPFD	photosynthetically-active photon flux density
PBR	photobioreactor
PSII	Photosystem II
rETR	relative electron transport rate through PSII
RLC	rapid light-response curve

References

An J-Y, Kim B-W (2000) Biological desulfurization in an optical-fiber photobioreactor using an automatic sunlight collection system. J Biotechnol 80:35–44

Baker NR (2008) Chlorophyll fluorescence: a probe of photosynthesis in vivo. Annu Rev Plant Biol 59:89–113

Blanken W, Cuaresma M, Wijffels RH, Janssen M (2013) Cultivation of microalgae on artificial light comes at a cost. Algal Res 2:333–340

Chen C-Y, Lee C-M, Chang J-S (2006a) Hydrogen production by indigenous photosynthetic bacterium *Rhodopseudomonas palustris* WP3–5 using optical fiber-illuminating photobioreactors. Biochem Eng J 32:33–42

Chen C-Y, Lee C-M, Chang J-S (2006b) Feasibility study on bioreactor strategies for enhanced photohydrogen production from *Rhodopseudomonas palustris* WP3-5 using optical-fiber-assisted illumination systems. Int J Hydrog Energy 31:2345–2355

Chen C-Y, Lee C-M, Chang J-S (2006c) Enhancing phototropic hydrogen production by solid-carrier assisted fermentation and internal optical-fiber illumination. Process Biochem 41:2041–2049

Chen C-Y, Saratale GD, Lee C-M, Chen P-C, Chang J-S (2008) Phototrophic hydrogen production in photobioreactors coupled with solar-energy-excited optical fibers. Int J Hydrog Energy 33:6886–6895

Chiang CL, Lee CM, Chen PC (2011) Utilization of cyanobacteria Anabaena sp. CH1 in biological carbon dioxide mitigation processes. Bioresour Technol 102:5400–5405

Chini Zittelli G, Pastorelli R, Tredici MR (2000) A modular flat panel photobioreactor (MFPP) for indoor mass cultivation of *Nannochloropsis* sp. under artificial illumination. J Appl Phycol 12:521–526

Choi YE, Yeoung-Sang Y, Jong MP, Ji-Won Y (2011) Multistage operation of airlift photobioreactor for increased production of astaxanthin from *Haematococcus pluvialis*. J Microbiol Biotechnol 21:1081–1087

Csögör Z, Kiessling B, Perner I, Fleck P, Posten C (2001) Growth and product formation of *Porphyridium purpureum*. J Appl Phycol 13:317–324

Cuaresma M, Janssen M, Vílchez C, Wijffels RH (2009) Productivity of *Chlorella sorokiniana* in a short light-path (SLP) panel photobioreactor under high irradiance. Biotechnol Bioeng 104:352–359

Genty B, Briantais JM, Baker NR (1989) The relationship between the quantum yield of photosynthesis electron transport and quenching of chlorophyll fluorescence. Biochim Biophys Acta 990:87–92

Grobbelaar JU (2013) Inorganic algal nutrition. In: Richmond A, Hu Q (eds) Handbook of microalgal culture: applied phycology and biotechnology. Wiley Blackwell, Oxford, pp 123–133

Hirata S, Hayashitani M, Taya M, Tone S (1996) Carbon dioxide fixation in batch culture of *Chlorella* sp. using a photobioreactor with a sunlight-collecting device. J Biosci Bioeng 81:470–472

http://en.wikipedia.org/w/index.php?search=biorefinary&title=Special%3ASearch&fulltext=1

http://www.bioreal.se

http://www.microlife.biz/it/prodotti/treelife

http://www.psi.cz

Jacobi A, Steinweg C, Rosello Sastre R, Posten C (2012) Advanced photobioreactor LED illumination system: scale-down approach to study microalgal growth kinetics. Eng Life Sci 12:621–630

Juneau P, Green BR, Harrison PJ (2005) Simulation of Pulse-Amplitude-Modulated (PAM) fluorescence. Limitations of some PAM-parameters in studying environmental stress effects. Photosynthetica 43:75–83

Koller M, Muhr A, Braunegg G (2014a) Microalgae as versatile cellular factories for valued products. Algal Res 6:52–63

Koller M, Salerno A, Strohmeier K, Schober S, Mittelbach M, Illieva V, Chiellini E, Braunegg G (2014b) Novel precursors for production of 3-hydroxyvalerate-containing poly[(*R*)-hydroxyalkanoate]s. Biocatal Biotransform 32:161–167

Kromkamp JC, Barranguet C, Peene J (1998) Determination of microphytobenthos PSII quantum efficiency and photosynthetic activity by means of variable chlorophyll fluorescence. Mar Ecol Prog Ser 162:45–55

Lee CG, Palsson BO (1994) High-density algal photobioreactors using light emitting diodes. Biotechnol Bioeng 44:1161–1167

Lee CG, Palsson BO (1995) Light emitting diode-based algal photobioreactor with external gas exchange. J Ferment Bioeng 79:257–263

Malapascua JRF, Jerez CG, Sergejevová M, Figueroa FL, Masojídek J (2014) Photosynthesis monitoring to optimize growth of microalgal mass cultures: application of chlorophyll fluorescence techniques. Aquat Biol 22:124–140

Markou G (2014) Effect of various colors of light-emitting diodes (LEDs) on the biomass composition of *Arthrospira platensis* cultivated in semi-continuous mode. Appl Biochem Biotechnol 172:2758–2768

Masojídek J, Torzillo G (2008) Mass cultivation of freshwater microalgae. Ecological engineering, vol 3, Encyclopaedia of ecology. Elsevier, Oxford, pp 2226–2235

Masojídek J, Papáček Š, Sergejevová M, Jirka V, Červený J, Kunc J, Korečko J, Verbovikova O, Kopecký J, Štys D, Torzillo G (2003) A closed solar photobioreactor for cultivation of microalgae under supra-high irradiance: basic design and performance. J Appl Phycol 15:239–248

Masojídek J, Sergejevová M, Rottnerová K, Jirka V, Korečko J, Kopecký J, Zaťková I, Torzillo G, Štys D (2009) A two-stage solar photobioreactor for cultivation of microalgae based on solar concentrators. J Appl Phycol 21:55–63

Masojídek J, Vonshak A, Torzillo G (2011a) Chlorophyll fluorescence applications in microalgal mass cultures. In: Suggett DJ, Prášil O, Borowitzka MA (eds) Chlorophyll a fluorescence in aquatic sciences: methods and applications. Springer, Dordrecht, pp 277–292

Masojídek J, Kopecký J, Giannelli L, Torzillo G (2011b) Productivity correlated to photobiochemical performance of *Chlorella* mass cultures grown outdoors in thin-layer cascades. J Ind Microbiol Biotechnol 38:307–317

Masojídek J, Koblížek M, Torzillo G (2013) Photosynthesis in microalgae. In: Richmond A, Hu Q (eds) Handbook of microalgal culture: applied phycology and biotechnology. Wiley Blackwell, Oxford, pp 20–39

Nedbal L, Tichy V, Xiong FH, Grobbelaar JU (1996) Microscopic green algae and cyanobacteria in high-frequency intermittent light. J Appl Phycol 8:325–333

Nedbal L, Trtílek M, Červený J, Komárek O, Pakrasi HB (2012) A photobioreactor system for precision cultivation of photoautotrophic microorganisms and for high-content analysis of suspension dynamics. Biotechnol Bioeng 100:902–910

Ogbonna JC, Yada H, Masui H, Tanaka H (1996) A novel internally illuminated stirred tank photobioreactor for large-scale cultivation of photosynthetic cells. J Ferment Bioeng 82:61–67

Ogbonna JC, Soejima T, Tanaka H (1999) An integrated solar and artificial light system for internal illumination of photobioreactors. J Biotechnol 70:289–297

Pegallapati AK, Nirmalakhandan N (2011) Energetic evaluation of an internally illuminated photobioreactor for algal cultivation. Biotechnol Lett 33:2161–2167

Pegallapati AK, Arudchelvam Y, Nirmalakhandan N (2012a) Energy-efficient photobioreactor configuration for algal biomass production. Bioresour Technol 126:266–273

Pegallapati AK, Koushik A, Nirmalakhandan N (2012b) Modeling algal growth in bubble columns under sparging with CO2-enriched air. Bioresour Technol 124:137–145

Pegallapati AK, Koushik A, Nirmalakhandan N (2013) Internally illuminated photobioreactor for algal cultivation under carbon dioxide-supplementation: performance evaluation. Renew Energy 56:129–135

Pegallapati AK, Nirmalakhandan N, Dungan B, Holguin FO, Schaub T (2014) Evaluation of internally illuminated photobioreactor for improving energy ratio. J Biosci Bioeng 117:92–98

Plaza M, Cifuentes A, Ibáñez E (2008) In the search of new functional food ingredients from algae. Trends Food Sci Technol 19:31–39

Pulz O, Broneske J, Waldeck P (2013) IGV GmbH experience report, industrial production of microalgae under controlled conditions: innovative prospects. In: Richmond A, Hu Q (eds) Handbook of microalgal culture: applied phycology and biotechnology. Wiley Blackwell, Oxford, pp 445–460

Radmer RJ (1989) Using a lamp having higher intensity and efficiency, US patent 4952511 B

Ralph PJ, Gademann R (2005) Rapid light curves: a powerful tool to assess photosynthetic activity. Aquat Bot 82:222–237

Richmond A (2013) Biological principles of mass cultivation of photoautotrophic microalgae. In: Richmond A, Hu Q (eds) Handbook of microalgal culture: applied phycology and biotechnology. Wiley Blackwell, Oxford, pp 171–204

Rorrer GL, Mullikin RK (1999) Modeling and simulation of a tubular recycle photobioreactor for macroalgal cell suspension cultures. Chem Eng Sci 54:3153–3162

Šetlík I, Komárek J, Prokeš B (1967) Short account of the activities from 1960 to 1965 and some future prospects. Annu Rep Algolog Lab for 1966, Knihtisk Praha

Suh IS, Lee SB (2001) Cultivation of a cyanobacterium in an internally radiating air-lift photobioreactor. J Appl Phycol 13:381–388

Suh IS, Lee SB (2003) Optimization of radiator position in an internally radiating photobioreactor: a model simulation study. J Microbiol Biotechnol 13:789–793

Torzillo G, Accolla P, Pinzani E, Masojídek J (1996) *In situ* monitoring of chlorophyll fluorescence to assess the synergistic effect of low temperature and high irradiance stresses in *Spirulina* grown in outdoor photobioreactors. J Appl Phycol 8:283–291

Torzillo G, Bernardini P, Masojídek J (1998) *On-line* monitoring of chlorophyll fluorescence to assess the extent of photoinhibition of photosynthesis induced by high oxygen concentration and low temperature and its effect on the productivity of outdoor cultures of *Spirulina platensis* (Cyanobacteria). J Phycol 34:504–510

Tredici M (2004) Mass production of microalgae: photobioreactors. In: Richmond A (ed) Handbook of microalgal mass cultures. Wiley Blackwell, Oxford, pp 178–214

Vonshak A, Torzillo G, Masojidek J, Boussiba S (2001) Sub-optimal morning temperature induces photoinhibition in dense outdoor cultures of the alga *Monodus subterraneus* (Eustigmatophyta). Plant Cell Environ 24:1113–1118

Wang S-K, Stiles AR, Guo C, Liu C-Z (2014) Microalgae cultivation in photobioreactors: an overview of light characteristics. Eng Life Sci 14:550–559

White AJ, Critchley C (1999) Rapid light curves: a new fluorescence method to assess the state of the photosynthetic apparatus. Photosynth Res 59:63–72

Wijffels RH, Barbosa MJ (2010) An outlook on microalgal biofuels. Science 329:796–799

Zarmi Y, Bel G, Aflalo C (2013) Theoretical analysis of culture growth in flat-plate bioreactors: the essential role of timescales. In: Richmond A, Hu Q (eds) Handbook of microalgal mass culture: applied phycology and biotechnology. Wiley Blackwell, Oxford, pp 205–224
Zittelli GC, Biondi N, Rodolfi L, Tredici MR (2013) Photobioreactors for mass production of microalgae. In: Richmond A, Hu Q (eds) Handbook of microalgal mass culture: applied phycology and biotechnology. Wiley Blackwell, Oxford, pp 225–266

Thin-Layer Systems for Mass Cultivation of Microalgae: Flat Panels and Sloping Cascades

Jiří Masojídek, Magda Sergejevová, José Romel Malapascua, and Jiří Kopecký

Abstract Two basic approaches to microalgae biomass production are used: one applies to cultivation in closed or semi-closed vessels – photobioreactors, while the other involves open reservoirs with direct contact of the microalgal culture with the environment. The most crucial variable for phototrophic growth is light availability. The amount of photon energy received by each cell is a combination of several factors: irradiance intensity, cell density, length of optical path (thickness of culture layer), rate of mixing as well as cultivation unit design. In practice, this should form a part of the considerations when designing cultivation systems.

The highest growth rate and productivity have been achieved in cultivation systems with microalgae layer thickness lower than 50 mm. The advantage of these thin-layer systems is that high biomass density is reached, which is advantageous for harvesting and processing. Basically, two thin-layer cultivation systems are being used that guarantee high areal or volumetric productivity due to high surface-to-volume ratio: vertical or inclined flat panels, and near-horizontal sloping cascades or raceways. The first type, flat-panel photobioreactors represent closed or semi-closed systems. In the other system, thin-layer sloping cascades – microalgae culture is grown on open, inclined-surface platforms that – by some means – combine the advantages of open systems (direct sun irradiance, easy heat dissipation, simple cleaning and maintenance, lower construction and biomass costs and efficient degassing) with positive features of closed systems (operation at high biomass densities achieving high volumetric productivity). Among the limitations of these systems, there can be a possibility of contamination by other microalgae strains which allows growing preferentially fast-growing strains or those cultivated in selective environments.

J. Masojídek (✉) • J.R. Malapascua • J. Kopecký
Centre Algatech, Laboratory of Algal Biotechnology, Institute of Microbiology, Czech Academy of Science, Opatovický mlýn, CZ-379 81 Třeboň, Czech Republic

Faculty of Science, University of South Bohemia, CZ-370 05 České Budějovice, Czech Republic
e-mail: masojidekj@seznam.cz

M. Sergejevová
Centre Algatech, Laboratory of Algal Biotechnology, Institute of Microbiology, Czech Academy of Science, Opatovický mlýn, CZ-379 81 Třeboň, Czech Republic

A. Prokop et al. (eds.), *Algal Biorefineries*, DOI 10.1007/978-3-319-20200-6_7

Keywords Biomass • Cascades • *Chlorella* • Cyanobacteria • Flat-panel • Growth rate • Mass culture • Microalgae • Nutrition • Open pond • Photobioreactor • Productivity • Thin-layer • Light-dark cycle

1 Introduction

In microalgal biotechnology, suitable species can be grown as productive strains in extensive aquacultures (algacultures) facilitating the efficient manipulation of cultivation processes. Although many microalgal strains are cultivated worldwide for different purposes, the bulk of annual biomass production is represented by only a few species. Algacultures are an ideal platform for the large-scale production of biomass because they are fast-growing, solar-powered 'biofactories' with low nutrient requirements. Their substantial benefits over plants are based on their short life cycles and metabolic plasticity that offers the possibility of modifying their biochemical pathways and cellular composition by varying culture conditions.

Over the last 60 years, microalgal biotechnology has shown a range of applications: from the traditional extensive biomass production in human and animal nutrition, health food products, soil conditioning in agriculture, aquaculture colorants, technologies for waste-water treatment, products for cosmetics, pharmaceuticals, and most recently the possible production of a 'third' generation biofuels. Dense, well-mixed *mass culture of microalgae*[1] (>0.5 g biomass per litre) with sufficient nutrition and gas exchange represents an artificial system that is a suitable model for a *biorefinery*.[2]

1.1 Cultivation Systems

Various cultivation systems and technologies have been developed to grow microalgal mass cultures. The choice of a suitable cultivation system and the adjustment of the cultivation regime must be worked out for each individual microalgal strain and production purpose. The key problem to solve in a cultivation unit design is how to use the photon flux at a maximum rate, i.e. how to allow each single microalgae cell to get access to an optimum number of photons every time. Two basic approaches to mass production are used: one applies to cultivation in open reservoirs (with direct contact of the microalgal culture with the environment), while the other involves closed or semi-closed vessels – photobioreactors (PBRs) with no direct contact between the culture and the atmosphere (for a recent review, see Zittelli et al. 2013). Large-scale outdoor PBRs for commercial production are

[1] The term – microalgae – is used by phycologists pragmatically for oxygenic phototrophic microorganisms, which include prokaryotic cyanobacteria and various eukaryotic algae and diatoms; it has no taxonomic significance.

[2] A biorefinery is a facility that integrates equipment and biomass conversion processes to produce fuels, power, heat, and value-added chemicals from biomass.

usually designed as modules. There are major operational differences between open reservoirs and PBRs and, consequently, the growth physiology of the microalgae is different between the two systems (Grobbelaar 2009, 2012).

The systems are used for specific purposes and this will determine which cultivation system is the most suitable since there is no universal all-purpose unit. Crucial variables are the irradiance intensity, temperature, optical depth, turbulence, light acclimated state of the organism, nutrient availability and gas exchange (supply of CO_2 and O_2 degassing). From a commercial point of view, the price of the final product is often an important consideration. At present open reservoirs are the only feasible culture systems for the production of thousands of tons of biomass as production is cheaper than a culture from a closed PBR. Unfortunately, the use of open ponds is restricted to a relatively small number of microalgal species due to the limited control of cultivation conditions and contamination. Hence, open systems are suitable for "robust" microalgal strains (e.g., *Chlorella* or *Scenedesmus*) that grow rapidly, or under very selective conditions (e.g., *Spirulina* or *Dunaliella*).

Compared to open systems, photobioreactors have certain advantages: reproducible cultivation conditions with regard to environmental influences; reduced risk of contamination; low CO_2 losses; lower cost of biomass down-stream processing; and smaller area requirements. On the downside, closed systems are: more difficult to clean; the construction material might partially decrease sunlight penetration; and the system must be cooled and degassed effectively since excessive oxygen produced by the growing cultures can reduce growth. Furthermore, the cost of construction is about one order of magnitude higher than that of open ponds.

The total ground area (i.e. including the ground area between panels) for the vertical flat plate PBR is significantly lower than that occupied by an open reservoir (e.g. raceway or cascade). Finally, the harvested cell density is close to one order of magnitude higher in the flat plate PBR than that in open ponds or raceways, which carries economic significance.

From a practical point of view, flat-panel PBRs have one serious disadvantage: biofouling at higher biomass density, especially of the channels, due to reduced turbulence in their narrow, rectangularly shaped channels.

1.1.1 Open Outdoor Systems

Open cultivation systems are usually artificial ponds, tanks, raceways (shallow racetracks mixed by paddle wheels) and sloping cascades (i.e. inclined-surface platforms). An overview of open culture systems used for the mass cultivation of microalgae outdoors has been presented recently (Tredici 2004; Zittelli et al. 2013). Productivity in these open systems is usually low (~1 g DM m^{-2} day^{-1}) due to the lack of mixing and CO_2 supply. To improve productivity, open systems are mixed by impellers, rotating arms, paddle wheels, or by a stream of CO_2-enriched air supplied into the culture. The culture depth may vary between 10 and 30 cm. The cultures are usually grown at a biomass concentration ranging between 0.5 and 1 g L^{-1} depending on the culture depth. Outdoor cultures are considered a photo-limited system as they are operated at an optimum concentration rather than at a maximum growth rate.

1.1.2 Closed and Semi-closed Photobioreactors

Compared to open systems closed PBRs are more flexible and can be better optimized according to the biological and physiological characteristics of a selected microalgal strain. A variety of PBRs (using either natural or artificial illumination) has been designed consisting of glass or transparent plastic tubes, columns or panels, positioned horizontally or vertically, arranged as serpentine loops, fences, flexible coils, a series of panels or column 'gardens'; these act as a photostage in which the microalgal suspension is continuously mixed. Necessary cooling is maintained by submerging the tubes in a pool of water, by heat exchangers, or by spraying water onto the PBR surface. In PBRs, a much greater biomass density can be maintained than in open systems. At present, panel or tubular PBRs are often mounted in greenhouses to maintain culture conditions for all-year functioning. Slowly-growing strains sensitive to contamination are grown in PBRs, e.g. *Nannochloropsis, Haematococcus, Tetraselmis, Phaeodactylum, Skeletonema, Pavlova, Thalassiosira, Nostoc, Navicula, Isochrysis, Chaetoceros,* etc.

Despite the higher biomass yields attainable with PBRs (as compared to open systems), their high construction and maintenance costs still make them uncompetitive for the industrial production of microalgal biomass. Their use can be foreseen for the production of high-value bioactive substances, which require the adoption of sterile conditions.

2 Culture Monitoring and Maintenance

Microalgal mass cultures grown in a cultivation unit should also have its physiological status checked operatively in order to optimize photosynthetic activity and growth. Successful cultivation requires a continuous monitoring of a culture's physicochemical parameters, namely its pH, temperature, dissolved oxygen concentration, and nutrient status. One method of observation is to use basic biological examination under the microscope: in order to detect morphological changes of cells and contamination by other microorganisms (microalgae, bacteria, fungi and protozoa). The nutrient status can be followed by monitoring the concentration of nitrogen, and then using this as a measure for adding proportional amounts of other nutrients. Sufficient carbon (CO_2 or bicarbonate) supply is a crucial point as the ambient CO_2 concentration is very low (about 0.04 %; v/v) and is the limiting factor in extensive microalgal mass cultures exposed to high irradiances.

Photobiochemical monitoring methods reflect the general status of the cells' photosynthetic apparatus and are thus often used to adjust the appropriate cultivation conditions. Oxygen production and chlorophyll (Chl) fluorescence have been used as reliable and sensitive techniques to monitor the photosynthetic activity of various photosynthetic organisms including microalgae (Bradbury and Baker 1984; Krause and Weis 1984, 1991; Walker 2009; Flameling and Kromkamp 1998; Gilbert et al. 2000; Figueroa et al. 2003; Wilhelm et al. 2004; Figueroa et al. 2013). Although providing analogous information, Chl fluorescence techniques are, as compared with measurements of O_2 production, considerably faster, more sensitive and moreover, can give information on absorbed energy distribution between photochemical and dissipative (protective) processes and the balance between photosynthetic electron transport and the Calvin-Benson cycle (Schreiber et al. 1986, 1995; Baker and Oxborough 2004; Suggett et al. 2011).

Since the 1990s Chl fluorescence measurement has become one of the most common and useful approaches used for monitoring the physiological status of microalgal mass cultures due to its sensitivity, ease of use, as well as its prompt provision of results (Ting and Owens 1992; Büchel and Wilhelm 1993; Vonshak et al. 1994, 1996; Torzillo et al. 1996, 1998; Baker 2008; Enriquez and Borowitzka 2011; Masojídek et al. 2011b). One, direct approach is to measure photosynthesis *on-line/in-situ* during the diel cycle to monitor the actual situation in a culture. The other possibility is to measure Chl fluorescence *off-line* using dark-adapted microalgal samples taken from a cultivation unit at selected times (Masojídek et al. 2011a).

Chl fluorescence measurements in our experiments showed that changes of some fluorescence variables can be well correlated with changes of cultivation conditions, physiological status and growth of a given microalgal culture and/or the suitability of a selected cultivation system (Torzillo et al. 1996, 1998; Masojídek et al. 2000, 2003, 2009, 2011a; Malapascua et al. 2014). Using pulse-amplitude-modulation (PAM) technique to carry out saturation pulse analysis of fluorescence quenching some of the fluorescence variables can be calculated (for recent reviews see Maxwell and Johnson 2000; Schreiber 2004; Baker 2008; Masojídek et al. 2011b; Malapascua et al. 2014). For example the maximum photochemical yield of PSII (F_V/F_M), actual photochemical yield of PSII, $\Delta F'/F_M'$ (= $[F_M'-F']/F_M'$) and the relative electron transport rate rETR through PSII (the product of multiplication $\Delta F'/F_M'$ by the photosynthetically active radiation E_{PAR} in the culture) reflect photosynthetic activity and can be correlated with analogous changes in the daily productivities of cultures grown under different conditions (Torzillo et al. 1996, 1998; Masojídek et al. 2000, 2011a). Namely, rETR proved to be a simple and reliable parameter to estimate growth and productivity in both indoor and outdoor mass cultures of microalgae (Malapascua et al. 2014). The so-called Stern-Volmer non-photochemical quenching NPQ (= $[F_M-F_M']/F_M'$) is, in principle, inversely related to photochemistry ($\Delta F'/F_M'$). It indicates an increased futile heat dissipation of absorbed energy and is considered a safety valve protecting PSII reaction centres from damage by excess irradiance (Bilger and Björkman 1990). It has been experimentally proven that a midday-depression of actual PSII photochemical yield (F_v/F_m) of between 20 and 30 % compared to morning values at high-cell density is compatible with well-performing cultures. A lower or higher depression of F_v/F_m indicated low-light acclimated or photoinhibited cultures, respectively (Masojídek et al. 2003, 2011a).

Although the theory is well described at present (Maxwell and Johnson 2000; Schreiber 2004, Strasser et al. 2004, Baker 2008), the interpretation of Chl fluorescence signals may not be straightforward, particularly when dealing with microalgae (Schreiber et al. 1995; Strasser et al. 1995, Campbell et al. 1998). Care must be taken when measuring fluorescence and evaluating data in cyanobacteria. This is because the fluorescence emission of phycobilisomes, as well as state transition effects and PSI fluorescence, can contribute significantly to the total signal, and this affects the correct determination of certain variables (Ting and Owens 1992; Büchel and Wilhelm 1993; Schreiber et al. 1995).

Culture growth might be estimated as changes in the optical density (OD) at 750 nm, the dry mass (biomass), or the number of cells. Pigment content is determined in solvent extracts using spectroscopy or liquid chromatography. Biomass productivity can be expressed as the areal or volumetric yield per unit time, for example in [g m^{-2} day^{-1}] or in [g l^{-1} day^{-1}].

Basically, two cultivation regimes are used for the growth of microalgal cultures. In the batch regime, the culture is inoculated and at a certain point of growth it is harvested. In the continuous regime, the culture is harvested continuously according to its growth rate and fresh medium is added to replace nutrients. In biotechnological practice, semi-continuous or semi-batch regimes are usually adopted, that is, where a part of the culture is harvested at regular intervals.

3 Biological Principles: Light-Regime, Biomass Density, Optical Path and the Importance of Time-Scales

The energy needed for the photosynthetic conversion of carbon dioxide into organic substances is delivered by photons, which come from the sun under natural conditions. Except light, the growth of microalgae biomass is further influenced by physicochemical conditions for growth: a suitable temperature and pH, and a sufficient supply of carbon and nutrients in the growth medium. Since microalgal mass cultures grow in dense suspensions (as compared with natural populations of phytoplankton), turbulent mixing is critical to expose cells to light and to allow for an efficient mass transfer. In outdoor cultures, solar photosynthetically-active radiation (PAR) represents a major growth-limiting factor in well-maintained mass microalgal cultures (where temperature and nutrients are not limiting). The kinetic reply of microalgal cells to irradiance intensity is shown in a model of photosynthetic light-response curve (Fig. 1), provided that the irradiance intensity is the sole growth-limiting factor. At low, light-limited levels the photosynthetic rate is approximately linearly proportional to irradiance. At light saturation, the maximum rate of photosynthesis P_{max} is reached. At irradiance values beyond the plateau region in Fig. 1, the rate of photon absorption exceeds the rate of electron turnover in the photosynthetic apparatus and it eventually leads to a decrease of photosynthesis, commonly referred to as photoinhibition. The ultimate rate limiting processes are the photosynthetic dark reactions (Gordon and Polle 2007; Richmond 2013; Masojídek et al. 2013).

From a practical point of view, flux requirement in commercial microalgal photobioreactors is typically ~200–400 μmol photons m^{-2} s^{-1} corresponding to the irradiance saturating photosynthesis (E_S in Fig. 1) that is only about 10–20 % of the maximum photosynthetic photon flux density (PPFD) (Gordon and Polle 2007; Richmond 2013; Masojídek et al. 2013). The requirements for efficient utilization

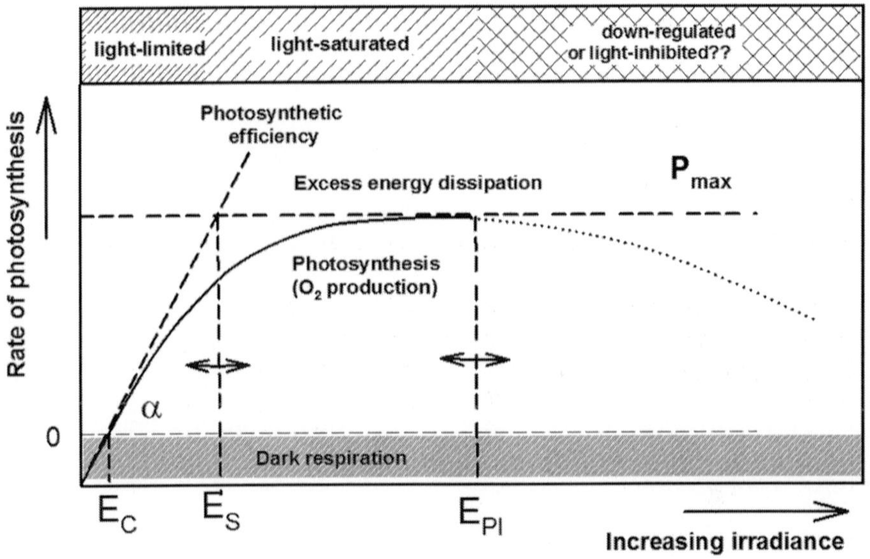

Fig. 1 A model of the light-response curve of photosynthesis. Three regions are present in the curve: light-limited, light-saturated and light inhibited. The intercept in the x-axis E_C designate the compensation irradiance intensity between dark respiration and photosynthesis; and E_S is an approximate irradiance level between photosynthesis limitation and saturation. At low, light-limited, levels the photosynthetic rate is approximately linearly proportional to irradiance. At light saturation, the maximum rate of photosynthesis P_{max} is reached. By reaching E_S the rate of photon absorption exceeds the rate of electron turnover in the PSII complex and excess energy is dissipated. Further irradiance increase eventually leads to a light-induced drop of photosynthesis, commonly referred as photoinhibition

of high light fluxes in microalgal cultures have been elucidated: the most important of these are a *short light-path combined with a highly turbulent flow at high cell densities* (i.e. >5 g DM L^{-1}) (Hu and Richmond 1996; Hu et al. 1998; Richmond 2003; Grobbelaar 2012).

Since early reports in the 1930s, it has been clear that intermittent (pulsed) light is an important issue for microalgae growth (Emerson and Arnold 1932). Microalgal cells may utilize strong light only if it is delivered intermittently, in 'pulses'. The so called 'flashing light' effect on photosynthesis in *Chlorella* was studied by Kok (1953). Later, the effect of L/D cycles was investigated by several research groups (e.g. Laws et al. 1983; Tennessen et al. 1995; Gordon and Polle 2007; Grobbelaar 2009; Zarmi et al. 2013). In the 1990s, the introduction of high-intensity LEDs to scientific use made it possible to measure the effect of intermittent illumination more preciously in the microsecond range (Matthijs et al. 1996; Nedbal et al. 1996).

In mass microalgal culture it is possible to achieve high photosynthetic yields in full sunlight when the turbulence and density of cells are adjusted to produce the proper pattern of light intermittence, i.e. the L/D cycles are sufficiently short in the order of tens to hundreds of microseconds (10–100 Hz), close to the time scale of the rate-limiting dark reactions of photosynthesis. The influence of L/D cycles of several seconds to tens of seconds does not appear to result in an improvement of the photosynthetic efficiency (Janssen et al. 2000, 2001, 2003). It was concluded that the averaged amount of photon energy received by each cell is a combination of several variables: irradiance intensity, cell population density, length of optical path (thickness of culture layer), spectral quality, light absorption, and the rate of mixing (Richmond 2003, 2004, 2013). As shown in Fig. 1, the maximum photochemical efficiency is achieved in the light-limited region, but maximum rates of photosynthesis are reached in microalgal cultures in which averaged cell irradiances are close to saturation E_S (and energy losses are still low).

4 Thin-Layer Systems – Layer Thickness, High Surface-to-Volume Ratio and Biomass Productivity

The averaged irradiance intensity of a microalgal cell is modulated not only by ambient irradiance, but also by culture density, mixing, culture depth (light path), light-dark cycle frequency as well as cultivation unit design and spatial setting with respect to exposure to the sun. In practical terms, this should form part of the considerations when designing cultivation systems.

As discussed above, principally two thin-layer cultivation systems are being used that guarantee high areal or volumetric productivity due to their high exposed surface-to-volume ratio (S/V ratio): vertical or inclined flat panels, and sloping raceways or cascades. *The higher the surface for light incidence and the smaller the volume for the microalgae culture (S/V ratio), the better the light supply.* A crucial point is the sufficient mixing of microalgal culture to induce fast L/D cycling of cells in 'short' light-path (<50 mm) cultivation systems. For a given system the culture exhibits the highest photosynthetic efficiency at optimal cell density. The other aspect concerns the overall photic volume that should comprise ~5–10 % of optical path (Tredici 2010; Richmond 2013) as the depth the light penetrates into the culture is a function of cell density. The operation regime – suitable biomass density, culture layer (optical path), cell movement patterns (averaged light/dark cycles for cells) and mass exchange – has to be developed to maximise/photo-optimise the use of high photon flux densities reaching the surface of cultivation systems. In general, the shorter the length of the light-path, the smaller the areal volume and the higher the volumetric productivity (Richmond and Cheng-Wu 2001).

L/D cycle considerations have indicated that the cell travel time begins to represent a relevant parameter for enhancement in photosynthetic productivity when the optical path is reduced to about 10 mm. In such a system, e.g. flat-panel PBR, the time range of L/D cycles is hundreds of milliseconds, assuming a photic volume of 5 % and a fluid velocity of 30 cm s^{-1} (Richmond 2003). In another experiment, light penetration was measured in a sloping cascade using a culture with a biomass density of about 11 g L^{-1}. As the culture depth was about 6 mm, a hydrodynamic model demonstrated highly turbulent flow allowing rapid L/D cycles (with a frequency of 0.47 s^{-1}) in a culture layer (Masojídek et al. 2011a).

Higher cell density cultures require a higher Reynolds number (an indication of the extent of turbulence) in order to induce short L/D cycles. Efficient use of strong light requires high frequency of L/D cycles which in turn facilitates higher biomass productivity. Applied intermittently to the individual cells in the turbulent culture, high irradiance is *diluted* by being available in smaller doses to more cells within a given time span. Thus, the light is used more effectively, compared with light use of cells illuminated continuously at low-density, or in poorly stirred cultures. Therefore the increased L/D cycle frequency can be considered a form of *light dilution* (Richmond 2013).

4.1 Closed Systems – Flat-Panel Photobioreactors

The first type of thin-layer systems used for microalgae cultivation, flat-panel PBRs represent closed or semi-closed systems. They usually consist of vertical or inclined transparent rectangular vessels with a relatively short light-path of 1–5 cm, made of firm material, i.e. glass, Plexiglas or polycarbonate in which baffles can be mounted to create a labyrinth of channels (Fig. 2). The other possibility is to use flat-panels, 3–5 cm thick, made of flexible, polyethylene bags enclosed in a rigid framework. The flat-panel systems are placed either outdoors or in a greenhouse exploiting sun light.

The height and width of individual panels can be varied. In practice, panels can be connected to modules of several metres in length with a total volume of hundreds to thousands litres. The panels are in series or parallel, arranged vertically some distance apart to avoid self-shading. Flat-panel PBRs are mixed with air (+ CO_2) introduced via a perforated tube at the bottom of panels to create a high degree of turbulence (air-bubble or air-lift), or the culture is circulated by a pump (*for recent review see* Zittelli et al. 2013; Pulz et al. 2013).

Fig. 2 Thin-layer photobioreactors for cultivation of microalgae. (**a**) Outdoor inclined flat-panel photobioreactors were arranged in series (Sde Boqer campus, J. Blaustein Institutes for Desert Research, Ben Gurion University, Israel,; courtesy of Prof. Amos Richmond); (**b**) Vertical-panel photobioreactors arranged in parallel and mounted in a greenhouse (Institut für Getreideverarbeitung, Potsdam-Rehbrücke, Germany)

Flat-panel PBRs with high biomass density have been one of the devices to establish the potential for massive improvements in bioproductivity as biomass yield and optimum biomass density (OBD) in closed systems is generally much higher than in open systems. The other advantage is the possibility of growing some strains sensitive to contamination by other fast-growing microalgae (*Nannochloropsis, Haematococcus, Tetraselmis, Isochrysis, Phaeodactylum, Skeletonema, Pavlova, Thalassiosira, Nostoc, Chaetoceros*, etc.). Experiments in inclined flat-panel PBRs (30°–60° tilt angle to sun) connected in series (Fig. 2a) for mass cultivation of fast-growing cyanobacterium *Spirulina (Arthrospira) platensis* showed the interrelationship between light path and OBD: the stepwise decreasing thickness of flat-panel from 104, 52, 26 to 13 mm corresponded to an increase of OBD from 1.7, 3.1, 8.4 to 15.8 g L^{-1} resulting in the biomass productivity of 33.6, 38.9, 49.4 to 51.1 g m^{-2} day^{-1}, respectively (Hu et al. 1996a, b). Naturally, biomass productivity depends on strain physiology and conditions. As compared to *Arthrospira*, productivities were much lower for slowly-growing marine microalga *Nannochloropsis* sp. which is the eminent producer of PUFA. In the same flat-panel PBRs with the culture layer thickness stepwise decreasing from 104, 52, 26 to 13 mm, the corresponding biomass productivities were 5.5, 7.3, 9.3 and 12.1 g m^{-2} day^{-1}, respectively (Zou and Richmond 1999) which were almost an order of magnitude lower than those in *Arthrospira*.

In laboratory experiments, *Arthrospira* was grown in flat panels of only 7.5 mm (!) thick illuminated continuously by 900 μmol photons m^{-2} s^{-1} (provided by 1500 W halogen lamps) which were placed either on one or on both sides). In this case, it was possible to work with ultra-high biomass densities of about 27 g L^{-1} reaching biomass productivity of about 100 g m^{-2} day^{-1} (Hu et al. 1998). From a technical point of view the scaling up of this system is hardly feasible as the use of very thin flat-plate panels would be rather difficult for maintenance.

Most of industrial flat-panel PBRs have light-path between 20 and 40 mm which determines a high surface-to-volume ratio. Several commercial large-scale systems for microalgae production have been developed working on the principle of flat-plate panels. In one example, a large-scale flat plate photobioreactor was constructed employing vertically oriented plastic plates in a greenhouse, in which microalgae were flowed horizontally in narrow rectangular channels created by baffles (Fig. 2b). This system was a pioneering project of microalgae-based industrial CO_2 fixation from flue gas produced by a lime kiln in Elbigerode (Germany). The productivity of this 6,000-L system was rather high, in the range of 30–50 g DM biomass m^{-2} day^{-1} using microalgae strains *Chlorella* and *Scenedesmus* (Pulz et al. 2013).

The concept of 'disposable panels' for large-scale applications was developed in the early 2000s by two groups working in Italy (University of Florence) and Israel (Ben Gurion University). The vertical photobioreactor called 'Green Wall Panel (GWP I)' consists of 100-litre bags (~4.5 × 100 × 250 cm; 800 L) made of a polyethylene foil enclosed in a rigid framework (Fig. 3) (Rodolfi et al. 2009; Zittelli et al. 2013). The GWP I modules (4.5 cm × 1 m × 20 m) can be connected, placing them in a single row or in parallel, 1-m apart to avoid mutual shading. For the outdoor experiments, air-flow is maintained for culture mixing and CO_2 is supplied during daylight hours keeping pH in the range 7.5–8.0. The cooling is generally provided by water (even seawater) spraying, or by insertion of cooling loops inside the panels to prevent the culture overheating. This low-cost, easy-to-operate and low-contamination system with good scalability has been used to produce various microalgae (*Tetraselmis, Nannochloropsis, Isochrysis, Cylindrotheca*) although biomass densities are in the scale of grams. The GWP PBRs (developed at the University of Florence and commercialised by Fotosintetica & Microbiologica Srl) have been successfully used in several large-scale demonstration projects worldwide to grow microalgae for various purposes. The GPW design has been continuously modified in order to improve functioning and to reduce costs. The most recent model of these PBRs, GWP III PBRs have east-west orientation, shorter light-path and adjustable tilt from vertical to inclined position to use sun light effectively and decrease energy input by integration with photovoltaics.

The technology provided by Subitec GmbH (Germany) enables the cultivation of microalgae at an industrial scale with an enclosed system based on flat-panel airlift photobioreactors to produce microalgae biomass as a source of feed, bulk chemicals and energy – clean biofuels from microalgae (Fig. 3).

A large-scale flat-panel photobioreactor 'Hanging Gardens' was developed and demonstrated by Ecoduna GmbH (Bruck a/L, Austria). One module unit (4.3 m^3) consist of 12 parallel flat panels (3 cm × 2 m × 6 m) which are placed 15 cm apart in

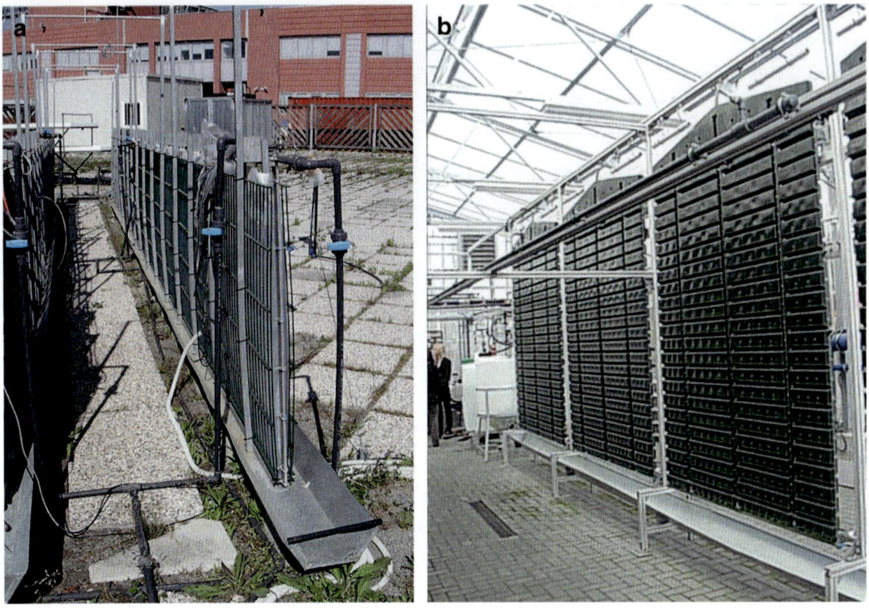

Fig. 3 Flat-plate photobioreactors for cultivation of microalgae. (**a**) Outdoor vertical flat-panel photobioreactor 'Green Wall Panel' arranged in a single row or in parallel (developed at the University of Florence and commercialised by Fotosintetica & Microbiologica Srl., Italy). (**b**) Vertical alveolar flat-panel photobioreactors produced by Subitec GmbH were arranged in series and mounted in a greenhouse at Vattenfall power plant in Senftenberg (Germany)

Fig. 4 Large-scale flat-panel photobioreactor 'Hanging Gardens' (developed and demonstrated by Ecoduna GmbH in Bruck a/L, Austria). One module unit consisting (see insert) of 12 closely spaced parallel panels (3 cm × 2 m × 6 m) which are placed 15 cm apart in a movable frame that allows tracking of the sun movements. The panels are internally partitioned by vertical baffles to allow culture circulation using a gas-lift effect. The insert shows details of the arrangements of the panels in the module

a movable frame that allows tracking the sun movements (Fig. 4). The panels are internally partitioned by baffles to allow culture circulation as air and CO_2 are injected from the bottom to generate a gas-lift effect.

4.2 *Open System – Sloping Cascades*

The other type of thin-layer systems used for microalgae cultivation represent sloping cascades which are known worldwide as the Třeboň's or Šetlík's type (Šetlík et al. 1967; 1970). In these cultivation units microalgae flow in thin-layer over open, inclined-surface platforms which – by some means – combine the advantages of open systems (direct sun irradiance, easy heat derivation, simple cleaning and maintenance, lower construction and biomass costs, efficient degassing) with positive features of closed systems (operation at high biomass densities achieving high volumetric productivity). The unique cultivation plant of 900 m^2 was constructed in 1962–1963 which was one of the first large-scale research facilities for mass microalgae production (Fig. 5). Later, in the 1970s as a part of collaborative projects some thin-layer cascades (TLC) of the Třeboň's type were also constructed in Bulgaria,

Fig. 5 Outdoor large-scale cascades for cultivation of microalgae (50 and 900 m^2) built in the 1960s. One of the first large-scale research facilities for mass microalgae production was located on the campus of the Opatovický mlýn, Institute of Microbiology, Třeboň (mid 1960s). The units had a plain glass surface, with a slope of 3 %, framed by a steel structure. The transverse baffles 3.5 cm high and 15 cm apart were fitted on the surface to create intensive turbulence in the microalgae layer of about 50 mm

Italy, Poland and Cuba, in order to compare microalgae cultivation under various climatic conditions (Bartoš 1967; Zahradník 1967; Vendlová 1969). A large-scale facility of 900 m^2 was operated in Rupite, Bulgaria until the 2000s. In mass cultivations, green microalgae *Scenedesmus* and *Chlorella* were mostly used. Recently, outdoor TLC were used in pilot trials to study the growth of the cyanobacterium *Arthrospira platensis* (Torzillo et al., unpublished results) and the freshwater microalga *Trachydiscus (Eustigmatophyceae)* (Malapascua et al. 2014).

Since the 1960s, the concept of thin-layer has been developed at the Laboratory of Algal Biotechnology of the Institute of Microbiology at Třeboň (for review see Masojídek and Prášil 2010). Originally, the design of microalgae cultivation in a relatively thin layer (<50 mm) has been based on turbulent flow using corrugated surfaces, or a plane fitted with transversal baffles (Fig. 5). As compared to open reservoirs (ponds, raceways) with the depth of suspension in the range of 100–300 mm where diluted cultures of microalgae (0.5–1 g DM L^{-1}) are grown under limited light, poor mixing and gas exchange, the main advantage of TLCs was to grow well-mixed, thick microalgae culture with a much higher biomass density (>10 g DM L^{-1}). Thus, a much lower volume of dense microalgae suspension can be handled during biomass processing.

Fig. 6 Outdoor large-scale cascades of 650 m^2 with a working volume of 6,500 l built in the 1990s. Microalgae are grown in a smooth thin-layer of 6–8 mm and the total surface-to-volume ratio is about 100 m^{-1}

In the 1990s, when the scientific atmosphere in the Czech Republic became more favourable, microalgae were put back on stage, and another generation of large-scale (650 m^2) outdoor TLCs for microalgae cultivation was built and tested at the Institute of Microbiology in Třeboň (Fig. 6) (Doucha et al. 1993; Lívanský et al. 1995; Doucha and Lívanský 1995; Grobbelaar et al. 1995). Pilot units of 25–50 m^2 of the same principle are also used in several institutions in the country (Institute of Botany at Třeboň, an agricultural farm at Dublovice, etc.). As compared to the TLCs used in the 1960–1970s, the second generation of TLCs employs a much thinner layer of microalgae – less than 10 mm. Instead of densely spaces baffles, plastic rods with a diameter of 13 mm diameter were placed 1.5 m apart and thus the flow velocity could be increased to 0.4–0.5 m s^{-1}. It was just a small step to realize that the inclined-surface system could work best if operated as a smooth inclined surface (glass plates framed by an angle steel structure) without any baffles where the layer of microalgae is only 6–8 mm. First and foremost, the cleaning and maintenance of smooth surface units has been much simpler, as compared with the baffled system. Another advantage of TLCs was easy heating-up by solar irradiance, but on the other hand microalgae suspension was also spontaneously cooled by water evaporation avoiding overheating.

The cell layer thickness below 10 mm in combination with high flow speed (0.4–0.5 m s^{-1}) generates the turbulent flow (Reynolds number of about 4,500) which prevents cell self-shading. Due to the short optical path, light utilisation is more efficient and high optimum biomass densities (15–35 g DM L^{-1}) can be operated in semi-continuous regime, enabling cheaper harvesting (Masojídek et al. 2011a, Masojídek and Torzillo 2014, Doucha and Lívanský 2014). These units are characterized by their high ratio of exposed surface to total culture volume (S/V of ~100 m^{-1}), which enables high volumetric and areal productivity as compared with that of open ponds (S/V ~10 m^{-1}). The short light path in combination with the high cell density and intensive turbulence enables cells to be exposed to intermittent light with short light/dark cycles (10–100 Hz), thus avoiding over-reduction of photosynthetic electron carriers (Hu and Richmond 1996; Richmond 2004; Masojídek et al. 2004). This set-up has allowed achievement of high growth rates, up to biomass concentration of 40–50 g DM L^{-1}. A 100 m^2 pilot system was also tested in the Mediterranean climate where summer productivities were as high as 32 g DM m^{-2} day^{-1} as compared to Central Europe with productivity maxima of about 23 g DM m^{-2} day^{-1} (Doucha and Lívanský 2006). Improved construction of the retention tank caused a significant reduction of the dark phase to about 20 % of total volume; such high productivities as 50 g DM m^{-2} day^{-1} could be achieved in cascade cultivation units in summer days, even in temperate climate zones (Masojídek et al. 2011a). These TLCs have been used for research and biomass production until now.

Recently, thin-layer cascades have been used for the pilot cultivation of various microalgae strains in several countries: Italy (Torzillo et al. 2010), Spain (Jerez et al. 2014; Ihnken et al. 2014), Switzerland (University of Applied Sciences, Zürich) and Greece (Doucha and Lívanský 2006). A large-scale plant consisting of 2 cascade raceway modules of 1,500 m^2 each (total volume of 180,000 L) was installed in Pataias, Portugal for the BIOFAT project (designed and built by the company A4F EU). The unit consists of two sloped platforms (declining of 0.5 %), 10 m wide and 75 m long which form a cascade-like system running in opposite directions. This facility is a hybrid technology between raceway pond and sloping cascade since the layer thickness is 40 mm, resulting in the S/V ratio of about 15 m^{-1}. In this case, the operating biomass density is about 4 g DM L^{-1}.

4.2.1 Latest Innovations of TLC Set-Up

TLCs are constructed in a way that the microalgae culture flows from the top to the bottom over sloping platforms and ends in a retention tank, from where it is pumped back to the top. The units are made up of five parts: cultivation surface – photostage, retention tank, pump, CO_2 supply and aeration, and measurement and control sensors. The module consists of two sloped platforms (divided into lanes separated by bent edges) where the lower end of the upper platform is connected by a trough to the beginning of the lower platform, which is declined in the opposite direction (Fig. 7). The operation cycle starts in a retention tank (degasser) from where the microalgal suspension is circulated by a pump via a return pipe (riser) to the upper part of the cultivation area. Then, the suspension flows back into a retention tank which helps in degassing of excess oxygen produced by the microalgae. Pure CO_2 is supplied directly into the microalgal suspension in the riser. Fast flow in a thin layer suspension shows great importance for growth of microalgae since the aquaculture is well mixed with sufficient light and nutrient availability and produced oxygen is released into the surrounding atmosphere. The culture is circulated over the surface only during the day; it is kept in a retention tank at night to reduce heat loss, or during rainfall to avoid dilution by rainwater. After collection in the retention tank, the culture is mixed by aeration to preserve biological activity. Special software has been designed to enable automatic control and data acquisition of the culture parameters in the experimental unit. The culture's behaviour is monitored by temperature, pH, dissolved oxygen and Chl fluorescence sensors.

As compared with the TLC units used in the 1990–2000, the latest generation of TLCs, model Dahlia designed and constructed in 2012–2013 as a modular system in the Institute of Microbiology in Třeboň. It has several innovations (see below) to improve cultivation process and ease of maintenance (Fig. 7). The module has an area of 90 m^2 and is made up of two identical platforms where microalgal culture is exposed to sunlight in a north-south orientation. Compared to previous units made of fragile glass plates glued to metal frames, the cultivation surface is made of stainless steel which is easily cleaned and maintained, avoiding any problems with winter damage and corrosion. The slope of each platform is (independently) vertically adjustable between 0.5° and 3° which in combination with a variable-flow (20–60 cm s^{-1}) using an open-impeller pump make it possible to set-up suspension layer thickness between 5 and 15 mm. Vertically-adjustable platforms and pumping speed make it possible to change layer thickness in order to study the optimal conditions for microalgal strains at varying biomass density. This is advantageous to regulate light supply to the microalgae culture according to its physiological demand to maintain optimum irradiance regime. The shape of the retention tank was designed to minimise the dark volume of the microalgal suspension which can be as low as 10 %. The S/V ratio can be operated in the range of 60–180 m^{-1}. Hydraulic properties, suspension distribution and flow were improved to reduce energy demand and CO_2 losses.

The cultivation unit is controlled and regulated via sensors to measure photosynthetic activity, temperature, dissolved oxygen concentration and pH. Thermoregulation of the culture can be partially controlled by a heat exchanger in the trough and heating cable in the retention tank. The CO_2 supply is regulated as a pH-stat according to the demand of the microalgal culture. Easy access to all parts of the cultivation device is necessary for cleaning and maintenance purposes.

In order to lower the cultivation unit height (maximum height of 1.7 m above ground) the retention tank is buried of about 0.5 m below ground. The unit is supported by a lightweight scaffolding structure made of rectangular profiles with ground anchors which enhances axial and torsional stiffness. All materials used for construction are biocompatible (PVC, PE, zinc-galvanised parts); the cultivation area is made of stainless steel. This construction is durable to climate conditions and corrosive

Fig. 7 (continued) set-up suspension layer thickness between 5 and 15 mm. The lower end of the upper platform is connected by a trough (8) to the beginning of the lower platform, which is declined in the opposite direction. The operation cycle starts in a retention tank (2; degasser) from where the microalgal suspension is circulated by a pump (3) via a riser (6; a return pipe) to the upper part of the cultivation area where it is distributed by a perforated tube (7; flow direction is indicated by an arrow). The lower end of the upper platform is connected by a trough (8) to the beginning of the lower platform, which is inclined in the opposite direction. Then, the suspension flows back into a retention tank via a screen (10) which helps in degassing of excess oxygen produced by the microalgae. Pure CO_2 is supplied directly into the microalgal suspension in the riser (4). A three-way valve (11) is used for harvesting. Measurement and control sensors (pH, dissolved oxygen, temperature and liquid-level) are mounted in the degasser and in the connecting trough (5). The circulation cycle takes about 60–80 s which can be varied by the pump velocity. The suspension can be harvested via a three-way valve (11). The whole system is controlled by computer software which enables regulation of the cultivation process and data acquisition

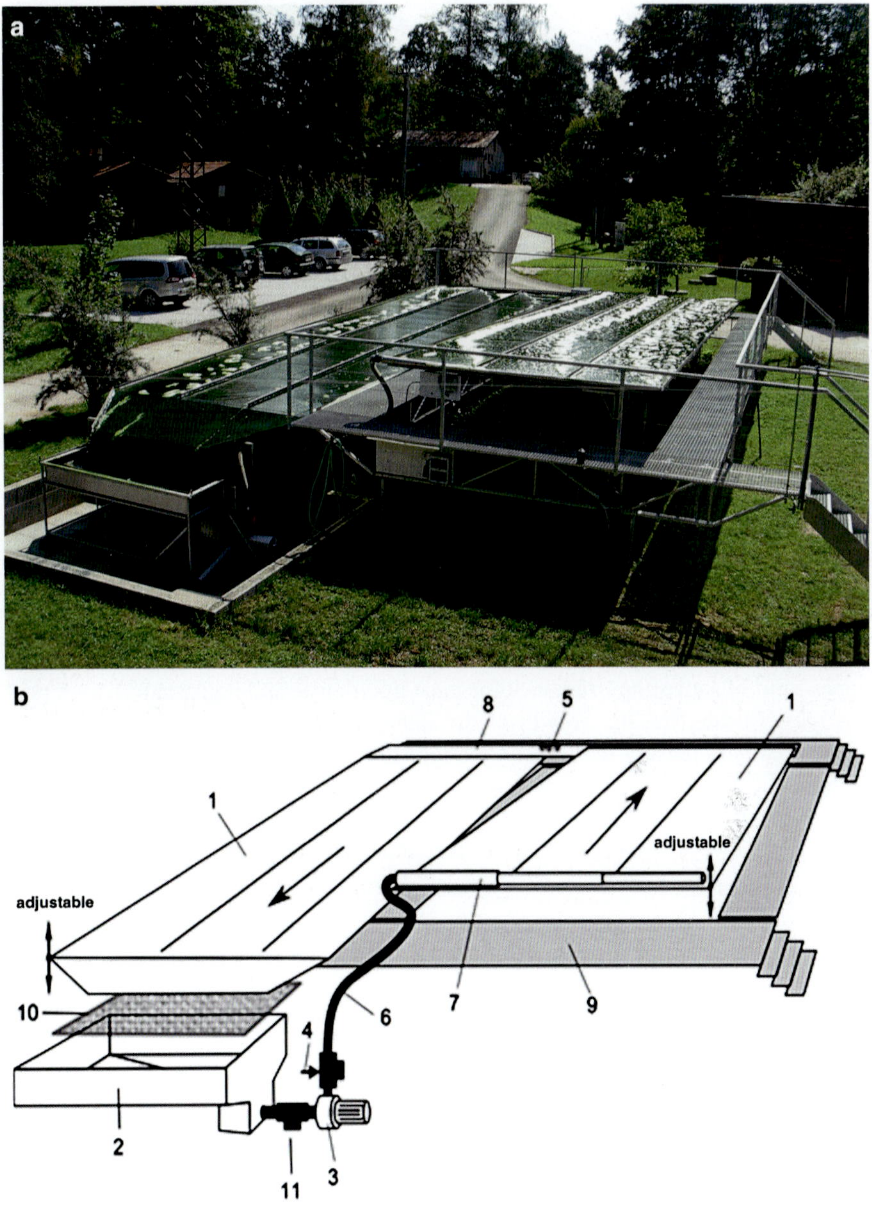

Fig. 7 (Panel **a**) Outdoor thin-layer cascade, model Dahlia for cultivation of microalgae built in 2013. It has an area of 90 m^2 and can contain a total volume of 500–1,500 l. The surface-to-total-volume ratio can be operated in the range between 60 and 180 m^{-1} corresponding to the layer thickness of suspension between 5 and 15 mm. (Panel **b**) Schematic diagram of the 90 m^2 cascade. The module consists of two identical platforms (1) divided into lanes separated by bent edges made of stainless steel. They are supported by scaffolding and exposed to sunlight in a north-south orientation. The slope of each platform is adjustable (between 0.5 and 3°; see two-way arrows) which in combination with a variable-flow maintained by an open-impeller pump make it possible to

environmental factors for tens of years. The unit was made using standardized parts as construction modules connected by joints which ease disassembly of the system. These features make it transportable with a long working life and easy to repair.

This unit for cultivation of microalgae was registered at the Industrial Property Office of the Czech Republic (patent pending PV 2013-803; utility design CZ 27021U1). The use of this demonstration unit is intended for microalgae production as food and feed additives, especially enriched in certain bioactive compounds (e.g. carotenoids, polyunsaturated fatty acids, etc.) or chemical elements (Se, Cr, Fe, Zn), for biodegradation and waste water treatment or CO_2 sequestration.

In model trials a culture of *Chlorella sorokiniana* was grown in an outdoor TLC of 90 m^2 (model Dahlia; Fig. 7) during late summer (September). The S/V ratio was operated between 120–160 m^{-1} and the culture layer thickness was between 5 and 6 mm. The maximum daily irradiance was about 1800 μmol photons m^{-2} s^{-1} and usually cultivation temperature rose from about 15–19 °C in the morning to the midday maximum of 24–29 °C. The growth optimum of this fast growing microalga is rather broad, between 20 and 40 °C. The dissolved oxygen concentrations were between 9 and 10 g L^{-1} at the start of the cultivation area rising to 18–32 g L^{-1} (the variability is caused by ambient irradiance and temperature) before flowing to the retention tank. The measurement of irradiance intensity close to the surface in the photic zone by a spherical microsensor (US-SQS/B; H. Walz, Germany) showed the mean light intensity of about 400 μmol photons m^{-2} s^{-1} which is within the usual upper-limit of saturating irradiance for most microalgae and it guarantees high productivity. In one trial the growth of the thinner culture was rather fast since the culture biomass density increased from the starting point of 2 g L^{-1} to 18 g L^{-1} in 8 days. The starting biomass concentration was about 2 g L^{-1} which was relatively diluted culture in this thin-layer set-up; it resulted in an initial lag-phase for 2 days (Fig. 8, curve with open circles). It corresponded to a chlorophyll concentration of

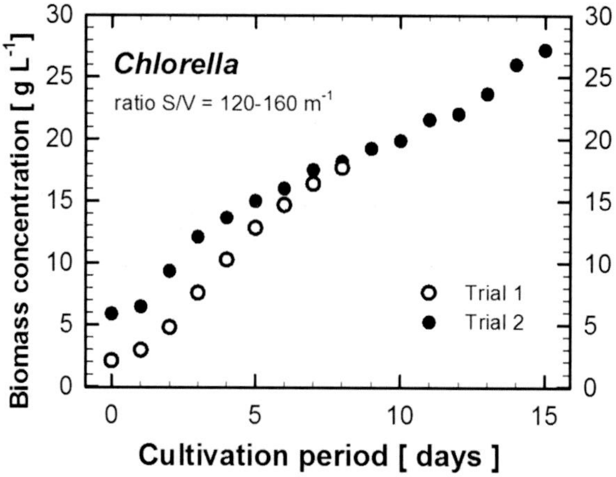

Fig. 8 Growth curves of the *Chlorella sorokiniana* culture. Two trials – 8 and 15-day long – were carried out to measure biomass concentration changes (g L^{-1}) using the 90-m^2 cascade (S/V ratio = 120–160 m^{-1}) in late summer (September). The starting biomass density was about 2 g L^{-1} (*curve with open circles*) and about 5.8 g L^{-1} (*curve with closed circles*), respectively

50–550 mg L^{-1}. In the exponential phase (5–15 g L^{-1}) the specific growth rate was about 0.27 day^{-1}. The diel course of photosynthetic activity of the *Chlorella* culture was monitored from 08:00 to 18:00 h as in samples taken from the culture. Maxima of the relative electron transport rate rETR and the maximum photochemical yield of PSII, F_v/F_m were calculated from rapid light-response curves (using saturation-pulse analysis of fluorescence quenching) (Fig. 9) as described in Malapascua et al. (2014). The so-called rapid light-response curve shows the dependency of photosynthetic electron transport (rETR) on the irradiance intensity E (Fig. 1; see also Kromkamp et al. 1998; White and Critchley 1999; Ralph and Gademann 2005) and provides detailed information on the saturation characteristics of electron transport, as well as the actual performance of a microalgal culture. The value of $rETR_{max}$ at midday was about 350, 2.3-times higher than that in the morning which indicated that the culture was rather active as it responded well to high irradiance. The values of the F_V/F_M ratio usually range between 0.7 and 0.8 in normal non-stressed green microalgae (Masojídek et al. 2013). In this case, the morning value of F_v/F_m was 0.75 which indicated a 'healthy' culture; this variable decreased to 0.6 at 13:00 h, i.e. by about 20 %. The experiments in closed photobioreactors as well as TLCs showed that a midday-depression of PSII photochemical yields between 20 and 30 % as compared with maximal morning values is essential for well-performing cultures (Masojídek et al. 2003, 2011a). A lower or higher depression of photochemical yields indicated low-light acclimated or photoinhibited cultures,

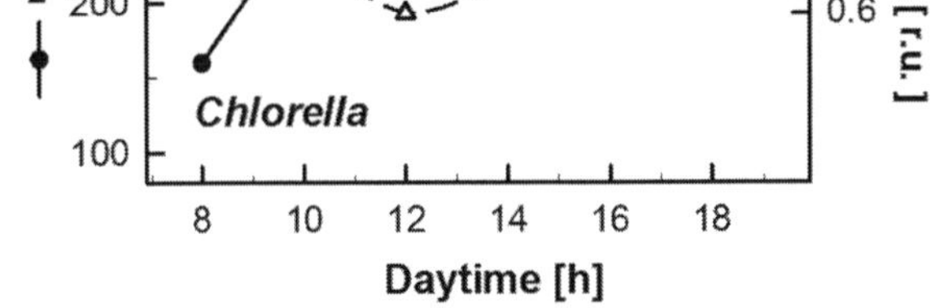

Fig. 9 Diel changes in the maximum relative electron transport rate $rETR_{max}$ and the maximum photochemical yield of PSII, F_v/F_m in *Chlorella sorokiniana* mass culture. The diel course of photosynthetic activity of the *Chlorella* culture was monitored from 08:00 to 18:00 h in samples taken from the culture at exponential phase of growth (see Fig. 7; Day 5). Maxima of the relative electron transport rate $rETR_{max}$ and the maximum photochemical yield of PSII, F_v/F_m were estimated from rapid light-response curves (using saturation-pulse analysis of fluorescence quenching) as described in Malapascua et al. (2014). One typical experiment was taken for presentation of data in this graph

respectively. The night temperatures below 20 °C minimized the respiration losses of biomass to less than 10 %.

In the other trial the starting biomass concentration was about 5.8 g L^{-1} which was about 3-times denser than the previous one. In this case the lag phase was seen only during Day 1, but the growth of this culture was slower during the first week as compared with the lower starting biomass density (Fig. 8, curve with closed circles).

TLCs have, however, some advantages, among which the much higher operating cell concentration, very high daylight productivities, and the possibility to quickly store the culture at night or in case of unfavourable weather conditions. These results are important from a biotechnological point of view in order to optimize the growth of outdoor microalgae mass cultures under varying climatic conditions.

5 Future Prospects of Thin-Layer Systems

Two principles have to be considered, namely reducing layer thickness and using vertical extensions in the configuration of cultivation modules. In thin-layer systems the culture layer usually varies between 10 and 40 mm. Empirically, a further reduction of layer thickness seems to be possible and opens a way to higher cell densities and productivities. An ultrathin-layer system where the microalgae suspension flows by gravity in a thin vertical coating of 0.5–2 mm between two plastic foils to assure a uniform and optimal photon supply was patented in 1994 (DE 4411486 CI, 1994) (Pulz et al. 2013). This system uses the adhesion forces between hydrophilic materials such as foils, plastics, or glass to grow microalgae culture in a thin flowing layer. Subject to research, various configurations of ultrathin layers in the space of a cultivation system are accessible, both for the formation of static (immobilized) and dynamic (flowing) biofilms. The next improvement is the use of light-penetrable and gas-permeable materials such as transparent textile tissues or meshes. These materials allow controlled light and gas supply (O_2, CO_2) within the culture suspension. The ultra-thin layer units with their uniform distribution of microalgae in the photostage allow high culture densities and high productivities.

Recently, a similar principle was used to design the so called 'accordion' PBR which consists of two adjacent transparent sheets, sealed together along to form thin-layer vessels holding microalgae culture arranged like a 'pleated sheet' (US 8709808 B2, 2014 by J.L Cuello & J.W Ley). The PBR includes a support structure and a base reservoir from which the culture is pumped to the upper edge of the photobioreactor between the two sheets.

The most recent development patented by IGV GmdH (Nuthetal, Bergholz-Rehbrücke, Germany) is the so called 'rain PBR' where microalgae are grown in ultrathin-layer (DE 10 2009 027, WO 2010/14154) (Pulz et al. 2013). The key issue of this approach is that a dynamic biofilm in the form of droplets and films is created using meshes or grids, distributing the high-cell-density microalgae culture into tiny fog-to-rain-like droplets in PBR space. While in the present PBR technology, biomass concentrations of 1–5 g DM L^{-1} are usually achievable, this system allows biomass concentrations of 20–40 g DM L^{-1}. Biomass productivity (footprint based) 80 g DM m^{-2} day^{-1} can be estimated for this system. Another advantage of the new system is the predictable reduction of investment costs.

Acknowledgements The authors thank Mr Pavel Souček, Mr Petr Novotný, Ms. Karolina Ranglová and Ms. Soňa Pekařová for technical assistance and Mr Jason Dean for language corrections. The Ministry of Education, Youth and Sports and the Technology Agency of the Czech Republic supported this work through the project Algatech CZ.1.05/2.1.00/03.0110, Algain CZ.1.07/2.3.00/30.0059 and Algaman CZ.1.07/2.3.00/20.0203 and project TA03011027.

Abbreviations

Chl	chlorophyll
DM	dry mass
E	irradiance
F_0, F_v, F_m	minimum, variable and maximum fluorescence in dark-adapted state
F', F_v', F_m'	minimum, steady-state, variable and maximum fluorescence in light-adapted state
F_v/F_m, $\Delta F'/F_m'$	maximum, resp. actual photochemical yield of PSII
GWP	green-wall panel
L/D	light-dark
LED	light-emitting diode
OBD	optimum biomass density
OD	optical density
PAM	pulse-amplitude-modulation
PAR	photosynthetically active radiation
PBR	photobioreactor
P_{max}	maximum rate of photosynthesis
PPFD	photosynthetic photon flux density
PSII	Photosystem II
rETR	relative electron transport rate through PSII
RLC	rapid light-response curve
S/V	surface-to-volume ratio
TLC	thin-layer cascade

References

Baker NR (2008) Chlorophyll fluorescence: a probe of photosynthesis in vivo. Annu Rev Plant Biol 59:89–113

Baker NR, Oxborough K (2004) Chlorophyll fluorescence as a probe of photosynthetic productivity. In: Papageorgiou GC, Govindjee (eds) Chlorophyll-*a* fluorescence: a signature of photosynthesis. Springer, Dordrecht, pp 65–82

Bartoš J (1967) Outdoor algae culture and selection of strains in Cuba. In: Nečas J, Lhotský O (eds) Ann Rep Algolog Lab Třeboň for 1966, pp 123–125
Bilger W, Björkman O (1990) Role of the xanthophyll cycle in photoprotection elucidated by measurements of light-induced absorbance changes, fluorescence and photosynthesis in leaves of *Hedera canariensis*. Photosynth Res 25:173–185
Bradbury M, Baker NR (1984) A quantitative determination of photochemical and nonphotochemical quenching during the slow phase of the chlorophyll fluorescence induction curve of bean leaves. Biochim Biophys Acta 765:275–281
Büchel C, Wilhelm C (1993) In vivo analysis of slow chlorophyll fluorescence induction kinetics in algae: progress problems and perspectives. Photochem Photobiol 58:137–148
Campbell D, Hurry V, Clarke A, Gustafsson P, Öquist G (1998) Chlorophyll fluorescence analysis of cyanobacterial photosynthesis and acclimation. Microbiol Mol Biol Rev 62:667–683
Doucha J, Lívanský K (1995) Novel outdoor thin-layer high density microalgal culture system: productivity and operation parameters. Arch Hydrobiol Algol Stud 76:129–147
Doucha J, Lívanský K (2006) Productivity, CO_2/O_2 exchange and hydraulics in outdoor open high density microalgal (*Chlorella* sp.) photobioreactors operated in a Middle and Southern European climate. J Appl Phycol 18:811–826
Doucha J, Lívanský K (2014) High density outdoor microalgal culture. In: Bajpai R, Prokop A, Zappi M (eds) Algal biorefinery, vol 1. Springer, Dordrecht, pp 147–173
Doucha J, Lívanský K, Bínová J, Kubičko P, Novotný P (1993) Thin-layer high density microalgal culture systém: productivity and operational energy costs. In: Masojidek J, Šetlík, I (eds) Progress in biotechnology of phototrophic microorganisms. Proceedings of 6th international conference on apple algol, Institute of Microbiology, Třeboň, p 40
Emerson R, Arnold W (1932) The photochemical reactions in photosynthesis. J Gen Physiol 16:191–205
Enriquez S, Borowitzka MA (2011) The use of the fluorescence signal in studies of seagrasses and macroalgae. In: Suggett DJ, Prášil O, Borowitzka MA (eds) Chlorophyll a fluorescence in aquatic sciences: methods and applications. Springer, Dordrecht, pp 187–208
Figueroa FL, Conde-Álvarez R, Gómez I (2003) Relations between electron transport rates determined by pulse amplitude modulated chlorophyll fluorescence and oxygen evolution in macroalgae under different light conditions. Photosynth Res 75:259–275
Figueroa FL, Jerez CG, Korbee N (2013) Use of *in vivo* chlorophyll fluorescence to estimate photosynthetic activity and biomass productivity in microalgae grown in different culture systems. Lat Am J Aquat Res 41:801–819
Flameling IA, Kromkamp J (1998) Light dependence of quantum yields for PSII charge separation and oxygen evolution in eukaryotic algae. Limnol Oceanogr 43:284–297
Gilbert M, Wilhelm C, Richter M (2000) Bio-optical modelling of oxygen evolution using in vivo fluorescence: comparison of measured and calculated photosynthesis/irradiance (P-E) curves in four representative phytoplankton species. J Plant Physiol 157:307–314
Gordon JM, Polle JEW (2007) Ultrahigh bioproductivity from algae. Appl Microbiol Biotechnol 76:969–975
Grobbelaar JU (2009) Factors governing algal growth in photobioreactors: the "open" versus "closed" debate. J Appl Phycol 21:489–492
Grobbelaar JU (2012) Microalgae mass culture: the constrains of scaling-up. J Appl Phycol 24:315–318
Grobbelaar JU, Nedbal L, Tichy L, Šetlík I (1995) Variation in some photosynthetic characteristics of microalgae cultured in outdoor thin-layered sloping reactors. J Appl Phycol 7:175–184
http://en.wikipedia.org/w/index.php?search=biorefinary&title=Special%3ASearch&fulltext=1
Hu Q, Richmond A (1996) Productivity and photosynthetic efficiency of *Spirulina platensis* as affected by light intensity, algal density and rate of mixing in a flat plate photobioreactor. J Appl Phycol 8:139–145
Hu Q, Guterman H, Richmond A (1996a) A flat inclined modular photobioreactor for outdoor mass cultivation of photoautotrophs. Biotechnol Bioeng 51:51–60
Hu Q, Guterman H, Richmond A (1996b) Physiological characteristics of *Spirulina platensis* (cyanobacteria) cultured at ultrahigh cell densities. J Phycol 32:1066–1073

Hu Q, Zarmi Y, Richmond A (1998) Combined effects of light intensity, light-path and culture density on output rate of *Spirulina platensis* (Cyanobacteria). Eur J Phycol 33:165–171

Ihnken S, Beardall J, Kromkamp JC, Gómez Serrano C, Torres MA, Masojídek J, Malpartida I, Abdala R, Jerez CG, Malapascua JR, Navarro E, Rico RM, Peralta E, Ezequil JPF, Figueroa FL (2014) Light acclimation and pH perturbations affect photosynthetic performance in Chlorella mass culture. Aquat Biol 22:95–110

Janssen M, de Bresser L, Baijens T, Tramper J, Mur LR, Snel JFH, Wijffels RH (2000) Scale-up aspects of photobioreactors: effects of mixing-induced light/dark cycles. J Appl Phycol 12:225–237

Janssen M, Slenders P, Tramper J, Mur LR, Wijffels RH (2001) Photosynthetic efficiency of *Dunaliella tertiolecta* under short light/dark cycles. Enzyme Microb Techol 29:298–305

Janssen M, Tramper J, Mur LR, Wijfels RH (2003) Enclosed outdoor photobioreactors: light regime, photosynthetic efficiency, scale-up and future prospects. Biotechnol Bioeng 81:193–210

Jerez CG, Navarro E, Abdala R, Malpartida I, Rico RM, Masojídek J, Figueroa FL (2014) Hydrodynamics and photosynthesis performance of *Chlorella fusca* grown in a thin-layer cascade. Aquat Biol 22:111–122

Kok B (1953) Experiments on photosynthesis by *Chlorella* in flashing light. In: Burlew JS (ed) Algal culture: from laboratory to pilot plant, vol 600. Carnegie Institution of Washington Publication, Washington, pp 63–75

Krause GH, Weis E (1984) Chlorophyll fluorescence as a tool in plant physiology. 2. Interpretation of fluorescence signals. Photosynth Res 5:139–157

Krause GH, Weis E (1991) Chlorophyll fluorescence and photosynthesis – the basics. Annu Rev Plant Physiol 42:313–349

Kromkamp JC, Barranguet C, Peene J (1998) Determination of microphytobenthos PSII quantum efficiency and photosynthetic activity by means of variable chlorophyll fluorescence. Mar Ecol Prog Ser 162:45–55

Laws EA, Terry KL, Wickman J, Challup MS (1983) A simple algal production system designed to utilize the flashing light effect. Biotechnol Bioeng 25:2319–2335

Lívanský K, Kajan M, Pilarski PS (1995) Productivity, respiration and chemical composition of the green alga Scenedesmus incrassatulus grown in outdoor cultivation units with and without baffles. Arch Hydrobiol Algol Stud 76:111–128

Malapascua JRF, Jerez CG, Sergejevová M, Figueroa FL, Masojídek J (2014) Photosynthesis monitoring to optimize growth of microalgal mass cultures: application of chlorophyll fluorescence techniques. Aquat Biol 22:124–140

Masojídek J, Prášil O (2010) The development of microalgal biotechnology in the Czech Republic. J Ind Microbiol Biotechnol 37:1307–1317

Masojídek J, Torzillo G, Kopecký J, Koblížek M, Nidiaci L, Komenda J, Lukavská A, Sacchi A (2000) Changes in chlorophyll fluorescence quenching and pigment composition in the green alga Chlorococcum sp grown under nitrogen deficiency and salinity stress. J Appl Phycol 12:417–426

Masojídek J, Papáček Š, Sergejevová M, Jirka V, Červený J, Kunc J, Korečko J, Verbovikova O, Kopecký J, Štys D, Torzillo G (2003) A closed solar photobioreactor for cultivation of microalgae under supra-high irradiance: basic design and performance. J Appl Phycol 15:239–248

Masojídek J, Koblížek M, Torzillo G (2004) Photosynthesis in microalgae. In: Richmond A (ed) Handbook of microalgal mass cultures, 1st edn. Wiley Blackwell, Chichester, pp 20–39

Masojídek J, Sergejevová M, Rottnerová K, Jirka V, Korečko J, Kopecký J, Zaťková I, Torzillo G, Štys D (2009) A two-stage solar photobioreactor for cultivation of microalgae based on solar concentrators. J Appl Phycol 21:55–63

Masojídek J, Kopecký J, Giannelli L, Torzillo G (2011a) Productivity correlated to photobiochemical performance of *Chlorella* mass cultures grown outdoors in thin-layer cascades. J Ind Microbiol Biotechnol 38:307–317

Masojídek J, Vonshak A, Torzillo G (2011b) Chlorophyll fluorescence applications in microalgal mass cultures. In: Suggett DJ, Prášil O, Borowitzka MA (eds) Chlorophyll a fluorescence in aquatic sciences: methods and applications. Springer, Dordrecht, pp 277–292

Masojídek J, Koblížek M, Torzillo G (2013) Photosynthesis in microalgae. In: Richmond A, Hu Q (eds) Handbook of microalgal culture: applied phycology and biotechnology. Wiley Blackwell, Chichester, pp 20–39
Masojídek J, Torzillo G (2014) Mass cultivation of freshwater microalgae, 2nd edn. On-line database Earth Systems and Environmental Sciences, Elsevier, 13 p. http://dx.doi.org/10.1016/B978-0-12-409548-9.09373-8
Matthijs HCP, Balke H, VanHes UM et al (1996) Application of light-emitting diodes in bioreactors: flashing light effects and energy economy in algal culture (*Chlorella pyrenoidosa*). Biotechnol Bioeng 50:98–107
Maxwell K, Johnson GN (2000) Chlorophyll fluorescence – a practical guide. J Exp Bot 51:659–668
Nedbal L, Tichy V, Xiong FH, Grobbelaar JU (1996) Microscopic green algae and cyanobacteria in high-frequency intermittent light. J Appl Phycol 8:325–333
Pulz O, Broneske J, Waldeck P (2013) IGV Gmb H experience report, industrial production of microalgae under controlled conditions: innovative prospects. In: Richmond A, Hu Q (eds) Handbook of microalgal culture: applied phycology and biotechnology. Wiley Blackwell, Chichester, pp 445–460
Ralph PJ, Gademann R (2005) Rapid light curves: a powerful tool to assess photosynthetic activity. Aquat Bot 82:222–237
Richmond A (2003) Growth characteristics of ultrahigh-density microalgal cultures. Biotechnol Bioprocess Eng 8:349–353
Richmond A (2004) Biological principles of mass cultivation. In: Richmond A (ed) Handbook of microalgal mass cultures. Wiley Blackwell, Chichester, pp 125–177
Richmond A (2013) Biological principles of mass cultivation of photoautotrophic microalgae. In: Richmond A, Hu Q (eds) Handbook of microalgal culture: applied phycology and biotechnology. Wiley Blackwell, Chichester, pp 171–204
Richmond A, Cheng-Wu Z (2001) Optimization of a flat plate glass reactor for mass production of *Nannochloropsis* sp. Outdoors. J Biotech 85:259–269
Rodolfi L, Chini Zittelli G, Bassi N, Padovani G, Biondi N, Bonini G, Tredici MR (2009) Microalgae for oil: strain selection, induction of lipid synthesis and outdoor mass cultivation in a low-cost photobioreactor. Biotechnol Bioeng 102:100–112
Schreiber U (2004) Pulse-amplitude-modulation (PAM) fluorometry and saturation pulse method: an overview. In: Papageorgiou GC, Govindjee (eds) Chlorophyll a fluorescence: a signature of photosynthesis, Advances in photosynthesis and respiration, vol 19. Springer, Dordrecht, pp 279–319
Schreiber U, Schliwa U, Bilger W (1986) Continuous recording of photochemical and nonphotochemical fluorescence quenching with a new type of modulation fluorometer. Photosynth Res 10:51–62
Schreiber U, Endo T, Mi H, Asada K (1995) Quenching analysis of chlorophyll fluorescence by the saturation pulse method: particular aspects relating to the study of eukaryotic algae and cyanobacteria. Plant Cell Physiol 36:873–882
Šetlík I, Komárek J, Prokeš B (1967) Short account of the activities from 1960 to 1965 and some future prospects, Annu. Rep. Algolog. Lab for 1966. Institute of Microbiology, Třeboň
Šetlík I, Šust V, Málek I (1970) Dual purpose open circulation units for large scale culture of algae in temperate zones. I. Basic design considerations and scheme of a pilot plant. Algol Stud 1:111–164
Strasser RJ, Srivastava A, Govindjee (1995) Polyphasic chlorophyll a fluorescence transient in plants and cyanobacteria. Photochem Photobiol 61:33–42
Strasser RJ, Tsimili-Michael M, Srivastava A (2004) Analysis of the chlorophyll a fluorescence transient. In: Papageorgiou GC, Govindjee (eds) Chlorophyll a fluorescence: a signature of photosynthesis. Advances in photosynthesis and respiration, vol 19. Springer, pp 321–362
Suggett DJ, Moore CM, Geider RJ (2011) Estimating aquatic productivity from active fluorescence measurements. In: Suggett DJ, Prášil O, Borowitzka MA (eds) Chlorophyll a fluorescence in aquatic sciences: methods and Applications. Springer, Dordrecht, pp 103–128

Tennessen DJ, Bula RJ, Sharkey TD (1995) Efficiency of photosynthesis in continuous and pulsed light emitting diode irradiation. Photosynth Res 44:261–269

Ting CS, Owens TG (1992) Limitation of the pulse-modulated technique for measuring the fluorescence characteristics of algae. Plant Physiol 100:367–373

Torzillo G, Accolla P, Pinzani E, Masojídek J (1996) *In situ* monitoring of chlorophyll fluorescence to assess the synergistic effect of low temperature and high irradiance stresses in *Spirulina* cultures grown outdoors in photobioreactors. J Appl Phycol 8:283–291

Torzillo G, Bernardini P, Masojídek J (1998) On-line monitoring of chlorophyll fluorescence to assess the extent of photoinhibition of photosynthesis induced by high oxygen concentration and low temperature and its effect on the productivity of outdoor cultures of *Spirulina platensis* (Cyanobacteria). J Phycol 34:504–510

Torzillo G, Giannelli L, Martinez-Roldan AJ (2010) Microalgae culturing in thin-layer photobioreactors. In: Bardone E, Viglia A (eds) 2nd international conference on industrial biotechnology. Chemical engineering transactions 20, pp 265–270

Tredici M (2004) Mass production of microalgae: photobioreactors. In: Richmond A (ed) Handbook of microalgal mass cultures. Wiley Blackwell, Chichester, pp 178–214

Tredici M (2010) Photobiology of microalgae mass cultures: understanding the tools for the next green revolution. Biofuels 1:143–162

Vendlová J (1969) Outdoor cultivation in Bulgaria. Mass culture of Scenedesmus in outdoor units. In: Nečas J, Lhotský O (eds) Ann Rep Algolog Lab Třeboň for 1968, Institute of Microbiology, Třeboň, pp 143–152

Vonshak A, Torzillo G, Tomaselli L (1994) Use of chlorophyll fluorescence to estimate the effect of photoinhibition in outdoor cultures of *Spirulina platensis*. J Appl Phycol 6:31–34

Vonshak A, Torzillo G, Accolla P, Tomaselli L (1996) Light and oxygen stress in *Spirulina platensis* (Cyanobacteria) grown outdoors in tubular reactors. Physiol Plant 97:175–179

Walker DA (2009) Biofluels, facts, fantasy, and feasibility. J Appl Phycol 21:509–517

White AJ, Critchley C (1999) Rapid light curves: a new fluorescence method to assess the state of the photosynthetic apparatus. Photosynth Res 59:63–72

Wilhelm C, Becker A, Vieler A, Rautenberger R (2004) Photophysiology and primary production of phytoplankton in freshwater. Physiol Plant 120:347–357

Zahradník J (1967) Bioengineering. Mass culture of Scenedesmus in outdoor units. In: Nečas J, Lhotský O (eds) Ann Rep Algolog Lab Třeboň for 1966, Institute of Microbiology, Třeboň, pp 103–122

Zarmi Y, Bel G, Aflalo C (2013) Theoretical analysis of culture growth in flat-plate bioreactors: the essential role of timescales. In: Richmond A, Hu Q (eds) Handbook of microalgal mass culture: applied phycology and biotechnology. Wiley Blackwell, Chichester, pp 205–224

Zittelli GC, Biondi N, Rodolfi L, Tredici MR (2013) Photobioreactors for mass production of microalgae. In: Richmond A, Hu Q (eds) Handbook of microalgal mass culture: applied phycology and biotechnology. Wiley Blackwell, Chichester, pp 225–266

Zou N, Richmond A (1999) Effect of light-path length in outdoor flat plate reactors on output rate of cell mass and of EPA in *Nannochloropsis* sp. J Biotech 70:351–356

Gas Balances and Growth in Algal Cultures

Marcia Morales, Juan Cabello, and Sergio Revah

Abstract This chapter reviews some of the most important aspects related to the growth of microalgae with their main substrate, carbon dioxide (CO_2), and product, oxygen (O_2). Both gases exist in the aqueous nutrient medium in dynamic equilibrium with the incoming CO_2 contained in air and in the case of open systems, with the exterior environment. The concentrations of both gases, along with other environmental parameters including light, temperature, nutrients, etc. affect photosynthesis and, consequently, the growth rates and productivities. The effect of environmental conditions and medium composition is reviewed in terms of mass balances initially and then kinetic growth models based on these conditions. Transfer of CO_2 and O_2 is described and then integrated with cell kinetics highlighting recent developments in models describing dynamic system, along with advanced computational tools used in solving and representing the hydrodynamics in photobioreactors. Finally, an example with experimental results is presented. Detailed procedures to obtain parameters used in microalgal kinetics, mass transfer and hydrodynamics expressions are described and used to validate a dynamic model simulating the performance of an airlift photobioreactor inoculated with *Scenedesmus* sp.

Keywords Microalgae • Photobioreactors • Gas balances • Growth kinetics • Hydrodynamics • Dynamic models • Scenedesmus • Photosynthesis • P-I curves • Photoadaptation • Phooinhibition • Arrenhius • K_La • Mixing • Dispersion coefficient • Mass transfer • Gas holup • Computational fluid dynamic modeling (CFD) • Intrinsic photosynthetic parameters • Model parameters

---

Authors thank Dr. Rakesh K. Baipai, Editor, for inputs and comments that greatly improved this chapter.

M. Morales (✉) • S. Revah
Departamento de Procesos y Tecnología, Universidad Autónoma
Metropolitana-Cuajimalpa, Colonia Santa Fe Cuajimalpa, México, DF, Mexico
e-mail: mmorales@correo.cua.uam.mx

J. Cabello
Departamento de Procesos e Hidráulica, Universidad Autónoma
Metropolitana-Iztapalapa, México, DF, Mexico

A. Prokop et al. (eds.), *Algal Biorefineries*, DOI 10.1007/978-3-319-20200-6_8

1 Introduction

Microalgae can grow autotrophically with light and carbon dioxide (CO_2) from the atmosphere or from other more concentrated CO_2 sources such as combustion gases producing biomass and valuable molecules (Pulz 2001; Pulz and Gross 2004; Chisti 2007). Microalgae can biosynthesize, accumulate, and secrete a wide range of primary and secondary metabolites, many of which are valuable substances with potential applications in the food, pharmaceutical, environmental and cosmetics industries. Some of the most biotechnologically relevant microalgae are the green algae (*Chlorophycea*) such as *Chlorella vulgaris, Haematococcus pluvialis, Dunaliella salina* and the cyanobacteria including diverse strains of *Spirulina.* Many of these microalgae are commercialized mainly as nutritional supplements for humans and as animal feed additives (Borowitzka 1999; Olaizola 2003; Pulz and Gross 2004; Spolaore et al. 2006; Harun et al. 2010; Chu 2012; Priyadarshani and Rath 2012). Biofuels are also produced from microalgal biomass (biodiesel, bioethanol, and biogas) or by reactions promoted by them (biohydrogen) (Mata et al. 2010; Milledge 2011; Varfolomeev and Wasserman 2011). Microalgae have been used as unconventional sources of proteins (Becker 2007), in wastewater treatment facilities for pollution abatement, and proposed for photosynthetic gas exchange and water recycling in space travel (Cogne et al. 2005). Microalgal cultivation systems can be installed near energy power plants which may provide a continuous supply of high CO_2 emissions to decrease greenhouse gases.

There have been extensive studies on microalgal cultivation process improvement to overcome the technological and economic issues faced in their industrial-scale production. These studies include screening and isolation of high CO_2 tolerant strains, optimization of biological and physicochemical parameters, development of new reactor configuration for cultivation, search for new valuable products, and energy saving strategies (Chisti 2007; Brennan and Owende 2010; Harun et al. 2010; Mata et al. 2010; Ratha and Prasanna 2012; Singh and Singh 2014; Zhao and Su 2014). Microalgae selection is based on their specific final use, but commercial viability of algae cultivation depends strongly on the biomass and final product yields and productivities. To achieve the highest performance, the key parameters such as light intensity, temperature, pH and nutrient concentration that require close monitoring and control if possible. Proper selection, design, and operation of a culture system play a critical role in fostering proper growth and cost-effective product formation conditions. One of the main design challenges includes efficient mass transfer of CO_2 from the gas phase (air, CO_2 enriched air, flue gas, etc.) to the liquid phase as microalgae consume mainly the dissolved CO_2 molecules to form the cellular organic carbon components. Reliable mass transfer correlations help to determine the transfer rate of CO_2 from gas to the liquid phase and the amounts of the gas supply required to support maximum possible growth and product formation.

This chapter presents the main mechanisms involved in gas transport, microalgal-CO2 fixation, and biomass production, specifically focusing on the effect of physicochemical and hydrodynamic processes. Recent advances on dynamic models based on differential gas mass balances are presented along with a detailed case study on kinetic and hydrodynamic characterization of an airlift photobioreactor.

2 Photosynthesis and CO_2 Concentrating Mechanism

Microalgal photosynthesis is a biophysicochemical process that converts CO_2 into organic compounds using photonic energy and releases molecular O_2. The photosynthesis involves light dependent and light independent reactions. Light dependent reactions capture the photonic energy and convert ADP and $NADP^+$ into the energy carriers ATP and $NADPH{\cdot}H^+$ via electron transport chain producing O_2. The light independent reactions then capture CO_2 and produce the precursors of carbohydrates using ATP and $NADPH{\cdot}H^+$ by the Calvin–Benson cycle. Absorption of photons by the carotenoids and chlorophyll antenna complex occurs in a time scale of picoseconds (Luo and Al-Dahhan 2011). The resulting excitation energy can be reradiated into fluorescence, dissipated as heat, or transferred to other pigment molecules and ultimately to the reaction centers (Falkowski and Raven 1997). The energy distribution between these processes depends on the fraction of reaction centers in the oxidative state. At the reaction centers, the excitation energy is utilized to split water molecules evolving O_2 and producing chemical reductants ($NADPH{\cdot}H^+$) and chemical energy (ATP) through a complex electron transport chain (e.g., the known Z-scheme) over a time scale of 10 ms (Falkowski and Raven 1997). These chemical reductants and energy are subsequently transferred into the stroma (surrounding aqueous phase), and used to assimilate inorganic carbon into organics through the Calvin-Benson cycle over a time scale of 10 ms (Luo and Al-Dahhan 2011). The turnover time of the overall electron transfer chain, i.e., the minimal time required to transfer an electron from water to inorganic carbon, is in the range of 1–50 ms (Falkowski and Raven 1997) although it is assumed constant under steady state. Finally, the produced organics are further utilized during cell metabolism and duplication.

The Calvin–Benson cycle involves redox reactions using the photosynthetic energy and electrons to convert CO_2 into reduced organic molecules. In unicellular microalgae, CO_2 concentrating mechanism (CCM) plays a vital role during the carbon fixation process as it can enhance the level of CO_2 at the active site of the ribulose bisphosphate carboxylase-oxygenase (Rubisco) in the cells. The CCM increases the rate of photosynthesis and decreases that of photorespiration (Zhao and Su 2014). Dissolved inorganic carbon (DIC) is present in the aqueous nutrient medium in the form of dissolved CO_2, carbonic acid (H_2CO_3), bicarbonate (HCO_3^-), and carbonate (CO_3^{2-}) under dynamic ionization equilibrium, described by the equations below.

1. CO_2 dissolves in water (slow reaction)	$CO_2(g) \leftrightarrows CO_2(aq)$
2. $CO_2(aq)$ reacts with water to form carbonic acid (slow reaction)	$CO_2(aq) + H_2O \leftrightarrows H_2CO_3$
3. Carbonic acid reacts with water to form bicarbonate and H_3O^+ ions (fast reaction)	$H_2CO_3 + H_2O \leftrightarrows HCO_3^- + H_3O^+$
4. Bicarbonate finally reacts with water to form carbonate and H_3O^+ (fast reaction)	$HCO_3^- + H_2O \leftrightarrows CO_3^{2-} + H_3O^+$

Dissolved CO_2 needs to be replenished continuously through mass transfer from air, flue gas or directly injected CO_2 as DIC is incorporated into the cells. Some microalgae species can use only the dissolved CO_2 for photosynthesis, whereas others can use HCO_3^- also. Others may utilize carbonate also, though this form may be toxic for other species (Moss 1973).

HCO_3^- utilization by cells occurs directly by active transport and cation exchange, as well as indirectly through action of carbonic anhydrases (CA) that catalyze conversion of HCO_3^- to CO_2 and OH^-. CAs enable high CO_2 flux rates in the chloroplast stroma (Raven et al. 2012; Meyer and Griffiths 2013). Three types of CAs have been reported; these include the periplasmic carbonic anhydrase (pCA), the cytosolic carbonic anhydrase (cyCA), and the chloroplast carbonic anhydrase (chCA). The function of pCA is to balance the CO_2 and HCO_3^- and continuously supply CO_2 for cells. The cyCA may accelerate the transport of CO_2 and HCO_3^- through plasma membrane to chloroplasts. The chCA is considered a key CA in CCM (Zhao and Su 2014) as the inorganic carbon transport system located on the chloroplast envelope delivers HCO_3^- to the stroma (Aizawa and Miyachi 1986; Badger and Price 1994). The uncatalyzed conversion rate of HCO_3^- into CO_2 is 10,000 times slower than the rates of CO_2 fixation by the primary CO_2 fixing enzyme (Rubisco) (Badger and Price 1994). Hence, ability of CAs to catalytically convert HCO_3^- into CO_2 is critical in microalgal CCM.

2.1 *Microalgal Growth*

Microalgae are fast-growing photosynthetic microorganisms, most dividing every 1–2 days under favorable growing conditions although some may have doubling times as little as 3–4 h. Due to their simple cell structure and fast growth rate, microalgae have CO_2 biofixation efficiencies that are 10–50 times higher than those of terrestrial plants (Richmond 2004). Certain microalgae can also grow heterotrophically using reduced organic carbon sources; this ability comes handy for certain applications as in the case of water treatment systems (Pittman et al. 2011). Some microalgae have the ability to grow mixotrophically with both light, CO_2 and organic carbon, showing an increase in cell production (Alcántara et al. 2013). Similarly to plants, the microalgal chemical composition varies among species and even within the same species, depending on the cultivation conditions (e.g., temperature, pH, CO_2, dissolved O_2, light intensity, nutrient concentration, etc.), and on the growth phase at harvest time (Sánchez Mirón et al. 2003; Juneja et al. 2013).

As shown in Table 1, the main cell components of microalgae include proteins, lipids, and water-soluble carbohydrates. Based on growth conditions, especially the composition of the medium with respect to C:N or C:P ratio, most microalgae will accumulate lipids or carbohydrates.

Table 1 General composition of various microalgae (% dry matter)

Algae	Protein	Carbohydrates	Lipids
Anabaena cylindrical	43–56	25–30	4–7
Chaetoceros calcitrans	36	27	15
Chlamydomonas rheinhardii	48	17	21
Chlorella pyrenoidosa	57	26	2
Chlorella vulgaris	51–58	12–17	14–22
Dunaliella salina	57	32	6–25
Euglena gracilis	39–61	14–18	14–20
Haematococcus pluvialis	10	40	41
Porphyridium cruentum	28–39	40–57	9–14
Scenedesmus obiquus	50–56	10–17	12–14
Spirogyra sp.	6–20	33–64	11–55
Spirulina platensis	46–63	8–14	4–9
Synechococcus sp.	63	15	11
Scenedesmus dimorphus	8–18	21–52	16–40
Spirulina maxima	60–71	13–16	6–7
Scenedesmus obtusiusculus[a]	11–25	28–63	20–56
Scenedesmus dimorphus	8–18	21–52	16–40

Adapted from Becker (2007)
[a]Toledo-Cervantes et al. (2013)

2.2 *Effect of Nutrients*

2.2.1 Carbon Source

Under the premise that cells contain about 50 % carbon, approximately 1.8 ton of CO_2 are required to produce 1 ton of algal biomass (Chisti 2007; Alcántara et al. 2013). The usual sources of CO_2 include: (i) atmospheric CO_2, (ii) industrial exhaust gases (e.g. flue gas), and (iii) soluble carbonates (e.g. $NaHCO_3$ and Na_2CO_3). Although atmospheric CO_2 can be used, its low concentration (approximately 400 ppmv) cannot sustain the high transfer rates required for intensive biomass production; hence, gas streams containing higher CO_2 content are usually employed for cultivation of algae. The requirements of CO_2 vary according to the microalgal species and the cultivation methods. While some species can survive in media containing very high dissolved CO_2 concentration, lower dissolved CO_2 concentrations are desirable for maximum biomass productivity. When flue gases are used as CO_2 source, other flue gas components may turn out to be toxic to cells and flue-gas conditioning may be required. Negoro et al. (1993) found that growth productivity of *Nannochloropsis* sp. and *Phaeodactylum* sp. was barely influenced by the content of SO_x and NO_x in flue gases. Similarly, Kao et al. (2014) also reported that low levels of NO_x typically present in scrubbed flue gas did not inhibit *Chlorella* growth.

2.2.2 Nutrient Source

Growth media must provide the inorganic elements that constitute the algal cell (Chisti 2007). These include nitrogen (N), phosphorus (P), magnesium (Mg), sulfur (S) and trace elements. Minimal nutritional requirements can be determined using the approximate molecular formula of microalgal biomass composition which was reported to be $CO_{0.48}H_{1.83}N_{0.11}P_{0.01}$ by Chisti (2007). Alcántara et al. (2013) reported an approximate composition for *Chlorella sorokiniana* as $CO_{0.43}H_{1.63}N_{0.14}P_{0.006}S_{0.005}$, that did not vary significantly when grown mixotrophically.

As a constituent of both nucleic acids and proteins, nitrogen is directly associated with the primary metabolism of microalgae (Kumar et al. 2010). Biomass production depends on the type of nitrogen source selected and its concentration. Common nitrogen sources include nitrate (NO_3^-), nitrite (NO_2^-), ammonium salts, or urea. When *Isochrysis galbana* was cultured in media containing nitrate or nitrite or urea, it was found that urea favored significantly higher growth rates and higher accumulation of lipids (Fidalgo et al. 1998). It should be noted that the N source has an interactive influence with other process factors, particularly with the C source (Bilanovic et al. 2009). A high N/P ratio helps increase the growth rate and biomass production under low atmospheric CO_2 concentrations. Phosphates are part of the backbone of DNA and RNA, essential energy carrier molecules such as ATP and phospholipids, and are generally supplied as inorganic $H_2PO_4^{1-}$ or HPO_4^{2-}. Limitation of these key nutrients (N, P deplete conditions) shifts the cellular metabolism to accumulation of energy storage compounds such as lipids or carbohydrates (Takagi et al. 2000; Goldberg and Cohen 2006; Rodolfi et al. 2009; Xin et al. 2010; Juneja et al. 2013). Sulfur is included in the culture medium as inorganic sulfates and is crucial for the formation of the amino acids methionine and cysteine. Other required components, usually include trace metal (<4 ppm) added as inorganic salts, such as calcium (Ca^{2+}), potassium (K^+) and magnesium (Mg^{2+}). Trace elements, iron (Fe^{3+}), manganese (Mn^{2+}), cobalt (Co^{2+}), zinc (Zn^{2+}), copper (Cu^{2+}) and nickel (Ni^{2+}) in the range of micro or nanograms, are also needed to sustain growth of cells. Excess metal concentrations, on the other hand, may inhibit growth, impair photosynthesis, deplete antioxidants, and damage the cell membrane (Juneja et al. 2013). Medium formulation based on cell requirements is a necessary step to increase growth rates and productivity (Danquah et al. 2010).

Sodium chloride (NaCl) is a relevant medium component, especially for marine species, and its control may regulate biomass growth and product formation as reported by Takagi and Yoshida (2006) for a marine *Dunaliella* strain. In another example, (Ben Amotz and Tornabene 1985) reported that lipid content of *Botryococcus braunii* grown in 0.50 M NaCl was higher than in cells grown in medium containing no NaCl; but these cells had reduced protein, carbohydrates, and pigment contents. Besides its influence in providing the appropriate growth environment, increased NaCl content affects the solubility of CO_2 and O_2 in medium also.

2.3 *Effect of Environmental Factors*

2.3.1 CO_2 Concentration

The concentration of DIC has a direct influence on growth and product formation in autotrophic microalgae. This concentration is the result of CO_2 input through gas-liquid mass transfer from the incoming gaseous stream, the chemical equilibria described previously, and its uptake rate by the microalgae. Gas-liquid mass transfer of CO_2 will be discussed in detail in the following section. CO_2 uptake rate depends on the microalgal strain, cell density, light intensity, temperature, and other environmental factors. Optimal CO_2 content in gas stream for cell growth and the maximum CO_2 fractions tolerated differ significantly between different strains (Solovchenko and Khozin-Goldberg 2013). Most microalgae grow only at low CO_2 concentrations and may be inhibited at levels higher than 5 % CO_2, (Cheng et al. 2006). Inhibition of photosynthesis at higher CO_2 fractions may be related to the inactivation of key enzymes of the Calvin-Benson cycle due to acidification of the stromal compartment of the chloroplast (Solovchenko and Khozin-Goldberg 2013). *Nannochloropsis oculata* grew better in 2 % CO_2 than in air, but growth was suppressed above 5 % CO_2 (Yun et al. 1997). Similarly, Chiu et al. (2008) showed that for a *Chlorella* sp. the optimum level of CO_2 for growth was 2 % and cells were inhibited by CO_2 levels beyond 10 %. This work also showed that higher cell concentrations in medium allowed sustaining growth under higher CO_2 contents. Some microalgae can grow with 10–15 % CO_2, which are the CO_2 concentrations found in flue-gases, but both carbon fixation and biomass productivity are reduced. For example, *Chlorella* sp. KR-1, a fresh water microalga, showed maximum growth at 10 % (v/v) CO_2 and tolerated up to 70 % CO_2, but with an increased lag time and suboptimal productivity (Sung et al. 1998). According to Singh and Ahluwalia (2013) even the most CO_2 tolerant strains such as *Chlorella* sp. T-1 (100 % CO_2), *Scenedesmus* sp. (80 % CO_2), and *Euglena gracilis* (45 %) exhibited their maximum biomass productivity at CO_2 concentrations of 10 %, 10–20 % and 5 % CO_2, respectively. Several authors (Farrelly et al. 2013; Singh and Singh 2014; Zhao and Su 2014; Cuellar-Bermudez et al. 2015) recently reported CO_2 removal efficiencies, CO_2 fixation rates, and biomass productivities for different microalgal strains at different operation conditions.

2.3.2 O_2 Concentration

During photosynthesis, O_2 is produced by photolysis of water and released to the culture medium. High cell density cultivation may be limited not only by the photosynthesis rate but also by the effects of O_2 accumulation (Molina-Grima et al. 1999, Pulz 2001, Suali and Sarbatly 2012). Photosynthetically-generated dissolved O_2 (DO) in the medium is also recognized as a reliable and sensitive indicator of cell culture activity in relation to growth and productivity. Unexpected declines in DO indicate a reduction in activity, which may be associated to an operational problem to be addressed. On the other hand, high DO values in actively growing cultures of photoautotrophic microorganisms result in decreased productivity by photo-inhibition and photo-oxidation, and eventual rapid cell death. DO concentrations over 35 mg L^{-1} may be easily reached in dense algal cultures at midday, which may lead to the photo-oxidative death of the culture under prolonged exposure to full sunlight (Fernández et al. 2012). Concentrations between 35 and 45 mg $O_2 L^{-1}$, representing super saturation conditions (up to 5-times the saturation levels with atmospheric air), have been observed in cultures of *Spirulina* in large open systems (1000 m^2) with insufficient mixing. DO concentrations affect cell protein contents also; addition of 45 % O_2 to the gas phase in a culture of *Spirulina* reduced cell protein content from 48 % (dry weight) to 22 % (Richmond 2004).

In open systems, strong mixing promotes a significant reduction in DO, providing an additional advantage to maintaining a turbulent flow when cells are actively growing, particularly at high cell densities. Since volumetric O_2 production relates to volumetric biomass productivity, O_2 build-up becomes a significant problem in photobioreactors with a high surface to volume (S/V) ratio.

Several design options have been proposed to reduce O_2 accumulation by increasing degassing in photobioreactors. These include keeping the tube length as short as economically possible to minimize O_2 build-up and to replace this volume with fresh mineral medium (Camacho-Rubio et al. 1999); and supplying the gas at higher flow rates, promoting a turbulent regime to favor O_2 equilibration with the incoming air (Grobbelaar 1994). Hydrophobic membranes have also been tested under the extreme conditions of long-term space missions (Cogne et al. 2005).

2.3.3 Culture Temperature

Temperature has a strong impact on microalgae growth and productivity. Low temperatures affect overall metabolic activity, specifically the enzymatic system associated with carbon fixation (Ras et al. 2013). High temperatures promote increased rates while inducing protein denaturation and inactivation, reducing the global anabolic activity and influencing respiration and photorespiration more strongly than photosynthesis. This effect is further exacerbated by the reduced solubility of CO_2 as per Henry's law (Zhao and Su 2014). The effect of temperature on chemical or enzymatic reactions is typically modeled by the modified Arrhenius equation (Table 2). The activation energies of enzymatic reactions (E_a) range typically from 10 to 20 kcal mol^{-1} (Cabello et al. 2015). On the other hand, activation energies of thermal death (E_d) are higher than E_a. Hence, beyond an optimal temperature range, the substantially increased thermal death rate results in a steep decrease in growth rate with increasing temperature until growth ceases completely. An example of this behavior is shown in Sect. 6, Fig. 6c. Most microalgae species are capable of carrying out photosynthesis and cellular division over a wide temperature range, generally between 15 and 30 °C, with optimal conditions between 20 and 25 °C although wide variations exist depending on the selected microalgal species. Temperatures under 16 °C will slow down

Table 2 Kinetic models for estimating the photosynthetic rate and microalgae growth

Kinetic expression	Reference/Microalgae/Conditions or parameter values
Light-inhibition model with temperature dependence	Cabello et al. (2014)
$P_{O_2}(I_{av},T)=\left(k_0\exp\left(\frac{-E_a}{T\,R}\right)-k_1\exp\left(\frac{-E_d}{T\,R}\right)\right)\left(\frac{I_{av}}{K_S+I_{av}+\frac{I_{av}^2}{K_I}}\right)$	*Scenedesmus obtusiusculus*
Light model: Beer-Lambert law	T: 5–40 °C, I_0: 8–2640 μmolm^{-2} s^{-1}
Arrhenius expression:	K_s: 75.7 μmolm^{-2} s^{-1}, K_I: 4970 μmolm^{-2} s^{-1}, E_a: 16.1 kcal mol^{-1}, E_d: 30 kcal mol^{-1}
$p_{O_2,\max}=k_0\exp\left(\frac{-E_a}{T\,R}\right)-k_1\exp\left(\frac{-E_d}{T\,R}\right)$	
Model with nutrients- and oxygen dependence	Concas et al. (2010)
$\zeta_m(m,z,I_{av},C_j,C_{O_2})=\left[\mu_{\max}\cdot\frac{I_{av}^n}{K_S^n+I_{av}^n}\cdot\prod_{j=1}^{2}1-\frac{C_j}{K_j+C_j}\cdot\left(1-\frac{C_{O_2}}{C_{O_2,\max}}\right)-\mu_c\right]\cdot m$	*Spirulina platensis*
Light model: Beer-Lambert law	K_S: 160 μmolm^{-2} s^{-1} , n: 1.49, $\mu_{\max}$: 0.074 h^{-1}, K_{NO_3} : 5.314 g m^{-3}, $K_{H_2PO_4}$: 0.028 g m^{-3}
	$C_{O_2,\max}$: 47.9 g m^{-3}
Model with irradiance- and oxygen dependence	Fernández et al. (2012)
$P_{O_2}(I_{av},C_{O_2})=p_{O_2,\max}\left(\frac{I_{av}^n}{K_S^n+I_{av}^n}\right)\cdot\left(\frac{K_{IO_2}}{K_{IO_2}+C_{O_2}}\right)-r_a\cdot P_{O_2,\max}$	*Scenedesmus almeriensis*
Light model: Beer-Lambert law	$P_{O_2\max}$: 2.2×10^{-3} molO_2 Kg^{-1} s^{-1}K_S: 69 μmolm^{-2} s^{-1} n: 0.87 K_{IO_2} : 1.13 molO_2 m^{-3}r_a: 0.01
Model with pH- oxygen- and irradiance dependence	Fernández et al. (2014)

(continued)

Table 2 (continued)

Kinetic expression	Reference/Microalgae/Conditions or parameter values
$P_{O_2}\left(I_{av},C_{O_2},pH\right)=p_{O_2,\max}\left(\frac{I_{av}^{n}}{K_i\exp\left(I_{av}m_1\right)+I_{av}^{n}}\right)\cdot\left(1-\left(\frac{C_{O_2}}{K_{IO_2}}\right)^2\right)\times$ $\times\left(B1\exp\left(\frac{-C1}{pH}\right)-B2\exp\left(\frac{-C2}{pH}\right)\right)-r_1\cdot C_b$	*Scenedesmus almeriensis*
Light model: Beer-Lambert law	$P_{O_2\max}: 4.9E-05 Kg_{O_2}\ Kg^{-1}s^{-1}$
	$K_{IO_2}: 0.7202\ mol\ m^{-3}$
	B1: 2.4, B2: 533, C1: 6.26
	C2: 68.80, K_i: 174 μmolm^{-2} s^{-1}
	m_1: 0.0022
	n: 0.9892
Light-inhibition model with temperature dependence	Bernard and Rémond (2012)
$\mu(T,I)=\mu_{opt}(I)\cdot\phi(T)\quad;\ \mu_{opt}(I)=\mu_{\max}\frac{I}{I+\frac{\mu_{\max}}{\alpha}\left(\frac{I}{I_{opt}}-1\right)^2}$	*Chlorella pyrenoidosa*
$\mu_{\max}=\{0 \text{ for } T<T_{\min};\mu_{opt}\cdot\phi(T) \text{ for } T_{\min}<T\langle T_{\max};0 \text{ for } T\rangle T_{\max}$	T_{min} :5.2 °C, T_{opt} : 38.7 °C
$\phi(T)=\frac{\left(T-T_{\max}\right)\left(T-T_{\min}\right)^2}{\left(T_{opt}-T_{\min}\right)\left[\left(T_{opt}-T_{\min}\right)\left(T-T_{opt}\right)-\left(T_{opt}-T_{\max}\right)\left(T_{opt}+T_{\min}-2T\right)\right]}$	T_{max} : 45.8 °C, μ_{max} : 2.0 d^{-1}
Light model: Beer-Lambert law	I_{opt}: 275 μmol m^{-2} s^{-1}
	I : 20–577 μmol m^{-2} s^{-1}
	T : 15–43 °C

	α: 0.05 d^{-1}
Power model[a]	Molina-Grima et al. (1994)
$\mu = \mu_{max}\left(\frac{I_{av}^{n}}{K_S^{n} + I_{av}^{n}}\right) - \lambda$	*Isochrysis galbana*
Light model: Beer-Lambert law	μ_{max} : 0.046 h^{-1}, n: 1.7
	K_S : 9.67×10^{15} quanta cm^{-2} s^{-1}
	λ: 0.00385 h^{-1}
Monod model[a]	Jeon et al. (2005)
$P_{O_2} = p_{O_2,max}\left(\frac{I_0}{K_S + I_0}\right) - \lambda$	*Haematococcus pluvialis*
	λ:: 3.66 mgO_2g^{-1} h^{-1}
	$P_{O_2,max}$:43.5 mgO_2g^{-1} h^{-1}
	K_S : 177.2 μmol m^{-2} s^{-1}
	I : 0–2000 μmol m^{-2} s^{-1}
	T : 25 °C
Light-inhibition model with first-order decay[a]	Pruvost et al. (2008)
$\mu = \mu_{max}\left(\frac{I}{K_S + I + \frac{I_{av}^{2}}{K_I^{2}}}\right) - \lambda$	*Chlamydomonas reinhardtii*
Light model: Two-flux model with back-scattering	μ_{max} : 0.2479 h^{-1}
	K_S : 69.75 μmol m^{-2} s^{-1}

(continued)

Table 2 (continued)

Kinetic expression	Reference/Microalgae/Conditions or parameter values
	K_I : 2509.66 μmol m^{-2} s^{-1}
	λ:: 0.0531 h^{-1} , E_{ab} = 172 m^2 Kg^{-1} E_s : 868 m^2 Kg^{-1},
	I : 0–236 μmol m^{-2} s^{-1}
	T : 25 °C
Kinetics of biomass growth based on the three-state model of a PSF	Wu and Merchuk (2004)
$\frac{dx_1}{dt} = -\alpha_1 I x_1 + \gamma_1 x_2 + \delta_1 x_3$ $\frac{dx_2}{dt} = -\alpha_1 I x_1 - \gamma_1 x_2 - \beta_1 x_3$ $x_1 + x_2 + x_3 = 1$ $\mu = k_1 \gamma_1 x_2 - \mathrm{Me}$	*Porphyridium* sp.
Biomass production rate	I : 110–550 μmol m^{-2} s^{-1}, T: 24 °C
$\frac{dC_b}{dt} = (\mu - \mathrm{Me}) C_b = (k_1 \gamma_1 x_2 - \mathrm{Me}) C_b$	γ_1: 0.1460 s^{-1},
$\mathrm{Me} = \overline{\mathrm{Me}} \cdot e^{k_m(\tau_1 - \tau_c)}$	β_1: $5.78 \times 10^{-7} m^2 \mu E^{-1}$
Light model: Beer-Lambert law in non-transparent culture medium	α1: 1.93×10^{-3} $m^2 \mu E^{-1}$
	δ1: 4.796×10^{-4} s^{-1}
	τc: 2400 Pa
	k_1: 3.65×10^{-4}
	km: 1.6×10^{-3} Pa^{-1}
	Me: 0.05908 h^{-1}
	$\overline{\mathrm{Me}}$: 1.64×10^{-5} s^{-1}

Model with various limiting factors (irradiance, dissolved CO_2 and metabolite concentration)	Hu et al. (2012)
$\mu = \mu_{\max} \frac{h I \left(1 - e^{-\delta C_b}\right)}{C_b \left(I + r_b\right)} \cdot \left(\frac{C_{L,CO_2}}{K_c + C_{L,CO_2}}\right) \cdot \left(\frac{K_P}{K_P + C_P}\right)$	*Spirulina plantesis*
	h:7.22 gL^{-1}, r_b: 287.42 μmol m^{-2} s^{-1}, δ : 0.72 Lg^{-1}, K_c : 2.38 gL^{-1}, K_p: 1.2×10^{-2} g L^{-1}

[a]Reported by Béchet et al. (2013)

growth, whereas those over 35 °C are lethal for a number of species (Briassoulis et al. 2010). Recently, Ras et al. (2013) presented a review of the effect of temperature on different species, emphasizing the biochemical effect and adaptation to high temperatures. Among the 15 species reported, *Chlorella pyrenoidosa* showed an optimum growth temperature at 38.7 °C and tolerating up to 45 °C. *Scenedesmus obtusiusculus* showed an optimum growth temperature at 35 °C, (Cabello et al. 2015) with a good response for outdoor cultivation.

Microalgae production rates are also impacted by temperature. This is of special interest for large-scale outdoor production systems, which are liable to undergo extreme temperatures. Due to variable conditions, microalgae production in outdoor photo-bioreactors experience temperature fluctuations between 10 and 45 °C. Tolerance and adaptability of microalgae to high temperatures can be improved by induced acclimation. Since flue gases from point sources, such as power plants, are at high temperatures (exiting at around 120 °C), use of temperature-tolerant algae in production systems would achieve significant savings in gas cooling costs. Although most microalgal species considered for carbon mitigation or industrial use are mesophilic (optimum growth temperatures of 20–35 °C), there have been some studies using thermophilic cyanobacteria (temperatures of 42–75 °C) also, in terms of maximum CO_2 tolerance, to address the problem associated with flue gas cooling. *Chlorogleopsis* sp. (or SC2), a thermophilic cyanobacterial species collected from the Yellowstone National Park was able to grow at 50 °C and showed potential for CO_2 biofixation (Ono and Cuello 2007).

2.3.4 Culture pH

Media pH plays an important role in cell growth and product accumulation. Optimal ranges are between 7.9–8.3 for marine-water algae and 6.0–8.0 for fresh-water microalgae (Azov 1982; Olaizola 2003; Pandey et al. 2010; Ying et al. 2014). Depending on the natural habitat of microalgal used, the optimal pH range may vary widely (Moss 1973). This is the case of the filamentous cyanobacteria *Spirulina,* adapted to the alkaline environment (pH 9–11) of soda lakes in Africa and Mexico (Koru 2012; Pandey et al. 2010; Sharma et al. 2014). In contrast, microalgal species that tolerate pH values as low as 4.0 have also been reported (Sung et al. 1998).

The pH of the medium determines the distribution of the different DIC species also, particularly that of CO_2 and HCO_3^- (de Morais and Costa 2007). Furthermore, pH variations may occur because of changes in dissolved CO_2 induced by changing CO_2 input concentrations or by alterations in CO_2 uptake by the cells due to growth rate or increased biomass. There are no significant pH changes in the medium when atmospheric air containing only the natural CO_2 concentration (around 400 ppmv) is fed to the reactors. On the other hand, high CO_2 levels such as those found in flue gases shift the medium pH towards acidic conditions. The medium pH usually decreases because of CO_2 dissolution, which in turn gradually increases as growth proceeds (Valdés et al. 2012; Kao et al. 2014). When flue gases containing SO_2 or NO_x are used, the pH may also vary due to the solubilization of these acid-forming gases (Negoro et al. 1993; Cuellar-Bermudez et al. 2015; Kao et al. 2014). Zhao and Su (2014), reported that pH dropped to about 5.5 in the culture medium when flue gas containing 10–20 % CO_2 was sparged with a rate of 0.25 VVM. Further reductions to 2.5–3.5 occurred when the gas additionally contained 100–250 ppm SO_2. Besides the DIC equilibrium, the uptake of ionic species, such as ammonia, nitrate, phosphates, etc. by the microalgae may also produce major pH changes in the medium. Studies have shown significant pH increases with increasing biomass concentration, and therefore careful management is critical to avoid pH inhibition in the early growth stages. This may be achieved by medium design or by the addition of neutralizing agents.

2.3.5 Irradiance

Light intensity is one of the most important limiting factors in photosynthetic growth of microalgae (Fernández et al. 2012; Costache et al. 2013). The fundamental relationship between photosynthesis and light intensity for individual cells is often represented by the P-I curves (Béchet et al. 2013). Examples of these curves can be found in Sect. 6, Fig. 6b, depicting three distinct light regimes – limited, saturated, and inhibited. At low light intensities, the photosynthesis rate is usually proportional to light intensity since it is limited by the rate of capture of photons. As light intensity increases, microalgae approach 'light-saturation' when their photosynthetic rate becomes limited by the reaction rates following the capture of photons. Under this condition, the rate of photosynthesis is usually maximal and independent of light intensity (Béchet et al. 2013). Microalgae respond to light changes through photo-acclimation to balance the light reactions in the chloroplasts, the energy demand for CO_2 fixation and other metabolic reactions (Vonshak and Torzillo 2004). Short-term adaptation involves state transition and non-photochemical mechanisms that operate to adjust the amount of light energy delivered to the Photosystem II, PSII, on a timescale of seconds to minutes. Long-term photoacclimation starts when the short-term variant is not enough to cope with the adjustments. The long-term strategies include changes in enzymatic activity and gene expression, leading to changes in concentrations of photosynthetic complexes and photosystem stoichiometry. The acclimation periods differ between species.

At higher irradiances, an inhibitory threshold may be reached and the photosynthesis rate decreases due to the deactivation of key proteins in the photosynthetic units (Camacho-Rubio et al. 2003; Fan et al. 2007). This phenomenon, known as photo-inhibition, occurs at light intensities only slightly greater than the level at which the specific growth rate peaks (Pulz 2001; Camacho-Rubio et al. 2003). Photo-inhibitory processes are time-dependent wherein irreversible 50 % damage to the photosynthetic system will occur after 10–20 min (Pulz 2001).

P-I relationships aim to accurately describe the intrinsic relationship between photosynthesis and light intensity, which vary with microalgal species. For this reason, the light response must be experimentally determined at low cell concentrations to ensure that all cells are exposed to uniform light intensity (Béchet et al.

2013). However, commercial cultivations aim to have high-cell concentrations to maximize algal productivity and minimize costs (Molina-Grima et al. 2003). These high concentrations cause light gradients to occur in the medium with individual cells experiencing different light intensities depending on their location in the medium (Béchet et al. 2013). Artificial light, solar light, or both can be used to illuminate algal culture systems. Laboratory-scale culture systems are usually illuminated with fluorescent lamps or other well-controlled artificial light systems at 400–700 nm. In outdoor cultivations, the ultimate source of light is solar, which cannot be controlled and therefore, studies on the optimization of light are usually done indoors with artificial illumination. A comparison of artificial light sources including cool fluorescent, incandescent, halogen, and aluminum-indium-gallium-phosphide (AllnGap II, 643 nm peak wavelength) light emitting diodes (LEDs) was conducted by Kommareddy and Anderson (2003). AllnGap II LEDs, emitting more than 98 % of their light between 400 and 700 nm, were the most efficient and economical light source.

2.4 *Kinetic Models for Microalgal Growth*

Reaction kinetic models describe, in general, how the rates depend on key measurable environmental variables. In the case of microalgae, they are irradiance, temperature, pH, CO_2, O_2 and nutrients. Both empirical and mechanistic kinetic equations have been developed to describe microalgal growth, CO_2 uptake and O_2 produced from photosynthesis or the accumulation of a product. Microalgal activity can be described by linear, exponential, saturation, logistic and fast-acceleration / slow-deceleration equations (Table 2). Modified Monod kinetics is the most widely used empirical model to describe cell growth dependency and inhibition on light, substrate and product. The effect of temperature on the microbial growth kinetics can be represented by the Arrhenius expression (Costache et al. 2013; Cabello et al. 2014). Alternatively, Bernard and Rémond (2012) expressed the maximum specific growth rate as a function of the minimum, maximum, and optimum temperatures for photosynthesis (T_{min}, T_{max}, and T_{opt}, respectively) (Table 2). Béchet et al. (2013) described that algal biomass productivity is the net result of photosynthesis and endogenous respiration. Predicting the rate of these mechanisms during outdoor cultivation is challenging because algal activity is influenced by factors such as light intensity, temperature, pH, DO, and nutrient concentration. Alternatively, experimental data of microbial growth can be fitted into empirical equations by non-linear regressions. Logistic models, such as the modified Gompertz equation, and simple saturation models (Monod) can be used to fit the experimental data of cell growth and product formation curves in order to estimate the maximum specific growth/production rates, lag phase and maximum cell/product concentration (Béchet et al. 2013).

The prediction of the P-I curve is the minimal requirement for mathematical modeling. As it was previously stated, these relationships have been established empirically for specific cases. In response to the limitations of the fixed-parameter empirical P-I models, dynamic models of photosynthesis have been proposed where individual steps of photosynthesis are represented, including as a minimum one photochemical energy capture step and a metabolic consumption step. Wu and Merchuk (2004) included in the dynamic model, the representation of the photosynthetic growth in an alternating light-dark regime, which is based on a three-state model (resting state, x_1, activated state, x_2, and inhibited state, x_3) of a photosynthetic factory model (PSF), furthering the work of Eilers and Peeters (1988). Furthermore, approaches considering photoadaptation and photoinhibition in a multi-step enzyme mediated process have been developed by Eilers and Peeters (1993) and Camacho-Rubio et al. (1989 and 2003).

The models in Table 2 represent intrinsic rates, referring only to the biological effects assuming no transport-limitations, no light attenuation, and uniform irradiance. In other words, these models account only for biological phenomena as affected by the amount of light energy at the cells. Maintaining high growth rates requires that the cells grow at their optimal conditions, which may need implementing heating or cooling, pH control, and proper nutrient management. However, observable production rates combine effects of biological processes, mass transport, and light attenuation and concentration gradients that occur under poor mixing conditions and cause local nutrient limitation (e.g. CO_2) or O_2 inhibition (Béchet et al. 2013). These phenomena are reviewed in the Sects. 4 and 5.

3 Mass Transfer

Microalgal cultivation systems can be classified as open systems (e.g., raceway ponds, lakes, etc.) or closed systems (e.g., column, tubular, flat plate, etc.). Table 3 compares the main photobioreactor configurations including mass transfer aspects, which are described in the following subsections.

These microalgae cultivation systems should be designed to deliver the required mixing and to provide turbulent flow in the multiphasic system (i.e. CO_2-medium-microalgae) to enhance mass transfer, to achieve light, temperature, pH and nutrients homogeneity, to remove O_2, and to prevent microalgal aggregation and sedimentation. Beyond the economics, proper mixing and flow turbulence design is also important since excessive shear stress may lead to microalgal cell damage (Sánchez Mirón et al. 2003; Gallardo-Rodríguez et al. 2011).

Gas-liquid mass transfer may become the limiting step to optimal performance of microalgae systems under two principal situations, both involving solubility of gases in the medium. In the first place, CO_2 needs to be constantly and sufficiently supplied because limited availability of CO_2 will affect both cell growth and product formation. Secondly, photosynthetically produced O_2, which is dissolved in the medium, must be removed as it is produced since its accumulation may inhibit growth. Most of the other nutrients required for cell growth and metabolism are highly soluble in water so their delivery to cells can be easily controlled.

Table 3 Comparison of open pond and different close photobioreactors for algal cultivation

Photobioreactors	Advantages	Disadvantages
Open pond	Low O_2 inhibition	Low biomass concentration and productivity
	Low operation costs	High contamination risk
	Low construction costs	High CO_2 losses
	Experience with the use of large scale systems	High water losses for evaporation
		Poor mixing and mass transfer
		Larger footprint than closed systems
Tubular	Large illumination surface area	Fouling and biofilm formation on reactor walls.
	High biomass productivity	Poor mass transfer
	Reduced loss of CO_2	High O_2 accumulation
	Low water losses from evaporation	Photoinhibition risk
		Difficult temperature control
		pH, O_2 and CO_2 gradients of in liquid phase
Flat panel	Easy to scale up	High hydrodynamic stress
	Good light path	Hard to control temperature
	High illumination surface area	Low mixing
	High biomass productivity	Biofilm formation on reactor walls
	Relatively low O_2 accumulation	
	High photosynthetic efficiency	
Column	Low energy consumption	Small illumination surface area
	Medium cost	High liquid holdup
	Easy operation	Complex mixing.
	Good mixing and low shear stress	
	Easy-to-control variables	
	High mass transfer	
	Low photoinhibition and photooxidation	
	High heat capacity (good temperature control)	
	Potentially high reaction rate per unit volume of reactor	

Adapted from Brennan and Owende (2010) and Ho et al. (2011)

Figure 1 depicts the CO_2 and O_2 mass balance issues in a photobioreactor. CO_2 transport from the bulk gas to the suspended microalgal cells involves gas-liquid mass transfer of CO_2 and ionic equilibrium between the different components of DIC, and uptake of DIC by the microalga. In the case of O_2, it is a microalgal byproduct that diffuses from the cell into the medium and may attain super saturation concentration before being transferred to the gas through the gas-liquid mass transfer to bubbles or at the water surface. Oversaturation conditions may lead to O_2 inhibition.

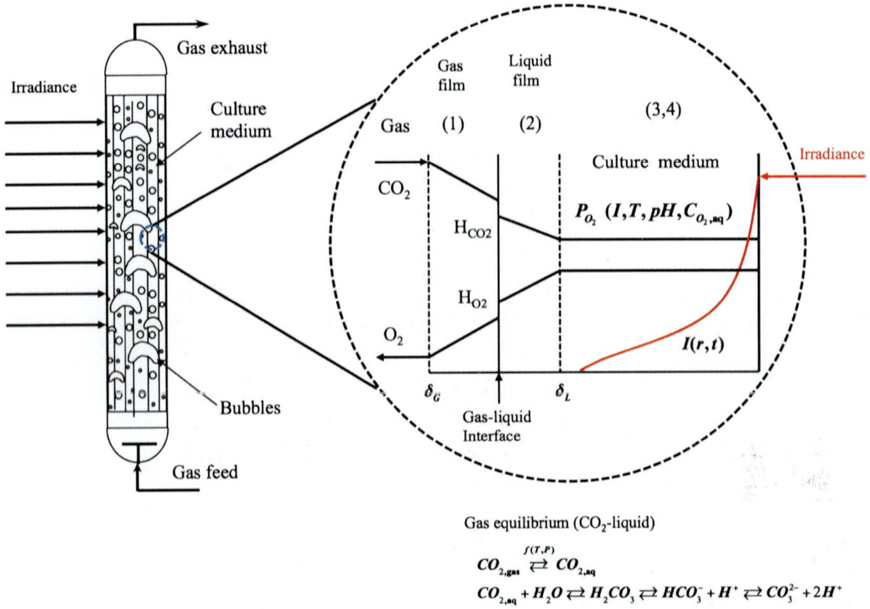

Fig. 1 Concentration distribution of CO_2 and O_2. (*1*) Mass transfer from the bulk concentration in the gas phase to the gas-liquid interface, (*2*) Mass transfer from the gas-liquid interface to the bulk-liquid-liquid phase, (*3*) Mixing, molecular and turbulent diffusion in the bulk liquid, (*4*) Reaction of photosynthesis in the microalga

The gas-liquid mass transfer rate of CO_2 can be expressed as $dC_{L,CO_2}/dt = K_L a_{CO_2}\left(C_{L,CO_2}{}^* - C_{L,CO_2}\right)$. In this equation, K_L is the mass transfer coefficient based on the liquid phase, a is the specific transfer area, and $C_{L,CO_2}{}^*$ and C_{L,CO_2} are the CO_2 concentration at saturation and in the culture medium (bulk-liquid), respectively. As can be seen, the gas-liquid volumetric mass transfer depends on both the K_La and the gradient, which is related to the solubility (absorbing and/or desorbing) of the gas and their variations with temperature. Factors influencing the global mass transfer process are reviewed in the following sections.

3.1 Gas Equilibrium (CO_2-Liquid)

Aqueous CO_2 solubilization typically involves six different reactions as represented in the Fig. 2 (Sugai-Guérios et al. 2014).

The microalgae consumption of CO_2(aq) by the solubilization reactions affects the driving force of CO_2 gas-liquid transfer. Therefore, even though the reactions do not directly affect K_La, it is important to consider them in the determination of the mass transfer rate. Erickson et al. (1987) developed a model based on five reactions valid for processes at 25 °C. Camacho-Rubio et al. (2003) and Nedbal et al. (2010)

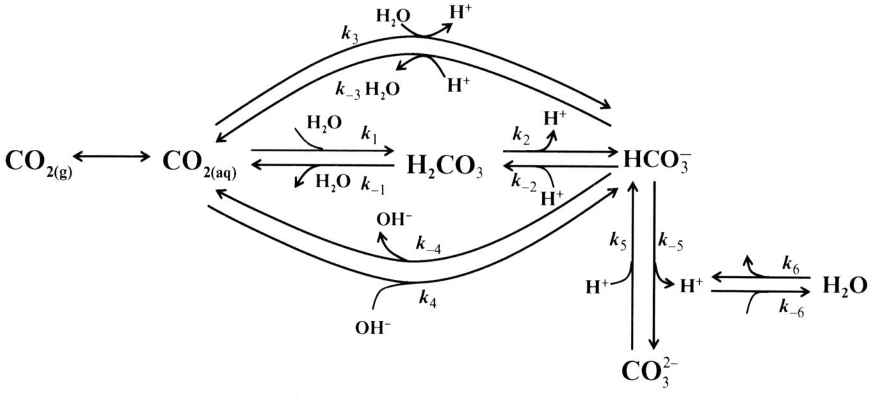

Fig. 2 Reactions involved in the solubilization of CO_2 in aqueous solution (Adapted from Sugai-Guérios et al. 2014)

only considered three reactions of solubilization. Sugai-Guérios et al. (2014) presented a mathematical model describing the complete reaction scheme shown in Fig. 2. The model included correlations to calculate reaction rate constants over the temperature range of 5–40 °C, for any value of pH. The model was validated using literature data for CO_2 mass transfer in two systems, a bubble column absorption unit and a flatpanel photobioreactor.

The solubility of gaseous CO_2 in water is low, but still about 26 times higher than that of O_2. The equilibrium between gaseous CO_2 partial pressure and liquid CO_2 concentration can be described by the Henry's law as $p_a = k_H\ C_a$. The k_H, the Henry coefficient, is related directly with gas solubility and has a value of (k_H =) 3.91×10^{-2} mol atm^{-1} L^{-1} for CO_2 at 20 °C in pure water. The solubility of CO_2 in water reduces with increased temperature and salinity; for example, k_H for CO_2 at 25 °C is 3.39×10^{-2} mol atm^{-1} L^{-1} which corresponds to a 13 % reduction compared to 20 °C. In seawater at 20 °C of 10 % and 35 % salinity, the k_H values are 3.73×10^{-2} mol atm^{-1} L^{-1} (−4.6 % compared to pure water) and 3.32×10^{-2} mol atm^{-1} L^{-1}, (−15 %), respectively (Stumm and Morgan 1996). Furthermore, microalgae preferably uptake dissolved CO_2 whose diffusivity in water is several orders of magnitude slower than in air. Hence, the rate of CO_2 absorption may be a limiting factor for algae cultivation especially when it is found diluted in the input stream or when there is high demand, as in the case of concentrated cultures.

3.2 Hydrodynamic and Mass Transfer Characteristics

Hydrodynamics and mass transfer in photobioreactors involve concepts such as the overall mass transfer coefficient (K_La), mixing, liquid velocity, gas bubble velocity and gas holdup. These are described in following sections.

3.2.1 Volumetric Gas Mass Transfer Coefficient K_La

The overall volumetric mass transfer coefficient (K_La) is the most commonly used parameters for evaluating the CO_2 transfer in photobioreactors. It depends on factors such as agitation rate, superficial gas velocity, sparger type, presence of surfactants/antifoam agents and temperature.

Table 4 presents some of the correlations for the CO_2 mass transfer coefficient. It is important to note that the physicochemical properties, such as viscosity and therefore the flow dynamics and mass transfer in the reactor, of the culture media are not invariant in photobioreactors and these can change due to excreted secondary metabolic products. However, most of the works do not consider these aspects as seen in Table 4.

3.2.2 Mixing

Physical processes such as interphase, interparticle, and intraparticle mass transfer occurring within a multiphase reactor depend significantly upon the mixing characteristics of the various phases involved. The mixing process is conventionally divided in macromixing (which gives information about the retention times of elementary volumes) and micromixing (describing the communication between elementary volumes). Macromixing occurs at the macroscopic scale, i.e., on the scale of the vessel and corresponds to the large-scale flow processes causing distributions of the fluid elements represented by residence time distribution (RTD) (Shah et al. 1978) or that of the local concentrations. Micromixing refers to the phenomena at cell and molecular scales.

Overall macromixing in the reactors is generally characterized by mixing time, axial dispersion coefficient (D_{az}), or the Peclet number $Pe = u_{Le} \, L \, D_{az}^{-1}$. Mixing time is usually defined as the time required for concentration of a tracer to reach 99 % of its equilibrium value from injection (Levenspiel 1999). The axial dispersion coefficient, D_{az}, can be calculated from empirical correlations or estimated from experimental residence time distribution (RTD) data (see Table 4 and Sect. 6).

The RTD curves permit quantitative evaluation of the nature and degree of macromixing as well as the dynamic holdup of each fluid phase in the reactor. RTD is normally measured by the so-called “stimulus-response” techniques that involve injecting a tracer in the inlet stream or at some point within a reactor and monitoring tracer concentration/signal in the exit stream or at some downstream point within the reactor (Levenspiel 1999). Luo and Al-Dahhan (2008) used pulse response technique to study macromixing in an airlift bioreactor. They injected NaOH solution at the bottom of the column and monitored pH values in the liquid phase. An axial dispersion model (ADM) was used to estimate the Peclet number (hence the axial dispersion coefficient) from the pulse response data. Alternatively, particle trajectories measured by the Computer-Automated Radioactive Particle Tracking (CARPT) technique may be used for RTD analysis (Luo and Al-Dahhan 2004). This technique uses a sophisticated detection system to track a specially

Table 4 Empirical and semi-empirical correlations for gas-liquid mass transfer coefficient, axial dispersion coefficients, and gas hold up

Correlations	Reference/Reactor/Conditions
Mass transfer correlations	
$\frac{K_L a D_C^2}{D_L} = 0.6\left(\frac{\mu_L}{\rho_L D_L}\right)^{0.5}\left(\frac{g D_C^2 \rho_L}{\sigma}\right)^{0.62}\left(\frac{g D_C^3 \rho_L^2}{\mu_L^2}\right)^{0.31} \varepsilon_G^{1.1}$	Akita and Yoshida (1973) Bubble column: u_G: 0.003–0.4 m s^{-1}; Dc: 0.152–0.6 m , Hc: 0.126–0.35 m
$K_L a = \frac{14.9g}{u_G}\left(\frac{u_G \mu_L}{\sigma}\right)^{1.76}\left(\frac{\mu_L^4 g}{\rho_L \sigma^3}\right)^{-0.248}\left(\frac{\mu_G}{\mu_L}\right)^{0.243}\left(\frac{\mu_L}{D_L \rho_L}\right)^{-0.604}$	Hikita et al. (1981)[a] Bubble column, u_G: 0.042–0.38 ms^{-1}, Dc: 0.10–0.19 m, Hc: 0.13–0.22 m.
$K_L a = 3.31\left(\frac{D_L \varepsilon_G}{d_b^2}\right)^{1.76}\left(\frac{\mu_L}{\rho_L D_L}\right)^{1/3}\left(\frac{d_b \rho_L u_G}{\mu_L \varepsilon_G}\right)^{1/2}$	Fair (1967)[a] Bubble column, theoretical equation
Axial dispersion correlations	
$D_{az} = 1.23 D_c^{1.5}(u_G)^{0.5}$	Shah et al. (1978) Bubble columns: System: Air-water
$D_{az} = 0.678 D_c^{1.4}(u_G)^{0.3}$	u_G: 0–0.152 ms^{-1}, Dc: 0.4–1 m.
$D_{az} = 0.344(u_G)^{0.32}\rho_L^{0.07}$	u_G: 0–0.05 ms^{-1}, Dc: 0.15–0.2 m.
$D_{az} = \left(0.114 + 0.523(u_G)^{0.77}\right)D_c^{1.25}(1/\mu_L)^{0.12}$	u_G: 0.0051–0.173 ms^{-1}, Dc: 0.019–0.045 m.
	u_G: 0.043–0.338 ms^{-1}, Dc: 0.1–0.19 m.
Axial dispersion model (ADM) for macromixing in bubble column reactor or airlift photobioreactor	
Mass balance for tracer	Luo and Al-Dahhan (2008)
$\frac{\partial c}{\partial \theta} = \frac{1}{\mathrm{Pe}}\frac{\partial^2 c}{\partial z^2} - \frac{\partial c}{\partial z}$	Peclet number, Pe, is obtained by adjusting the ADM to experimental data of residence-time distribution (RTD). The RTD for a flowing fluid is normally obtained by the so-called "Stimulus-response" technique.
$\theta = 0; c(z,\theta) = 0; z = 0; c = 1 + \frac{1}{\mathrm{Pe}}\frac{\partial c(0^+,\theta)}{\partial z}; z = 1; \frac{\partial c(1^-,\theta)}{\partial z} = 0$	
$c = \frac{C}{C_0};\ z = \frac{Z}{L};\ \theta = \frac{t\,u_{Le}}{L};\ \mathrm{Pe} = \frac{u_{Le} L}{D_{az}}$	

Gas hold up correlations	
$$\frac{\varepsilon_G}{(1-\varepsilon_G)^4}=0.2\left(\frac{g\,D_C^2\,\rho_L}{\sigma}\right)^{1/8}\left(\frac{g\,\rho_L^2 D_C^3}{\mu_L^2}\right)^{1/12}\left(\frac{u_G}{\sqrt{g\,D_C}}\right)$$	Akita and Yoshida (1973)[a] Bubble column; u_G: 0.003–0.4 m s^{-1}, u_L: 0–0.044 ms^{-1}, Dc: 0.15–0.6 m, Hc: 1.26–3.5 m. External loop ; column; $A_r/A_d\sim 1$, u_G: 0.2–12 cm/s
$$\varepsilon_{Gr}=0.16\left(1+\frac{A_d}{A_r}\right)\left(\frac{u_{Gr}}{u_{Lr}}\right)^{0.56}$$	Bello (1981)[b] External loop column; Dc ~ 0.15 m (riser) , A_r/A_d ~0.11–0.68
$$\varepsilon_{Gr}=1.07\left(\frac{u_G^2}{gD_C}\right)^{1/3}$$	Kawase and Moo-Young (1987)[b] Concentric cylinder; Dc ~ 0.23 m, Hc ~ 1.22 m; $A_d/A_r\sim 0.33$
$$\varepsilon_{Gr}=0.203\left(\frac{u_{Lr}+u_{Gr}}{gD_{cr}}\right)^{0.31}\left(\frac{\sigma_L^3\rho_L^2}{g(\rho_L-\rho_G)}\right)^{0.012}\left(\frac{u_{Gr}}{u_{Lr}}\frac{A_r}{A_d}\right)^{0.74};Mo=\frac{g(\rho_L-\rho_G)}{\sigma_L^3\rho_L^2};Fr=\frac{(u_{Lr}+u_{Gr})}{gD_{cr}}$$	Kemblowski et al. (1993)[b]
$$\varepsilon_{Gr}=\left(\frac{u_G}{\sqrt{gD_c}}\right)\left(0.415+4.27\left(\frac{u_{Gr}+u_{Lr}}{\sqrt{gD_c}}\right)\left(\frac{g\,d_C^2\rho_L}{\sigma_L}\right)^{-0.188}+1.13\left(\frac{u_{Gr}}{\sqrt{g\,D_c}}\right)^{1.22}\left(\frac{g\mu_L^4}{\rho_L\sigma_L^3}\right)^{0.0386}\right)$$	Koide et al. (1988)[b] Concentric cylinder; A_r/A_d=0.3–1.2, aspect ratio: 5–16
$$\varepsilon_{Gr}=2.47u_G^{0.97}$$	Chisti (1989)[b] Bubble column under bubbly flow regime
$$\varepsilon_G=0.505u_G^{0.47}\left(\frac{0.072}{\sigma}\right)^{2/3}\left(\frac{0.001}{\mu_L}\right)^{0.05}$$	Hikita and Kikukawa (1974)[a] Bubble columns; u_G: 0.042–0.38 ms^{-1}, Dc: 0.1–0.19 m, Hc: 0.6–1.35 m
$$\frac{\varepsilon_G}{(1-\varepsilon_G)^4}=0.14u_G\left(\frac{\rho_L^2}{\sigma(\rho_L-\rho_G)g}\right)^{1/4}\left(\frac{\rho_L^2\sigma^3}{\mu_L^4(\rho_L-\rho_G)g}\right)^{1/24}\left(\frac{\rho_L}{\rho_G}\right)^{5/72}\left(\frac{\rho_L}{\rho_L-\rho_G}\right)^{1/13}$$	Mersmann (1978)[a] Ranges are not defined (semi theoretical equation)
$$\varepsilon_G=0.0672\left(\frac{u_G\mu_L}{\sigma}\right)^{0.578}\left(\frac{\mu_L^4 g}{\sigma^3\rho_L}\right)^{-0.131}\left(\frac{\rho_G}{\rho_L}\right)^{0.062}\left(\frac{\mu_G}{\mu_L}\right)^{0.107}$$	Hikita et al. (1980)[a] u_G,: 0.042–0.38 ms^{-1}, Dc: 0.1 m, Hc: 0.65 m

[a]Reported by Shah et al. (1982)
[b]Reported by Luo and Al-Dahhan (2010)

made small radioactive particle that follows the liquid or slurry phase of interest in the reactors. Three-dimensional Lagrangian trajectories of the tracer particle can be obtained for a statistically long period, and thus provide intrinsic information for classical RTD analyses.

Various macromixing models have been used to correlate the RTD of various phases in a three-phase photobioreactor (Sánchez Mirón et al. 2000; Camacho-Rubio et al. 2004; Luo and Al-Dahhan 2008). Liquid/slurry mixing in these column reactors has been studied by pulse response techniques and analyzed by the classical axial dispersion model (ADM) (Chisti 1998; Sánchez Mirón et al. 2000). However, the ADM model is recommended only for flows with small deviation from plug flow (Peclet number < 20).

3.2.3 Liquid and Gas Bubble Velocities

The mixing characteristics and transport processes within photobioreactors depend strongly on the prevailing flow regime which depends on gas and liquid flow rates, nature of gas distribution, and the fluid properties. Gas bubble size and velocity are dependent on the liquid flow rate. Fine spargers are usually employed to increase gas dispersion inside photobioreactors. Nevertheless, bubbles coalesce reducing the contact area between liquid and gas and the mass transfer rates. At higher gas flow rates, larger bubbles are formed increasing the gas bubble velocity and reducing mass transfer. Baffles or static mixers inside the reactors help break down the larger bubbles into finer ones increasing gas dispersion, thereby improving the mass transfer rates. Three regimes generally occur in bubble columns (homogeneous bubbly, heterogeneous bubbly and slug flows). A schematic representation of these flow regimes is shown in Fig. 3.

The homogeneous bubbly flow occurs at low gas superficial velocities. Its bubble size distribution is monomodal and coalescence and break-up phenomena are negligible. In large diameter columns, a heterogeneous flow regime develops as the superficial gas velocity is increased, due to frequent bubble coalescence and break-up. Large bubbles travel in the center whereas smaller bubbles move along the reactor walls. The undesirable slug flow regime happens at even higher superficial gas velocity and/or in particular when the column diameter is smaller than 0.15 m. In this regime, very large bubbles, i.e., slugs, span the entire cross section of the bubble column.

Sánchez Mirón et al. (2003) reported that superficial gas velocities >0.01 m s^{-1} had a strong effect on cell integrity in the cultures of *Phaeodactylum tricornutum*. As expected, shear stress was reduced with the addition of carboxymethyl cellulose (CMC) in the medium due to increased viscosity of the medium. In CMC-containing medium, increased aeration rates had a positive effect on cell productivity in bubble columns but not in airlift reactors. The difference has been attributed to the increased radial mixing in bubble column reactors.

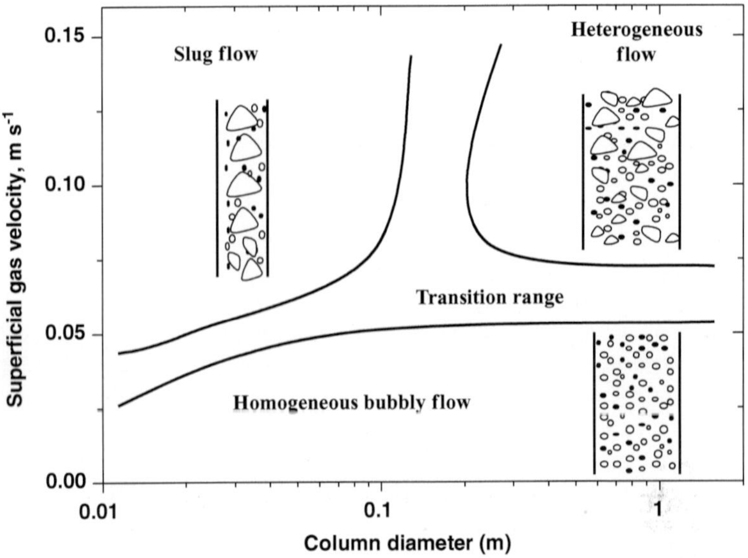

Fig. 3 Approximate dependency of flow regime on superficial gas velocity and column diameter (Adapted of Shah et al. 1982)

3.2.4 Gas Hold Up

Gas holdup (the fraction of the reactor volume occupied by the gas phase) is an important aspect of reactor hydrodynamics, and a critical parameter in photobioreactor design. It can be estimated from the expansion of reactor operating volume due to aeration. In photobioreactors, the gas holdup directly influence the overall gas-liquid mass transfer coefficient (K_La) through its effect on gas circulation rate and the gas residence time in reactor (Chisti 1998; Sánche Mirón et al. 2000).

The relationship between gas holdup and system conditions for bubble columns reactors have been described by Luo and Al-Dahhan (2010) and Shah et al. (1982). In airlift reactors, gas holdups under various operating conditions have been reported for the whole column, as well as separately in the riser and the downcomer. The overall gas holdup is usually calculated from the liquid volumes before and after the gas injection. Gas holdups in the riser and downcomer regions are calculated from measurements of differential pressure drops across these regions (Merchuk et al. 1998; Krishna and van Baten 2003). Empirical or semi-empirical correlations have been developed to predict the overall gas holdups in bioreactors (see Table 4; Shah et al. 1982; Chisti and Moo-Young 1987; Joshi et al. 1990, and Chisti 1998). Gas hold-up values from these correlations can be used in other reactor performance parameters such as gas-liquid mass transfer coefficients for use in reactor designs.

4 Mathematical Modeling of Transport Phenomena in Photobioreactors

Integration of the transport steps within cell kinetics allows expressing growth and product formation in terms of abiotic parameters (i.e. temperature, gas flow rates, nutrient and carbon concentration in the liquid, etc.). Computational Fluid Dynamic (CFD) modeling has been used to study fluid flows and associated mass transfer and kinetic processes in bioreactors. In CFD calculations, the governing equations for continuity, momentum balance, and energy conservation are numerically solved to predict velocity, temperature, shear, pressure profiles, and other parameters such as chemical species concentrations in a fluid flow system (Wang et al. 2015). The primary phase is treated as a continuum and its behavior is predicted using time averaged momentum-lance equations along with continuity and energy balance equations. Trajectories in the particulate phase are calculated using the primary phase flow field forces acting on a large number of the particulate elements. The CFD transport equations can be applied to laminar (Navier-Stokes equations) and turbulent flows (using Reynolds stresses in momentum balance equations). Turbulent flows are characterized by fluctuating velocity, which affects the heat and mass transfer. The increase in turbulence can be represented by an increase in effective fluid viscosity. The Reynolds-average Navier-Stokes model is widely used to determine the effective viscosity by the average and fluctuating components.

Numerous CFD studies have been conducted for different types of photobioreactors. Bitog et al. (2011) and Wang et al. (2015) have presented these studies in detail. COMSOL in the Chemical Engineering module of the commercial Multiphysics (Burlington MA, USA) and FLUENT are generally used in CFD studies investigating the hydrodynamics of bubble-column photobioreactors. Incorporating the reactor geometry and hydrodynamic characteristics in the kinetic models of growth and product formation allows determination of optimal conditions for growth while reducing the dead zones and the energy consumption. Depending on the type of the microalgae, turbulent flow may improve reactor productivity by enhancing mass transfer rates and algae cell exposure to the light, decreasing the effect of photo-inhibition, preventing cell settling, reducing exchange transfer barriers around the cells, and maintaining uniform pH and temperature (Zhang et al. 2002; Liffman et al. 2013). Liffman et al. (2013) used a CFD model to investigate fluid velocity distributions, pressure loss, and hydrodynamic power in high-rate algal raceway ponds. These reactors had different bend configurations to minimize the loss of energy required to circulate the fluid around the raceway and the new design of the raceways generated by the CFD model reduced energy requirements by 87 %, compared to those of conventional raceway ponds. Hadiyanto et al. (2013) studied the fluid flow, shear stress and the dead zone areas in different open raceway pond designs using a CFD model. Based on the model predictions, the optimal ratio of length to width of the channels was suggested to be greater than 10 (Hadiyanto et al. 2013).

Mixing in reactors prevents algae sedimentation and reduces cell attachment to the reactor wall also. The CFD model can be used to track algal cells and generate their trajectories in the reactor. Luo and Al-Dahhan (2011) developed a model to simulate the flow dynamics in a draft tube airlift photobioreactor. They used the Lagrangian equation to track trajectory of algal particles. Perner et al. (2003) carried out theoretical flow calculations with the CFD in order to characterize pressure losses and insufficiently mixed zones in plate photobioreactors with different inner structures, and experimentally validated their calculations. Pruvost et al. (2006) applied CFD to study the fluid trajectory in torus photobioreactors with different impeller configuration. Hekmat et al. (2010) studied the hydrodynamic of an airlift reactor using CFD simulation. The authors suggested that, in the field of photosynthetic microorganism cultivation, the mixing condition along the light gradient was of critical relevance, and that CFD has a great potential for the design of well-mixed photobioreactors. Sato et al. (2006) used CFD to evaluate the mixing performance and light absorption capacity in different kinds of photobioreactors. Optimal reactor configuration was obtained using CFD together with the data from *Chlorococum littorale* cultivation experiments. However, the relationship among parameters of inner structure and hydrodynamics, and algal cultivation efficiency were not analyzed quantitatively in these mentioned studies. Therefore, there is no generic structure optimization method for photobioreactors.

The CFD modeling accounts for column geometry and scale effects (Bitog et al. 2011). However, the success of the CFD simulation strategy is dependent on proper modeling of the momentum exchange and drag coefficients between the gas and liquid phases. Although the air–water system has several drag correlations, there are no general guidelines available to estimate drag coefficients for systems other than air–water systems.

5 Dynamic Models Coupling Mass Transfer and Microalgal Growth

Improved productivity in microalgal systems requires accurate quantitative comprehension of the relationship between performance and design, and operational elements. In photobioreactors under transient conditions, identification and characterization of the dynamic response of key variables is required to implement robust control strategies, leading to increased performance and safety. Models estimating the effect of exogenous variables on growth and productivity have different levels of complexity depending on their applications. Table 5 compiles current dynamic models describing microalgal growth in photobioreactors.

Concas et al. (2010) developed a model for a tubular photobioreactor (BIOCOIL) that involves the cell growth and transient mass balances for nutrients (nitrate and phosphate) and oxygen concentration in liquid phase. The model was validated with a 21 L bench-top photobioreactor operated under steady-state conditions with *Spirulina platensis* under artificial illumination and continuous CO_2 sparging. Fernández et al. (2012) has also developed a microalgal-culture dynamic model that factors in fluid dynamics, CO_2 and O_2 gas-liquid mass transfer, and O_2

Table 5 Mathematical models of microalgae cultures in photobioreactors that consider the gas-liquid mass transfer characteristics, hydrodynamic parameters and light intensity

Model	Reference/Photobioreactor
Mass balances for liquid phase	Camacho-Rubio et al. (1999):
$Q_L dC_{L,O_2} = K_L a_{O_2}\left(C^*_{L,O_2} - C_{L,O_2}\right)Sdz + R_{O_2}(\varepsilon_L)Sdz$ $Q_L dC_{CI} = K_L a_{CO_2}\left(C^*_{L,CO_2} - C_{L,CO_2}\right)Sdz + R_{CO_2}(\varepsilon_L)Sdz$	Predicts axial concentration profiles of dissolved O_2 and CO_2 in continuous outdoor tubular reactor.
Mass balances for gas phase	
$dF_{G,O_2} = -K_L a_{O_2}\left(C^*_{G,O_2} - C_{G,O_2}\right)Sdz$ $dF_{G,CO_2} = -K_L a_{CO_2}\left(C^*_{G,CO_2} - C_{G,CO_2}\right)Sdz$	
Mass balance for nutrients in the liquid phase	Concas et al. (2010):
$\frac{\partial C_j}{\partial t} = D_{az}\frac{\partial^2 C_j}{\partial z^2} - u_{Le}\frac{\partial C_j}{\partial z} - \frac{1}{Y_{X/j}}\int_0^\infty \zeta_m\left(m,z,I,C_j,C_{O_2}\right)\cdot\psi(m,z)\cdot dm$ $j = 1,2;\ 1 = NO_3^-;\ 2 = H_2PO_4^-$	Simulates growth in the tubular cross-section of a recirculating helical reactor (BIOCOIL). Describes the spatial concentration of nutrients and dissolved O_2 in the liquid phase.
Mass balance for oxygen in the liquid phase	
$\frac{\partial C_{O_2}}{\partial t} = D_{az}\frac{\partial^2 C_{O_2}}{\partial z^2} - u_{Le}\frac{\partial C_{O_2}}{\partial z} + \frac{1}{Y_{X/O_2}}\int_0^\infty \zeta_m\left(m,z,I,C_j,C_{O_2}\right)\cdot\psi(m,z)\cdot dm$	
Mass balance for O_2 in the liquid phase	Hu et al. (2012)
$\frac{dC_{L,O_2}}{dt} = K_L a_{O_2}\left(C^*_{L,O_2} - C_{L,O_2}\right) + Y_{O_2/b}\,\mu C_b(t-\tau) + 0.21 k_2 \upsilon_a / V_L$	Estimates the dynamic O_2 concentration in a closed-loop system
Mass balance for CO_2 in the liquid phase	Includes a feed-back routine to regulate and control the O_2 concentration in the gas phase. Uses linear-quadratic gaussian servo controller.
$\frac{dC_{L,CO_2}}{dt} = K_L a_{CO_2}\left(C^*_{L,CO_2} - C_{L,CO_2}\right) + k_{1r} C_{H^+} C_{CO_3^{2-}} - R_a Y_{O_2/b}\,\mu C_b(t-\tau) + 0.03 k_C \upsilon_a / V_L$	

Mass balance for O_2 in the gas phase	
$\frac{dC_{G,O_2}}{dt} = \left\{-K_L a_{O_2}\left(C^*_{L,O_2} - C_{L,O_2}\right)V_L - Q_G C_{G,O_2}\right\} / V_G$	
Mass balance for CO_2 in the gas phase	
$\frac{dC_{G,CO_2}}{dt} = \left\{-K_L a_{CO_2}\left(C^*_{L,CO_2} - C_{L,CO_2}\right)V_L - Q_G C_{G,CO_2}\right\} / V_G$	
Mass balance for the biomass concentration	
$\frac{dC_b}{dt} = \mu C_b\left(t-\tau\right) - K_d C_b$	
Mass balances for O_2 in the liquid phase	Fernández et al. (2012)
$dV\left(1-\varepsilon_G\right)\frac{dC_{O_2}}{dt} = \left(Q_L C_{O_2,\text{inlet}} - Q_L C_{O_2,outlet}\right) + P_{O_2}\left(1-\varepsilon_G\right)C_b dV - K_L a_{O_2}\left(C^*_{O_2} - C_{O_2}\right)_{ml}\left(1-\varepsilon_G\right)dV$	Predicts dissolved O_2 and CO_2 profiles, photosynthesis rate and biomass concentration. Includes mass balance and biological phenomena
Mass balances for CO_2 in the liquid phase ($P_{CO_2} = -P_{O_2}$)	
$dV\left(1-\varepsilon_G\right)\frac{dC_T}{dt} = \left(Q_L C_{T,\text{inlet}} - Q_L C_{T,outlet}\right) + P_{CO_2}\left(1-\varepsilon_G\right)C_b dV - K_L a_{CO_2}\left(C^*_{CO_2} - C_{CO_2}\right)_{ml}\left(1-\varepsilon_G\right)dV$	
Mass balances for the biomass concentration	
$dV\left(1-\varepsilon_G\right)\frac{dC_b}{dt} = \left(Q_L C_{O_2,\text{inlet}} - Q_L C_{O_2,outlet}\right) + P_{O_2}\left(1-\varepsilon_G\right)C_b\, dV\, Y_{b/O_2}$	
Mass balances in the gas phase	
$\frac{F_{N_2} dV}{Q_L}\frac{dY_{O_2}}{dt} = \left(F_{N_2} Y_{O_2,\text{inlet}} - F_{N_2} Y_{O_2,outlet}\right) - K_L a_{O_2}\left(C^*_{O_2} - C_{O_2}\right)_{ml}\left(1-\varepsilon_L\right)dV$	

(continued)

Table 5 (continued)

Model	Reference/Photobioreactor
$\frac{F_{N_2} dV}{Q_L}\frac{dY_{CO_2}}{dt} = \left(F_{N_2} Y_{CO_2,\text{inlet}} - F_{N_2} Y_{CO_2,outlet}\right) - K_L a_{CO_2}\left(C^*_{CO_2} - C_{CO_2}\right)_{ml}(1-\varepsilon_L)dV$	
Mass balance for O_2 in the liquid phase	Cabello et al. (2014):
$\varepsilon_L \frac{\partial C_{L,O_2}}{\partial t} = \varepsilon_L D_{az}\frac{\partial^2 C_{L,O_2}}{\partial z^2} - \varepsilon_L u_{Le}\frac{\partial C_{L,O_2}}{\partial z} - K_L a_{O_2}\left(C_{L,O_2} - \frac{C_{G,O_2}}{H}\right) + P_{O_2} C_b$ $t = 0;\ C_{L,O_2} = C^*_{L,O_2}\ z = 0;\ C_{L,O_2} = C^*_{L,O_2} + \frac{D_{az}}{u_{Le}}\frac{\partial C_{L,O_2}}{\partial z}\ z = L;\ \frac{\partial C_{L,O_2}}{\partial z} = 0$	Predicts O_2 profiles and biomass productivity in an airlift reactor. Includes hydrodynamics and mass transfer and intrinsic oxygen as a function of temperature, light intensity and biomass concentration under N-deplete conditions.
Mass balance for O_2 in the gas phase	
$\varepsilon_G \frac{\partial C_{G,O_2}}{\partial t} = -u_{Ge}\varepsilon_G \frac{\partial C_{G,O_2}}{\partial z} + K_L a_{O_2}\left(C_{L,O_2} - \frac{C_{G,O_2}}{H}\right)$ $t = 0; C_{G,O_2} = C^0_{G,O_2}\ z = L; \frac{\partial C_{G,O_2}}{\partial z} = 0$	
Mass balance for the biomass concentration	
$\frac{dC_b}{dt} = P_b = P_{O_2} C_b Y_{b/O_2} - K_d C_b$	

Hydrodynamic model	
$h_L \cdot \rho_L \frac{\partial u_L}{\partial t} + h_L \cdot \rho_L u_L \cdot \nabla u_{Le} = -\nabla p + \nabla \cdot \left[h_L \cdot \eta_L \left(\nabla u_L + \nabla u_L^T \right) \right] + h_L \cdot \rho_L g$	
$\nabla \cdot (u_L) = 0$; $u_G = u_L + u_{\text{slip}}$; $h_G + h_L = 1$	
$\frac{\partial \rho_G h_G}{\partial t} + \nabla \cdot (h_G \cdot \rho_G \cdot u_G) = 0$	
$\frac{3}{4} \frac{C_d}{d_b} \rho_L \lvert u_{\text{slip}} \rvert u_{\text{slip}} = -\Delta p$; $C_D = \frac{16}{\text{Re}_b}$; $\text{Re}_b = \frac{d_b \rho_L \lvert u_{\text{slip}} \rvert}{\eta_l}$	
$u_{Le} = \frac{\oint_A r \cdot u_L(r,z) \cdot dA}{\oint_A r \cdot dA}$; $u_{Ge} = \frac{\oint_A r \cdot u_G(r,z) \cdot dA}{\oint_A r \cdot dA}$; $\varepsilon_G = \frac{\oint_A r \cdot h_G(r,z) \cdot dA}{\oint_A r \cdot dA}$; $\varepsilon_L = \frac{\oint_A r \cdot h_L(r,z) \cdot dA}{\oint_A r \cdot dA}$	

production. The model was validated using data from a tubular photobioreactor under outdoor conditions with *Scenedesmus almeriensis*. It predicts the dynamic evolution of dissolved CO_2 and O_2 concentrations in the culture in response to solar radiation with estimated values of physicochemical and biological parameters. They highlight that adequate dynamic models are needed to design more advanced predictive control strategies for improved operational efficiency and effectiveness. Hu et al. (2012) validated a dynamic model with a 1.5 L plate-type laboratory photobioreactor with *Spirulina platensis* culture and indoor conditions. This model predicts the O_2 concentration with respect to light intensity and includes a controller routine/module for O_2 regulation. Other models (Camacho-Rubio et al. 1999; Rebolloso-Fuentes et al. 1999; Concas et al. 2010; Fernández et al. 2012; Cabello et al. 2014) have been developed to predict oxygen production during photosynthesis as a function of light intensity.

Microalgae culture dynamic models permit analysis of short-term physiological adaptations of microalgae within photobioreactors when key process parameters are modified. Spadiut et al. (2013) reviewed a range of dynamic changes including shifts, pulses, ramps and oscillations. Dynamic modeling may generate abundant data points, therefore, it is necessary to have a proper data-mining strategy, which entails, a robust experimental procedure with proper on-line monitoring of the response variables. This approach has been used for microalgal culture by Melnicki et al. (2013) following the response of the dissolved oxygen after a series of successive light ramps. The stabilized concentration of dissolved gas for each light step was converted into net photosynthetic production rates to determine parameters such as the maximum photosynthetic capacity ($P_{O2,max}$), the saturating irradiance (I_k), and the apparent quantum yield of PSII. Cabello et al. (2014) analyzed the short-term effect of incident light fluctuations on the oxygen production by the microalga *Scenedesmus obtusiusculus*. In the following section, this case study is presented in detail focusing on the kinetic and hydrodynamic characterization and dynamic response.

6 Case Study

This section presents a dynamic model to describe the growth and O_2 production of *Scenedesmus obtusiusculus* cultivated in an airlift photobioreactor (Cabello et al. 2014, 2015; Toledo-Cervantes et al. 2013). The model equations are presented in Tables 2 and 4 (referred as Cabello et al. 2014). This model includes a kinetic expression for the intrinsic oxygen production rate as function of temperature and irradiance, effect of light attenuation by the biomass and the representation of the hydrodynamics and mass transfer within the system. The model parameters were experimentally determined from the results of short-term experiments carried out in a mini-photobioreactor with no substrate or light limitations. Additional experiments were conducted to obtain the main coefficients for hydrodynamics and mass transfer in the airlift photobioreactor.

6.1 *Description of the Experimental System*

6.1.1 Microalga

The microalga *Scenedesmus obtusiusculus,* a promising strain for carbon dioxide sequestration and lipid storage (Toledo-Cervantes et al. 2013) is an indigenous microalga isolated from the springs in Cuatro Ciénegas, a highly restrictive environment, in the state of Coahuila in Mexico. The microalga, exhibited CO_2 fixation rates and cell productivities higher than those reported for other species used for CO_2 removal. This strain has the potential to be grown with flue gas since no inhibition was observed with 10 % CO_2 in gas phase. It may also be cultivated for biofuels since its lipid profile is suitable for biodiesel production, and it could grow to maximum biomass concentration of 6,000 gDW m^{-3} with biomass productivity of ~500 gDW m^{-3} d^{-1}, and lipid productivity of 200 g m^{-3} d^{-1}. Under conditions of nitrogen starvation, it was able to store lipid content as high as 55.7 % of DW.

6.2 *Experimental Setup*

6.2.1 Airlift Photobioreactor

S. obtusiusculus was cultivated in an internal loop 20 L airlift photobioreactor (8.3 cm inner diameter concentric tube; 105 cm effective height). Figure 4 shows a picture of the reactor as well as schematics of the reactor set-up. The illumination system consisted of high intensity white light LEDs and fluorescent lamps. The intensity of light incident on the surface and in the center of photobioreactor was measured with 2π and 4π quantum sensors, respectively

Liquid pH and the concentrations of dissolved oxygen (DO) and those of CO_2 in the liquid and the gas phases were continuously monitored and recorded on-line by a data acquisition module connected to a computer. The reactor was operated in batch mode with 16.8 L of a mineral medium that was inoculated with 1.8 L of *S. obtusiusculus* and exposed to continuous light intensity of 117 μmol m^{-2} s^{-1}. Air containing 3.8 % CO_2 was constantly sparged at a rate of 3.4 L min⁻1 (corresponding to gas superficial velocity of 0.0104 m s^{-1} based on the cross sectional area of the riser). More detailed description of experimental system can be found in Cabello et al. (2014).

6.2.2 Mini-photobioreactor

The optimal operating conditions and intrinsic kinetic parameters were determined through estimation of photosynthetic activity (Po_2) and rapid assays in a batch 2.75 mL glass mini-photobioreactor (Fig. 5). This mini-reactor had magnetic agitation and an external jacket for temperature control. It was placed inside a cylindrical

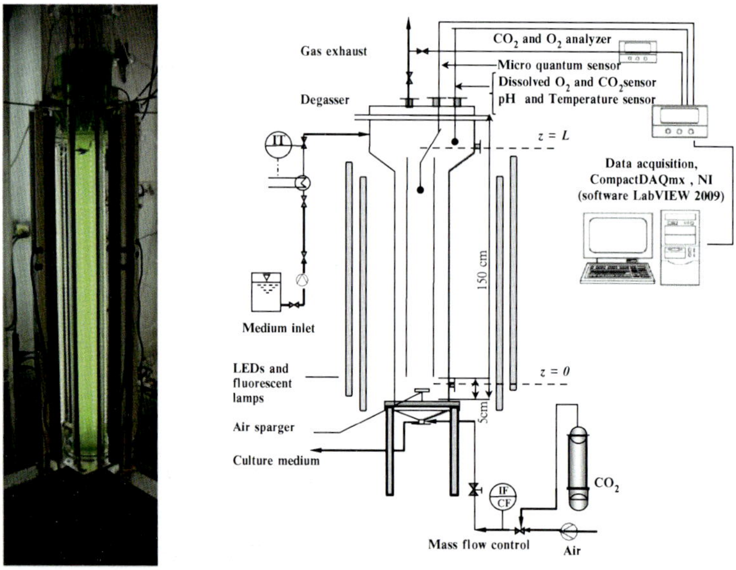

Fig. 4 Schematic diagram of the experimental airlift photobioreactor used for growth of the microalga *S. obtusiusculus*

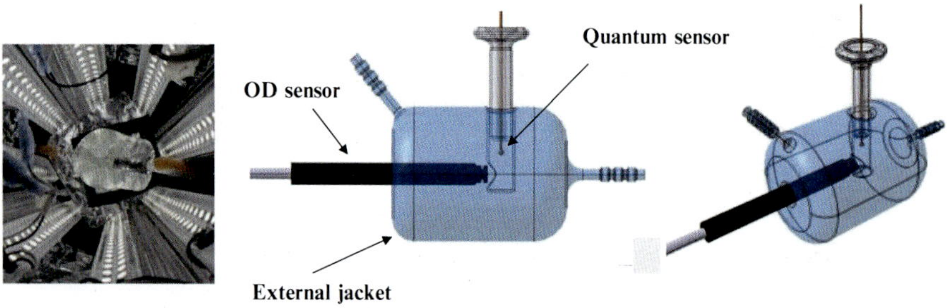

Fig. 5 Controlled mini-photobioreactor used to obtain the P-I curves at different temperatures

panel of high intensity white lights consisting of dimmable LEDs to regulate the irradiance. The irradiance values were measured with a spherical micro quantum sensor located at the center of the mini-photobioreactor. DO was measured, logged, and used to calculate the Po_2 as the biomass-concentration specific rate of evolution of DO (see Fig. 6a). Detailed information about sensor specifications and calculations can be found in Cabello et al. (2015).

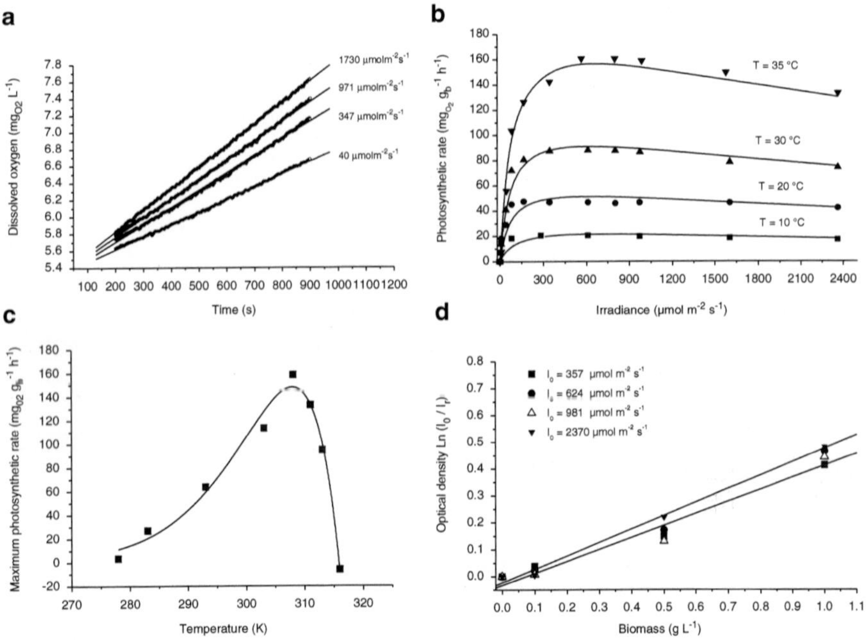

Fig. 6 (**a**) Evolution of the dissolved oxygen at different irradiances in the mini-reactor at 30 °C and 0.1 g DW L^{-1} biomass concentration. (**b**) Photosynthesis rates as function of irradiance at different temperatures (10, 20, 30 and 35 °C). (**c**) Dependence of photosynthetic activity on temperature. The continuous lines represent the calculated values using the double Arrhenius expression. (**d**) Light attenuation of the biomass at low concentrations

6.3 Intrinsic Photosynthetic Parameters

Experiments in the mini-photobioreactor were performed at irradiances between 4 and 2360 μmol m^{-2} s^{-1}, temperatures between 5 and 40 °C (Fig. 6b). For each condition, the Po_2 was determined as described above using constant slopes of DO vs time in the mini-photobioreactor (Fig. 6a)

The photosynthesis-irradiance (P-I) curves at different temperatures are shown in Fig. 6b. These values were fitted using a kinetic expression for the production of oxygen as function of the maximum production rate and an expression considering an average irradiance (I_{av}) inhibition effect (Table 2).

$$P_{O_2} = p_{O_2,\max}\left(\frac{I_{av}}{K_s + I_{av} + \dfrac{I_{av}^2}{K_I}}\right)$$

$$I_{av} = \frac{I_0}{D_c K_a C_b}\left[1-\exp\left(-D_c K_a C_b\right)\right]$$

Values of the parameters K_s, K_I and $Po_{2,max}$ (Table 6) obtained by Cabello et al. (2014) are consistent with those reported in the works by Fernández et al. (2012) and Béchet et al. (2013).

The values of the maximum oxygen production rates (Fig. 6c), $Po_{2,max}$, were used to evaluate the effect of the temperature on the photosynthetic activity. The data of the maximum oxygen production rate, and the temperature were fitted using a double Arrhenius expression to obtain E_a, E_d, k_0 and k_1.

$$p_{O_2,\max} = k_o \exp\left(-\frac{E_a}{T\,R}\right) - k_1 \exp\left(-\frac{E_d}{T\,R}\right)$$

Table 6 Parameters used in the model by Cabello et al. (2014)

Parameters	Value	Units
D_{az}	0.027	$m^2\ s^{-1}$
D_{CO_2}	1.9×10^{-5}	$cm^2\ s^{-1}$
D_{O_2}	2.7×10^{-5}	$cm^2\ s^{-1}$
E_a	16.1	kcal mol^{-1}
E_d	30	kcal mol^{-1}
ε_G	0.02	m^3 gas m^{-3}gas-liquid
K_a	0.096	$m^2\ g_b^{-1}$
K_d	0.005	h^{-1}
K_I	4970	μmol $m^{-2}\ s^{-1}$
$K_La_{O_2}$	12.3	h^{-1}
$K_La_{CO_2}$	10.3	h^{-1}
K_s	75.7	μmol $m^{-2}\ s^{-1}$
k_0	8.60×10^{13}	$g_{O_2}Kg_b^{-1}h^{-1}$
k_1	3.63×10^{23}	$g_{O_2}Kg_b^{-1}h^{-1}$
u_{Ge}	0.74	$m\ s^{-1}$
u_{Le}	0.07	$m\ s^{-1}$
Y_{b/O_2}	0.65	$g_b g_{O_2}^{-1}$

6.4 Biomass Light Absorption Coefficient

The biomass light absorption coefficient, K_a, was determined with the absorbance of cultures at different biomass concentrations (between 0.1 and 1 g L^{-1}) and the Beer-Lambert law equation. Figure 6d represents the light attenuation resulting from biomass increase in the mini-photobioreactor, and K_a was calculated using linear regression between the representation of data of optical density, I_0/I_r, and biomass concentration $I_r = I_0 \exp(-K_a C_b)$.

Experiments were also performed to determine the yield coefficient, Y_{b/O_2} , and the metabolic coefficient, K_d. These parameters were calculated from O_2 production and biomass evolution in the airlift photobioreactor. A detailed explanation is shown at the end of the next section. Values of the parameters are listed in Table 6.

6.5 Hydrodynamic and Mass Transfer Characterization

The mixing degree was determined by the RTD curves obtained by injection of a tracer (1 M NaOH, 15 ml) at the bottom of the column and registering the response of a pH sensor at the top of the column. An axial dispersion model (ADM) (Camacho-Rubio et al. 2004) of a single parameter was used to describe the behavior of RTD curves. The dimensionless form of this model includes the Peclet number (Pe), which depends on the axial dispersion coefficient, D_{az}.

$$\frac{\partial c}{\partial \theta} = \frac{1}{Pe}\frac{\partial^2 c}{\partial z^2} - \frac{\partial c}{\partial z}$$

The first approximation of D_{az} or Pe values was done from the dimensionless variance defined by Levenspiel (1999).

$$\sigma_\theta^2 = \frac{\sigma_t^2}{\bar{t}^2} = 2\left(\frac{D_{az}}{u_{Le}L}\right) - 2\left(\frac{D_{az}}{u_{Le}L}\right)^2\left[1 - \exp\left(\frac{u_{Le}L}{D_{az}}\right)\right]$$

Afterwards, the value was adjusted to reproduce RTD curve (Fig. 7a) using the ADM solved with the *software FlexPDE* transforming the description of a partial differential equations system into a finite element model to represent the mass balance of the tracer. The simulation was done for a gas superficial velocity of 0.0104 m s^{-1}. The value of the axial dispersion coefficient, D_{az} was 0.027 m^2s^{-1} and the Peclet number (Pe) at these conditions was 3.3; it represents macro mixing and indicates the deviation of a completely mixed reactor (Levenspiel 1999). These deviations might result from the non-uniformity of the velocity profiles, recirculation, channeling flow, turbulent flow or molecular diffusion, shape and geometry of the reactor, reflux due to the difference of velocity between different stages and recirculation due to the macroscopic agitation.

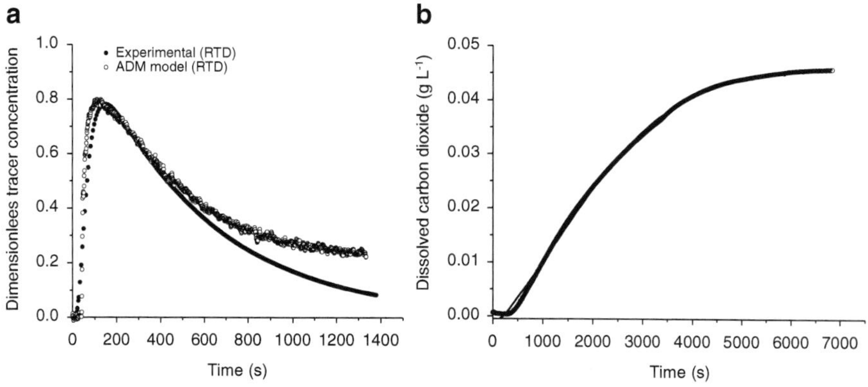

Fig. 7 (**a**) Experimental curve of the residence time distribution (RTD) and ADM model prediction. (**b**) Evolution of the CO_2 concentration in the liquid phase to obtain the volumetric coefficient of mass transfer CO_2

The overall gas-liquid mass transfer coefficient of carbon dioxide was determined at pH of 4 using the equation representing the CO_2 absorption in liquid phase.

$$\frac{dC_{L,CO_2}}{dt} = K_L a_{CO_2}\left(C^*_{L,CO_2} - C_{L,CO_2}\right)$$

The experimental dissolved CO_2 data (Fig. 7b) were used to estimate the value of $K_L a_{CO_2}\left(10.3\,h^{-1}\right)$. $K_L a_{O_2}\left(12.3\,h^{-1}\right)$ was calculated from $K_L a_{CO_2}$ using the following expression:

$$K_L a_{O_2} = K_L a_{CO_2}\left(\frac{D_{O_2}}{D_{CO_2}}\right)^{1/2}$$

Sanchez Mirón et al. (2000) used a similar methodology to calculate the volumetric mass transfer coefficient of oxygen. The values were 28 and 90 h^{-1} for superficial gas velocities of 0.01–0.03 m s^{-1}. At these conditions, the hold-up of the gas phase was between 0.018 and 0.04 m^3 (gas volume) m^{-3} (reactor volume). The gas hold-up, ε_G, effective gas velocity, u_G, and effective liquid velocity, u_{Le}, in our airlift photobioreactor were estimated with CFD (Cabello et al. 2014), using the momentum transport equations (two-fluid Euler-Euler model) for bubble column flow for the axial symmetry (2D) geometry (mesh of 12,704 triangular elements). The simulation was done with a gas superficial velocity of 0.0104 m s^{-1}.

Figure 8a shows the results for the simulation of the hydrodynamics of an airlift column using CFD. Gas entered the reactor from the lower part developing velocity profiles in the liquid phase. Figure 8 represents the steady-state profiles for liquid and gas velocity and the gas holdup. The gas flows upward through the draft tube and liquid velocity increases in the central zones close to the gas inlet (Fig. 8b), and

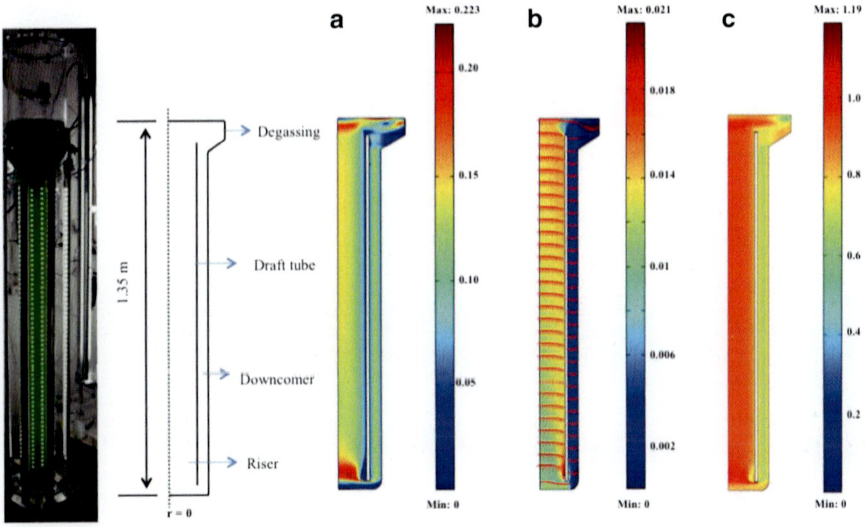

Fig. 8 Simulations of (**a**) the liquid phase velocity profile (m s^{-1}), (**b**) gas hold up (m^3 m^{-3}) and (**c**) gas phase velocity (m s^{-1}), for the airlift geometry operated at 0.0104 m s^{-1}. The scale of low (*blue*) to high values (*red*) of the hydrodynamic quantities is represented by a color scale (*right side*) to the reactor representation. The total height of the airlift column is 1.05 m and a radius of 0.06 m. The draft tube radius was 0.04 m

it is approximately constant in the middle-upper zone. In the degasser zone, the liquid accelerates due to the release of the gas bubbles and moves into the downcomer at a slower velocity. The velocity close to the inner side of the draft tube and in the lower zones of the reactor indicate the presence of a stagnant liquid film above the draft tube. Gas velocity contours show, as expected, that velocities in the riser zone are higher than in the downcomer (Fig. 8d) and the gas fraction in this zone is close to zero (Fig. 8c). From these, average values of ε_G, u_L and u_G were calculated and used for simulations and predictions of biomass concentration or productivity.

6.6 *Biomass Yield Coefficient and Metabolic Coefficient*

Two additional parameters (Table 6) have to be obtained to validate the model and evaluate its performance: the growth yield coefficient of oxygen to biomass $\left(Y_{b/O_2}\right)$ and the maintenance coefficient (K_d). A Y_{b/O_2} of $0.65\, g_b\, g_{O_2}^{-1}$ was obtained from growth experiments, from the ratio of the biomass ($g_b L^{-1}$) and the DO $\left(g_{O_2} L^{-1}\right)$ concentrations.

In order to determine K_d, a growth experiment was performed at specific operating conditions and the data were fitted using the following expression:

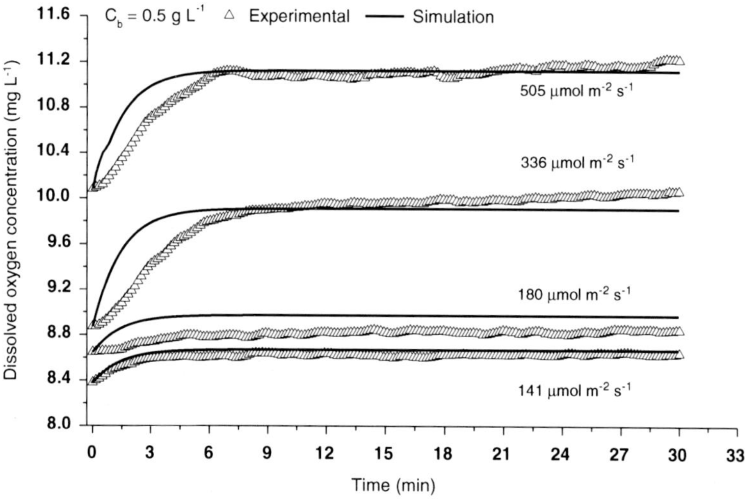

Fig. 9 Experimental data and predictions for dynamic simulation oxygen concentration in the airlift photobioreactor. The initial irradiance was 117 µmol m^{-2} s^{-1} each curve represents the increase in DO after the irradiance change

$$\frac{dC_b}{dt} = P_{O2}C_bY_{b/O2} - K_dC_b$$

This equation considers the term K_dC_b corresponding to the endogenous respiration rate. The value for K_d of 0.005 h^{-1} was obtained by fitting this equation to the experimental data shown in Fig. 9.

6.7 Model Validation

Once all parameters were obtained, validation of the model was done by comparison of experimental data for light step-changes with the predicted O_2 concentration. At the end of Sect. 5, it was mentioned that dynamic processes might generate abundant information when changes including shifts, pulses, ramps and oscillations in light were applied. Figure 10 shows the oxygen response for step changes in irradiances starting from 117 µmol m^{-2} s^{-1} to 180, then to 336 and to 505 µmol m^{-2} s^{-1}. As can be observed, good agreement (±5 %) was found at steady state between experimental and predicted data. Nonetheless, deviations in the initial stages were observed, mainly because the model does not consider the biological delay in the transient period.

Predictions of the evolution of biomass concentration at different irradiances (150, 200 and 350 µmol m^{-2} s^{-1}) are shown in Fig. 9 as solid lines. It can be seen that the model adequately predicts the changes from exponential growth stage to a

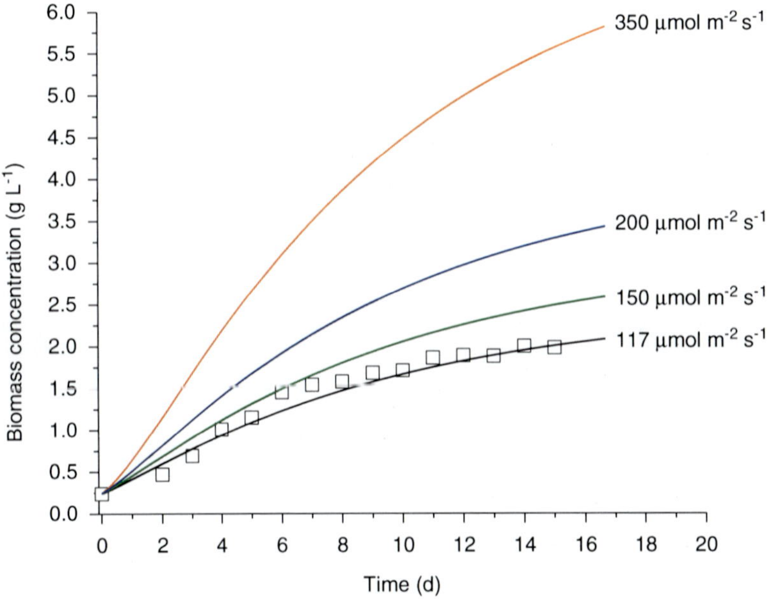

Fig. 10 Simulation (▬) and experimental (□) values obtained for the biomass concentrations at irradiance of 117 μmol m^{-2} s^{-1} and optimum temperature 35 °C

steady state biomass content associated with the contribution of the kinetic term representing specific endogenous respiration rate of the microalgae.

Predictions of biomass productivity ($P_b = dC_b/dt$) for conditions assayed in Fig. 9 are plotted in Fig. 11. The biomass productivity was predicted to peak on the third day of operation and then decreased by 65 % over 17 days of operation due to increase in biomass concentration and the biomass-associated increased light attenuation and endogenous respiration rate.

7 Final Remarks

Development of models representing the non-biotic and biotic dynamics of microalgal activity and growth is important for design, optimization and control of industrial-scale microalgae culture systems. Several aspects influencing the microalgae growth were reviewed and models were presented in this chapter, which jointly use kinetics and hydrodynamics phenomena. Further, detailed explanation of a dynamic model and experimental determination of parameters was presented. The use of short term experiments using a mini-photobioreactor with adequate CO_2, nutrients, and light at DO levels below photoinhibition to determine the intrinsic parameters was a useful and rapid tool to evaluate the effect of environmental factors on biological activity and to establish optimal

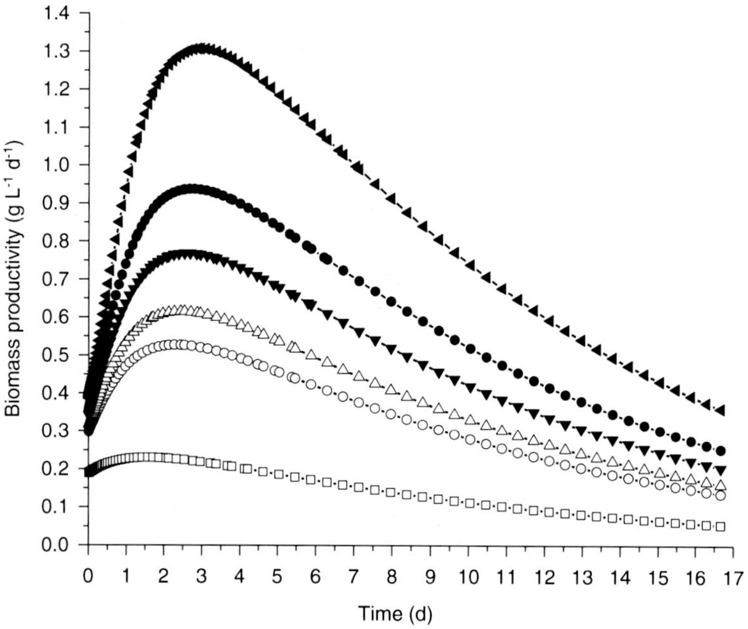

Fig. 11 Predicted biomass productivity for irradiances of 150(□), 350(○), 400(Δ), 500(▼), 610(●) and 850 μmol^{-1} s^{-1}(◀)

growth conditions. The proposed model predicted the effect of temperature and light on the dynamic O_2 concentration and consequently CO_2 uptake and biomass productivity. However, additional research is needed to further improve the model's dynamic response. This could include a better representation of local light distribution, allow proper microalgae adaptation to changing conditions and the establishment of optimal conditions to maximize biomass productivity. Further work could focus on effectively representing the effect of nutrients and their limitation on specific conditions for the accumulation or excretion of lipids, carbohydrates, pigment or other value-added products.

8 Nomenclature

a	Specific gas-liquid interfacial area, m^{-1}
A	Cross-section of the column, m^2
A_d	Cross-section of the downcomer, m^2
A_r	Cross-section of the riser, m^2
B1, B2	Pre exponential factors, dimensionless
C1, C2	Activation energy, dimensionless
c	Trace concentration, defined by C/C_0, dimensionless

C_b	Biomass concentration, g L^{-1}
C_{CI}	Dissolved total inorganic carbon concentration, mol m^{-3}
$C_{O_2}^{*}$	Equilibrium concentration, mg L^{-1}
C_{G,O_2}	O_2 bulk-gas concentration, mg L^{-1}
C_{G,CO_2}	CO_2 bulk-gas concentration, mg L^{-1}
C_{G,O_2}^{0}	Initial O_2 concentration, mg L^{-1}
$C_{O_2,max}$	Maximum O_2 concentration in culture medium, mg L^{-1}
C_{L,O_2}	O_2 bulk-liquid concentration, mg L^{-1}
C_{L,CO_2}	CO_2 bulk-liquid concentration, mg L^{-1}
C_j	Concentration of *j*th nutrient in culture medium, mg L^{-1}
C_{O_2}	Dissolved O_2 concentration in culture medium, mg L^{-1}
C_P	Culture metabolite concentration, g L^{-1}
C_{H+}	H^+ concentration, g L^{-1}
$C_{CO_3^{2-}}$	CO_3^{2-} concentration, g L^{-1}
C_D	Drag coefficient, dimensionless
C^*_{L,O_2}	Equilibrium O_2 concentration, mg L^{-1}
C^*_{L,CO_2}	Equilibrium CO_2 concentration, mg L^{-1}
D_{az}	Axial dispersion coefficient, m^2 s^{-1}
D_c	Column diameter, m
D_L	Molecular diffusivity of solute in liquid phase, m^2s^{-1}
D_{O_2}	Molecular diffusivity of O_2 in liquid phase, m^2s^{-1}
D_{CO_2}	Molecular diffusivity of CO_2 in liquid phase, m^2s^{-1}
d_t	Tube diameter, m
d_b	Bubble diameter, m
dz	Differential distance in the solar tube, m
E_a	Activation energy, kcal mol^{-1}
E_d	Deactivation energy, kcal mol^{-1}
E_{ab}	Mass absorption coefficient, m^2 Kg^{-1}
E_S	Mass scattering coefficient, m^2 Kg^{-1}
F_G	Gas flow, L min^{-1}
F_{CO_2}	CO_2 molar flow rate in the gas phase, mol s^{-1}
F_{O_2}	O_2 molar flow rate in the gas phase, mol s^{-1}
F_{N_2}	Nitrogen molar flow rate in the gas phase, mol s^{-1}
g	Gravitational acceleration, m^2s^{-1}
h	experimental constant, g L^{-1}
h_G	Local volume fraction of the gas phase, m^3 gas m^{-3} gas-liquid
h_L	Local volume fraction of the liquid phase, m^3 liquid m^{-3} gas-liquid
Hc	Height column, m
H	Partition coefficient from Henry law, defined by $H=k_H/RT$, dimensionless
HCO_2	Partition coefficient for carbon dioxide, dimensionless
HO_2	Partition coefficient for oxygen, dimensionless
I	Light intensity, μmol m^{-2} s^{-1}
I_0	Incident irradiance on reactor surface, μmol m^{-2} s^{-1}
I_{av}	Average irradiance within the reactor, μmol m^{-2} s^{-1}

I_k	Saturating irradiance, μmol m^{-2} s^{-1}
I_r	$I_r = I_0 exp(-K_a C_b)$
k_1	kinetic constant for growth, dimensionless
k	Kinetic energy of turbulence, m^2 s^{-2}
k_0, k_1	Frequency factors of Arrhenius, $mg_{O_2} g_b^{-1} s^{-1}$
k_2	Conversion coefficient between volume and mass of O_2, g L^{-1}
k_{1r}	First-order kinetic coefficient (CO_2 hydrolysis), d^{-1}
k_c	Conversion coefficient between volume and mass of CO_2, g L^{-1}
k_H	Henry coefficient, atm-m^3 mol^{-1}
k_m	constant coefficient, Pa^{-1}
K	Mass transfer coefficient, ms^{-1}
K_a	Biomass light absorption coefficient, m^2 gb^{-1}
K_c	CO_2 half-saturation constant, g L^{-1}
K_d	Metabolic coefficient, h^{-1}
K_I	Inhibition constant for irradiance, μmol m^{-2} s^{-1}
K_i	Form parameter, μmol m^{-2} s^{-1}
$K_L a_{O_2}$	Volumetric mass transfer coefficient for O_2, h^{-1}
$K_L a_{CO_2}$	Volumetric mass transfer coefficient for CO_2, h^{-1}
K_s	Irradiation saturation constant, μmol m^{-2} s^{-1}
K_j	Saturation constant of *j*th nutrient, g m^{-3}
$K_{H_2PO_4}$	Saturation constant of H_2PO_4, g m^{-3}
K_{NO_3}	Saturation constant of NO_3, g m^{-3}
K_{IO_2}	Inhibition constant for O_2, molO_2 m^{-3}
K_p	Metabolite halft-saturation constant, g L^{-1}
L	Reactor length, m
m	Single cell mass, g
m_1	Form parameter, dimensionless
$\underline{Me}$	Maintenance term, h^{-1}
$\overline{Me}$	Maintenance without shear effects, h^{-1}
n	Form exponent
p	Pressure, Pa
P_b	Biomass productivity, $g_b L^{-1} h^{-1}$
$P_{O_2,max}$	Maximum O_2 production rate per biomass unit, $mg_{O_2} g_b^{-1} s^{-1}$
P_{O_2}	O_2 production rate per biomass unit, $mg_{O_2} g_b^{-1} s^{-1}$
Pe	Liquid phase Peclet number, defined by $u_{Le} L D_{az}^{-1}$
Q_G	Volumetric gas flow rate, $m^3 s^{-1}$
Q_L	Volumetric liquid flow rate, $m^3 s^{-1}$
R_{CO_2}	CO_2 consumption rate of the culture, mol $CO_2 m^{-3}$ s^{-1}
R_{O_2}	O_2 production rate of the culture, mol $O_2 m^{-3}$ s^{-1}
Re_b	Liquid phase Reynolds number, dimensionless
R	gas constant, Pa m^3 Kg^{-1} mol^{-1} K^{-1}
Ra	Assimilatory quotient of *S. platensis*, dimensionless
r	radial distance, m
r_b	experimental constant, μmol m^{-2} s^{-1}
r_a	respiration rate, dimensionless

r_1	Losses caused by respiration, $g_{O_2} g_b^{-1}$
S	Cross-section of the tube, m^2
T	Temperature, K
T	Time, min
u_G	Superficial gas velocity, m s^{-1}
u_{Ge}	Effective gas velocity, m s^{-1}
u_L	Superficial liquid velocity, m s^{-1}
u_{Le}	Effective liquid velocity, m s^{-1}
u	Superficial fluid phase velocity, m s^{-1}
u_{slip}	Relative velocity between gas and liquid, m s^{-1}
x_1	Fraction of PSF in the open state, dimensionless
x_2	Fraction of PSF in the closed state, dimensionless
x_3	Fraction of PSF in the inhibited state, dimensionless
Y_{O_2}	O_2 molar fraction
Y_{CO_2}	CO_2 molar fraction
Y_{b/O_2}	Biomass yield on O_2, $g_b g_{O_2}^{-1}$
$Y_{O_2/b}$	O_2 yield on biomass, $g_{O_2} g_b^{-1}$
$Y_{X/j}$	ratio of weight of dry biomass produced to weight of *j*th nutrient consumed
Y_{X/O_2}	ratio of weight of dry biomass produced to weight of O_2 produce
V	Reactor volume, m^3
V_L	Volume of liquid phase, L
V_G	Volume of gas phase, L
Z	Axial direction, m

Greek Letter

σ	Interfacial tension, N m^{-1}
ρ	phase density, Kg m^{-3}
μ_G	Gas dynamic viscosity, Pa s
μ_L	Liquid dynamic viscosity, Pa s
μ	Specific growth rate, d^{-1}
ε_G	Gas phase hold up
ε_L	Liquid phase hold up
ε	rate of dissipation of the kinetic energy, Kgm^{-3} s^{-1}
ζ_m	Time rate of change of cell mass *m*, g s^{-1}
ψ	Cell distribution function in a generic spatial position, g^{-1} m^{-3}
μ_C	Catabolic growth rate
μ_{max}	Maximum specific rate of cell growth, s^{-1}
η_L	Liquid dynamic viscosity, Pa s
λ	Maintenance coefficient, h^{-1}
$\phi(T)$	Function of temperature, °C
τ	Constant time delay of *S. platensis* growth, d
τ_1	Shear stress, Pa
τ_c	Critical shear stress, Pa

υa Flow rate in aeration pipe and a control input to reactor, L d^{-1}
δ Coefficient reflecting *S. platensis* sensitivity to red-blue light, Lg^{-1}
δ_1 Kinetic constant for $x_3 \rightarrow x_1$, s^{-1}
δ_G Thickness of the gas film, m
δ_L Thickness of the liquid film, m
θ Time, defined by t $u_{Le/}L$,dimensionless
α Constant, d^{-1}
α_1 Kinetic constant for $x_1 \rightarrow x_2$, $m^2\ \mu E^{-1}$
β_1 Kinetic constant for $x_2 \rightarrow x_3$, $m^2\ \mu E^{-1}$
γ_1 Kinetic constant for $x_2 \rightarrow x_1$, s^{-1}

Subscript
G Gas phase
L Liquid phase
r riser
j nutrient
max maximum value
min minimum value
opt optimal value
r riser

List of Acronyms

ADM	Axial dispersion model
ADP	Adenosine diphosphate
AllnGap II	Aluminum indium gallium phosphide
ATP	Adenosine triphosphate
CA	Carbonic anhydrase
CARPT	Computer-automated radioactive particle
CCM	Carbon dioxide concentration mechanism
CFD	Dynamic Fluid Computational
chCA	Chloroplast carbonic anhydrase
cyCA	Cytosolic carbonic anhydrase
DIC	Dissolved inorganic carbon
DNA	Deoxyribonucleic acid
DO	Dissolved oxygen
DW	Dry weight
LEDs	Light emitting diodes
$NADP^+$	Nicotinamide adenine dinucleotide phosphate
NADPH	Nicotinamide adenine dinucleotide phosphate hydrogen
pCA	Periplasmic carbonic anhydrase
PSF	Photosynthetic factory
PSII	Photosystem II

RNA	Ribonucleic acid
RTD	Residence time distribution
Rubisco	Ribulose bisphosphate carboxylase-oxygenase
STR	Stirred tank reactor
VVM	Gas volume flow per unit of liquid volume per minute

References

Aizawa K, Miyachi M (1986) Carbonic anhydrase and CO_2 concentrating mechanisms in microalgae and cyanobacteria. FEMS Microbiol Rev 39:215–233

Akita K, Yoshida F (1973) Gas holdup and volumetric mass transfer coefficient in bubble columns. Effects of liquid properties. Ind Eng Chem Process Des Dev 12(1):76–80

Alcántara C, García-Encina PA, Muñoz R (2013) Evaluation of mass and energy balances in the integrated microalgae growth-anaerobic digestion process. Chem Eng J 221:238–246

Azov Y (1982) Effect of pH on inorganic carbon uptake in algal cultures. Appl Environ Microbiol 43:1300–1306

Badger MR, Price GD (1994) The role of carbonic acid anhydrase in photosynthesis. Annu Rev Plant Physiol Plant Mol Biol 45:369–392

Béchet Q, Shilton A, Guieysse B (2013) Modeling the effects of light and temperature on algae growth: state of the art and critical assessment for productivity prediction during outdoor cultivation. Biotechnol Adv 31:1648–1663

Becker EW (2007) Micro-algae as a source of protein. Biotechnol Adv 25(2):207–210

Bello RA (1981) A characterization study of airlift contactors for application to fermentations. Ph D thesis. University of Waterloo, Ontario, Canada

Ben Amotz B, Tornabene TG (1985) Chemical profile of selected species of microalgae with emphasis on lipids. J Phycol 21:72–81

Bernard O, Rémond B (2012) Validation of a simple model accounting for light and temperature effect on microalgal growth. Bioresour Technol 123:520–527

Bilanovic D, Andargatchew A, Kroeger T, Shelef G (2009) Fresh water and marine microalgae sequestering of CO_2 at different C and N concentrations – response surface methodology analysis. Energy Convers Manag 50:262–267

Bitog JP, Lee IB, Lee CG, Kim KS, Hwang HS, Hong SW, Mostafa E (2011) Application of computational fluid dynamics for modeling and designing photobioreactors for microalgae production: a review. Comput Electron Agric 76(2):131–147

Borowitzka MA (1999) Commercial production of microalgae: ponds, tanks, tubes and fermenters. J Biotechnol 70:313–321

Brennan L, Owende P (2010) Biofuels from microalgae – a review of technologies for production. Renew Sust Energ Rev 14:557–577

Briassoulis D, Panagakis P, Chionidis M, Tzenos D, Lalos A, Tsinos C, Berberidis K, Jacobsen A (2010) An experimental helical-tubular photobioreactor for continuous production of *Nannochloropsis* sp. Bioresour Technol 101:6768–6777

Cabello J, Morales M, Revah S (2014) Dynamic photosynthetic response of the microalga *Scenedesmus obtusiusculus* to light intensity perturbations in an airlift photobioreactor. Chem Eng J 252:104–111

Cabello J, Toledo A, Sánchez L, Revah S, Morales M (2015) Effect of the temperature, pH and irradiance on the photosynthetic activity by *Scenedesmus obtusiusculus* under optimal conditions and nitrogen starvation. Bioresour Technol 181:128–135

Camacho-Rubio F, García Camacho F, Fernández Sevilla JM, Chisti Y, Molina Grima E, Chisti Y (1989) A mechanistic model of photosynthesis in microalgae. Biotechnol Bioeng 81(4):459–473

Camacho-Rubio F, Acién Fernández FG, Sánchez Pérez JA, García Camacho F, Molina Grima E (1999) Prediction of dissolved oxygen and carbon dioxide concentration profiles in tubular photobioreactors for microalgal culture. Biotechnol Bioeng 62(1):71–86

Camacho-Rubio F, García Camacho F, Fernández Sevilla JM, Chisti Y, Molina Grima E (2003) A mechanistic model of photosynthesis in microalgae. Biotechnol Bioeng 81(4):459–473

Camacho-Rubio F, Sánchez Mirón A, Cerón García MC, García Camacho F (2004) Mixing in bubble columns: a new approach for characterizing dispersion coefficients. Chem Eng Sci 59:4369–4376

Cheng L, Zhang L, Chen H, Gao C (2006) Carbon dioxide removal from air by microalgae cultured in a membrane-photobioreactor. Sep Purif Technol 50(3):324–329

Chisti Y (1989) Airlift bioreactors. Elsevier, London

Chisti Y (1998) Pneumatically agitated bioreactors in industrial and environmental bioprocessing: hydrodynamics, hydraulics and transport phenomena. Appl Mech Rev 51:33–112

Chisti Y (2007) Biodiesel from microalgae. Biotechnol Adv 25:294–306

Chisti MY, Moo-Young M (1987) Airlift-reactors: characteristic, applications and design considerations. Chem Eng Commun 60:195–242

Chiu SY, Kao CY, Chen CH, Kuan TC, Ong SC, Lin CS (2008) Reduction of CO_2 by a high-density culture of Chlorella sp. in a semicontinuous photobioreactor. Bioresour Technol 99(9):3389–3396

Chu W-L (2012) Biotechnological applications of microalgae. IeJSME 6(Suppl 1):S24–S37

Cogne G, Cornet JF, Gros JB (2005) Design, operation, and modeling of a membrane photobioreactor to study the growth of the cyanobacterium *Arthrospira platensis* in space conditions. Biotechnol Prog 21(3):741–750

Concas A, Pisu M, Cao G (2010) Novel simulation model of the solar collector of BIOCOIL photobioreactors for CO_2 sequestration with microalgae. Chem Eng J 157:297–303

Costache TA, Acién FG, Morales MM, Fernández-Sevilla JM, Stamati I, Molina-Grima E (2013) Comprehensive model of microalgae photosynthesis rate as a function of culture conditions in photobioreactor. Appl Microbiol Biotechnol 97(17):7627–7637

Cuellar-Bermudez SP, Garcia-Perez JS, Rittmann BE, Parra-Saldivar R (2015) Photosynthetic bioenergy utilizing CO_2: an approach on flue gases utilization for third generation biofuels. J Clean Prod 98:53–65. doi:10.1016/j.jclepro.2014.03.034

Danquah MK, Harun R, Halim R, Forde GM (2010) Cultivation medium design via elemental balancing for *Tetraselmis suecica*. Chem Biochem Eng Q 24(3):361–369

de Morais MG, Costa JA (2007) Carbon dioxide fixation by *Chlorella kessleri*, *C. vulgaris*, *Scenedesmus obliquus* and *Spirulina* sp. cultivated in flasks and vertical tubular photobioreactors. Biotechnol Lett 29:1349–1352

Eilers PHC, Peeters JCH (1988) A model for the relationship between light intensity and the rate of photosynthesis in phytoplankton. Ecol Model 42:199–215

Eilers PHC, Peeters JCH (1993) Dynamic behaviour of a model for photosynthesis and photoinhibition. Ecol Model 69:113–133

Erickson LE, Curless CE, Lee HY (1987) Modeling and simulation of photosynthetic microbial growth. Ann N Y Acad Sci 506:308–324

Fair JR (1967) Designing gas-sparged reactors. Chem Eng 74(14):67–74

Falkowski PG, Raven JA (1997) Aquatic photosynthesis. Blackwell Science, Malden

Fan LH, Zhang YT, Cheng LH, Zhang L, Tang D-S, Chen H-L (2007) Optimization of carbon dioxide fixation by *Chlorella vulgaris* cultivated in a membrane-photobioreactor. Chem Eng Technol 30(8):1094–1099

Farrelly DJ, Everard CD, Fagan CC, McDonnell KP (2013) Carbon sequestration and the role of biological carbon mitigation: a review. Renew Sust Energ Rev 21:712–727

Fernández I, Acién FG, Fernández JM, Guzmán JL, Magán JJ, Berenguel M (2012) Dynamic model of microalgal production in tubular photobioreactors. Bioresour Technol 126:172–181

Fernández I, Acién FG, Berenguel M, Guzmán JL, Andrade GA, Pagano DJ (2014) A lumped parameter chemical–physical model for tubular photobioreactors. Chem Eng Sci 112:116–129

Fidalgo JP, Cid A, Torres E, Sukenik A, Herrero C (1998) Effects of nitrogen source and growth phase on proximate biochemical composition, lipid classes and fatty acid profile of marine microalgae *Isochrysis galbana*. Aquaculture 166:105–116
Gallardo-Rodriguez JJ, García-Camacho F, Sánchez Mirón A, López-Rosales L, Chisti Y, Molina-Grima R (2011) Shear-induced changes in membrane fluidity during culture of a fragile dinoflagellate microalga. Biotechnol Prog 28(2):467–473
Goldberg IK, Cohen Z (2006) The effect of phosphate starvation on the lipid and fatty acid composition of the fresh water eustigmatophyte *Monodus subterraneus*. Phytochemistry 67:696–701
Grobbelaar JU (1994) Turbulence in mass algal cultures and the role of light/dark fluctuations. J Appl Phycol 6:331–335
Hadiyanto H, Elmore S, Gerven TV, Stankiewicz A (2013) Hydrodynamic evaluations in high rate algae pond (HRAP) design. Chem Eng J 217:231–239
Harun R, Singh M, Forde GM, Danquah MK (2010) Bioprocess engineering of microalgae to produce a variety of consumer products. Renew Sust Energ Rev 14:1037–1047
Hekmat A, Amooghin AE, Moraveji MK (2010) CFD simulation of gas-liquid behavior in an air-lift reactor: determination of the optimum distance of the draft tube. Simul Model Pract Theory 18(7):927–945
Hikita H, Kikukawa H (1974) Liquid-phase mixing in bubble columns: effect of liquid properties. Chem Eng J 8(3):191–197
Hikita H, Asai S, Tanigawa K, Segawa K, Kitao M (1980) Gas hold-up in bubble columns. Chem Eng J 20(1):59–67
Hikita H, Asai S, Tanigawa K, Segawa K, Kitao M (1981) The volumetric liquid-phase mass transfer coefficient in bubble columns. Chem Eng J 22(1):61–69
Ho S-H, Chen C-Y, Lee D-J, Chang J-S (2011) Perspectives on microalgal CO_2-emissions mitigation systems-A review. Biotechnol Adv 29:189–198
Hu D, Li M, Zhou R, Sun Y (2012) Design and optimization of photo bioreactor for O_2 regulation and control by system dynamics and computer simulation. Bioresour Technol 104:608–615
Jeon J-C, Cho C-W, Yun Y-S (2005) Measurement of microalgal photosynthetic activity depending on light intensity and quality. Biochem Eng J 27:127–131
Joshi JB, Ranade VV, Gharat SD, Lele SS (1990) Sparged loop reactors. Can J Chem Eng 68:705–741
Juneja A, Ceballos RM, Murthy GS (2013) Effects of environmental factors and nutrient availability on the biochemical composition of algae for biofuels production: a review. Energies 6(9):4607–4638
Kao C-Y, Chen T-Y, Chang Y-B, Chiu T-W, Lin H-Y, Chen C-D, Chang J-S, Lin C-S (2014) Utilization of carbon dioxide in industrial flue gases for the cultivation of microalga Chlorella sp. Bioresour Technol 166:485–493
Kawase Y, Moo-Young M (1987) Theoretical prediction of gas holdup in bubble columns with Newtonian fluids. Ind Eng Chem Res 26:933–937
Kemblowski Z, Przywarski J, Diab A (1993) Average gas holdup and liquid circulation velocity in airlift reactors with external loop. Chem Eng Sci 48:4023–4035
Koide K, Kimura M, Nitta H, Kawabata H (1988) Liquid circulation in bubble column with draught tube. J Chem Eng Jpn 21:393–399
Kommareddy AR, Anderson GA (2003) Study of light as a parameter in the growth of algae in a photo-bio reactor (PBR). ASAE Paper No. 034057. ASAE, St. Joseph
Koru E (2012) Earth food *Spirulina* (*Arthrospira*): production and quality standards. In: Yehia E-S (ed) Food additive. InTech Europe, Rijeka, Croatia, ISBN 978-953-51-0067-6
Krishna R, van Baten JM (2003) Mass transfer in bubble columns. Catal Today 79–80:67–75
Kumar A, Yuan X, Sahu AK, Dewulf J, Ergas SJ, Van Langenhove H (2010) A hollow fiber membrane photo-bioreactor for CO_2 sequestration from combustion gas coupled with wastewater treatment: a process engineering approach. J Chem Technol Biotechnol 85(3):387–394
Levenspiel O (1999) Chemical reaction engineering. Wiley, New York, pp 260–327

Liffman K, Paterson DA, Liovic P, Bandopadhayay P (2013) Comparing the energy efficiency of different high rate algal raceway pond designs using computational fluid dynamics. Chem Eng Res Des 91:221–226

Luo HP, Al-Dahhan MH (2004) Analyzing and modeling of photobioreactors by combining first principles of physiology and hydrodynamics. Biotechnol Bioeng 85(4):382–393

Luo HP, Al-Dahhan MH (2008) Macro-mixing in a draft-tube airlift bioreactor. Chem Eng Sci 63(6):1572–1585

Luo HP, Al-Dahhan MH (2010) Local gas holdup in a draft tube airlift bioreactor. Chem Eng Sci 65(15):4503–4510

Luo HP, Al-Dahhan MH (2011) Verification and validation of CFD simulations for local flow dynamics in a draft tube airlift bioreactor. Chem Eng Sci 66:907–923

Mata TM, Martins AA, Caetano NS (2010) Microalgae for biodiesel production and other applications: a review. Renew Sust Energ Rev 14:217–232

Melnicki MR, Pinchuk GE, Hill EA, Kucek LA, Stolyar SM, Fredrickson JM, Konopka JM, Beliaev AS (2013) Feedback-controlled LED photobioreactor for photophysiological studies of cyanobacteria. Bioresour Technol 134:127–133

Merchuk JC, Contreras A, García F, Molina-Grima E (1998) Studies of mixing in a concentric tube airlift bioreactor with different spargers. Chem Eng Sci 53(4):709–719

Mersmann A (1978) Design and scale up of bubble and spray columns. Ger Chem Eng 1:1–11

Meyer M, Griffiths H (2013) Origins and diversity of eukaryotic CO_2-concentrating mechanisms: lessons for the future. J Exp Bot 64(3):769–786

Milledge JJ (2011) Commercial application of microalgae other than as biofuels: a brief review. Rev Environ Sci Biotechnol 10:31–41

Molina-Grima E, García Camacho F, Sánchez Pérez JA, Fernández Sevilla J, Acién Fernández FG, Contreras Gómez A (1994) A mathematical model of microalgal growth in light limited chemostat cultures. J Chem Technol Biotechnol 61:167–173

Molina-Grima E, Acién Fernández FG, García Camacho F, Chisti Y (1999) Photobioreactors:light regime, mass transfer, and scaleup. J Biotechnol 70:231–247

Molina-Grima E, Belarbia EH, Acién Fernández FG, Robles Medina A, Chisti Y (2003) Recovery of microalgal biomass and metabolites: process options and economics. Biotechnol Adv 20:491–515

Moss B (1973) The influence of environmental factors on the distribution of freshwater algae: an experimental study: II. The role of pH and the carbon dioxide-bicarbonate system. J Ecol 61:157–177

Nedbal L, Cervený J, Keren N, Kaplan A (2010) Experimental validation of a non-equilibrium model of CO_2 fluxes between gas, liquid medium, and algae in a flat-panel photobioreactor. J Ind Microbiol 37(12):1319–1326

Negoro M, Hamasaki A, Ikuta Y, Makita T, Hirayama K, Suzuki S (1993) Carbon dioxide fixation by microalgae photosynthesis using actual flue gas discharged from a boiler. Appl Biochem Biotechnol 39:643–653

Olaizola M (2003) Commercial development of microalgal biotechnology: from the test tube to the marketplace. Biomol Eng 20:459–466

Ono E, Cuello JL (2007) Carbon dioxide mitigation using thermophilic cyanobacteria. Biosyst Eng 96(1):129–134

Pandey JP, Pathak N, Tiwari A (2010) Standardization of pH and light intensity for the biomass production of *Spirulina platensis*. J Algal Biomass Util 1(2):93–102

Perner I, Posten C, Broneske J (2003) CFD Optimization of a plate photobioreactor used for cultivation of microalgae. Eng Life Sci 3(7):287–291

Pittman JK, Dean AP, Osundeko O (2011) The potential of sustainable algal biofuel production using wastewater resources. Bioresour Technol 102(1):17–25

Priyadarshani I, Rath B (2012) Commercial and industrial applications of microalgae – a review. J Algal Biomass Util 3(4):89–100

Pruvost J, Pottier L, Legrand J (2006) Numerical investigation of hydrodynamic and mixing conditions in a torus photobioreactor. Chem Eng Sci 61:4476–4489

Pruvost J, Cornet JF, Legrand J (2008) Hydrodynamics influence on light conversion in photobioreactors: an energetically consistent analysis. Chem Eng Sci 63:3679–3694

Pulz O (2001) Photobioreactors: production systems for phototrophic microorganisms. Appl Microbiol Biotechnol 57:287–293

Pulz O, Gross W (2004) Valuable products from biotechnology of microalgae. Appl Microbiol Biotechnol 65:635–648

Ras M, Steyer JP, Bernard O (2013) Temperature effect on microalgae: a crucial factor for outdoor production. Rev Environ Sci Biotechnol 12(1):53–164

Ratha SK, Prasanna R (2012) Bioprospecting microalgae as potential sources of "Green Energy"-Challenges and perspectives (Review). Appl Biochem Biotechnol 48(2):109–125

Raven JA, Giordano M, Beardall J, Maberly SC (2012) Algal evolution in relation to atmospheric CO_2: carboxylases, carbon-concentrating mechanisms and carbon oxidation cycles. Philos Trans R Soc Lond B Biol Sci 367:493–507

Rebolloso-Fuentes MM, García JL, Fernández JM, Acién FG, Sánchez JA, Molina E (1999) Outdoor continuous culture of *Porphyridium cruentum* in a tubular photobioreactor: quantitative analysis of the daily cyclic variation of culture parameters. J Biotechnol 70:271–288

Richmond A (2004) Handbook of microalgal culture: biotechnology and applied phycology. Blackwell Science, Oxford

Rodolfi L, Zittelli GC, Bassi N, Padovani G, Biondi N, Bonini G, Tredici MR (2009) Microalgae for oil: strain selection, induction of lipid synthesis and outdoor mass cultivation in a low-cost photobioreactor. Biotechnol Bioeng 102(1):100–112

Sánchez Mirón A, García Camacho F, Contreras Gomez A, Molina Grima E, Chisti Y (2000) Bubble-column and airlift photobioreactors for algal culture. AIChE J 46(91):1872–1877

Sánchez Mirón A, Cerón Garcıa MC, Contreras Gómez A, García Camacho F, Molina Grima E, Chisti Y (2003) Shear stress tolerance and biochemical characterization of *Phaeodactylum tricornutum* in quasi steady-state continuous culture in outdoor photobioreactors. Biochem Eng J 16:287–297

Sato T, Usui S, Tsuchiya Y, Kondo Y (2006) Invention of outdoor closed type photobioreactor for microalgae. Energy Convers Manag 47:791–799

Shah YT, Stiegel GJ, Sharma MM (1978) Backmixing in gas-liquid reactors. AIChE J 24(3):369–400

Shah Y, Kelkar B, Delker W (1982) Design parameters estimations for bubble column reactors. AIChE J 28(3):353–379

Sharma G, Kumar M, Ali MI, Jasuja ND (2014) Effect of carbon content, salinity and pH on *Spirulina platensis* for phycocyanin, allophycocyanin and phycoerythrin accumulation. J Microb Biochem Technol 6:202–206

Singh UB, Ahluwalia AS (2013) Microalgae: a promising tool for carbon sequestration. Mitig Adapt Strateg Glob Chang 18(1):73–95

Singh SP, Singh P (2014) Effect of CO_2 concentration on algal growth: a review. Renew Sust Energ Rev 38:172–179

Solovchenko A, Khozin-Goldberg I (2013) High-CO_2 tolerance in microalgae: possible mechanisms and implications for biotechnology and bioremediation. Biotechnol Lett 35:1745–1752

Spadiut O, Rittmann S, Dietzsch C, Herwig C (2013) Dynamic process conditions in bioprocess development. Eng Life Sci 13:88–101

Spolaore P, Joannis-Cassan C, Duran E, Isambert A (2006) Commercial applications of microalgae. J Biosci Bioeng 101:87–96

Stumm W, Morgan JJ (1996) Aquatic chemistry: chemical equilibria and rates in natural waters, 3rd edn. Wiley-Interscience, New York, p 1022

Suali E, Sarbatly S (2012) Conversion of microalgae to biofuel. Renew Sust Energ Rev 16:4316–4342

Sugai-Guérios MH, Mariano AB, Vargas JVC, de Lima Luz LF, Mitchell DA (2014) Mathematical model of the CO_2 solubilisation reaction rates developed for the study of photobioreactors. Can J Chem Eng 92(5):787–795

Sung KD, Lee JS, Shin CS, Park SC (1998) Isolation of a new highly CO_2 tolerant fresh water microalgae *Chlorella* sp. KR-1. Kor J Chem Eng 15:449–450
Takagi M, Yoshida T (2006) Effect of salt concentration on intracellular accumulation of lipids and triacylglyceride in marine microalgae *Dunaliella* cells. J Biosci Bioeng 101:223–226
Takagi M, Watanabe K, Yamaberi K, Yoshida Y (2000) Limited feeding of potassium nitrate for intracellular lipid and triglyceride accumulation of *Nannochloris* sp. UTEX LB1999. Appl Microbiol Biotechnol 54:112–117
Toledo-Cervantes A, Morales M, Novelo E, Revah S (2013) Carbon dioxide fixation and lipid storage by *Scenedesmus obtusiusculus*. Bioresour Technol 130:652–658
Valdés FJ, Hernández MR, Catalá L (2012) Estimation of CO2 stripping/CO2 microalgae consumption ratios in a bubble column photobioreactor using the analysis of the pH profiles. Application to Nannochloropsis oculata microalgae culture. Bioresour Technol 119:1–6
Varfolomeev SD, Wasserman LA (2011) Microalgae as source of biofuel, food, fodder, and medicines. Appl Biochem Biotechnol 47(9):789–807
Vonshak A, Torzillo G (2004) Environmental stress physiology. In: Richmond A (ed) Handbook of microalgal culture: biotechnology and applied phycology. Blackwell Science, Oxford, p 57
Wang L, Agyemang SA, Amini H, Shahbazi A (2015) Mathematical modeling of production and biorefinery of energy crops. Renew Sust Energ Rev 43:530–544
Wu X, Merchuk JC (2004) Simulation of algae growth in a bench scale internal loop airlift reactor. Chem Eng Sci 59:2899–2912
Xin L, Hong-ying H, Ke G, Ying-xue S (2010) Effects of different nitrogen and phosphorus concentrations on the growth, nutrient uptake, and lipid accumulation of a freshwater microalga *Scenedesmus* sp. Bioresour Technol 101:5494–5500
Ying K, James Gilmour D, Zimmerman WB (2014) Effects of CO_2 and pH on growth of the microalga *Dunaliella salina*. J Microb Biochem Technol 6:167–173
Yun YS, Lee SB, Park JM, Lee CI, Yang JW (1997) Carbon dioxide fixation by algal cultivation using wastewater nutrients. J Chem Technol Biotechnol 14:1297–1300
Zhang K, Kurano N, Miyachi S (2002) Optimization aeration by carbon dioxide gas for microalgal production and mass transfer characterization in a vertical flat-plate photobioreactor. Bioprocess Biosyst Eng 25:97–101
Zhao B, Su Y (2014) Process effect of microalgal-carbon dioxide fixation and biomass production: a review. Renew Sust Energ Rev 31:121–132

Beneficial or Toxic Effects of Selenium on Green Algae and Their Application as Nutrient Supplements or Bio-remediators

Milada Vítová, Kateřina Bišová, Jiří Doucha, and Vilém Zachleder

Abstract Selenium is an essential element in many organisms, with beneficial roles for animal health, but it can also be a dangerous toxin. Research into these two paradoxical faces of selenium is reviewed in relation to unicellular species of microalgae. Recent information on algal selenoproteins in general, and selenium-containing enzymes in particular, will be described, as well as the bioaccumulation of different forms of selenium in algae. In this review, we will also consider the impact of selenium on algal growth and other cellular events: firstly from the point of its toxicity, including research into selenium-resistant algal strains and their possible use for remediation of selenium contaminated environments. Secondly, we will examine case examples of selenium-enriched algae as a source of organic selenium with health benefits for different domestic animals.

Keywords Algae • Bioaccumulation of selenium • Detoxication • Glutathione peroxidase • Selenate • Selenite • Selenium • Selenium-enriched algae • Selenium in animal feeding • Selenium in human health • Selenium resistant strains • Selenoenzymes • Selenoproteins • Thioredoxin reductase • Toxicity of selenium

1 Introduction

Intensive and long-term research into selenium (Se), comprising thousands of papers, have been carried out since its discovery until the most recent times, due to its influence on a wide range of organisms, from plants to mammals including humans. The history of Se research is described in a detailed and impressive review by Flohé (2009). Interestingly, Se was, for a long time, known as an industrial hazard with a negative environmental impact on vertebrates, causing extinction of local fish populations and teratogenesis in birds, or fish, and reproductive toxicity in egg-laying vertebrates. Mortality, mass wasting in adults, reduced juvenile growth and immune suppression

M. Vítová • K. Bišová • J. Doucha • V. Zachleder (✉)
Laboratory of Cell Cycles of Algae, Centre Algatech, Institute of Microbiology, Czech Academy of Sciences (CAS), Opatovický mlýn, 37981 Třeboň, Czech Republic
e-mail: vitova@gmail.com; bisova@alga.cz; doucha@alga.cz; zachleder@gmail.com

A. Prokop et al. (eds.), *Algal Biorefineries*, DOI 10.1007/978-3-319-20200-6_9

were additional negative effects of environmental selenium (Skorupa 1998). The pivotal change in research into Se was the discovery of its positive impact as an essential trace element in mammals (Schwarz and Foltz 1957). Selenium deficiency was found to reduce growth, productivity, and reproduction and even cause death in fish, birds, animals, and humans. A detailed description and discussion is outside of the scope of this review and is already described in numerous reviews that deal with a wide range of negative effects of Se deficiency in the mammalian diet as well as with potential positive effects on nutrition and health (Brown and Arthur 2001; Flohé 2009; Hatfield et al. 2009; Letavayova et al. 2006; Novoselov et al. 2005; Patrick 2004; Qi et al. 2010; Rayman 2009; Surai 2006; Whanger 2004; Young et al. 2010; Zhuo and Diamond 2009). Lu and Holmgren (2009) reviewed biological functions of selenoproteins in oxidoreductions, redox signaling, antioxidant defense, thyroid hormone metabolism, and immune responses. They described a strong correlation between Se and human diseases such as cancer, Keshan disease, virual infections, male infertility, and abnormalities in immune responses and thyroid hormone function.

One of the reasons that Se has attracted researchers for so long is the fact that Se has large positive and negative environmental impacts with a narrow distinction between the two. This can lead to controversial findings and conclusions on the impact of Se compounds on animal and particularly human health, which was succinctly expressed by Flohé (2009): "More recent insights have revealed that Se in the context of tumor development remains a two-edged sword".

From this enormously extensive topic of selenium research, we will review a narrow field of Se research related to algae. We will concentrate on various microalgae comprising unicellular species from different taxa that have been extensively studied. Recent knowledge on algal selenoproteins in general and selenium-containing enzymes in particular, will be described, as well as the bioaccumulation of different forms of Se in algae.

2 Selenium Properties, Classification and Compounds

2.1 Element Classification

Although Se is sometimes referred to as a metal, or at least as a borderline metalloid, it should be stressed that the main classification of Se is as a non-metal. Selenium, with an atomic number of 34 and an atomic mass of 78.9, belongs to the chalcogen group, which includes oxygen, sulfur, tellurium, and polonium (Lide 1994). Elemental Se has several different allotropes that display either non-metal (red Se, black Se) or borderline metalloid or non-metal behavior (grey Se) (Lide 1994; McQuarrie and Rock 1991). In contrast to metals that typically form cations in aqueous solution, Se forms the anions selenite (SeO_3^{2-}) and selenate (SeO_4^{2-}). Oxyanions increase solubility and mobility with increasing pH, while metals show the opposite behavior.

2.2 Selenoproteins

Selenoproteins are proteins containing the rare amino acid selenocysteine (SeCys), encoded by a UGA codon and covalently linked in the protein primary structure (Kryukov et al. 2003; Stadtman 1996). They have been identified in such diverse organisms as bacteria, archea, and mammals (Burk et al. 2003).

Typical representatives of selenoproteins are antioxidant enzymes, a member of which is thioredoxin reductase (TR), an important protector of cells against Se toxicity. It is also a key enzyme in Se metabolism as it provides active selenide for the synthesis of all selenoproteins. Other important selenoproteins are glutathione peroxidases (GPXs), antioxidant enzymes protecting various organisms from oxidative stresses by catalyzing the reduction of hydroperoxides at the expense of glutathione (Roy et al. 2005). Selenoproteins also include enzymes such as deiodinases, selenophosphate-synthetase 2, selenoprotein H, I, K, M and Sep15, N, O, P, R, S, T, V, W (Kryukov et al. 2003; Reeves and Hoffmann 2009). Interestingly, most known selenoproteins are animal proteins, while their homologs in yeast and land plants are not selenoproteins (Stillwell and Berry 2005). Yeast and land plants thus contain seleno-independent glutathione peroxidases, also named non-selenium GPX (NS-GPX)/GPx7 and GPx5 (Herbette et al. 2007).

However, selenoproteins were detected in algae. In the green alga *Chlamydomonas reinhardtii*, its repertoire is almost comparable to that of mammalian models (Novoselov et al. 2002). A survey of the *Chlamydomonas* genome led to the identification of the complete selenoproteome, defined by 12 selenoproteins representing 10 families (Grossman et al. 2007; Lobanov et al. 2007). Among these selenoproteins, five GPXs could be identified in the *Chlamydomonas reinhardtii* genome. The selenite-induced GPX in *Chlamydomonas reinhardtii* has been purified (Shigeoka et al. 1991) and an active Se -dependent GPX containing the expected amino acid selenocysteine has been identified (Fu et al. 2002). Selenoproteins were also discovered and studied in detail in other green algae. The chlorococcal alga *Scenedesmus quadricauda* produces biomass with a high content of organically bound Se and it was used to study the relationship between the presence of inorganic Se compounds and thioredoxin reductase activity (Vítová et al. 2011). Bioinformatic approaches have identified other selenoproteins in the green algae *Ostreococcus tauri* and *O. lucimarinus* (Prasinophyceae) and other species (Lobanov et al. 2007, 2009). Furthermore, selenoenzymes were also found in the haptophyte *Emiliania huxleyi* and the diatom *Thalassiosira pseudonana* (Araie et al. 2008; Obata and Shiraiwa 2005; Price and Harrison 1988).

In general, eukaryotes have highly variable sets of selenoproteins, varying from zero in higher plants and fungi to more than 30 in some fish and algae.

The mechanism of selenoprotein synthesis has been studied intensively and has been described in detail in several reviews. Thus, reviewing these topics would be duplicating the excellent work of others and is outside of the scope of this paper. We therefore refer readers to the following reviews:

Lobanov et al. (2009) provide an overview of eukaryotic selenoproteins and selenoproteomes. Low and Berry (1996) describe molecular mechanisms in the regulation of translation, including alternative decoding of UGA, typically a stop codon, as selenocysteine. Two RNA structures, the mRNA selenocysteine insertion sequence and a unique selenocysteyltRNA, are required for this process.

3 Bioaccumulation

In natural waters, Se is present in its inorganic forms in three oxidation states: selenate (+VI), selenite (+IV) and elemental Se (0). Selenate is the dominant dissolved form, representing more than 67 % of the total dissolved Se (Gojkovic et al. 2014).

Selenate is the Se form primarily absorbed by algae and translocated to the chloroplast, where it follows the sulfur assimilation pathway, i.e. it is activated by ATP sulphyrylase, reduced to selenite and incorporated into other Se-containing compounds (Neumann et al. 2003). Both selenates and selenites are taken up by microalgae and converted to protein-bound SeCys and SeMet, soluble inorganic forms, non toxic organic compounds such as SeMeSeCys, several free amino acids, and volatile organoselenium compounds (dimethylselenide) (Neumann et al. 2003).

Kinetic analysis of the marine alga *Emiliania huxleyi*, using [^{75}Se]selenite (Obata et al. 2004) revealed that the selenite uptake process consists of two phases, one saturable and one linearly related to substrate concentration. The uptake activity of the first phase was suppressed by inhibitors of ATP biogenesis, suggesting that selenite uptake is driven by an active transport system. Inhibition of amino acid and protein synthesis by cycloheximide caused a decrease in Se incorporation into low molecular compounds and proteins. These results suggested that *E. huxleyi* rapidly absorbs selenite, filling a small intracellular pool and low molecular compounds are immediately synthesized and then metabolized to selenoproteins.

Selenoproteins were also found in the unicellular green alga, *Chlamydomonas reinhardtii* and the dinoflagellate *Oxyrrhis marina*, suggesting that these algae also possess translation machinery corresponding to that in animals (Fu et al. 2002; Novoselov et al. 2002; Osaka et al. 2003).

From a comparison of uptake of four Se species (selenite, selenate, SeCys and SeMet) in the green alga *C. reinhardtii* it was clear that organic Se was taken up in higher amounts than inorganic forms. The most rapid uptake was that of SeMet, resulting in SeCys production (Zhang 2013). Thus, SeMet seems to be a favorable selenoprotein precursor, and this is further supported by the fact that it is the most readily available Se species in aquatic environments. Interestingly, Se-Cys was taken up and released back into the medium while selenite produced by the cells was reabsorbed, indicating that this is an important intermediate compound in Se biochemistry. These results were also confirmed in other green algae. In the chlorococcal alga *Chlorella sorokiniana,* selenate promotes the accumulation of large amounts of SeMet. The SeMet enrichment of biomass also depends on sulfur and Se nutritional conditions (see Sect. 4.1). Apart from SeMet, free amino acids and proteins of *C. sorokiniana* also contained selenocysteine and selenocystine. Enrichment of *C. sorokiniana* biomass in SeMet could be scaled up to produce Se-enriched algal biomass as a bioactive food supplement (Gojkovic et al. 2014).

Bioaccumulation and biotransformation of selenate was studied in detail in the chlorococcal alga *C. sorokiniana*. Selenium concentration in culture medium and Se inside *Chlorella* showed inverse time-course trends. Inorganic selenate was transformed into organic forms: SeCys, SeMeSeCys and SeMet. In the first 48 h of incubation, an increase in the bioaccumulation of all selenocompounds was observed, accompanied by a rapid increase in the production of SeCys inside the cells, reaching a maximum concentration of about 6 μg Se/L. In a long-term batch experiment, SeMet concentration inside the cells rapidly increased to a final concentration of about 25 μg/L. The increase observed in SeMet occurred in parallel with a decrease in the intracellular content of SeCys (Gojkovic et al. 2014).

Similarly, in *Scenedesmus quadricauda*, the total amount of Se and SeMet in biomass increased with increasing concentrations of Se compounds in the culture medium and SeMet made up 30–40 % of the total biomass of Se (Umysová et al. 2009)

Exposure experiments with selenite have been carried out in the unicellular marine algae *Dunaliella primolecta*, *Porphyridium cruentum*, and *Chlorella* sp. grown in the presence of selenite. The species accumulated Se in several biochemical fractions, including proteins, amino acids (*Dunaliella primolecta* and *Chlorella* sp.) and soluble carbohydrates (*Dunaliella primolecta* and *Chlorella* sp.) (Moreno et al. 2014).

4 Selenium Toxicity

Selenium toxicity in different organisms, particularly vertebrates (Creighton and Twining 2010; Hamilton 2004; Schrauzer 2000; Spalholz 1994), is well known and has been intensively studied. The concentration at which Se becomes toxic is, however, very different depending on the organisms as well as the oxidative state of Se compounds to which the organisms are exposed (Table 1).

Selenate (SeO_4^{2-}) and selenite (SeO_3^{2-}) usually dominate in aquatic ecosystems (Cutter 1989). Selenate toxicity was found to be higher than the toxicity of selenite in some organisms. This has been observed for freshwater phytoplanktonic communities (Riedel et al. 1996), marine algae (Price et al. 1987; Wong and Oliveira 1991), and for microalgae (Umysová et al. 2009; Vítová et al. 2011). On the other hand, some authors (Pastierova et al. 2009) observed higher toxicity of selenite than selenate in *Desmodesmus (Scenedesmus) quadricauda*, *D. subspicatus*, *Chlorella vulgaris* and *Pseudokircheriella subcapitata*. Higher toxicity of selenite than selenate was also found in *Chlamydomonas reinhardtii* (Morlon et al. 2005b).

Selenium in the aquatic environment comes from both natural and anthropogenic sources, such as irrigation of agricultural lands, coal mining and combustion. Typical freshwater concentrations range from 0.13 to 2.50 nmol/L (equivalent to 0.01–0.5 μg Se/L); higher concentrations up to 5 μmol/L (equivalent to 400 μg Se/L) have been observed in contaminated areas (Conde and Sanz Alaejos 1997). However, concentrations may reach 1600 μg/L (micromolar range) in waters draining seleniferous soils (Hamilton 2004).

Table 1 EC_{50} values expressed as Se concentration (in mg/L and μmol/L) as determined for different algal species grown in the presence of selenite or selenate

	EC_{50}				
	Selenite		Selenate		
Species, strain	mg/L	μmol/L	mg/L	μmol/L	References
Scenedesmus quadricauda, wild type	4	50	12	417	Umysová et al. (2009)
Scenedesmus quadricauda, strain SeIV	300	3800	33	151	Umysová et al. (2009)
Scenedesmus quadricauda, strain SeVI	50	632	180	2270	Umysová et al. (2009)
Chlamydomonas reinhardtii			0.032	0.4	Fournier et al. (2010)
Chlamydomonas reinhardtii			0.245	3.1	Fournier et al. (2010)
Chlamydomonas reinhardtii	6.3	80			Morlon et al. (2005b)
Chlamydomonas reinhardtii			0.355	4.5	Geoffroy et al. (2007)
Chlorella pyrenoidosa			0.79	10	Bennett (1988)
Chlorella sorokiniana			45	238.2	Gojkovic et al. (2013)

4.1 *Targets of Selenium Toxicity in Algae*

Depending on its concentration, Se can act as an essential micro-nutrient protecting against damage by reactive oxygen species, or as a toxic compound. It has devastated wildlife populations in several large scale incidents, such as at Lake Belews in North Carolina or at the Kesterson Reservoir and San Joaquin Valley in California. Its toxic effects led to deformities and massive reproductive failures in fish and birds (Hamilton 2004). It has been proposed that Se substitution for sulfur in sulfur-containing proteins may be responsible for its toxic effects. During protein synthesis, Se substitution for sulfur could modify protein structure and function leading to teratogenic effects in fish (Lemly 1997).

Selenium toxicity in the alga *Spirogyra* and some species of Cyanophyta (*Anabaena ambigua, Anabaena subcylindrica, Nostoc commune, Nostoc muscorum, Spirulina* sp.), expressed as the effect on total chlorophyll, total protein, total carbohydrate, total starch and total free amino acids, was in the EC_{50} range of 3–5 mg Se/L. All parameters tested gradually decreased with increasing concentrations of Se. In agreement with the reported dual effects of Se, low concentrations of Se had stimulatory rather than toxic effects (Mane et al. 2013).

The species most sensitive to the presence of Se compounds seems to be the green alga *Chlamydomonas reinhardtii*, the growth of which was inhibited at 4.5 ± 0.2 μM (0.355 mg/L) selenate, which is lower than that commonly found in environmental concentrations. Estimated benchmark 50 % doses (EC_{50}) were 14 μM (1.104 mg/L) of selenate (Geoffroy et al. 2007). The effects of both selenite and selenate on *C. reinhardtii* included ultrastructural damage to chloroplasts resulting in impaired photosynthesis (Morlon et al. 2005b); this seems to be an effect in other algae too. Chloroplast membranes of the chlorococcal alga *Scenedesmus quadricauda*, in the presence of selenite, were reorganized into thick bundles of thylakoids and the stroma became granulose. In the presence of selenate, the chloroplast had a fingerprint-like appearance, and the stroma became less dense. Both Se compounds caused an increase in starch production (Vítová et al. 2011).

Effects observed at the population and sub-cellular levels were strongly linked. Photosynthetic and respiratory rates may be reduced due to ultrastructural damage, leading to a decrease in energy available to the cells, and ultimately to a reduction in growth capacity. As a consequence in phytoplankton species, dense granules containing Se and rich in phosphorus were assumed to be the result of detoxification processes. It is not only growth that is affected by increased concentrations of Se but also cell division that is specifically blocked in *S. quadricauda* incubated in the presence of selenite or selenate. Of the two compounds, selenite was more toxic than selenate. This was probably due to an over-accumulation of SeMet that was more pronounced for selenite (29 % of SeMet in the case of selenate and 41 % of SeMet in the case of selenite) (Vítová et al. 2011).

The resistance of *Chlorella sorokiniana* (Gojkovic et al. 2014) and *S. quadricauda* (Umysová et al. 2009) to Se is an order of magnitude greater than resistance of *C. reinhardtii.* In a wild type strain of *S. quadricauda,* growth was reduced at a concentration of 50 mg Se/L and completely blocked at 100 mg Se/L (Fig. 1). However, strains of *S. quadricauda* have been selected that were able to grow at concentrations of Se as high as 400 mg Se/L (Fig. 1).

The selected strains differed, however, in their resistance to different Se compounds. In the strain resistant to high levels of selenite, toxicity of selenate was comparable to that of the wild type. On the other hand, the strain resistant to high levels of selenate was sensitive to high levels of selenite (Fig. 2). A strain resistant to both Se compounds was also selected but toxicity to both compounds was higher than in strains resistant to only one of the compounds (Fig. 2) (after Umysová et al. 2009).

Fig. 1 Time course of dry weight of wild type and a selenite resistant strain SelV of *Scenedesmus quadricauda* grown in different selenium concentrations from 0 to 400 mg Se/L solid lines, full symbols strain SelV; short dashes, empty symbols wild type (Original, previously unpublished graph)

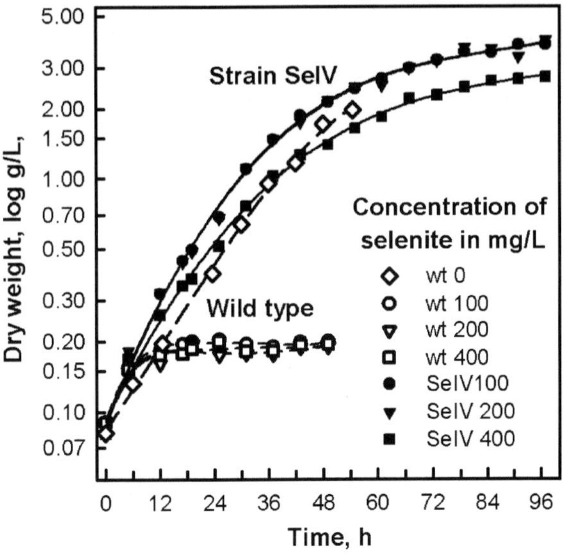

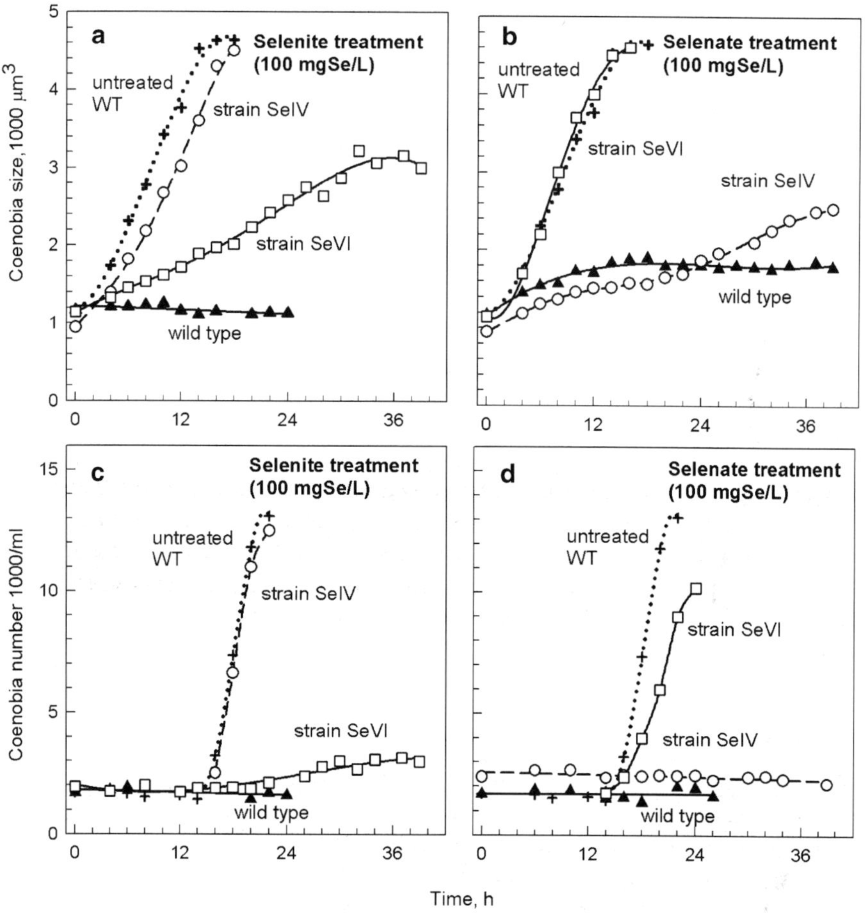

Fig. 2 Growth and cell division of *Scenedesmus quadricauda* under selenium treatment Growth, monitored as coenobia size (**a** and **b**) and cell division, monitored as coenobia number (**c** and **d**) during the cell cycle of synchronous cultures of wild type (*crosses*-untreated; *triangles*-treated with selenium), selenite resistant (*circles*), and selenate resistant (*squares*) strains of *Scenedesmus quadricauda* grown in the presence of selenite (**a** and **c**) or selenate (**b** and **d**) (100 mg Se/L (After Vítová et al. 2011)

4.2 Glutathione Peroxidases (GPXs)

One of the markers of stress caused by Se is glutathione peroxidase activity. GPXs are antioxidant enzymes that protect organisms against oxidative stress by catalyzing the reduction of hydroperoxides at the expense of glutathione (Roy et al. 2005).

The green alga *C. reinhardtii* responded to the presence of selenite by decreasing the level of ascorbate peroxidase and inducing that of glutathione peroxidase. The induced maximum activity of glutathione peroxidase and its enzymatic properties closely resembled those of animal glutathione peroxidases that contain Se (Yokota et al. 1988).

The green alga *Scenedesmus quadricauda* was used to investigate the relationship between Se toxicity and glutathione peroxidase activity. During growth of the untreated wild type, glutathione peroxidase activity increased slightly and then declined gradually until the end of the cell cycle. A similar pattern was observed in untreated resistant strains, and when resistant strains were grown in the presence of Se in the oxidation state to which they were resistant. In the wild type strain cultivated with 50 mg Se/L (selenite or selenate), activity increased to a high level and slowly declined until the end of the cell cycle. Similarly, activity increased in strains SeIV and SeVI when grown in the presence of Se in the oxidation state to which they were not resistant (Fig. 3) (Vítová et al. 2011). The activity of glutathione peroxidase in *Scenedesmus* was affected by Se in an oxidation state-dependent manner. In the wild type cultivated with 50 mg Se/L as selenite or selenate, activity increased dramatically during the growth phase; up to 150 and 50 rel. U/mg, respectively. GPX activity was higher in the presence of selenite than selenate (Vítová et al. 2011).

GPX activity in *S. quadricauda* was increased in the presence of high levels of Se, to values similar to those found in mammalian cells, where GPX is used as a biomarker. Whole blood GPX mRNA levels have been used as molecular biomarkers for assessing dietary Se requirements in rats (Sunde et al. 2009). Oxidative stress biomarker levels (GPX, vitamin A and E) were higher in rainbow trout after exposure to Se (Miller et al. 2009).

4.3 Thioredoxin Reductase

A similar trend as for GPX activity was shown for thioredoxin reductase (TR) activity in *S. quadricauda,* in both wild type and strains resistant to selenite (strain SeIV) and selenate (strain SeVI) (Umysová et al. 2009). TR activity was the same in all strains studied if grown in the absence of Se. However, TR was affected by Se treatment in a dose-dependent and toxicity-dependent manner. The findings implied that the increase in TR activity in algal cells was a stress response to Se cytotoxicity.

4.4 Detoxification of Selenium in Algal Cells

High levels of Se will lead to the metabolizing of toxic selenate to volatile dimethylselenide, which is thought to be one way that algal cells detoxify accumulated Se. In field studies, the largest proportion of Se was volatilized (e.g., Cook and Bruland 1987). A significant proportion of Se (more than 90 %) (Zhang 2013) can be cycled from the aquatic environment to the air through biological uptake, and this Se flux can be a fairly efficient way of detoxifying Se. Freshwater microalgae (eg. *Chlorella* sp.) were found to produce volatile dimethylselenide at exceptionally high rates in water containing 20 μM selenate. (3.5 μg Se/L) Up to 90 % of selenate supplied to

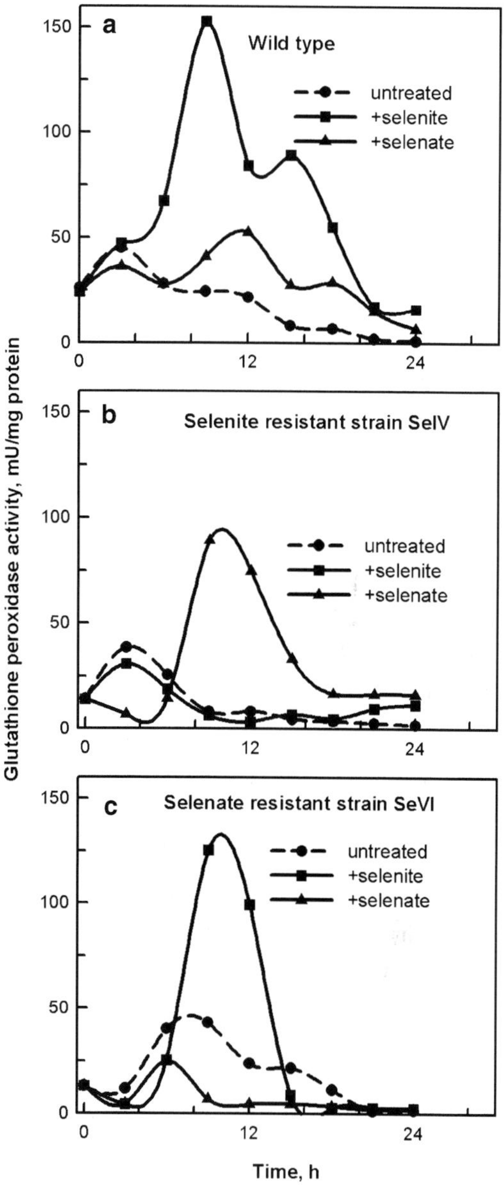

Fig. 3 Glutathione peroxidase activity during the cell cycle of *Scenedesmus quadricauda* Activity of glutathione peroxidase during the cell cycle of synchronous cultures of wild type (**a**), selenite resistant (**b**), selenate resistant (**c**) strains of the alga *Scenedesmus quadricauda* grown in the absence (*circles*) or presence (50 mg Se/L) of selenite (*squares*) or selenate (*triangles*) (After Vítová et al. 2011)

the microalgae were removed through accumulation and volatilization (Neumann et al. 2003). Organic Se is more bioavailable, taken up faster and effectively released from the cell because it can be transformed faster into the gaseous phase than inorganic Se, which has to be reduced within the cell (Zhang 2013).

4.4.1 Sulfate Role in Selenium Toxicity

Selenate and selenite differ in their mode of entry as well as following metabolism, causing different types of toxicity. Selenate is assumed to transfer through the cell wall similarly to sulfate, due to the fact that the enzymes of the trans-sulfuration pathway cannot discriminate between sulfur and Se (Birringer et al. 2002). Selenium and sulfur (S) share many chemical properties so competition of sulfates and selenates for transfer through the cell wall is an important factor determining toxicity of external selenate in algae. It was found in *Chlorella* (Neumann et al. 2003) that additions of 1 mM sulfate, but not nitrate, inhibited Se accumulation and volatilization so that only low amounts (1.8 %) of the supplied selenate were removed from the nutrient medium.

Competition with chemically analogous sulfate ions for sulfate/selenate transporters was shown in microalgae cultured in sulfate-free nutrient solution; sulfate transporter activity was increased and about 87 % of the accumulated selenate was metabolized to SeMet and selenocystine. Selenium was also found to compete with sulfur for assimilation pathways in *C. reinhardtii* (Morlon et al. 2005b) and in other organisms (Wheeler et al. 1982; Williams et al. 1994); SeCys was produced by the sulfate metabolism pathway (Pilon-Smits and Quinn 2010).

Interference of Se with S metabolism was also demonstrated in *Scenedesmus quadricauda*, where with sulfur deficiency, selenium toxicity increased (Umysová et al. 2009). Similarly in *C. reinhardtii*, the presence of sulfate markedly decreased the toxicity of selenate, as shown by increased EC_{50} values estimated in the presence of 8 and 80 μmol/L of sulfate ions. An increase in toxicity was accompanied by a decrease in selenate bioaccumulation, which was 10 times lower in the presence of 80 μmol/L, than in the presence of 8 μmol/L, sulfate ions (Fournier et al. 2010). Decreased Se toxicity with an increased concentration of sulfate was also observed in *Chlorella pyrenoidosa* (Bennett 1988) (Table 1). All these findings show that it is very important to know the sulfate ion concentration in the medium when estimating the EC_{50} of selenate. A significant relationship between acute selenate toxicity and medium sulfate concentration has also been observed for different freshwater organisms including *Cerodaphnia dubia*, *Pimephales promelas* (Brix et al. 2001) and *Selenastrum capricornatum* (Williams et al. 1994).

In contrast to the toxicity of selenate, selenite toxicity is not dependent on the concentration of sulfur, as is apparent by comparing EC_{50} values and lethal concentrations of selenite, which were similar in *S. quadricauda* (Umysová et al. 2009) and *C. reinhardtii* (Morlon et al. 2005a, 2005b) (Table 1).

5 Applications

As described above, Se compounds play a controversial role in the environment. On the one hand they are harmful products of human activity, such as coal combustion and oil refinery waters, coal, phosphate, and metal mining, causing direct health threats to a taxonomically wide range of animals, including humans, through contaminated water. On the other hand, considerable attention has been devoted to the effects of Se shortage in daily food intakes. A diet deficient in Se causes serious health problems, while Se sufficiency has marked beneficial effects on many serious diseases and improves the quality of animal products such as eggs and meat (for more details, see Sect. 5.2.2).

It is therefore not surprising that in relation to Se compounds, algae can be used in various ways. Algae can on the one hand be used to adsorb and accumulate Se compounds and to render harmless Se contaminated areas, and on the other hand, they can be used in the form of Se-enriched algal biomass as a source of organic Se in beneficial dietary supplements for animals and humans.

5.1 Detoxication

As discussed above, Se in both oxidation states can accumulate in microalgae and be converted to protein-bound SeCys and SeMet, soluble inorganic forms, non toxic organic compounds such as selenomethylselenocysteine (SeMeSeCys), several free amino acids, and volatile organoselenium compounds (dimethylselenide) (Neumann et al. 2003). These characteristics of microalgae make them convenient organisms for remediation of Se contaminated areas. These can include agricultural drainage waters resulting from the irrigation of Se-rich soils, or industrial wastewater generated during the processing of coal or oil, which may contain inorganic Se salts in such high concentrations that they act as toxic environmental contaminants.

5.1.1 Absorption of Selenium Compounds from Polluted Water Environments

A complicating factor for algal applications in contaminated water environments is Se toxicity, which prevents its more advantageous use at higher concentrations and consequently there is a relatively low uptake of Se by algal cells. The solution is to select strains of algae that are resistant to high concentrations of inorganic Se, and with the ability to accumulate Se to high levels.

Wild type strains and species of microalgae differ in their upper limits of Se resistance as well as in resistance to the type of Se compound. The highest tolerable limit in algae is up to 100 mg/L of sodium selenite (45 mg Se/L), and was found for *Chlorella zofingiensis* (Pelah and Cohen 2005). Growth of untreated and selenite-treated cells was similar, irrespective of the selenite concentration. Increasing the concentration of sodium selenite above 100 mg/L (45 mg Se/L), resulted in a toxic effect and the cells died after 4 days. Similarly, the fresh-water green microalga *Chlorella sorokiniana,* in a culture medium supplemented with selenate (SeO_4^{2-}) at concentrations ranging from 5 to 50 mg/L (2.9–29 mg Se/L), grew well at all tested selenate concentrations (Gojkovic et al. 2013; Neumann et al. 2003). In the wild type strain of *S. quadricauda,* growth was slowed with selenate or selenite at a Se concentration of 50 mg Se/L, and completely blocked at 100 mg Se/L (Umysová et al. 2009).

5.1.2 Utilization of Algae with Resistance to Very High Concentrations of Se Compounds

The upper concentration of Se observed in contaminated areas was about 400 μg Se/L (Conde and Sanz Alaejos 1997). Clearly, algal species studied up to now, were able to grow and accumulate Se compounds in these areas, and can do so even if the concentration of Se compounds in contaminated waters would be more than two orders of magnitude higher than those found in 1997 (Conde and Sanz Alaejos 1997). Extremely contaminated environments could therefore be cleaned by selected strains with extremely high resistance to Se (400 mg Se/L) (Umysová et al. 2009). Three such resistant strains of *Scenedesmus quadricauda* were isolated: strain SeIV was resistant to selenite (Na_2SeO_3) but not to selenate (Doušková et al. 2009a), while strain SeVI was resistant to selenate (Na_2SeO_4) but sensitive to selenite in the same way as wild type (Doušková et al. 2009b), the third strain was resistant to both selenium compounds (Doušková et al. 2009c). These strains grew even at very high concentrations of Se, at a rate similar to that of the untreated wild type, but absorbed significantly higher amounts of organically bound Se than the wild type (Umysová et al. 2009). The selected strain, Se(IV+VI), which was resistant to a combination of both selenite and selenate (Doušková et al. 2009c) had, however, the growth rate lower than in the untreated wild type. This strain's resistance to both compounds was also lower than that of the respective resistant strains (SeIV and SeVI). These differences between resistant mutants indicated that the effect of selenite and selenate on the cells is based on distinctly different mechanisms (Umysová et al. 2009). Selenate is chemically analogous to sulfate (the same charge and virtually the same chemical structure) and is likely to be accumulated via sulfate transporters. It appears that high rates of Se accumulation by sulfate-deprived microalgae results from reduced competition with chemically analogous sulfate ions for selenate uptake via up-regulated sulfate/selenate transporters and rapid reductive metabolism of selenate. Consequently the algae accumulated Se from selenate at much faster rates than from selenite (Neumann et al. 2003).

5.2 *Decreasing Toxicity of Other Toxic Material*

In contrast to direct toxicity, the presence of Se in cells can also suppress or decrease the toxicity of other compounds, particularly heavy metals. Detailed studies have been published on the detoxification of mercuric chloride and methylmercury chloride by *Chlorella sorokiniana*. Methylmercury was found to be extremely toxic to freshwater algae since it disturbed mitosis and cytokinesis, leading to the formation of giant multinucleated cells, polyploid nuclei and numerous alterations in nuclear and nucleolar structures. A protective role of Se against mercury toxicity was found in several organisms and was confirmed and studied in detail in *Chlorella sorokiniana* (Moreno et al. 2014).

In counteracting the toxicity of $MeHg^+$, algae pre-treated with SeMet markedly increased the total content of chlorophyll, carotenoids, as well as dry weight and light dependent oxygen production, compared to a control culture that was only exposed to $MeHg^+$. The levels of $MeHg^+$ measured in cells were lower in cultures pre-treated with SeMeCys than controls, and they were able to incorporate and transform these Se compounds, indicating that uptake of $MeHg^+$ into the cells was negligible when carried out in the presence of SeMet, or that SeMet enhanced the release of $MeHg^+$ (Moreno et al. 2014).

Similar work also revealed the protective role of Se against $K_2Cr_2O_7$-induced hepatotoxicity in rats (Soudani et al. 2011).

5.2.1 Production and Utilization of Selenium enriched Algae

There is a considerable body of evidence that Se-containing compounds are effective chemoprotective agents, reducing the incidence of breast, liver, prostate, and colorectal cancers in model systems (Brown et al. 2000; Burk et al. 2003, 2006; de Rosa et al. 2012; Duffield-Lillico et al. 2002; Hatfield et al. 2009; Rayman et al. 2006; Rayman 2000; 2002, 2009; Reilly 2006; Whanger 2004). When algae are exposed to environmental Se in the forms of selenite or selenate, they transform them into organic Se compounds. Such Se-enriched algae could be used as an alternative to Se-enriched yeast.

Selenium-enriched biomass was extensively tested as a feed additive on a wide range of farm animals (chickens, laying hens, sows, boars, ewes, newborn lambs) and compared with the effects of both inorganic Se compounds and in some cases with Se-enriched yeast (Skřivan et al. 2006, 2008, 2010; Travnicek et al. 2007, 2008). In general, it was found that addition of Se-enriched *Chlorella* or *Scenedesmus* biomass to the diet of all animals tested had statistically better effects on specific physiological and physical parameters than addition of a selenite salt. In some cases, Se-enriched algae showed some advantage over Se-enriched yeast, and it was assumed that the different selenoamino acid profiles between Se-enriched yeast and Se-enriched *Chlorella* may explain the differences. The effects of dietary sodium selenite, Se-enriched yeast and heterotrophically produced Se-enriched spray-dried *Chlorella* biomass on physical parameters of eggs and laying hen production were investigated (Skřivan et al. 2006). Utilization of the organic form of Se from Se-enriched yeast and Se-enriched *Chlorella* in laying hens was higher than that of the inorganic form of Se (Na_2SeO_3). Higher egg white thickness was observed only in the Se-*Chlorella* group. Hens receiving the diet with Se-*Chlorella* also had better egg production compared to the basal diet (Skřivan et al. 2008). In other experiments, Se from Se-yeast and Se-*Chlorella* (0.3 mg/kg) were added as a supplement to the diet for broiler chickens. Selenium addition increased the Se concentration in muscle and Se-enriched algae had the best feed conversion (Ševčíková et al. 2006) and enhanced oxidative stability of broiler meat compared to sodium selenite (Dlouhá et al. 2008). Se-enriched *Chlorella* supplementation of feed mixtures for poultry can therefore be used to improve poultry health as well being a source of organic Se in Se-enriched poultry meat for human consumption.

The effects of supplementation of inorganic and organic forms of Se on ewes and in newborn lambs were studied by other research groups (Travnicek et al. 2007, 2008). They used algal biomass produced by controlled cultivation in solar bioreactors, with a Se content of 255 mg per kg of dry matter.

Addition of Se-enriched spray-dried *Chlorella* biomass to the diet had a positive effect on feeding and can be explained by direct utilization of selenoamino acids during proteosynthesis, as was shown for Se-yeast (Qin et al. 2007). Likewise, further experiments confirmed a positive influence on sows and boars, regardless of the form of Se supplementation, and higher activity of GSH-Px in whole blood, and a higher content of Se in blood serum and the milk of ewes. The organic form of Se was, however, more efficient and higher fertility increase (by 38 %) of ewes receiving Se-enriched algae was reported. The levels of immunity in postnatal lambing ewes and postnatal proteosynthesis in the offspring of lambing ewes were enhanced (Rodinová et al. 2008). These positive effects corresponded to similar results obtained with Se-enriched yeast (Kim and Mahan 2001; Knowles et al. 1999). Maintenance of a higher Se output in milk in the second month of lactation reflects a higher level of utilization of organic forms of Se. The higher conversion of organic Se, such as Se-enriched yeast, which was reflected in a higher Se content in tissues and blood and higher activity of GSH-Px, was explained by direct utilization of selenoamino acids during proteosynthesis (Qin et al. 2007). The positive effect of feeding with Se-enriched *Chlorella* can be explained in a similar way.

Except for the unique utilization of Se-resistant strains described in *Scenedesmus quadricauda* (Umysová et al. 2009), most algal species or strains used for production of Se-enriched biomass are susceptible to toxicity by Se compounds at relatively low concentrations, probably due to damage to the thylakoid membrane structure leading to impaired PS II function and limited electron transport between PSII and PSI (Geoffroy et al. 2007). However, at sublethal levels, detoxification occurs through Se accumulation and transformation into SeMet, and selenate tolerance mechanisms are based on assimilation and biotransformation of selenate to less-toxic organic forms (Umysová et al. 2009). Several strategies for obtaining algal biomass highly enriched in Se were developed to grow algal cultures up to the point where overall detoxification activity competes with selenate toxicity.

5.2.2 Continuous Supply of Low Nontoxic Dosage of Selenium

A laboratory scale procedure for continuous cultivation of *Chlorella sorokiniana* was described for the production of biomass enriched in the high value amino acid, SeMet. The effect of dilution rate on biomass productivity, viability and SeMet content, at several selenate concentrations, was studied and maximal SeMet productivity of 21 μg/L/day was demonstrated with 40 mg/L of selenate (16.7 mg Se/L) in the culture medium at a low dilution rate of 0.49/day, calculated on a total daily effluent volume (Gojkovic et al. 2013). The absorbed Se was transformed into organic forms SeCys, SeMeSeCys and SeMet (Gojkovic et al. 2014), which are more anti-carcinogenic than inorganic Se (Stadtman 1996). SeMet has beneficial physiological effects that are not shared with other Se compounds and it cannot be synthesized in higher animals, including humans. It was also observed that in a long-term batch experiment, cellular SeMet increased to a final concentration of about 25 ppm (μg/L), which is sixfold higher than the 4 ppm (μg/L) that were achieved at exponential phase after 96 h of growth. The observed increase in SeMet was accompanied by a decrease in the content of intracellular SeCys (Gojkovic et al. 2014). The authors concluded that enrichment of *Chlorella sorokiniana* biomass in SeMet could be scaled up to produce Se-enriched algal biomass that could be used as a bioactive food supplement. Unfortunately, scaled up production of Se-enriched *Chlorella sorokiniana* biomass and its application in animal diets have not yet been tested.

5.2.3 Stepwise Supply of Low Dosage of Selenium

This strategy stems from the finding that algae growing in batch culture can detoxify low concentrations of Se compounds that are present in the nutrient medium by assimilation and biotransformation of inorganic Se compounds into intracellular, less-toxic, or non-toxic organic forms. With an increased growth of biomass, the tolerated dose of a Se compound can be increased without any toxic cellular effects. Stepwise addition of Se during the growth phase of *Spirulina platensis* avoided the growth inhibitory effects of high Se concentrations (Chen et al. 2006). This procedure was also successfully applied to commercial/industrial biomass production of *Chlorella vulgaris* enriched with organically bound Se (Doucha and Lívanský 2012). The Se-enriched algal biomass was produced both autotrophically (Doucha and Lívanský 2009), grown in scaled up outdoor thin layer photobioreactors (2 % of CO2 in air) and heterotrophically (Doucha and Lívanský 2008), in fermenters with glucose added as a source of carbon and energy (Doucha and Lívanský 2012). The heterotrophically grown cultures had a high volumetric productivity and high cell concentrations (up to 100 g of dry mass/L) (Doucha et al. 2009). The procedure for consecutive addition of selenite to heterotrophically grown *Chlorella vulgaris* is illustrated in Fig. 4. These results (Doucha et al. 2009) verified that the stepwise addition of Se to the medium could prevent a toxic effect and offers an effective and economical method for the production of Se-enriched algal products.

5.2.4 One-Shot Addition of a High Dose of Selenium to Resistant Strains of Algae

From the discussion above it is clear that the main problem faced by production of Se enriched algae is Se toxicity. The toxic concentration differs depending on the organism, and the highest resistance was found in freshwater *Chlorella zofingiensis,* which can tolerate sodium selenite up to a concentration of 100 mg/L; higher concentrations of exogenous Se were toxic (Pelah and Cohen 2005). An alternative approach is to use algae selected for high resistance to different Se compounds (Umysová et al. 2009);

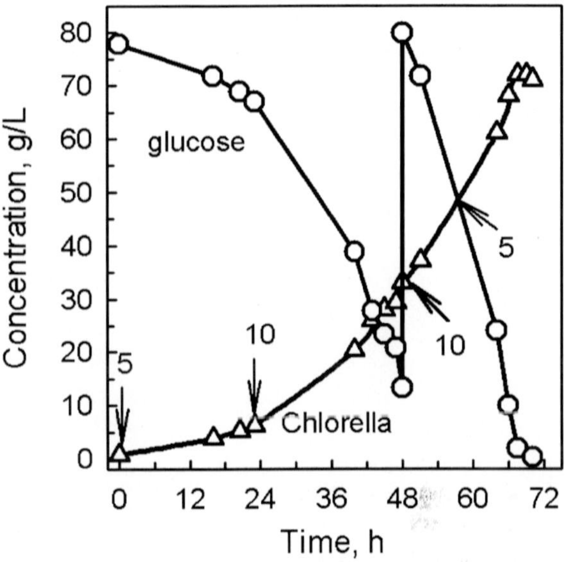

Fig. 4 Time course of *Chlorella* sp. dry mass and glucose concentration in the fermentation broth (culture volume 450 L). Secondary additions of glucose and nutrients were performed after 48 h of cultivation. Selenium additions (mg/L) in the form of sodium selenite solution are indicated by *arrows*) (After Doucha and Lívanský 2001)

this is discussed in the Sect. 4.4, where 'one shot' addition of a relatively high concentration of inorganic Se could support the production of Se-enriched algal biomass.

Therefore, apart from remediation of reservoirs containing toxic levels of Se compounds, Se-enriched biomass of such strains can also be used as a source of organic Se compounds for feeding domestic animals or as human dietary additives with the following benefits:

1. Algal biomass can be produced in scaled up outdoor open photobioreactors without any risk of contamination by other species because high concentrations of selenite or selenate in the nutrient medium will be toxic for most potential contaminating organisms.
2. In the case of resistant strains, Se can be added in a sufficiently high dose at the very beginning of the cultivation, without any toxic effects on algal cells during the time necessary for production of Se-enriched biomass. As an example, the Se-enriched biomass of the selenite resistant *S. quadricauda* strain, SeIV (Douskova et al. 2009a), was produced in the presence of 100 mg/L of selenite in an outdoor scaled up thin layer bioreactor (Doucha and Lívanský 2009). The enriched biomass contained 641 mg/kg of total Se. The amount of organically bound Se in the biomass was 23 %, which was similar to that attained in *Chlorella* (24 and 39 % of organic Se) as reported by (Neumann et al. 2003) and much higher than the SeMet content (in the range of ng/g) found by (Larsen et al. 2001).

In support of this idea, the application of Se-enriched biomass as an additive to chicken feed confirmed no toxic effects on body weight, feed conversion and survival rate in treated chickens. This means that it was an equally good source of organically bound Se for the production of Se-enriched chicken meat (Skřivan et al. 2010) as Se-enriched yeast or cereal grains (Medina et al. 2001).

6 Summary of Algal Application as Nutrient Supplements or Bio-remediators

The responses of algae to Se compounds can be different and contradictory due to potential toxic or beneficial properties of this element.

6.1 Refining of Selenium-Contaminated Aquatic Environments

In relation to Se toxicity it is clear that algae can accumulate inorganic selenium and convert this into non toxic organic compounds such as selenomethylselenocysteine (SeMeSeCys), several free amino acids, or volatile organoselenium compounds (dimethylselenide).

Therefore, algae can be used for remediation of areas such as agricultural drainage waters from irrigation of Se-rich soils, or industrial wastewater generated during the processing of coal, oil or other industrial activities where high concentrations of Se compounds become toxic environmental contaminants.

Wild types of several algal species, such as *Chlorella sorokiniana and Chlorella zofinginensis* (whose resistance to Se compounds is higher than the Se-concentration in contaminated areas) (see Sects. 4.1 and 5.1.1), can remediate moderately contaminated areas. Extremely contaminated environments can be cleaned by selected strains with increased resistance to Se. The properties and characteristics of three such resistant strains of *Scenedesmus quadricauda* are described (Sects. 5.1.2 and 5.2.4).

6.2 Selenium-Enriched Algae

Beneficial contributions of selenium stems from the fact that Se is an essential micronutrient in the diet of many organisms, including humans. Several epidemiological studies have shown its role in preventing or slowing the development of breast, prostate and colorectal cancer (for review, see Combs 2001a; b; Duffield-Lillico et al. 2004; Husbeck et al. 2006; Patrick 2004). Additional evidence suggests that selenium may have a positive role in mammalian development (Kohrle 2000; Minta et al. 2005), in immune functions (Arthur et al. 2003), and in slowing the aging process (Rayman 2000, 2009).

Algal cells enriched with selenium can be used as an excellent source of non-toxic organic selenium compounds for human and animal nutrition, compensating for a shortage of selenium in feed mixtures for animals, as well as in human food. The SeMet content is considered to be a measure of quality for this "organic selenium" due to its enhanced bioavailability.

The application of Se-enriched algal biomass was tested in several studies without toxic effects and with improvements in the health of domestic animals (chickens, laying hens, sows, boars, ewes, newborn lambs) (see Sect. 5.2.1). It was as good source of organically bound Se for the production of Se-enriched chicken meat (Skřivan et al. 2010) as Se-enriched yeast or cereal grains (Medina et al. 2001).

To produce sufficient amounts of algal selenium-enriched biomass the toxicity of inorganic selenium compounds must be overcome. Two approaches have been successfully applied:

6.2.1 Consecutive Addition of Low Doses of Inorganic Selenium Compounds

This approach was successfully used in the production of selenium-enriched algal biomass from species for which high doses of inorganic selenium compounds would be toxic.

It was particularly advantageous in heterotrophically grown cultures in scaled up fermenters. Selenite was added consecutively at defined time intervals, in lower than toxic concentrations. This had no toxic effect on cell growth but selenium compounds continuously accumulated and were transformed into non-toxic organic selenium compounds, particularly SeMet (see Sect. 2.2). Similarly to selenium enriched biomass in resistant strains, this algal biomass was shown to be beneficial when added to nutrients for domestic animals (Sect. 5.2.4).

6.2.2 Selenium-Resistant Algal Strains

Resistant strains of *Scenedesmus quadricauda* have been tested (Sect. 5.1.2), providing a 'one shot' addition of a relatively high concentration of inorganic Se to achieve Se-enriched algal biomass. This biomass was successfully tested as an additive to chicken feed mixtures (Sect. 5.2.4).

In addition to the technological advantageous, single dose application of a high concentration of external inorganic selenium at the beginning of cultivation in outdoor bioreactors, this approach also decreases the risk of contamination by other species of algae for which high doses of selenium would be toxic.

Acknowledgments The work was supported by the National Programme of Sustainability I, ID: LO1416.

List of Abbreviations

EC_{50}	half maximal effective concentration
DMSe	dimethylselenide
GPX	glutathione peroxidase

SeCys	selenocysteine
SeMeCys	selenomethylcystein
SeMeSeCys	selenomethylselenocysteine
SeMet	selenomethionine
TR	thioredoxin reductase
ICP-MS	inductively coupled plasma mass spectrometry

References

Araie H, Suzuki I, Shiraiwa Y (2008) Identification and characterization of a selenoprotein, thioredoxin reductase, in a unicellular marine haptophyte alga, *Emiliania huxleyi*. J Biol Chem 283:35329–35336

Arthur JR, McKenzie RC, Beckett GJ (2003) Selenium in the immune system. J Nutr 133:1457–1459

Bennett WN (1988) Assessment of selenium toxicity in algae using turbidostat culture. Water Res 22:939–942

Birringer M, Pilawa S, Flohe L (2002) Trends in selenium biochemistry. Nat Prod Rep 19:693–718

Brix KV, Volosin JS, Adams WJ, Reash RJ, Carlton RG, McIntyre DO (2001) Effects of sulfate on the acute toxicity of selenate to freshwater organisms. Environ Toxicol Chem 20:1037–1045

Brown KM, Arthur JR (2001) Selenium, selenoproteins and human health: a review. Public Health Nutr 4:593–599

Brown KM, Pickard K, Nicol F, Beckett GJ, Duthie GG, Arthur JR (2000) Effects of organic and inorganic selenium supplementation on selenoenzyme activity in blood lymphoctyes, granulocytes, platelets and erythrocytes. Clin Sci (Lond) 98:593–599

Burk RF, Hill KE, Motley AK (2003) Selenoprotein metabolism and function: evidence for more than one function for selenoprotein P. J Nutr 133:1517–1520

Burk RF, Norsworthy BK, Hill KE, Motley AK, Byrne DW (2006) Effects of chemical form of selenium on plasma biomarkers in a high-dose human supplementation trial. Cancer Epidemiol Biomark Prev 15:804–810

Chen T, Zheng W, Yang F, Bai Y, Wong Y-S (2006) Mixotrophic culture of high selenium-enriched *Spirulina platensis* on acetate and the enhanced production of photosynthetic pigments. Enzyme Microb Technol 39:103–107

Combs GF Jr (2001a) Considering the mechanisms of cancer prevention by selenium. Adv Exp Med Biol 492:107–117

Combs GF Jr (2001b) Impact of selenium and cancer-prevention findings on the nutrition-health paradigm. Nutr Cancer 40:6–11

Conde JE, Sanz Alaejos M (1997) Selenium concentrations in natural and environmental waters. Chem Rev 97:1979–2003

Cook TD, Bruland KW (1987) Aquatic chemistry of selenium: evidence of biomethylation. Environ Sci Technol 21:1214–1219

Creighton N, Twining J (2010) Bioaccumulation from food and water of cadmium, selenium and zinc in an estuarine fish, Ambassis jacksoniensis. Mar Pollut Bull 60:1815–1821

Cutter GA (1989) The estuarine behaviour of selenium in San Francisco Bay. Estuar Coastal Shelf Res 28:13–34

de Rosa V, Erkekoglu P, Forestier A, Favier A, Hincal F, Diamond AM, Douki T, Rachidi W (2012) Low doses of selenium specifically stimulate the repair of oxidative DNA damage in LNCaP prostate cancer cells. Free Radic Res 46:105–115

Dlouhá G, Ševčiková S, Dokoupilová A, Zita L, Heindl J, Skřivan M (2008) Effect of dietary selenium sources on growth performance, breast muscle selenium, glutathione peroxidase activity and oxidative stability in broilers. Czech J Anim Sci 53:265–269

Doucha J, Lívanský K (2001) Method of controlled cultivation of algae in heterotrophic mode of nutrition. Institute of Microbiology ASCR, Czech Republic, Czech Patent 288638

Doucha J, Lívanský K (2008) Production strain of the alga *Chlorella vulgaris* BEIJ. strain Doucha et Lívanský 1996/H 14. Institute of Microbiology ASCR, Czech Republic, Czech Patent No. 299352

Doucha J, Lívanský K (2009) Outdoor open thin-layer microalgal photobioreactor: potential productivity. J Appl Phycol 21:111–117

Doucha J, Lívanský K (2012) Production of high-density *Chlorella* culture grown in fermenters. J Appl Phycol 24:35–43

Doucha J, Lívanský K, Kotrbáček V, Zachleder V (2009) Production of *Chlorella* biomass enriched by selenium and its use in animal nutrition: a review. Appl Microbiol Biotechnol 83:1001–1008

Doušková I, Hlavová M, Umysová D, Vítová M, Zachleder V (2009a) Industrial strain *Scenedesmus quadricauda* SeIV of the green chlorococcal alga *Scenedesmus quadricauda* (Turp.) Bréb. Institute of Microbiology ASCR, Czech Republic, Czech Patent No. 300 861

Doušková I, Hlavová M, Umysová D, Vítová M, Zachleder V (2009b) Industrial strain Scenedesmus quadricauda SeVI of the green chlorococcal alga *Scenedesmus quadricauda* (Turp.) Bréb. Institute of Microbiology ASCR, Czech Republic, Czech Patent No. 300 809

Doušková I, Hlavová M, Umysová D, Vítová M, Zachleder V (2009c) Industrial strain Scenedesmus quadricauda SeIV+VI of the green chlorococcal alga *Scenedesmus quadricauda* (Turp.) Bréb. Institute of Microbiology ASCR, Czech Republic, Czech Patent No. 300 808

Duffield-Lillico AJ, Reid ME, Turnbull BW, Combs GF Jr, Slate EH, Fischbach LA, Marshall JR, Clark LC (2002) Baseline characteristics and the effect of selenium supplementation on cancer incidence in a randomized clinical trial: a summary report of the nutritional prevention of cancer trial. Cancer Epidemiol Biomarkers Perv 11:630–639

Duffield-Lillico AJ, Shureiqi I, Lippman S (2004) Can selenium prevent colorectal cancer? A signpost from epidemiology. J Natl Cancer Inst 96:1645–1647

Flohé L (2009) The labour pains of biochemical selenology: the history of selenoprotein biosynthesis. Biochim Biophys Acta 1790:1389–1403

Fournier E, Adam-Guillermin C, Potin-Gautier M, Pannier F (2010) Selenate bioaccumulation and toxicity in *Chlamydomonas reinhardtii*: influence of ambient sulphate ion concentration. Aquat Toxicol 97:51–57

Fu LH, Wang XF, Eyal Y, She YM, Donald LJ, Standing KG, Ben-Hayyim G (2002) A selenoprotein in the plant kingdom. Mass spectrometry confirms that an opal codon (UGA) encodes selenocysteine in *Chlamydomonas reinhardtii* glutathione peroxidase. J Biol Chem 277:25983–25991

Geoffroy L, Gilbin R, Simon O, Floriani M, Adam C, Pradines C, Cournac L, Garnier-Laplace J (2007) Effect of selenate on growth and photosynthesis of *Chlamydomonas reinhardtii*. Aquat Toxicol 83:149–158

Gojkovic Ž, Garbayo-Nores I, Gómez-Jacinto V, García-Barrera T, Gómez-Ariza J, Márová I, Vílchez-Lobato C (2013) Continuous production of selenomethionine-enriched *Chlorella sorokiniana* biomass in a photobioreactor. Process Biochem 48:1235–1241

Gojkovic Ž, Vílchez C, Torronteras R, Vigara J, Gómez-Jacinto V, Janzer N, Gómez-Ariza J-L, Márová I, Garbayo-Nores I (2014) Effect of selenate on viability and selenomethionine accumulation of Chlorella sorokiniana grown in batch culture. Sci World J 2014:1–13. http://dx.doi.org/10.1155/2014/401265

Grossman AR, Croft M, Gladyshev VN, Merchant SS, Posewitz MC, Prochnik S, Spalding MH (2007) Novel metabolism in *Chlamydomonas* through the lens of genomics. Curr Opin Plant Biol 10:190–198

Hamilton SJ (2004) Review of selenium toxicity in the aquatic food chain. Sci Total Environ 326:1–31

Hatfield DL, Yoo MH, Carlson BA, Gladyshev VN (2009) Selenoproteins that function in cancer prevention and promotion. Biochim Biophys Acta 1790:1541–1545

Herbette SP, Roeckel-Drevet P, Drevet JR (2007) Seleno-independent glutathione peroxidases. More than simple antioxidant scavengers. FEBS 274:2163–2180

Husbeck B, Nonn L, Peeh DM, Knox SJ (2006) Tumor-selective killing by selenite in patient-matched pairs of normal and malignant prostate cells. Prostate 66:218–225

Kim YY, Mahan DC (2001) Effects of high dietary levels of selenium-enriched yeast and sodium selenite on macro and micro mineral metabolism in growerfinisher swine Asian-australas. J Anim Sci 14:243–249

Knowles SO, Grace ND, Wurms K, Lee J (1999) Significance of amount and form of dietary selenium on blood, milk, and casein selenium concentrations in grazing cows. J Dairy Sci 82:429–437

Kohrle J (2000) The deiodinase family: selenoenzymes regulating thyroid hormone availability and action. Cell Mol Life Sci 57:1853–1863

Kryukov GV, Castellano S, Novoselov SV, Lobanov AV, Zehtab O, Guigo R, Gladyshev VN (2003) Characterization of mammalian selenoproteomes. Science 300:1439–1443

Larsen EH, Hansen M, Fan T, Vahl M (2001) Speciation of selenoaminoacids, selenonium ions and inorganic selenium by ion exchange HPLC with mass spectrometric detection and its application to yeast and algae. J Anal At Spectrom 16:1403–1408

Lemly AD (1997) Environmental implications of excessive selenium: a review. Biomed Environ Sci 10:415–435

Letavayova L, Vlckova V, Brozmanova J (2006) Selenium: from cancer prevention to DNA damage. Toxicology 227:1–14

Lide D (1994) CRC handbook of chemistry and physics: a ready-reference book of chemical and physical data. CRC Press, Boca Raton

Lobanov AV, Fomenko DE, Zhang Y, Sengupta A, Hatfield DL, Gladyshev VN (2007) Evolutionary dynamics of eukaryotic selenoproteomes: large selenoproteomes may associate with aquatic life and small with terrestrial life. Genome Biol 8:198

Lobanov AV, Hatfield DL, Gladyshev VN (2009) Eukaryotic selenoproteins and selenoproteomes. Biochim Biophys Acta 1790:1424–1428

Low SC, Berry MJ (1996) Knowing when not to stop: selenocysteine incorporation in eukaryotes. Trends Biochem Sci 21:203–208

Lu J, Holmgren A (2009) Selenoproteins. J Biol Chem 284:723–727

Mane PC, Kadam DD, Chaudhari RD (2013) Biochemical responses of some freshwater algal species to selenium: a laboratory study. Central Eur J Exp Biol 2:27–33

McQuarrie D, Rock E (1991) Chemistry of the main-group elements II. In: McQuarrie D, Rock P (eds) General chemistry. Freenan, New York, pp 1083–1097

Medina D, Thompson H, Ganther H, Ip C (2001) Se-methylselenocysteine: a new compound for chemoprevention of breast cancer. Nutr Cancer 40:12–17

Miller LL, Rasmussen JB, Palace VP, Hontela A (2009) The physiological stress response and oxidative stress biomarkers in rainbow trout and brook trout from selenium-impacted streams in a coal mining region. J Appl Toxicol 29:681–688

Minta M, Wilk I, Zmudzki J (2005) Cadmium and selenium: effect on proliferation and differentiation in hamster embryo cell culture. Bull Vet Inst Pulawy 49:329–335

Moreno F, Garcia-Barrera T, Gómez-Jacinto V, Gómez-Ariza JL, Garbayo-Nores I, Vilchez-Lobato C (2014) Antagonistic interaction of selenomethionine enantiomers on methylmercury toxicity in the microalgae *Chlorella sorokiniana*. Metallomics 6:347–355

Morlon H, Fortin C, Adam C, Garnier-Laplace J (2005a) Cellular quotas and induced toxicity of selenite in the unicellular green alga *Chlamydomonas reinhardtii*. Radioprotection 40:S101–S106

Morlon H, Fortin C, Floriani M, Adam C, Granier-Laplace J, Boudou A (2005b) Toxicity of selenite in the unicellular green alga *Chlamydomonas reinhardtii*: comparison between effects at the population and sub-cellular level. Aquat Toxicol 73:65–78

Neumann PM, De Souza MP, Pickering IJ, Terry N (2003) Rapid microalgal metabolism of selenate to volatile dimethylselenide. Plant 26:897–905

Novoselov SV, Rao M, Onoshko NV, Zhi H, Kryukov GV, Xiang Y, Weeks DP, Hatfield DL, Gladyshev VN (2002) Selenoproteins and selenocysteine insertion system in the model plant cell system *Chlamydomonas reinhardtii*. EMBO J 21:3681–3693

Novoselov SV, Calvisi DF, Labunskyy VM, Factor VM, Carlson BA, Fomenko DE, Moustafa ME, Hatfield DL, Gladyshev VN (2005) Selenoprotein deficiency and high levels of selenium compounds can effectively inhibit hepatocarcinogenesis in transgenic mice. Oncogene 24:8003–8011

Obata T, Shiraiwa Y (2005) A novel eukaryotic selenoprotein in the haptophyte alga *Emilliana huxleyi*. J Biol Chem 280:18462–18468

Obata T, Araie H, Shiraiwa Y (2004) Bioconcentration mechanism of selenium by a coccolithophorid, *Emiliania huxleyi*. Plant Cell Physiol 45:1434–1441

Osaka T, Beika A, Hattori A, Kohno Y, Kato KH, Mizutani T (2003) The protozoa dinoflagellate Oxyrrhis marina contains selenoproteins and the relevant translation apparatus. Biochem Biophys Res Commun 300:236–240

Pastierova J, Kramarova Z, Molnarova M, Fargasova A (2009) Comparison of the sensitivity of four freshwater microalgae to selenate and selenite. Fresenius Environ Bull 18:2029–2033

Patrick L (2004) Selenium biochemistry and cancer: a review of the literature. Altern Med Rev 9:239–258

Pelah D, Cohen E (2005) Cellular response of *Chlorella zofingiensis* to exogenous selenium. Plant Growth Regul 45:225–232

Pilon-Smits EAH, Quinn CF (2010) Selenium metabolism in plants. In: Hell R, Mendel RR (eds) Cell biology of metals and nutrients, Plant cell monographs 17. Springer, Berlin/Heidelberg, pp 225–241

Price NM, Harrison PJ (1988) Specific selenium-containing macromolecules in the marine diatom *Thalassiosira pseudonana*. Plant Physiol 86:192–199

Price NM, Thompson PA, Harrison PJ (1987) Selenium – an essential element for growth of the coastal marine diatom *Thalassiosira pseudonana* (Bacillariophyceae). J Phycol 23:1–9

Qi YM, Schoene NW, Lartey FM, Cheng WH (2010) Selenium compounds activate ATM-dependent DNA damage response via the mismatch repair protein hMLH1 in colorectal cancer cells. J Biol Chem 285:33010–33017

Qin SY, Gao JZ, Huang KH (2007) Effects of different selenium sources on tissue selenium concentrations, blood GSH-Px activities and plasma interleukin levels in finishing lambs. Biol Trace Elem Res 116:91–102

Rayman MP (2000) The importance of selenium to human health. Lancet 356:233–241

Rayman MP (2002) The argument for increasing selenium intake. Proc Nutr Soc 61:203–215

Rayman MP (2009) Selenoproteins and human health: insights from epidemiological data. Biochim Biophys Acta 1790:1533–1540

Rayman M, Thompson A, Warren-Perry M, Galassini R, Catterick J, Hall E, Lawrence D, Bliss J (2006) Impact of selenium on mood and quality of life: a randomized, controlled trial. Biol Psychiatry 59:147–154

Reeves MA, Hoffmann PR (2009) The human selenoproteome: recent insights into functions and regulation. Cell Mol Life Sci 66:2457–2478

Reilly C (2006) Selenium in food and health. Springer, New York, 222 p

Riedel GF, Sanders JG, Gilmour CC (1996) Uptake, transformation, and impact of selenium in freshwater phytoplankton and bacterioplankton communities. Aquat Microb Ecol 11:43–51

Rodinová H, Kroupová V, Trávníček J, Staňková M, Písek L (2008) Dynamics of IgG in the blood serum of sheep with different selenium intake. Vet Med 53:260–265

Roy G, Sarma BK, Phadnis PP, Mugesh G (2005) Selenium-containing enzymes in mammals: chemical perspectives. J Chem Sci 117:287–303

Schrauzer GN (2000) Selenomethionine: a review of its nutritional significance, metabolism and toxicity. J Nutr 130:1653–1656

Schwarz K, Foltz CM (1957) Selenium as integral part of factor-3 against dietary necrotic liver degeneration. J Am Chem Soc 79:3292–3293

Ševčíková S, Skřivan M, Dlouhá G, Koucký M (2006) The effect of selenium source on the performance and meat quality of broiler chickens. Czech J Anim Sci 51:449–457

Shigeoka S, Takeda T, Hanaoka T (1991) Characterization and immunological properties of selenium-containing glutathione-peroxidase induced by selenite in *Chlamydomonas-reinhardtii*. Biochem J 275:623–627

Skorupa JP (1998) Selenium poisoning of fish and wildlife in nature: lessons from twelve real-world examples. In: Frankenberger W, Engberg RA (eds) Environmental chemistry of slenium. Marcel Dekker, New York, pp 315–354

Skřivan M, Šimáně J, Dlouhá G, Doucha J (2006) Effect of dietary sodium selenite, Se-enriched yeast and Se-enriched *Chlorella* on egg Se concentration, physical parameters of eggs and laying hen production. Czech J Anim Sci 51:163–167

Skřivan M, Marounek M, Dlouhá G, Ševčíková S (2008) Dietary selenium increases vitamin E contents of egg yolk and chicken meat. Br Poult Sci 49:482–486

Skřivan M, Skřivanová V, Dlouhá G, Brányiková I, Zachleder V, Vítová M (2010) The use of selenium-enriched alga *Scenedesmus quadricauda* in a chicken diet. Czech J Anim Sci 55:565–571

Soudani N, Ben Amara I, Sefi M, Boudawara T, Zeghal N (2011) Effects of selenium on chromium (VI)-induced hepatotoxicity in adult rats. Exp Toxicol Pathol 63:541–548

Spalholz JE (1994) On the nature of selenium toxicity and carcinostatic activity. Free Radic Biol Med 17:45–64

Stadtman TC (1996) Selenocysteine. Annu Rev Biochem 65:83–100

Stillwell RJ, Berry MJ (2005) Expanding the repertoire of the eukaryotic selenoproteome. Proc Natl Acad Sci U S A 102:16123–16124

Sunde RA, Thompson KM, Evenson JK, Thompson BM (2009) Blood glutathione peroxidase-1 mRNA levels can be used as molecular biomarkers to determine dietary selenium requirements in rats. Exp Biol Med (Maywood) 234:1271–1279

Surai PF (2006) Selenium in nutrition and health. Nottingham University Press, Nottingham

Travnicek J, Pisek L, Herzig I, Doucha J, Kvicala J, Kroupova V, Rodinova H (2007) Selenium content in the blood serum and urine of ewes receiving selenium-enriched unicellular alga *Chlorella*. Vet Med 52:42–48

Trávníček J, Racek J, Trefil L, Rodinová H, Kroupová V, Illek J, Doucha J, Písek L (2008) Activity of glutathione peroxidase (GSH-Px) in the blood of ewes and their lambs receiving the selenium-enriched unicellular alga *Chlorella*. Czech J Anim Sci 53:292–298

Umysová D, Vítová M, Doušková I, Bišová K, Hlavová M, Čížková M, Doucha J, Machát J, Zachleder V (2009) Bioaccumulation and toxicity of selenium compounds in the green alga *Scenedesmus quadricauda*. BMC Plant Biol 9:58. doi:10.1186/1471-2229-9-58:1-16

Vítová M, Bišová K, Hlavová M, Zachleder V, Rucki M, Čížková M (2011) Glutathione peroxidase activity in the selenium-treated alga *Scenedesmus quadricauda*. Aquat Toxicol 102:87–94

Whanger PD (2004) Selenium and its relationship to cancer: an update. Br J Nutr 91:11–28

Wheeler AE, Zingaro RA, Irgolic K, Bottino NR (1982) The effect of selenate, selenite, and sulfate on the growth of 6 unicellular marine-species. J Exp Mar Biol Ecol 57:181–194

Williams MJ, Ogle RS, Knight AW, Burau RG (1994) Effects of sulfate on selenate uptake and toxicity in the green alga *Selenastrum capricornutum*. Arch Environ Contam Toxicol 27:449–453

Wong D, Oliveira L (1991) Effects of selenite and selenate on the growth and motility of 7 species of marine microalgae. Can J Fisher Aquat Sci 48:1193–1200

Yokota A, Shigeoka S, Onishi T, Kitaoka S (1988) Selenium as inducer of glutathione peroxidase in low-CO_2-grown *Chlamydomonas reinhardtii*. Plant Physiol 86:649–651

Young T, Finley K, Adams W, Besser J, Hopkins WD, Jolley DF, McNaughton E, Presser TS, Shaw D, Unrine J (2010) What you need to know about selenium. In: Chapman PM, Adams WJ, Brooks ML, Delos CG, Luoma SN, Maher W, Ohlendorf HM, Presser TS, Bradshaw P (eds) Ecological assessment of selenium in the aquatic environmental toxicology and chemistry. Society of Environmental Toxicology and Chemistry. SETAC Press, Pensacola, pp 7–45

Zhang X (2013) Aqueous speciation of selenium during its uptake by green algae *Chlamydomonas reinhardtii*. University of Waterloo, Waterloo, 61 p

Zhuo P, Diamond AM (2009) Molecular mechanisms by which selenoproteins affect cancer risk and progression. Biochim Biophys Acta 1790:1546–1554

Rare Earth Elements and Algae: Physiological Effects, Biorefinery and Recycling

Franz Goecke, Vilém Zachleder, and Milada Vítová

Abstract This chapter discusses new and multiple uses of rare earth elements (REEs) in modern commercial, industrial and military products and their ecological effects on the aquatic environment, with a particular focus on algae. Their unique physical and chemical properties, global locations and availability are described. Various applications for REEs, which may involve algae, are reviewed, including uses as fertilizers, markers, tracers, "bloom killers", or as biochemical agents such as tracers or bioindicators for physiological studies. We describe the ecological implications of increased release of REEs into the environment through mining, agricultural and industrial activities, as well as the possibility of using algae for bioremediation and recycling purposes. We conclude that a better understanding of the bioavailability, toxicity and uptake of REEs, as well as their physiological implications for algae at the molecular, enzymatic and life-cycle levels, are vital for environmentally-friendly production and use of these valuable resources.

Keywords Algae • Algal blooms • Bioaccumulation of metals • Bioindicators • Biosorption • Cerium • Environmental pollution • Fertilizers • Lanthanoids • Lanthanum • Metal enriched algae • Rare earth elements • Recycling of REEs • Toxicity of metals

1 Introduction

The name Rare Earths Elements (REEs) reflects the first observation of their existence in 1794 (Muraleedharan et al. 1994). The term 'rare' actually should not imply that they are uncommon in nature (Brown et al. 1990). While some of these elements are scarce, many REEs are present in the Earth's crust in similar proportions to elements in common use. They were named because of their scattered distribution and difficulties in refining (Liang et al. 2014).

REEs are integral to the development of several disciplines related to coordination chemistry, organo-metallic compounds, luminescent compounds, catalysis,

F. Goecke • V. Zachleder • M. Vítová (✉)
Laboratory of Cell Cycles of Algae, Institute of Microbiology, Czech Academy of Sciences (CAS), Opatovický mlýn, 37981 Třeboň, Czech Republic
e-mail: franz@alga.cz; zachleder@gmail.com; milada.vitova@gmail.com

A. Prokop et al. (eds.), *Algal Biorefineries*, DOI 10.1007/978-3-319-20200-6_10

solid-state chemistry, analytical and environmental chemistry, biology and medicine (Oliveira et al. 2012). They have applications in many areas and products e.g. audio systems, defense applications, liquid crystal displays, plasma, auto converters, family appliances, magnetic resonance imaging, polishing powders, automobiles, fertilizers, medicine, sports equipment, catalysts, glass additives, metallurgy, wind turbines, computers, lighting, NiMH batteries (for details see, Du and Graedel 2011; EPA 2012; Zepf 2013).

Becoming indispensable for a number of critical technologies, the use of these metals has increased, and demand is expected to rise in the near future. Consequently, their release into the biosphere will rise in parallel. There is therefore an urgent need to address the environmental effects of REEs. The affinity of algae for these metals may represent a serious environmental threat, or, on the other hand, may offer opportunities for bioremediation of contaminated areas.

REEs are non-essential elements that induce both positive and negative physiological responses. They are assumed not to be required for any known metabolic process, several reports have demonstrated that they can have beneficial effects under particular conditions (Goecke et al. 2015).

The aim of this review is to provide information about research covering the wide spectrum of impacts, both toxic and beneficial, in relation to algae, including possible applications for remediation of REE-contaminated areas as well as their recycling from industrial waste.

2 Characterization of Rare Earth Elements

The REE group includes scandium, yttrium, lanthanum and a series of 14 lanthanides (Ln) with atomic numbers of 58–71 (Table 1). They are members of Group IIIb in the periodic table and thus exhibit similar physical and chemical properties, but differ slightly in atomic number (Zhu et al. 2012). For example, Ln possess a partially filled f orbital, which is responsible for the uniformity in their oxidation states (Valcheva-Traykova et al. 2014). Another particular feature of lanthanides is their special electron configurations at the atomic level, in which with increasing atomic number the ionic radii get smaller; known as the lanthanide-contraction. This fact results in ionic radii of most trivalent lanthanides being similar to the radii of Ca^{2+}, Th^{4+}, and U^{4+}, with further implications to petrogenesis, toxicity and biological activities too (Zepf 2013). These different atomic structures and states especially lead to the unique and attractive properties of this group of elements.

REEs are widely dispersed and are generally common in nature, in terrestrial and marine environments with the exception of radioactive promethium (Brown et al. 1990, Table 1). In comparison with 'common' metals, the theoretical content of REEs in the Earth's crust is, on average, close to 0.015 % (Kastori et al. 2010), which matches that of copper, lead and zinc, and is much higher than that of tin, cobalt, silver and mercury (Hu et al. 2004). This does not imply that they are more readily exploitable as less abundant metals. Not all elements occur in concentrated,

Table 1 The lanthanide series and the so called REEs, their symbol, atomic number and atomic weight

Name	Symbol	Atomic Nr.	Atomic Weight	Origin	Earth's Crust (ppm)
Scandium	Sc	21	44.96	nat	7–14
Yttrium	Y	39	88.91	nat	17–24
Lanthanum	La	57	138.9	nat	30–71
Cerium	Ce	58	140.1	nat	57.5–66.4
Praseodymium	Pr	59	140.9	nat	6.3–7.1
Neodymium	Nd	60	144.2	nat	25.9–30.4
Promethium*	Pm	61	145.0	syn	0.0
Samarium	Sm	62	150.4	nat	4.5–5.09
Europium	Eu	63	152.0	nat	0.88–1.21
Gadolinium	Gd	64	157.3	nat	2.8–4.21
Terbium	Tb	65	158.9	nat	0.48–0.82
Dysprosium	Dy	66	162.5	nat	2.9–3.5
Holmium	Ho	67	164.9	nat	0.62–0.83
Erbium	Er	68	167.3	nat	2.3
Thulium	Tm	69	168.9	nat	0.30–0.33
Ytterbium	Yb	70	173.0	nat	1.47–2.26
Lutetium	Lu	71	175.0	nat	0.23–0.35

Radioactive element is indicated by (*)
Elemental crust concentration reviewed by Rudnick and Gao (2003)

pure or elemental forms, and in this case, REEs are reported to occur in more than 200 minerals such as carbonates, silicates, and phosphates, making them economically challenging to isolate (Loell et al. 2011). In fact, even though interest in REEs commenced during World War II, the use of individual rare earths only became possible in the 1950s when separation and metallurgical technologies improved (Diniz and Volesky 2005a).

REEs are traditionally considered as biologically non-essential elements, although recently Pol et al. (2014) suggested the reverse for certain methanotrophic and methylotrophic bacteria. Because of their unique physical and chemical properties e.g., high density, high melting point, high conductivity and high thermal conductance (see, Cockerill et al. 1973; Brown et al. 1990; Bünzli and Eliseeva 2011), they are used in a growing number of applications and have become indispensable for a number of critical technologies in modern commercial, industrial and military products (Du and Graedel 2011).

Prior to the 1960s, the relatively low demand for REEs worldwide was supplied by India, Brazil and South Africa. From the 1960s until the 1980s, the United States was the world leader in rare earth oxide production, especially due to the development of color television. In the 1980s, China began to produce important amounts of rare earth oxides (EPA 2012; USGS 2014). This country has abundant REEs reserves and became the world's leading producer in the early 1990s. The production of rare earth oxides was around 133,200 tonnes in 2009 (Polinares 2012). REEs have been used extensively in China as agricultural fertilizers (Zhang et al. 2006), an activity in which consumption, expressed as oxides, has reached 1100 tonnes per year, and is becoming one of the leading demands for REEs (see, Hu et al. 2004).

China's dominant position as the producer of over 95 % of the world output of REEs, and the rapid increase in consumption and demand for REEs due to new 'clean' technologies has resulted in concerns with respect to supply and price (Tse Pui-Kwan 2011; Schwabe et al. 2012). The recent downturn in Chinese production and exports, including an ongoing World Trade Organization export dispute, have led to re-opening of REE-mines and refining operations in the United States, India, Canada and Australia. Additionally, exploration efforts to develop new REE-projects have intensified in the last 4 years in countries like Australia, Brazil, Canada, China, Finland, Greenland, India, Kyrgyzstan, Madagascar, Malawi, Mozambique, South Africa, Sweden, Tanzania, Turkey, and Vietnam (USGS 2014).

As future demand for automobiles, electronics, microscopes, computers and portable electronic equipment grows, the use of REEs is expected to increase (Alonso et al. 2012; Das and Das 2013). Consequently, the release of these elements into the biosphere will rise in parallel (Loell et al. 2011). During the last decades, a large number of studies have focused on the geochemistry of heavy metals and metalloids such as As, Cd, Hg, and Pb, whose toxic effects are well-understood, whereas fewer researchers have focused on the behavior of REEs (Liang et al. 2014). The waste footprint and environmental impacts related to REE mining, processing, storage and transport are significant especially to surface and ground water qualities (EPA 2012). Toxic metal contamination in wastewaters is a worldwide problem. Metal recovery, including REEs, from industrial wastewater is important not only in view of environmental issues, but also on the technological aspects i.e. sustainability of supply by recycling (Kanchana et al. 2014).

3 REEs and Algae

The presence of REEs (Pr, Nd and Sm) was first reported in the calcareous red alga *Lithotamnium calcareum* in Roscoff, France (Servigne and Tchakirian 1939). Further information on REEs was then very dispersed.

More recently it was discovered that regardless of size (micro or macroalgae), structural organization (unicellular, filamentous, crustose or foliose), type of algal division (e.g. Chlorophyta, Ochrophyta, Rhodophyta, and Charophyta, plus Cyanobacteria), or geographical origin, algae contain a diverse spectrum of REEs (Hou and Yan 1998; Fu et al. 2000; Kano et al. 2001; Mashitah et al. 2012). From these analyses, it is known that concentrations of REEs in marine algae can be 10–20 times higher than those in terrestrial plants (Hou and Yan 1998; see, Table 2 for comparison), and 10^2–10^6 times higher than in seawater (Kano et al. 2001; Sakamoto et al. 2008b). The total amount of REEs can easily reach 1.3 μg g^{-1} of fresh algal material, which is considered high (Yan et al. 1998) compared to water

Table 2 Examples of REEs and their concentrations in different vegetables at different locations

REE	Tree[a]	Tea[b]	Moss[c]	Potato[d]	Alga[e]	Alga[f]	Alga[g]
Sc	nd	0.085	nd	nd	nd	nd	nd
Y	nd	0.360	0.127	0.011	nd	nd	nd
La	0.280	0.600	0.266	0.017	0.362	**3.990**	0.032
Ce	0.370	1.000	0.493	0.038	0.943	**9.080**	0.076
Pr	0.091	0.120	0.056	0.007	0.049	**0.910**	0.008
Nd	0.155	0.440	0.402	0.015	0.191	**4.910**	0.039
Sm	0.031	0.085	0.036	0.008	0.034	**0.900**	0.009
Eu	0.004	0.018	0.009	0.001	0.008	**0.090**	0.028
Gd	0.024	0.093	0.037	0.007	0.044	**1.020**	0.012
Tb	0.017	nd	0.005	0.001	0.006	**0.090**	0.001
Dy	0.021	0.074	0.024	0.002	0.030	**0.710**	0.012
Ho	0.004	0.019	0.004	0.000	0.006	**0.090**	0.002
Er	0.006	–	0.013	0.002	0.015	**0.350**	0.008
Tm	0.001	–	0.001	0.000	0.002	**0.020**	0.001
Yb	0.008	0.044	0.011	0.001	0.008	**0.290**	0.007
Lu	0.019	0.007	0.001	0.000	0.001	**0.020**	0.001
Total	**1.034**	**2.945**	**1.489**	**0.117**	**1.704**	**22.460**	**0.239**

The presented data corresponded to mean values established in μg g^{-1} dry weight
In **bold** are highlighted the highest value of the series
Pm is not considered because is radioactive and extremely rare (nd = not determinated)
[a]Samples of pine needles *Pinus silvestris*, Germany (Markert 1987)
[b]Certified reference material GBW07605 tea leaves, China (Zhang et al. 2007)
[c]*Hylocomium splendens* Sweden (Tyler 2004)
[d]*Solanum* sp. from a food market, China (Li et al. 2012)
[e]Red algae *Grateloupia filicina* Japan (Kano et al. 2001)
[f]Brown algae *Padina* sp., Malaysia (Mashitah et al. 2012)
[g]Green algae *Codium fragile*, Japan (Fu et al. 2000)

bodies (freshwater and marine seawater) that normally contain very low concentrations of REEs, ranging between 10^{-3} and 10^{-1} μg L^{-1} (Ogata and Terakado 2006; Sahoo et al. 2012; Richards and Mullins 2013; Liang et al. 2014, and references therein).

We are aware of only two studies that allow us to compare the REE content in different coexisting organisms, including algae. Both studies show clear differences between the coexisting groups of organisms and the special importance of REEs in macroalgae (Table 3).

These broad ranges in biotic concentrations of REEs can be due to: (i) the relative abundance of the element in water; (ii) the physical and metabolic processes particular to each algal species (cell wall components, enzymes, proteins, etc.); and (iii) by environmental factors relevant to each region e.g. temperature, light, pH, nitrogen availability that can affect the two previous factors (Jayasekera and Rossbach 1996; Vásquez and Guerra 1996; Sánchez-Rodríguez et al. 2001).

Table 3 Content of REEs in coexisting environmental samples in two studies in China

Organisms	Yao et al. (2007)	Shi et al. (2004)
Crustacea	0.15	0.15–0.81
Fish	0.067.06–0.23	nd
Macroalgae	**1.30–1.40**	**0.78–49.10**
Mollusks	3.32	0.37–21.60
Zooplankton	0.17	nd

Macroalgae in bold and values in $\mu g\ g^{-1}$ d.w

Topographical and climatic conditions, orogenetic processes, atmospheric deposition, submarine groundwater discharge, hydrothermal inputs and recent volcanic activities increase metal concentrations (Vásquez and Guerra 1996; Schacht et al. 2010). There are also important anthropogenic sources of REEs, primarily as phosphoric mineral fertilizers, phosphorus plaster, industrial wastewater, sewer mud, and mining processes and related atmospheric depositions (Volokh et al. 1990; Olmez et al. 1991; Elbaz-Poulichet and Dupuy 1999; Sahoo et al. 2012; Zhu et al. 2012; Liang et al. 2014).

As it will be discussed later, algae can concentrate these elements and can be used as bioindicators even when elements are no longer detectable in a particular environment. Despite the wide range of natural and anthropogenic loadings of REEs in the environment, the abundance of REEs in algae have always been measurable.

4 Biological Effects of Rare Earths

In general, as described by Valcheva-Traykova et al. (2014), the biological activities of REEs may be related to similarities in their ionic radii and coordination numbers with those of some essential elements (i.e., Ca, Mn, Mg, Fe and Zn), the variability of Ln ionic charges, and the ability of Ln ions to form stable complexes with organic molecules. Chemical interactions of Ln with biologically active molecules may result in altered enzymatic activities, substitution of essential metal ions from their ion-binding proteins, or polymerization of macromolecules. Their chelation with ion binding sites of proteins and ion channels may alter the specific permeability of the cellular membranes, resulting in a shortage or excess of ions in the intracellular and extracellular voids (see, Brown et al. 1990; Horovitz 2000; Bulman 2003; Wang et al. 2003). Surprisingly, despite being used as fertilizers for decades in Chinese agriculture, their effects on plants, particularly on algae, remain poorly understood.

4.1 *Effects on Algal Physiology*

As mentioned above, REEs are able to react with other molecules, penetrate cells, bioaccumulate, associate with receptors, and block specific ion-channels (Brown et al. 1990; Wang et al. 2003). Such special properties present advantages and opportunities for the study of algal physiology. Ln^{3+} ions were used as agents to study mechanisms of Ca^{2+} influx and Ca^{2+}-related downstream events (Wang et al. 2003). REEs have been used in the green freshwater microalga *Eremosphaera viridis* (Bauer et al. 1998; Schönknecht et al. 1998) and the charophyte *Nitella flexilis* (Ueda et al. 1974) as plasma-membrane Ca^{2+} channel-blockers for further characterization of the structure, polarization and permeability of the plasma membrane to divalent cations.

REEs have also played a role in movement responses, phototaxis, graviorientation and avoidance reactions in microalgae (Nultsch 1979; Yoshimura 1998). For example, by the use of less than 50 μM Gd^{3+}, Hill et al. (2000) was able to block the flagellar responses in the green freshwater microalgae *Spermatozopsis similis*. Kam et al. (1999) observed a reduction in swimming speed of *Chlamydomonas reinhardtii* without affecting graviorientation as in plants, and the authors suggested that calcium-mediated gravitaxis originated in organisms that were evolutionary more advanced than this flagellate. Also, in the marine non-flagellated cyanobacterium *Synechococcus* sp., Pitta et al. (1997) used Tb^{3+} as a calcium blocker to demonstrate that calcium was required for motility.

Other studies have involved Gd^{3+} as an actin inhibitor in rhizoid morphogenesis of *Spirogyra* sp. (Yoshida and Shimmen 2009), and a di-tripolyphosphate complex has been used as a sodium shift reagent to study intracellular Na^+ concentrations in the halotolerant microalga *Dunaliella salina* (Bental et al. 1988). In transmission electron microscopy studies, $La(NO_3)_3$ has been used as an electron-dense tracer to delineate apoplastic pathways for ion transport in the brown alga *Cystoseira nodicaulis* (Pellegrini et al. 1991).

Cerium chloride ($CeCl_3$) has diverse applications for studying algal physiology. It penetrates biological membranes, reacts with hydrogen peroxide (H_2O_2) and produces insoluble cerium perhydroxide electron-dense deposits, which can be visualized by conventional transmission electron microscopy. Reactive oxygen species are constantly generated as by-products of metabolic pathways and can be overproduced in response to stress, as studied in the unicellular green alga *Micrasterias denticulata* (Darehshouri and Lütz-Meindl 2010) or in studies of specific defense responses of rhodophyta *Gracilaria* spp. (Weinberger et al. 2005). The same salt has been used for localization of non-specific acid phosphatases in the filamentous freshwater alga *Stigeoclonium tenue* (Michetti et al. 2006).

5 Potential Positive Effects of REEs

Studies on the positive effects of REEs have been carried out mostly on terrestrial organisms. Exposure of agricultural plants as different as alfalfa, banana, cabbage, cotton, legumes, maize, mushroom, orange, peanut, potato, rape, rice, sugar beet, tobacco, tomato and wheat to REEs, have positive effects on growth and product quality (Brown et al. 1990; Hu et al. 2004; Tyler 2004).

As mentioned above in China, REEs have been widely used for decades as plant growth regulators for crops, for decreasing vulnerability to diseases and as feed additives for livestock, poultry and fish (Tyler 2004; von Tucher and Schmidhalter 2005; Zhang et al. 2006; Schwabe et al. 2012). These effects have been linked to the stimulation of absorption, the transfer and assimilation of nutrients, alleviation of metal deficiencies, increases in metabolism (by enzyme activity), influences on photosynthesis, or effects on stress resistance e.g., against drought, acid rain and toxic metals (reviewed by Pang et al. 2001; Hu et al. 2004; Tyler 2004; He and Xue 2005; Volland et al. 2014). However, these observations have not been linked to any specific cellular or molecular model and are therefore far from explaining the mechanisms of action of REEs in plants or algae (Wang et al. 2014).

REEs have been especially studied concerning metal deficiency alleviation of calcium limitation, because REEs and Ca^{2+} ions share many common properties including similarities in ionic radii, affinity for oxygen ligands, and participation in electrostatic interactions. Due to those similarities and by having a valence higher than that of calcium, REE-ions can replace Ca^{2+} and bind with a higher affinity to multiple receptors, having different consequences depending on the role of the native metal (Brown et al. 1990; Jegerschöld et al. 2000; Ono 2000; Wei and Zhou 2000; Wang et al. 2003).

Most of investigations carried out with REEs and algae have focused on microalgal (and cyanobacterial) growth, usually with no attempt to elucidate the mechanism(s) of beneficial effects (Table 4). It is not clear whether the positive effects of REEs are due to alleviation of a nutrient deficiency (such as Ca^{2+}, Mg^{2+} or Mn^{2+}), as suggested previously for plants (Wei and Zhou 2000; Huang et al. 2008; Yin et al. 2009; Gong et al. 2011; Qu et al. 2012; Goecke et al. 2015), or if the elements participate in some other physiological reactions e.g. scavenging of oxygen-free radicals (Peng and Pang 2002; Ippolito et al. 2010; Valcheva-Traykova et al. 2014) or as compensation for the inhibitory effects of heavy metals (Volland et al. 2014).

Alleviation of metal-deficiency in algae by REEs was studied by Li et al. (2011), who demonstrated that low concentrations of La^{3+} were able to partially compensate for Ca^{2+}-deficiency in the green macroalga *Chara corallina*, thus permitting cytoplasmic streaming. Goecke et al. (2015) showed that lanthanides can produce a stimulatory effect on the freshwater microalga *Desmodesmus quadricauda*. Adding low concentrations of five Ln partly alleviated the adverse effects of Ca^{2+}-deficiency, but not Mn^{2+}-deficiency. Physiological stress by nutrient limitation was specifically measured as a decline in cellular growth and division, and by changes in photosynthetic parameters using a pulse amplitude modulation (PAM) fluorimeter. Ln increased F_v/F_m, $rETR_{max}$, E_k and α under Ca^{2+}-deficient conditions establishing control and suggesting that REEs may act as a substitute where calcium is limited (Fig. 1).

Although there is still no certainty as to the environmental effects of REEs, their applications in agronomy and aquaculture continues to intensify. REEs are not, up to now, commercially used for increase of algal production, but there are a few studies related to their effect on algal pigments and lipids of economic interest. For example, effect of 1 mg L^{-1} Ce^{3+} on increasing growth and astaxanthin production of *Haematococcus pluvialis* has been described. This effect was, however, concentration-dependent and inhibition of growth at higher concentrations occurred (Li et al. 2008).

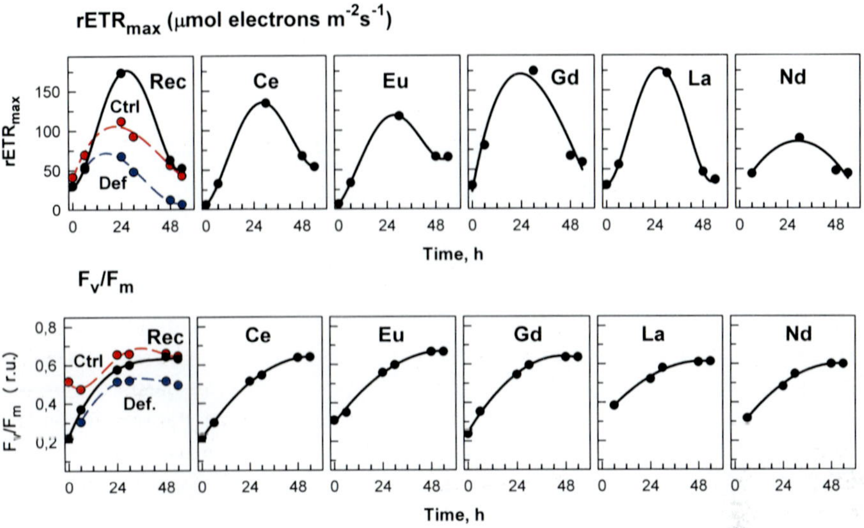

Fig. 1 Photosynthetic parameters expressed as maximum relative electron transport rates ($rETR_{max}$), and the maximal quantum yield (F_v/F_m), in cultures of the alga *Desmodesmus quadricauda* grown either in complete mineral medium (Ctrl, *red* symbols, *dashed* curve) or in calcium-deficient mineral medium (Def, *blue* symbols, *dashed* curves) are shown. To calcium deficient cultures either the complete mineral medium was added (Rec, *black* symbols, *solid* line) or different lanthanides (Ce, Eu, Gd, La, Nd) as marked in individual panels. The complete photosynthetic parameters are displayed in the original publication (Modified from Goecke et al. 2015)

6 Toxicity of REEs

According to the Hodge-Sterner classification system, REEs were generally considered to be of low toxicity, but this depends on their chemical form and route of administration (Hodge and Sterner 1949). However, excessive REEs in soil/water acted as pollutants and have negative effects on humans, plants and animals (Abramczuk 1985). There are several reports that describe different levels of REE toxicity in diverse organisms (Brown et al. 1990; Bulman 2003; Wang et al. 2003), but the level of toxicity, threshold limits and maximum permissible concentrations are poorly established in the literature (Thomas et al. 2014). Toxicity varies between types of REEs or organisms and species, but the exact mode of action still remains unknown (Barry and Meehan 2000, Table 4).

As a possible cause of REE toxicity, their ability to interfere with the metabolism of several essential elements has been mentioned (Pang et al. 2001). In this way, changes may occur in the normal physiological functions of several enzymes as demonstrated for ATPase and pectate lyase (Squier et al. 1990 and Yoder and Jurnak 1995, respectively), blocking of ionic-channels (Palasz and Czekaj 2000), or affecting the uptake and metabolism of minerals (Wang et al. 2003; Kastori et al. 2010).

Table 4 Examples of studies testing the effect of REEs on the growth, physiology and survival of microalgae, and specifying the concentrations in at which positive, neutral and negative effects were observed (values in μmol L^{-1})

Algae	REE	[Positive effect]	[Negative effect]	Reference
Arthrospira platensis[B]	La^{3+}	38.53–53	>53.94	Gong et al. (2011)
**Arthrospira platensis*[B]	$LaCl_3$	30–40	>40	Li et al. (1999)
Chlamydomonas reinhardtii[C]	Ce^3	5–20	–	Liu et al. (1986)
	La^{3+}	5–20	–	Liu et al. (1986)
Chlorella vulgaris[C]	Ce^{3+}	1.8	2.1	Evseeva et al. (2010)
**Ch. vulgaris v. autotrophica*[C]	12Ln	–	29.14	Su et al. (2005)
**Desmodesmus quadricauda*[C]	La^{3+}	<7.2	>72	Jin et al. (2009)
Euglena gracilis[E]	Dy^{3+}	50–100	180–1000	Fuma et al. (2005)
Isochrysis galbana[H]	La	7.28–87.4	–	Qu et al. (1998)
	Gd	6.36–57.23	–	Qu et al. (1998)
	Yb	5.78–17.34	–	Qu et al. (1998)
Microcystis aeruginosa[B]	La^{3+}	<7.2	>72	Jin et al. (2009)
Skeletonema costatum[O]	13Ln	–	28–30	Tai et al. (2010)
	Sc	–	21.88	Tai et al. (2010)
	Y	–	43.21	Tai et al. (2010)

Algal divisions are characterized as Chlorophyta (C), Haptophyta (H), and Ochrophyta (O). Cyanobacteria (B) and Euglenophyta (E) are also considered

If the algal species has a new name it is referred to using the actual name and an asterisk (*), names according to Algaebase (see, Guiry and Guiry 2014)

In consequence, REEs may influence the stability, permeability and functioning of cell membranes (Brown et al. 1990; Hu et al. 2004; Wang et al. 2011).

Even though biological toxicities of REEs on different microalgae have been reported (Table 4), information is still sparse and incomplete. Only a few members of Charophyta (Reid et al. 1996), Chlorophyta (Hu et al. 2001; Jin et al. 2009; Evseeva et al. 2010), Dinophyta (Yang and Kong 2002), Euglenophyta (Fuma et al. 2005), Bacillariophyceae (Xin et al. 1998; Tai et al. 2010), and Haptophyta (Qu et al. 1998) (plus Cyanobacteria: Singh and Subbaramaiah 1970; Li et al. 1999; Wang et al. 2012), have been investigated. Almost all studies, however, lacked measurements on the bioavailability of lanthanides, which could also explain contradictory results between the concentrations of REEs and stimulatory/inhibitory effects on the same species. In addition, their effect on many algal groups and species is unknown and there are no toxicity tests on macroalgae. On the other hand, some specific inhibitory properties are well established and have been frequently studied in various physiological studies (see, Sect. 4.1).

As algae are primary producers and ecologically important organisms, transfer of these elements through the food chain is expected (Wang et al. 1993; Thomas et al. 2014). Any detrimental effects of REEs may result in enhanced negative effects on organisms at higher trophic levels; therefore toxicity needs to be addressed from this point of view as well (see, Barry and Meehan 2000; Bao et al. 2001).

Recently, the toxic effects of REEs on algae were proposed to involve nutrient depletion rather than toxicity *per se* (Yuan et al. 2009; Lürling and van Oosterhout 2013; see subchamber 7.2). It was suggested that these elements sequester essential nutrients such as phosphates, which may produce an effect on growth (death by starvation). Besides possible applications, this is an important characteristic that should be further investigated, because it could influence the effective concentration (EC_{50}) of those metals, and thus influence environmental-decision making.

6.1 Algal Defenses Against Pollutants

In order to survive, algae have developed mechanisms to remove, sequester or tolerate toxic elements present in the environment. These mechanisms may involve either adhesive or barrier mechanisms to prevent uptake of toxic elements to cells or cell organelles, or physiological responses to localize, reduce or eliminate toxic concentrations of metal ions inside the cells (Sandau et al. 1996). These include the production of binding factors (like phytochelatins), proteins or peptides (like metalloproteins), ion-selective transporters, and excretion or compartmentalization (Pakrasi et al. 2001). Therefore, tolerance varies among organisms (Table 4). Studies on the toxicity of those metals not only indicated the most susceptible species in cases of contamination, but also the more tolerant and resistant ones, which may have interesting prospects for future studies on biosorption and bioaccumulation.

6.2 Tolerance, Biosorption and Bioaccumulation of Metals in Algae

The passive binding of elements to algal biomass "biosorption", or active processes whereby removal/uptake of the metals is metabolically controlled "bioaccumulation", has attracted attention over the last decades (Davis et al. 2003). The accumulation of elements in algal cells is a very complicated process, depending on properties of the species (type, size, form, state of development), the element (charge, chemical form, concentration) and the medium (pH, type and concentration of metal salts or complexing agents present) (Starý et al. 1983). There are several studies on the accumulation, biosorption and desorption of REEs, using both micro- and macroalgae; these involved brown, green and red algae, some algal flagellates, and also cyanobacteria (examples in Table 5). They demonstrate that algae, either dead or alive, can efficiently accumulate these metals due to their ability to form chelated metabolites, for example with proteins, sugars, nucleic acids, amino acids, nucleotides, etc. (Hu et al. 2004). Those elements can also combine with pigments and polysaccharides such as cellulose, alginic acid, carrageenan, fucoidan, etc.; these exist in great diversity and abundance in algae (Lunde et al. 1972; Diniz and Volesky 2005a, b; Gok and Aytas 2009; Okajima et al. 2010), making algae a realistic candidate for the development and testing of biosorption methods.

Table 5 Studies on accumulation, biosorption and/or desorption of REEs using algae

Algae	REEs	Reference
**Amphidinium carterae*$^{D(m)}$	Ce	Rice and Willis (1959)
Aphanothece sacrum$^{C(m)}$	14 Ln^{3+}, Y	Okajima et al. (2010)
Carteria sp.$^{C(m)}$	Ce	Rice and Willis (1959)
Chaetoceros muelleri$^{O(m)}$	Ce, La	Richards and Mullins (2013)
Chlorella vulgaris$^{C(m)}$	La	Wang et al. (1996)
**Cylindrotheca closterium*$^{O(m)}$	Ce	Rice and Willis (1959)
**Diacronema lutheri*$^{C(m)}$	Ce, La	Richards and Mullins (2013)
Euglena gracilis$^{E(m)}$	Nd	Kang et al. (2000)
Euglena gracilis$^{E(m)}$	Ce, Nd	Shen et al. (2002)
Microcystis aeruginosa$^{B(m)}$	Ce, La	Zhou et al. (2004)
Nannochloropsis gaditana$^{C(m)}$	Ce, La	Richards and Mullins (2013)
Platymonas sp.$^{C(m)}$	Ce	Rice and Willis (1959)
**Porphyridium purpureum*$^{R(m)}$	Ce	Rice and Willis (1959)
Sargassum polycystum^{O}	Eu, La, Yb	Diniz and Volesky (2005a)
Sargassum polycystum^{O}	Eu, La	Diniz et al. (2008)
Sargassum sp.^{O}	Eu, Gd, La, Nd, Pr, Sm	Oliveira et al. (2011, 2012)
Tetraselmis chui$^{C(m)}$	Ce, La	Richards and Mullins (2013)
Thalassiosira sp.$^{O(m)}$	Ce	Rice and Willis (1959)
Turbinaria conoides^{O}	Ce, Eu, La, Yb	Vijayaraghavan et al. (2010)
Ulva lactuca^{C}	14 Ln^{3+}, Y	Zoll and Schijf (2012)

Algal division, Chlorophyta (C), Ochrophyta (O), and Rhodophyta (R) are specified. Cyanobacteria (B), and the protist classes Dinophyceae (D) and Euglenophyceae (E) are also considered
If microalgae were utilized, they are specified with an (m). If an algal species has a new name, it is referred to with the actual name and an asterisk (*), names according to Algaebase (see Guiry and Guiry 2014)

However, there is still a huge gap in our knowledge regarding entry and internal distribution of REEs in algal cells and where those metals actually accumulate. Even in terrestrial plants, the activation of cellular processes such as endocytosis by Ln has only very recently been demonstrated (Wang et al. 2014). In a few published studies, it has been suggested that REEs may concentrate in chloroplasts (Guo et al. 2000; Kang et al. 2000; Shen et al. 2002; Ren et al. 2007, 2013), unfortunately these studies only covered one phototrophic protist (*Euglena gracilis* Klebs) and only a few REEs; thus it is unknown how other micro- or macroalgae may react. Thus, many questions remain open when considering the exact influence of REEs at molecular and physiological levels. Algae and cyanobacteria possess complex cell walls and the exact manner of REE incorporation, intracellular transport, and storage remain unknown. It is not even clear whether these elements are compartmentalized in specific areas or remain as free cytoplasmic components.

Information on resistant strains or natural hyper-accumulators is also sparse and toxicity tests involving marine macroalgae have not yet been carried out.

Furthermore, precise data on mechanisms underlying REE toxicity, the acquisition of tolerance, and effects on algal physiology and biochemistry at the cellular, subcellular and molecular levels are yet to be elaborated (Pakrasi et al. 2001; Hu et al. 2004).

7 Applications of REEs

7.1 Applications as Functional Foods and Fertilizers

For decades, algae have been added to feeds to promote growth and maintain health in live stock (Chapman and Chapman 1980), and more recently, REEs have been proposed as growth promoters in different animals like pigs and other domestic animals (He and Rambeck 2000). Algae rich in REEs just represent a potential alternative to dietary supplements or functional foods. There is, however, only one study conducted on young abalones where the use of REE-enriched algae enriched with REEs as feed was effective as a growth promoter (Bao et al. 2001). Thus, it would be important to expand the number of studies, acquire relevant data on carry-over effects of REEs, and to assess the risk of human exposure via animal-derived foods (Schwabe et al. 2012).

7.2 Application of REEs to Control Harmful Algal Blooms

A number of microalgal species, including cyanobacteria (e.g., *Alexandrium* spp., *Microcystis aeruginosa*), produce harmful algal blooms with implications for health, the ecology and economics. These blooms are not only related to anoxygenic conditions and bad odor, but also to the accumulation of potent toxins (e.g., microcystins, saxitoxins) with lethal effects on animals and humans (Lürling and van Oosterhout 2013). As described above, there are several studies demonstrating toxic effects of REEs on microalgae (see Table 4), and the special affinity of alga for REEs. Therefore, the idea of developing a product containing REEs to control algal blooms is not new, and there are a few products commercially available, although their mechanisms of action differ. Novel techniques involve the application of modified clays, such as LaC13-modified kaolinite (Yuan et al. 2009) and La^{3+} modified bentonite as a dephosphatisation method (van Oosterhout and Lürling 2013). For example, PHOSLOCK®, a commercial product developed by CSIRO, Australia (Phoslock Water Solutions Ltd.), has a strong binding affinity for ortho-phosphate (Lürling and Faassen 2012; van Oosterhout and Lürling 2013), potentially removing the source of nutrition for PO_4 utilizing microorganisms, and is therefore an interesting tool for effectively controlling outbreaks of harmful algal blooms (Li et al. 2004). These products seem quite promising and apparently have relatively low harmful effects on other organisms. Their effectiveness and environmental safety will, however, need further testing in long term field studies (see, Lürling and van Oosterhout 2013).

7.3 Application of REEs as Tracers and Bioindicators

Because of the unique chemical properties of REEs, which enable them to record subtle geochemical processes in natural systems (Fu et al. 2000), REEs have been extensively used in studies of provenance, petrogenesis, oceanic cycles and the chemical evolution of the Earth (Sakamoto et al. 2008a; Zhu et al. 2012). They offer a highly detectable tag without introducing radioactive material into marine environments. In real time, they have proved to be useful in studies to confirm the impact of cyanobacterial mats on deep waters off French Polynesia, providing evidence of an endo-upwelling flow (Jehl and Barsczus 1996). In other studies, based on stromatolites – sedimentary deposits formed by the interaction of benthic microbial communities (e.g. with cyanobacteria) and chemical and detrimental sediments – REEs have been proposed as useful elements to explore palaeo-environmental conditions (Allwood et al. 2010; Oliveri et al. 2010; Corkeron et al. 2012; Censi et al. 2015). Since fossils are one of the most ancient records of life on Earth, these structures may provide critical information about early microbial life and ancient environmental conditions (Johannesson et al. 2014). However, there is little knowledge related to microbial uptake, and potential biological fractionation of REEs that should be taken into consideration for the geological interpretation of bioaccumulation data (Corkeron et al. 2012; Johannesson et al. 2014).

REEs have also been proposed as an indicator of anthropogenic activities (Olmez et al. 1991). Coincidently, the use of macroalgae has been used as continuous sampling monitors of pollutants, to characterize coastal water quality (Jayasekera and Rossbach 1996; Vásquez and Guerra 1996). Due to their special affinity for algae, the REE pattern may be a useful tracer for investigating the surrounding marine environment (Kano et al. 2001), and can be used for monitoring sources of pollution from natural events such as volcanic activity (Schacht et al. 2010).

7.4 Application of REEs as Markers and Detection Tools

Both, the reactivity and the inert nature of different lanthanide compounds are attractive properties for developing markers for different experiments as detection tools (Brown et al. 1990). REEs have been used as non-toxic inert indigestible particulate markers in rate-of-passage and digestibility studies in animal and human nutrition (Sheng et al. 2005; Garatun-Tjeldstø et al. 2006). Related to algae, rare earth oxides have been used in the sea cucumber *Apostichopus japonicus* Liao, as markers in food choice experiments conducted with different macroalgal diets (Xia et al. 2012).

In the search for more sensitive, fast and inexpensive markers, some active chelates of REEs have already been obtained and tested in various biological assays. REEs have potential in the design of sensitive and specific immunoassays. The use of lanthanide chelate labels has been suggested as an effective way of minimizing background signals (Hagan and Zuchner 2011). In this case, the use of europium and terbium cryptates, as well as europium chelates, were proposed for labeling cyanotoxins e.g. microcystin (Oliveira et al. 2006; Santos et al. 2013). These methods can detect trace amounts of substances using highly specific immune responses and very sensitive, fluorescent tracers (Niu et al. 2012).

8 Applications of Algae in Relation to REEs

8.1 Applications of Algae for Bioremediation of REEs Pollution

REEs have been detected in runoff and waste water and in aquatic ecosystems as a consequence of industrial and agricultural practices, thus suggesting that REEs can reach groundwaters and migrate to rivers and lakes (Ippolito et al. 2010). Unfortunately, there is no certainty as to the ecological safety of REEs because there are no long lasting community studies on their effects of REEs, neither at toxic nor sub-lethal levels. Some studies have been recently published which focus on ecological effects of REEs as well ahas their potential threats as a consequence of bioaccumulation (see, Li et al. 2006, 2010). However, no corresponding regulations or standards have been established, up to now, regarding doses and threshold limits for the use of REEs in the environment (Wang et al. 2014).

Little is known about the impact sub-lethal levels of REEs on algal communities, which is important because concentrations of REEs in the biosphere are increasing. The waste footprints and environmental impacts from mining operations to extract rare earth mineral ores are expected to be significant (EPA 2012; Liang et al. 2014), emphasizing the urgency to support additional toxicity and ecological impact studies on REEs and to use this information in conducting risk assessments related to REE mining, processing, and recycling.

In areas, where REE contamination is likely, the slow accumulation of these elements in the environment could become problematic (Thomas et al. 2014). In China, REEs have been classified as a significant environmental pollutant and the elimination of excessive REEs from the environment is therefore considered a worthwhile goal (Ren et al. 2013).

The conventional treatment methods for metal removal from solutions such as chemical precipitation, electrochemical separation, membrane separation, reverse osmosis, ion-exchange or adsorption resins, present several disadvantages including either capital or high energy and operational costs. These traditional physical-chemical treatments become more expensive or even inefficient for treatment of effluents containing metal ions at low concentrations (Bhat et al. 2008). Therefore there is an increasing demand for eco-friendly technologies using low cost alternatives (Oliveira et al. 2011; Das and Das 2013, and references therein).

Among the most successful and widely used 'innovative technologies' is *in situ* bioremediation. Biological treatment is routinely used to remove many toxic wastes and biological nutrients such as nitrogen and phosphorus, from water and wastewater (Banaszak et al. 1999). Algae are of special interest for the development of new absorbent material due to their high sorption capacity and ready availability in practically unlimited quantities (Kanchana et al. 2014). They are naturally abundant, autotrophic, and are found in all kinds of aquatic bodies under different environmental conditions including pH (Bhat et al. 2008).

Das and Das (2013) recently reviewed the use of different biosorbents, including algae, bacteria, fungi, yeasts, resin, activated carbon, etc., for the recovery of REEs. Algae, due to their affinity for REEs, may be interesting options that warrant further study. Marine algae, in particular brown seaweeds, have been identified as potent biosorbents due to the presence of binding sites for chemical groups such as carboxyls, sulfonates, amines and hydroxyls (Davis et al. 2003). For example, Oliveira et al. (2011, 2012) evaluated the potential of *Sargassum* sp. biomass for the biosorption of Eu, Gd, La, Nd, Pr and Sm. They observed a fast and efficient recovery of the metals, although they were unable to separate them. The authors suggested that the carboxylic groups present in alginates (the main component of the brown algae's cell wall) were the main reactive functionalities. Similar results have been obtained with other brown macroalgae such as *Sargassum* spp. (Diniz and Volesky 2005a, b; Diniz et al. 2008; Sakamoto et al. 2008a), and *Turbinaria conoides* (Vijayaraghavan et al. 2010). Microalgae like *Chlorella* spp., *Nannochloropsis* spp. and the cyanobacterium *Microcystis* spp. have also been shown to be effective biosorbents of metals such as La^{3+} and Ce^{3+} as well as aluminum or iron, among others (see Zhou et al. 2004; Richards and Mullins 2013, Table 5).

The use of algal biosorbents has also been shown to have implications in secondary wastes. Any biological treatment for toxic waste streams cannot be a stand-alone system, but must be integrated into a more complex overall treatment strategy (Banaszak et al. 1999). The post-treatment of secondary solid wastes must also be addressed. The management of toxic solid waste is therefore very important due to high costs of disposal (Lee et al. 2014). From the results of comparative experiments with biosorbents (using the brown macroalga *Saccharina japonica*) and other chemical adsorbents and precipitants, such as mordenite, for the treatment of uranium wastes, a significant volume and weight reduction of waste biosorbents was demonstrated by subsequent ignition (Fig. 2). Interestingly, this also provides an opportunity for recycling of waste materials (see below).

8.2 Applications of Algae to Recycle REEs

A key concern regarding the commercial availability of REEs in the future is their abundance and accessibility (Tse Pui-Kwan 2011; Das and Das 2013; USGS 2014). The high prices for most of these elements, and rapid expansion in their applications

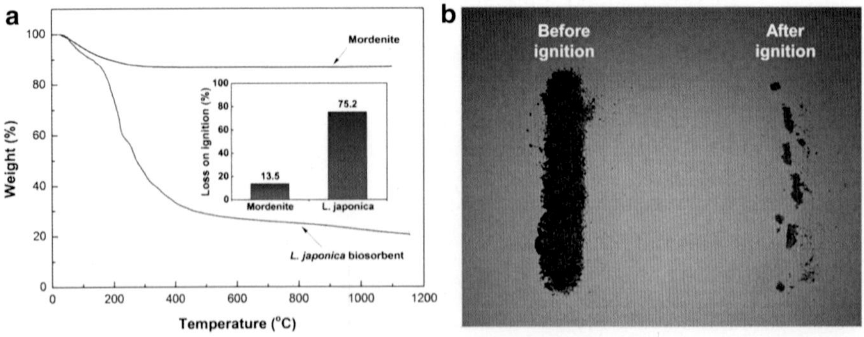

Fig. 2 (**a**) Weight reduction (in %) of mordenite (MOR), an alumino-silicate mineral, and *Saccharina* (as *Laminaria*) *japonica* biosorbent by ignition treatment. (**b**) Picture of *S. japonica* biosorbent before and after ignition treatment (After Lee et al. 2014)

has resulted in increased pressure on REE production; consequently, minimizing industrial production losses has become an important objective (EPA 2012). Research emphasis has been focused on the progressively more cost-effective removal and recovery of REEs for industrial processes (Diniz and Volesky 2005b; Oliveira and Garcia Jr. 2009). In addition to providing some limitations to supply risk, recycling could minimize the environmental challenges present in REE mining and processing (Du and Graedel 2011), by reducing the amount of those metals released in wastes.

No studies to date have investigated *in situ* whether relatively low concentrations of REEs stimulate algal growth and/or lipid production. Algal biomass has been investigated for the implementation of economic conversion processes producing different biofuels such as biodiesel, bioethanol, biogas, biohydrogen and other valuable co-products (Behera et al. 2015). If algae can also recycle REEs and/or bioremediate industrial wastes, investigations should urgently be focused in this direction.

9 Conclusions

Algae as primary producers and the basis of many trophic nets, are important and sensitive organisms. These ecologically important organisms have interesting relationships with REEs, and their affinity for these metals may represent a serious environmental threat as well as an attractive opportunity worthy of further investigation. As bioconcentration, stimulatory effects and toxicity of these elements vary among species, it is extremely difficult to predict an ecologically dangerous threshold. On the other hand, algae have many applications, and many of these are related to their mineral content e.g. foods, nutritional supplements, fertilizers, in medicine and industry (Moreda-Piñeiro et al. 2012). REEs have been used to detect algal toxins (cyanobacteria) and to measure defense and physiological reactions, and algae have been used to effectively bioaccumulate these metals, reflecting significant associations among them. This has led to the possibility of using algae for bioremediation of REEs and recycling purposes. A better understanding of bioavailability, toxicity and uptake of REEs, as well as physiological implications for algae at the molecular, enzymatic and life-cycle levels are vital for environmentally-friendly production and use of these valuable resources.

Acknowledgements We thank John Brooker for critical reading and language corrections of the manuscript. This work was supported by the Academy of Sciences of the Czech Republic (grant numbers M200201205, and RVO 61388971) and by the National Programme of Sustainability I, ID: LO1416.

List of Abbreviations

EC_{50}	half maximal effective concentration
ICP-MS	inductively coupled plasma mass spectrometry
Ln	lanthanide series of elements
REE	rare earth elements

References

Abramczuk JW (1985) The effects of lanthanum chloride on pregnancy in mice and on preimplantation mouse embryos in vitro. Toxicology 34:315–320

Allwood AC, Kamber BS, Walter MR, Burch IW, Kanik I (2010) Trace elements record depositional history of an Early Archean stromatolitic carbonate platform. Chem Geol 270:148–163

Alonso E, Sherman AM, Wallington TJ, Everson MP, Field FR, Roth R, Kirchain RE (2012) Evaluating rare earth element availability: a case with revolutionary demand from clean technologies. Environ Sci Technol 46:3406–3414

Banaszak JE, Rittmann BE, Reed DT (1999) Subsurface interactions of actinide species and microorganisms: implications for the bioremediation of actinide-organic mixtures. J Radioanal Nucl Chem 241:385–435

Bao YM, Zhang HJ, Zhang YS, Guo TW, Yang YX (2001) Study on rare earth compound feeds for young abalone. J Chin Rare Earth Soc 22:35–39. (in Chin.)

Barry MJ, Meehan BJ (2000) The acute and chronic toxicity of lanthanum to *Daphnia carinata*. Chemosphere 41:1669–1674

Bauer CS, Plieth C, Bethmann B, Popescu O, Hansen U, Simonis W, Schönknecht G (1998) Strontium-induced repetitive calcium spikes in a unicellular green alga. Plant Physiol 117:545–557

Behera S, Singh R, Arora R, Kumar Sharma N, Shukla M, Kumar S (2015) Scope of algae as third generation biofuels. Front Bioeng Biotechnol. doi:10.3389/fbioe.2014.00090

Bental M, Degani H, Avron M (1988) ^{23}Na-NMR studies of the intracellular sodium ion concentration in the halotolerant alga *Dunaliella salina*. Plant Physiol 87:813–817

Bhat SV, Melo JS, Chaugule BB, Souza SFD (2008) Biosorption characteristics of uranium (VI) from aqueous medium onto Catenella repens, a red alga. J Hazard Mater 158:628–635

Brown PH, Rathjen AH, Graham RD, Tribe DE (1990) Rare earth elements in biological systems. In: Gschneidner KA Jr, Eyring L (eds) Handbook on the physics and chemistry of rare earths, vol 13. Elsevier Science Publishers, Amsterdam, pp 423–452

Bulman RA (2003) Metabolism and toxicity of the lanthanides. In: Sigel A, Sigel H (eds) Metal ions in biological systems, vol 40, The lanthanides and their interrelations with biosystems. Marcel Dekker, New York, pp 39–67

Bünzli JCG, Eliseeva SV (2011) Basics of lanthanide photophysics. In: Hänninen P, Härmä H (eds) Lanthanide luminescence: photophysical, analytical and biological aspects, Springer series on fluorescence. Springer, Berlin/Heidelberg, pp 1–46

Censi P, Cangemi M, Brusca L, Madonia P, Saiano F, Zuddas P (2015) The behaviour of rare-earth elements, Zr and Hf during biologically-mediated deposition of silica-stromatolites and carbonate-rich microbial mats. Gondwana Res 27(1):209–215

Chapman VJ, Chapman DJ (1980) Seaweed as animal fodder, manure and for energy. In: Chapman VJ, Chapman DJ (eds) Seaweed and their uses. London and New York: Chapman & Hall, pp 30–61

Cockerill AF, Davies GLO, Harden RC, Rackham DM (1973) Lanthanide shift reagents for nuclear magnetic resonance spectroscopy. Chem Rev 73:553–588

Corkeron M, Webb GE, Moulds J, Grey K (2012) Discriminating stromatolite formation modes using rare earth element geochemistry: trapping and binding versus in situ precipitation of stromatolites from the Neoproterozoic Bitter Springs Formation, Northern Territory, Australia. Precambrian Res 212–213:194–206

Darehshouri A, Lütz-Meindl U (2010) H_2O_2 localization in the green alga *Micrasterias* after salt and osmotic stress by TEM-coupled electron energy loss spectroscopy. Protoplasma 239:49–56

Das N, Das D (2013) Recovery of rare earth metals through biosorption: an overview. J Rare Earths 31:933–943

Davis TA, Volesky B, Mucci A (2003) A review of the biochemistry of heavy metal biosorption by brown algae. Water Res 37:4311–4330

Diniz V, Volesky B (2005a) Biosorption of La, Eu and Yb using *Sargassum* biomass. Water Res 39:239–247

Diniz V, Volesky B (2005b) Effect of counterions on lanthanum biosorption by *Sargassum polycystum*. Water Res 39:2229–2236

Diniz V, Weber ME, Volesky B, Naja G (2008) Column biosorption of lanthanum and europium by *Sargassum*. Water Res 42:363–371

Du X, Graedel TE (2011) Global in-use stocks of the rare earth elements: a first estimate. Environ Sci Technol 45:4096–4101

Elbaz-Poulichet F, Dupuy C (1999) Behaviour of rare earth elements at the freshwater-seawater interface of two acid mine rivers: the Tinto and Odiel (Andalucia, Spain). Appl Geochem 14:1063–1072

EPA 2012 (2012) Rare earth elements: a review of production, processing, recycling, and associated environmental issues. United States Environmental Protection Agency EPA 600/R-12/572/ December 2012

Evseeva T, Geras'kin S, Majstrenko T, Brown J, Belykh E (2010) Comparative estimation of 232Th and stable Ce (III) toxicity and detoxification pathways in freshwater alga *Chlorella vulgaris*. Chemosphere 81:1320–1327

Fu FF, Akagi T, Yabuki S, Iwaki M, Ogura N (2000) Distribution of rare earth elements in seaweed: implication of two different sources of rare earth elements and silicon in seaweed. J Phycol 36:62–70

Fuma S, Takeda H, Takaku Y, Hisamatsu S, Kawabata Z (2005) Effects of dysprosium on the species-defined microbial microcosm. Bull Environ Contam Toxicol 74:263–272

Garatun-Tjeldstø O, Otterå H, Julshamn K, Austreng E (2006) Food ingestion in juvenile cod estimated by inert lanthanide markers effects of food particle size. ICES J Mar Sci 63:311–319

Goecke F, Jerez CG, Zachleder V, Figueroa FL, Řezanka T, Bišová K, Vitová M (2015) Use of lanthanides to alleviate the effects of metal ion-deficiency in *Desmodesmus quadricauda* (Sphaeropleales, Chlorophyta). Front Microbiol 6:2. doi:10.3389/fmicb.2015.00002

Gok C, Aytas S (2009) Biosorption of uranium(VI) from aqueous solution using calcium alginate beads. J Hazard Mater 168:369–375

Gong X, Hong M, Wang Y, Zhou M, Cai J, Liu C, Gong S, Hong FS (2011) Cerium relieves the inhibition of photosynthesis of maize caused by manganese deficiency. Biol Trace Elem Res 141:305–316

Guiry MD, Guiry GM (2014) AlgaeBase. World-wide electronic publication, National University of Ireland, Galway. http://www.algaebase.org. Searched on 6 Feb 2014

Guo P, Wang J, Li X, Zhu J, Reinert T, Heitmann J, Spemann D, Vogt J, Flagmeyer R-H, Butz T (2000) Study of metal bioaccumulation by nuclear microprobe analysis of algae fossils and living algae cells. Nucl Inst Methods Phys Res B 161–163:801–807

Hagan AK, Zuchner T (2011) Lanthanide-based time-resolved luminescence immunoassays. Anal Bioanal Chem 400:2847–2864

He ML, Rambeck WA (2000) Rare earth elements – a new generation of growth promoters for pigs? Arch Tierernahr 53:323–334

He Y, Xue L (2005) Biological effects of rare earth elements and their action mechanisms. Chin J Appl Ecol 16:1983–1989. (in Chin.)

Hill K, Hemmler R, Kovermann P, Calenberg M, Kreimer G, Wagner R (2000) A Ca^{2+} – and voltage-modulated flagellar ion channel is a component of the mechanoshock response in the unicellular green alga *Spermatozopsis similis*. BBA Rev Biomembr 1466:187–204

Hodge HC, Sterner JH (1949) Tabulation of toxicity classes. Am Ind Hyg Assoc Q 10:93–96

Horovitz CT (2000) Biochemistry and physiology of scandium and yttrium, Part 2: Biochemistry and applications, vol 13B, Biochemistry of the elements. Kluwer Academic/Plenum Publishers/Springer, New York, pp 39–163

Hou XL, Yan XJ (1998) Study on the concentration and seasonal variation of inorganic elements in 35 species of marine algae. Sci Total Environ 222:141–156

Hu QH, Zheng SP, Tang SM, Guan L (2001) Effects of Sm and Y on growth of *Chlorella ellipsoidea*. Agro-Environ Prot 20:398–404. (in Chin.)

Hu ZH, Richter H, Sparovek G, Schnug E (2004) Physiological and biochemical effects of rare earth elements on plants and their agricultural significance: a review. J Plant Nutr 27:183–220

Huang H, Chen L, Liu XQ, Liu C, Cao W, Lu Y, Hong FS (2008) Absorption and transfer of light and photoreduction activities of spinach chloroplasts under calcium deficiency: promotion by cerium. Biol Trace Elem Res 122:157–167

Ippolito MP, Fasciano C, d'Aquino L, Morgana M, Tommasi F (2010) Responses of antioxidant systems after exposition to rare earths and their role in chilling stress in common duckweed (*Lemna minor* L.): a defensive weapon or a boomerang? Arch Environ Contam Toxicol 58:42–52

Jayasekera R, Rossbach M (1996) Use of seaweeds for monitoring trace elements in coastal waters. Environ Geochem Health 18:63–68

Jegerschöld C, Rutherford AW, Mattioli TA, Crimi M, Bassi R (2000) Calcium binding to the photosystem II subunit CP29. J Biol Chem 275:12781–12788

Jehl C, Barsczus GH (1996) Origine des terres rares dans les mattes cyanobactériennes (kopara) de l'atoll de Tikehau (Tuamotu, Polynésie Francaise). C R Acad Sci 322:205–212

Jin X, Chu Z, Yan F, Zeng Q (2009) Effects of lanthanum(III) and EDTA on the growth and competition of *Microcystis aeruginosa* and *Scenedesmus quadricauda*. Limnologica 39:86–93

Johannesson KH, Telfeyan K, Chevis DA, Rosenheim BE, Leybourne MI (2014) Rare earth elements in stromatolites. 1. Evidence that modern terrestrial stromatolites fractionate rare earth elements during incorporation from ambient waters. In: Dilek Y, Furnes H (eds) Archean earth and early life. Springer, Dordrecht, pp 385–411

Kam V, Moseyko N, Nemson J, Feldman LJ (1999) Gravitaxis in *Chlamydomonas reinhardtii*: characterization using video microscopy and computer analysis. Int J Plant Sci 160:1093–1098

Kanchana S, Jeyanthi J, Kathiravan R, Suganya K (2014) Biosorption of heavy metals using algae: a review. Int J Pharm Med Bio Sci 3:1–9

Kang L, Shen Z, Jin C (2000) Neodymium cations Nd^{3+} were transported to the interior of *Euglena gracilis* 277. Chin Sci Bull 45:585–592
Kano N, Aoyagi Y, Imaizumi H (2001) Determination of rare earth elements in seaweed and seawater samples on the coast in Niigata Prefecture by ICP-MS after solvent extraction. Anal Sci 17:1011–1014
Kastori RR, Maksimović IV, Putnik-Delić MI, Zeremski-Škorić TM (2010) Rare earth elements: yttrium and higher plants. Matica Srpska Proc Nat Sci 118:87–98
Lee KY, Kim KW, Baek YJ, Chung DY, Lee EH, Lee SY, Moon JK (2014) Biosorption of uranium(VI) from aqueous solution by biomass of brown algae *Laminaria japonica*. Water Sci Technol 70:136–143
Li XF, Wang CH, Wen SH (1999) Study on culture conditions of *Spirulina platensis*. Food Ferment Ind 25:13–17. (in Chin.)
Li FR, Wang JT, Chen YJ (2004) Experimental research on treating the blue algae bloom with rare earth element in Dianchi Lake. Environ Sci (China) 23:101–103. (in Chin.)
Li G, Jiang J, Chen J, Zou Y, Zhang X (2006) Effects of rare earth elements on soil fauna community structure and their ecotoxicity to *Holotrichia parallela*. Chin J Appl Ecol 17:159–162. (in Chin.)
Li Z, Cai MG, Huang SY, Shi RG, Lu XX, Qi AX, Wu R (2008) Effects of cerium on cell growth and astaxanthin production of *Haematococcus pluvialis*. Mar Sci (China) 32:37–41. (in Chin.)
Li J, Hong M, Yin X, Liu J (2010) Effects of the accumulation of the rare earth elements on soil macrofauna community. J Rare Earths 28:957–964
Li Z, Zhang Z, Yu M, Zhou Y, Zhao Y (2011) Effects of lanthanum on calcium and magnesium contents and cytoplasmic streaming of internodal cells of *Chara corallina*. Biol Trace Elem Res 143:555–561
Li Y, Yang JL, Jiang Y (2012) Trace rare earth element detection in food and agricultural products based on flow injection walnut shell packed microcolumn preconcentration coupled with inductively coupled plasma mass spectrometry. J Agric Food Chem 60:3033–3041
Liang T, Li K, Wang L (2014) State of rare earth elements in different environmental components in mining areas of China. Environ Monit Assess 186:1499–1513
Liu YF, Tang RH, Zhang QX, Shi JY, Li XM, Liu ZQ, Zhao W (1986) Stimulation of cell growth of *Tetrahymena pyriformis* and *Chlamydomonas reinhardtii* by trace elements. Biol Trace Elem Res 9:89–99
Loell M, Reiher W, Felix-Henningsen P (2011) Contents and bioavailability of rare earth elements in agricultural soils in Hesse (Germany). J Plant Nutr Soil Sci 174:644–654
Lunde G, Smidsrød O, Haug A (1972) Selectively of polyuronates for lanthanide ions. Acta Chem Scand 26:3421–3426
Lürling M, Faassen EJ (2012) Controlling toxic cyanobacteria: effects of dredging and phosphorus-binding clay on cyanobacteria and microcystins. Water Res 46:1447–1459
Lürling M, van Oosterhout F (2013) Case study on the efficacy of a lanthanum-enriched clay (Phoslock®) in controlling eutrophication in Lake Het Groene Eiland (The Netherlands). Hydrobiologia 710:253–263
Markert B (1987) The pattern of distribution of lanthanide elements in soils and plants. Phytochemistry 26:3167–3170
Mashitah SM, Shazili NAM, Rashid MKA (2012) Elemental concentrations in brown seaweed, *Padina* sp. along the east coast of Peninsular Malaysia. Aquat Ecosyst Health 15:267–278
Michetti KM, Leonardi PI, Cáceres EJ (2006) Cytochemical localization of acid phosphatase in *Stigeoclonium tenue* (Chaetophorales, Chlorophyceae). Biocell 30:491–496
Moreda-Piñeiro A, Peña-Vázquez E, Bermejo-Barrera P (2012) Significance of the presence of trace and ultratrace elements in seaweeds. In: Kim SK (ed) Handbook of marine macroalgae: biotechnology and applied phycology, 1st edn. Wiley-Blackwell, Oxford, UK, pp 116–172
Muraleedharan TR, Philip L, Iyengar L, Venkobachar C (1994) Application studies of biosorption for monazite processing industry effluents. Bioresour Technol 49:179–186
Niu W, He E, Wu Q, Zhou W, Zhang Y, Huang B, Zhao X (2012) Use of fluorescent europium chelates as labels for detection of microcystin-LR in Taihu Lake, China. J Rare Earths 30:941–946

Nultsch W (1979) Effect of external factors on phototaxis of *Chlamydomonas reinhardtii*. III Cations. Arch Microbiol 123:93–99

Ogata T, Terakado Y (2006) Rare earth element abundances in some seawaters and related river waters from the Osaka Bay area, Japan: significance of anthropogenic Gd. Geochem J 40:463–474

Okajima MK, Nakamura M, Mitsumata T, Kaneko T (2010) Cyanobacterial polysaccharide gels with efficient rare-earth-metal sorption. Biomacromolecules 11:1773–1778

Oliveira RC, Garcia O Jr (2009) Study of biosorption of rare earth metals (La, Nd, Eu, Gd) by *Sargassum* sp. biomass in batch systems: physicochemical evaluation of kinetics and adsorption models. Adv Mater Res 71–73:605–608

Oliveira EJA, Vila Nova SP, Alves-Jr S, Santa-Cruz P, Molica RJR, Teixeira A, Malageño E, Lima Filho JL (2006) A fluorescent-labeled microcystin-LR terbium cryptate. J Braz Chem Soc 17:243–250

Oliveira RC, Jouannin C, Guibal E, Garcia O Jr (2011) Samarium(III) and praseodymium(III) biosorption on *Sargassum* sp.: batch study. Process Biochem 46:736–744

Oliveira RC, Guibal E, Garcia O Jr (2012) Biosorption and desorption of lanthanum(III) and neodymium(III) in fixed-bed columns with *Sargassum* sp.: perspectives for separation of rare earth metals. Biotechnol Prog 28:715–722

Oliveri E, Neri R, Bellanca A, Riding R (2010) Carbonate stromatolites from a Messinian hypersaline setting in the Caltanissetta Basin, Sicily: petrographic evidence of microbial activity and related stable isotope and rare earth element signatures. Sedimentology 57:142–161

Olmez I, Sholkovitz ER, Hermann D, Eganhouse RP (1991) Rare earth elements in sediments off Southern California: a new anthropogenic indicator. Environ Sci Technol 25:310–316

Ono T (2000) Effects of lanthanide substitution at Ca^{2+}-site on the properties of the oxygen evolving center of Photosystem II. J Inorg Biochem 82:85–91

Pakrasi H, Ogawa T, Bhattacharrya-Pakrasi M (2001) Chapter 14: Transport of metals: a key process in oxygenic photosynthesis. In: Aro EM, Andersson B (eds) Regulation of photosynthesis. Springer Netherlands, Dordrecht, pp 253–264

Palasz A, Czekaj P (2000) Toxicological and citophysiological aspects of lanthanides action. Acta Biochim Pol 47:1107–1114

Pang X, Li D, Peng A (2001) Application of rare-earth elements in the agriculture of China and its environmental behavior in soil. J Soils Sediments 1:124–129

Pellegrini L, Epiard-Lahaye M, Penot M (1991) De l'utilisation du lanthane comme traceur de la voie apoplastique chez *Cystoseira nodicaulis* (Fucales, Cystoseiraceae). Can J Bot 69:18–25. (in Franz.)

Peng A, Pang X (2002) The free radical mechanism of rare earth elements in anti-adversity for plants. Environ Chem 21:313–317. (in Chin.)

Pitta TP, Sherwood EE, Kobel AM, Berg HC (1997) Calcium is required for swimming by the nonflagellated cyanobacterium *Synechococcus* strain WH8113. J Bacteriol 179:2524–2528

Pol A, Barends TRM, Dietl A, Khadem AF, Eygensteyn J, Jetten MSM, Op den Camp HJM (2014) Rare earth metals are essential for methanotrophic life in volcanic mudpots. Environ Microbiol 16:255–264

Polinares (2012) Fact sheet: rare earths oxides (REO). Working paper no. 37, March 2012

Qu KM, Yuan XY, Xin FY (1998) Enhancement of 3 rare earth elements to *Isochrysis galbana*. J Fish Res 5:42–47. (in Chin.)

Qu CX, Gong X, Liu C, Hong M, Wang L, Hong FS (2012) Effects of manganese deficiency and added cerium on photochemical efficiency of maize chloroplasts. Biol Trace Elem Res 146:94–100

Reid RJ, Rengel Z, Smith FA (1996) Membrane fluxes and comparative toxicities of aluminium, scandium and gallium. J Exp Bot 47:1881–1888

Ren QG, Hua Y, Shen H, Zhong L, Jin CZ, Mi Y, Yao HY, Xie YN, Wei SQ, Zhou LW (2007) Cytochemical behavior of rare earth ions in *Euglena gracilis* studied by XAFS. J Radioanal Nucl Chem 272:359–362

Ren M, Chen X, Zheng Y, Shen H, Ren Q, Li Y, Watt F (2013) Sub-100-nm STIM imaging and PIXE quantification of rare earth elements in algae cells. X-Ray spectrom special issue. In:

European conference on X-Ray spectrometry, vol 42, 18–22 June 2012, Vienna, Austria, pp 237–241
Rice TR, Willis VM (1959) Uptake, accumulation and loss of radioactive cerium-144 by marine planktonic algae. Limnol Oceanogr 4:277–290
Richards RG, Mullins BJ (2013) Using microalgae for combined lipid production and heavy metal removal from leachate. Ecol Model 249:59–67
Rudnick RL, Gao S (2003) The composition of the continental crust. In: The crust (ed RL Rudnick), vol 3, Treatise on geochemistry (eds Holland HD, Turekian KK). Elsevier-Pergamon, Oxford, pp 1–64
Sahoo PK, Tripathy S, Equeenuddin SM, Panigrahi MK (2012) Geochemical characteristics of coal mine discharge vis-à-vis behavior of rare earth elements at Jaintia Hills coalfield, north-eastern India. J Geochem Explor 112:235–243
Sakamoto N, Kano N, Imaizumi H (2008a) Biosorption of uranium and rare earth elements using biomass of algae. Bioinorg Chem Appl 706240, 8 pp. doi:10.1155/2008/706240
Sakamoto N, Kano N, Imaizumi H (2008b) Determination of rare earth elements, thorium and uranium in seaweed samples on the coast in Niigata Prefecture by inductively coupled plasma mass spectrometry. Appl Geochem 23:2955–2960
Sánchez-Rodríguez I, Huerta-Diaz MA, Choumiline E, Holguín-Quiñones O, Zertuche-González JA (2001) Elemental concentration in different species of seaweeds from Loreto Bay, Baja California Sur, Mexico: implications for the geochemical control of metals in algal tissue. Environ Pollut 114:145–160
Sandau E, Sandau P, Pulz O (1996) Heavy metal sorption by microalgae. Acta Biotechnol 16:227–235
Santos JG, Dutra JDL, Alves S Jr, de Sá GF, da Costa JNB, Freire RO (2013) Theoretical spectroscopic study of the conjugate microcystin-LR-europium cryptate. J Braz Chem Soc 24:236–240
Schacht U, Wallmann K, Kutterolf S (2010) The influence of volcanic ash alteration on the REE composition of marine pore waters. J Geochem Explor 106:176–187
Schönknecht G, Bauer CS, Simonis W (1998) Light-dependent signal transduction and transient changes in cytosolic Ca2+ in a unicellular green alga. J Exp Bot 49:1–11
Schwabe A, Meyer U, Grün M, Voigt KD, Flachowsky G, Dänicke S (2012) Effect of rare earth elements (REE) supplementation to diets on the carry-over into different organs and tissues of fatng bulls. Livest Sci 143:5–14
Servigne M, Tchakirian A (1939) Sur la presence d'elements des terres rares dans les algues calcaires (*Lithothamnium calcareum*). C R Acad Sci 209:570–572
Shen H, Ren QG, Mi Y, Shi XF, Yao HY, Jin CZ, Huang YY, He W, Zhang J, Liu B (2002) Investigation of metal ion accumulation in *Euglena gracilis* by fluorescence methods. Nucl Inst Methods Phys Res B 189:506–510
Sheng XY, Hambidge KM, Krebs NF, Lei S, Westcott JE, Miller LV (2005) Dysprosium as a non-absorbable fecal marker in studies of zinc homeostasis. Am J Clin Nutr 82:1017–1023
Shi Q, Guo WD, Hu MH, Yang YP, Wu YM, Gong ZB (2004) The content of rare earth elements in benthic organisms from the Xiamen Bay and their distribution and environmental implications. Acta Oceanol Sin 26:87–94. (in Chin.)
Singh RN, Subbaramaiah K (1970) Effects of chemicals on *Fischerella muscicola* (Thuret) Gom. Can J Microbiol 16:193–199
Squier T, Bigelow DJ, Fernandez-Belda FJ, deMeis L (1990) Calcium and lanthanide binding in the sarcoplasmic reticulum ATPase. J Biol Chem 265:13713–13720
Starý J, Kratzer K, Prášilová J (1983) Systematic study of the cumulation of elements on alga. Toxicol Environ Chem 7:47–60
Su D, Li PJ, Tai PD (2005) Toxic effects of lanthanides on *Chlorella autotrophica*. Chin J Ecol 24:382–384. (in Chin.)
Tai P, Zhao Q, Su D, Li P, Stagnitti F (2010) Biological toxicity of lanthanide elements on algae. Chemosphere 80:1031–1035

Thomas PJ, Carpenter D, Boutin C, Allison JE (2014) Rare earth elements (REEs): effects on germination and growth of selected crop and native plant species. Chemosphere 96:57–66
Tse Pui-Kwan (2011) China's rare earth industry, U.S. Geological Survey Open-File Report 2011–1042, 11 pp
Tyler G (2004) Rare earth elements in soil and plant systems – a review. Plant Soil 267:191–206
Ueda T, Muratsugu M, Kobatake Y (1974) Roles of Ca^{2+}, phospholipids and proteins in the excitable membrane of protoplasmic droplet isolated from *Nitella*. Biochim Biophys Acta 373:286–294
USGS (2014) Mineral commodity summaries, Annual publication of the United States Geological Survey, February 2014, 128–129 pp
Valcheva-Traykova M, Saso L, Kostova I (2014) Involvement of lanthanides in the free radicals homeostasis. Curr Top Med Chem 14:2508–2519
van Oosterhout F, Lürling M (2013) The effect of phosphorus binding clay (Phoslock®) in mitigating cyanobacterial nuisance: a laboratory study on the effects on water quality variables and plankton. Hydrobiologia 710:265–277
Vásquez JA, Guerra N (1996) The use of seaweeds as bioindicators of natural and anthropogenic contaminants in northern Chile. Hydrobiologia 326:327–333
Vijayaraghavan K, Sathishkumar M, Balasubramanian R (2010) Biosorption of lanthanum, cerium, europium, and ytterbium by a brown marine alga, *Turbinaria conoides*. Ind Eng Chem Res 49:4405–4411
Volland S, Bayer E, Baumgartner V, Andosch A, Lütz C, Sima E, Lütz-Meindl U (2014) Rescue of heavy metal effects on cell physiology of the algal model system *Micrasterias* by divalent ions. J Plant Physiol 171:154–163
Volokh AA, Gorbunov AV, Gundorina SF, Revich BA, Frontasyeva MV, Pal CS (1990) Phosphorus fertilizer production as a source of rare earth elements pollution of the environment. Sci Total Environ 95:141–148
von Tucher S, Schmidhalter U (2005) Lanthanum uptake from soil and nutrient solution and its effects on plant growth. J Plant Nutr Soil Sci 168:574–580
Wang XR, Tu Q, Tian L, Dai L (1993) Simulation study on transport of rare earth elements along an aquatic food chain. Environ Chem 12:212–218. (in Chin.)
Wang XR, Sun H, Xu Z, Dai L, Li Z, Chen YJ (1996) The effects and bioconcentration of REE La and its EDTA complex on the growth of algae (*Chlorella vulgaris* beijerinck). J Nanjing Univ (Nat Sci) 32:460–465. (in Chin.)
Wang K, Cheng Y, Yang X, Li R (2003) Cell responses to lanthanides and potential pharmacological actions of lanthanides. In: Sigel A, Sigel H (eds) Metal ions in biological systems, vol 40, The lanthanides and their interrelations with biosystems. Marcel Dekker, New York/Basel, pp 707–751
Wang C, Lu X, Tian Y, Cheng T, Hu L, Chen F, Jiang C, Wang X (2011) Lanthanum resulted in unbalance of nutrient elements and disturbance of cell proliferation cycles in *Vicia faba* L. seedlings. Biol Trace Elem Res 143:1174–1181
Wang YJ, Li J, Lü Y, Jin HB, Deng SH, Zeng YM (2012) Effects of cerium on growth and physiological characteristics of *Anabaena flosaquae*. J Rare Earths 30:1287–1292
Wang L, Li J, Zhou Q, Yang G, Ding XL, Li X, Cai CX, Zhang Z, Wei HY, Lu TH, Deng WW, Huang XH (2014) Rare earth elements activate endocytosis in plant cells. Proc Natl Acad Sci U S A 111:12936–12941
Wei YZ, Zhou XB (2000) Effect of neodymium on physiological activities in oilseed rape during calcium starvation. J Rare Earths 18:57–61
Weinberger F, Leonardi P, Miravalles A, Correa JA, Lion U, Kloareg B, Potin P (2005) Dissection of two distinct defense-related responses to agar oligosaccharides in *Gracilaria chilensis* (Rhodophyta) and *Gracilaria conferta* (Rhodophyta). J Phycol 41:863–873
Xia S, Zhao P, Chen K, Li Y, Liu S, Zhang L, Yang H (2012) Feeding preferences of the sea cucumber *Apostichopus japonicus* (Selenka) on various seaweed diets. Aquaculture 344–349:205–209

Xin FY, Yuan YX, Qu KM (1998) Influence of lanthanum-amino acid complexes on *Chaetoceros muller*. Chin J Appl Ecol 9:206–208. (in Chin.)

Yan XJ, Hou XL, Sun B, Fan X, Han LJ (1998) Element composition of *Sargassum thunbergii*. Chinese J Oceanol Limnol 16:189–192

Yang G, Kong Q (2002) Effect of La^{3+} and Nd^{3+} on growth, DHA yield and nitrogenase activity of *Crypthecodinium cohnii*. J Chin Rare Earth Soc 20(S1):168–171

Yao JM, Gong ZB, Li YC, Wen YY, Li J, Wang T (2007) Determination of rare earth elements in marine organisms inductively coupled plasma-mass spectrometry with microwave digestion. J Instrum Anal 26:473–477. (in Chin.)

Yin S, Ze Y, Liu C, Li N, Zhou M, Duan Y, Hong FS (2009) Cerium relieves the inhibition of nitrogen metabolism of spinach caused by magnesium deficiency. Biol Trace Elem Res 132:247–258

Yoder MD, Jurnak F (1995) The refined three-dimensional structure of pectate lyase C from *Erwinia chrysanthemi* at 2.2 Angstrom resolution: implications for an enzymatic mechanism. Plant Physiol 107:349–364

Yoshida K, Shimmen T (2009) Involvement of actin filaments in rhizoid morphogenesis of *Spirogyra*. Physiol Plant 135:98–107

Yoshimura K (1998) Mechanosensitive channels in the cell body of *Chlamydomonas*. J Membr Biol 166:149–155

Yuan XZ, Pan G, Chen H, Tian BH (2009) Phosphorus fixation in lake sediments using $LaCl_3$-modified clays. Ecol Eng 35:1599–1602

Zepf V (2013) Rare earth elements. A new approach to the nexus of supply, demand and use: exemplified along the use of neodymium in permanent magnets. Springer thesis, Springer, Berlin/Heidelberg, 157 pp

Zhang SQ, Zhang AQ, Yan JC (2006) Study on the performance enhancing effect of rare earth elements on fattening pigs, broilers and laying hens. Rev China Agric Sci Technol 8:35–39

Zhang N, Huang CH, Hu B (2007) ICP-AES determination of trace rare earth elements in environmental and food samples by on-line separation and preconcentration with acetylacetone-modified silica gel using microcolumn. Anal Chem 23:997–1002

Zhou PJ, Lin J, Shen H, Li T, Song LR, Shen YW, Liu YD (2004) Kinetic studies on the combined effects of lanthanum and cerium on the growth of *Microcystis aeruginosa* and their accumulation by *M. aeruginosa*. Bull Environ Contam Toxicol 72:711–716

Zhu ZZ, Wang ZG, Li J, Li Y, Zhang ZG, Zhang P (2012) Distribution of rare earth elements in sewage-irrigated soil profiles in Tianjin, China. J Rare Earth 30:609–613

Zoll AM, Schijf J (2012) A surface complexation model of YREE sorption on Ulva lactuca in 0.05–5.0 M NaCl solutions. Geochim Cosmochim Acta 97:183–199

Part II
Biorefinery Design and Processing Steps

Utilization of Biorefinery Waste Proteins as Feed, Glues, Composites, and Other Co-Products

William M. Chirdon

Abstract In recent decades, there have been significant advances in bioprocessing technologies which have brought large-scale algal biorefineries closer to becoming environmentally and economically sustainable. Yet, due to the costs of algae culture and lipid extraction, it is difficult to design a profitable biorefinery operation, especially if the only valuable products produced are biofuels. Biorefineries typically generate large volumes of solid, proteinaceous biomass with little or no value. These residues are often referred to as algae cake or post-extracted algal residue (PEAR). If a company needs to pay to dispose of these byproducts, it will be nearly impossible to operate profitably.

New technologies will likely result in the gradual improvement of culturing and processing methods to make them more cost effective. Biorefineries may increase profitability by utilizing existing material streams that may have little or negative value including various nutrient-rich wastewater streams and carbon dioxide output from local industries. This chapter addresses the economic viability of biorefineries by considering options for converting proteinaceous waste streams from biorefineries into products. The processes and products discussed are generally applicable to most algal biomasses, whether they are the generated as lipid-extracted byproducts of biofuel production, wastewater treatment, or both. While there are a number of potential strategies for making biorefineries profitable in the long run, the invention of valuable coproducts would be immediately transformative to the viability of these industries.

Keywords Post-extraction algal residue (PEAR) • Algae • Algae cake • Sustainable materials • Algae protein • Waste utilization • Biorefinery • Particleboard • Glue • Coproducts

W.M. Chirdon (✉)
Department of Chemical Engineering, University of Louisiana at Lafayette,
Post Box 43675, Lafayette, LA 70504-4130, USA
e-mail: wmc8199@louisiana.edu

A. Prokop et al. (eds.), *Algal Biorefineries*, DOI 10.1007/978-3-319-20200-6_11

1 Need for Co-Product Development

While it has been thoroughly demonstrated that microalgae and other microorganisms can be used to create a variety of useful fuels and materials, the ultimate challenge is not to show that making products from these bioprocesses is scientifically possible. Rather, the true challenge lies in making biorefineries economically feasible. To this end, a large volume of research has been devoted to improving the cultivation, harvest, and extraction of useful fuels and chemicals from algae. Invariably, this body of research will steadily push the technology forward towards economic viability; however, Richardson et al. (2014) investigated the impacts of contemporary harvesting and lipid extraction technologies and has found that it would take dramatic reductions in the present capital and operating expenses for this technology to become profitable. While the development of culturing, harvesting, and extraction technologies has been gradual yet steady, this chapter focuses on technologies which have a potentially transformative effect on the profitability of biorefineries by discussing potential co-products that may be developed from the proteinaceous waste streams.

Many bioprocesses, including those producing lipids for biofuels, generate large volumes of proteinaceous byproducts. In many instances, the mass of protein produced by the algae exceeds that of the lipid mass. Bryant et al. (2012) estimate twenty pounds of PEAR is produced for each gallon of biodiesel, but of course this ratio is highly dependent on all aspects of the biorefinery operation. These proteinaceous biomasses at present have very little market value. With a typical value of $50–300 per ton, it may be difficult to find buyers willing to purchase the large volumes of biomass produced who are located in sufficient proximity so that the value of the biomass will exceed the transportation cost. In some instances, producers may need to pay to have the proteinaceous biomasses landfilled if that is cheaper than shipping the large volumes of byproducts to available buyers. When introducing this subject to students, I have often described the issue as the "lemonade problem." That is, when you squeeze a lemon to get the juice, there is a relatively large amount of solid byproduct that needs to be used for something. If these solids can be developed into marketable materials, they will be transformed from a problem to a product.

This type of problem is not unique to bioprocessing. Almost every chemical industry makes unintended chemicals and materials aside from their main products which could make their industry unprofitable if a company has to pay to dispose of these material streams. Even in undergraduate plant designs in chemical engineering, it can be seen that the most successful business plans are the ones that make intelligent use of their byproducts. Not only is this typically more profitable than paying for disposal, it is generally considered to be a wiser and more sustainable practice. The petroleum industry is actually a good example of this concept in that if gasoline was separated from the crude oil and the rest of the petroleum was disposed of as waste, then the industry would not be nearly as profitable as it is now producing a wide range of gasses, oils, waxes, asphalt, and a myriad of petrochemicals. Just as in the case of gasoline, the profitability of biofuels will depend on developing markets for co-products from post-extraction algal residue (PEAR). In this chapter, a variety of potential products from PEAR are discussed with an emphasis on the conversion of PEAR into solid products that can be sold in large volumes including feeds, glues, and composite materials.

2 Value of PEAR

The value of PEAR is contingent on a wide variety of factors. In addition to the species of algae and the conditions of the culture, the number and types of extraction processes after harvesting will also affect the value of the resulting PEAR. If multiple, aggressive lipid extractions are performed, the protein may be degraded. The value of the PEAR will also depend on the amount residual organic solvents if they were used to extract lipids. Some algae produce various toxins on their own which limit the applications and value. The PEAR may also suffer from heavy metal contamination if there are heavy metals in the water or the organic sources of the biorefinery, which may be concentrated inadvertently in the biorefinery operations, especially if there is significant evaporation in an open pond.

Bryant et al. (2012) developed a hedonic pricing model on the value of PEAR based on its potential for livestock feed. Due to a lack of commercial scale PEAR production and sale, the price of PEAR needs to be estimated from economic models. In the study conducted by Bryant et al. (2012), the price of PEAR was estimated by assuming that it would be used in livestock feed and would be valued according to its nutrient content. By analyzing the nutritional value through specific independent variables of twenty-two common animal feed ingredients including oilseed products, animal byproducts, grain products and other feedstuffs, the price of PEAR could be hedonically modeled based on its nutritional value compared to competitive feed products. Since PEAR has less protein and higher ash content than soybean meal, its value was estimated to be consistently less than soybean meal for livestock feed with an estimated value of $100–225 per ton over the years 2006–2010 based off of feed prices over the same time scale.

3 The Problem with PEAR

The need for co-product development from PEAR has been recognized for some time; however, there are several innate difficulties which arise when attempting to convert "proteinaceous biomass" into value-added products which make innovation in this area challenging. Firstly, the biomass is difficult to characterize to a level where chemical syntheses can be designed. For instance, a polymer chemist could easily utilize 1,4-butanediol in a polyester formulation or utilize the double bonds in an unsaturated oil using a cationic polymerization. (Ionescu and Petrovic 2009: US Patent 7501479), but there is very little a chemist can do with uncharacterized, proteinaceous biomass. With such a complex mixture, it can only be characterized statistically. That is, one may determine the molecular weight distribution of the constituents or conduct a functional group analysis, but it is nigh-impossible to identify well-defined chemicals which can be used in specific synthetic strategies to create specific products. Even if functional groups are identified and targeted for some synthetic strategy, the menagerie of other constituents is likely to interfere with the intended synthesis. Even if the PEAR could be completely characterized to know the exact structure of each macromolecule, the sheer complexity of the biomass would make plotting specific synthetic routes difficult. It should be noted that these biological residues are likely to be even more complex than the original biological system, because they will likely contain many of the original constituents of the biological system in addition to constituents in various levels of degradation. One may also consider the separation of PEAR to isolate chemicals valuable in themselves or for further synthesis, but separations are expensive processes and will result in additional byproducts. Yet, some components may justify the expense of separation. For instance, carotenoids, which are found in microalgae, have a market that has been projected to reach \$1.3 billion globally by the year 2017 due to their applications in food products, cosmetics, vitamin supplements, and animal feeds among others (Yaakob et al. 2014).

4 Human Consumption of PEAR

Since PEAR is a protein-rich byproduct, it has significant nutritional value. However, there are difficulties in using this as a food source for human consumption. One issue is that the lipids are often extracted using organic solvents which are toxic, carcinogenic, and highly-regulated. Removing these solvents to the point where the PEAR is fit for human consumption is expensive and could pose a legal liability if done improperly, or even if it was perceived to have been done improperly. However, the issues with solvent extraction should not be overstated, as the process has already been established for other food and beverage products such as olive oil and decaffeinated coffee. Some microalgae also produce compounds that are toxic, and therefore are poor candidates for a food source. This should be of particular concern when culturing algae in systems that are open to the environment as there is a potential for toxic microalgae to enter open systems as an "invasive" species. There is also a marketing challenge in convincing consumers to eat "post-extracted algal residue" even without the issue of residual organic solvents, especially in the western hemisphere. Yet, the PEAR typically contains large amounts of proteins, carbohydrates, and other constituents that also have potential as food ingredients. As such, it will be more likely to be added to food products incrementally as one of several ingredients, for instance as a protein supplement, in the near future. As with most applications, the microalgae species, the history of culturing environment, and the extraction and processing methods will affect the suitability of the PEAR for a given application or product, (Batista et al. 2013) so it is difficult to make generalizations about the food use of all PEAR. Some microalgae and other microbes may also produce other valuable compounds, such as pigments, that could be used to color food or cosmetics (Dufossé et al. 2005).

5 Use of PEAR in Agriculture

Historically, agricultural and bioprocessing byproducts have been used as animal feed or plant fertilizers. For instance, commercial brewers and distillers will often will give or sell their used grains to local farmers who raise pigs or cattle. This strategy may also be applied to PEAR. Considering the amount of proteins, carbohydrates, and minerals, it has the potential to be a nutritious feed for animals. Among the advantages of selling the PEAR to farmers is the large volume of food that the animals consume and the large amount of nutrients a farm needs on a commercial scale. There is some interest in extracting high-value materials from PEAR such as pigments, nutraceuticals, or pharmaceutical components, and these products would help the profitability of the biorefinery (Dufossé et al. 2005). However, after the high-value components are separated from the biomass, there would still be large volumes of PEAR remaining of which would need to be disposed. Using PEAR to feed animals and fertilize farmlands would allow for the large volumes to be fully utilized. While there are stringent regulations and consumer concerns about feeding foods that have been exposed to solvents to humans, there is far less risk and regulation to feeding these materials to livestock.

With the global population projected to reach nine billion within several decades (Lum et al. 2013), the world will need to simultaneously increase the food and energy supply. By using PEAR as an animal feed, it would allow more livestock to be farmed while having the potential to at least partially displace the need to feed grains such as corn to livestock, allowing these grains to be used by people. From 2007 to 2011, the worldwide production of ethanol dramatically increased over these 4 years from 50 to 85 billion, which is largely credited for a rise in price from \$163 to \$291 per metric ton over the same years with 38 % of the corn being utilized for bioethanol and co-products in 2011 (Lum et al. 2013). The use of microalgae on a commercial scale to create fuels and food through direct human consumption or as animal feed has the potential to simultaneously supplement the global food and energy supply.

Although livestock do not have any preconceived aversion to eating algal co-products, not all livestock will eat all types of protein. Rather, some livestock may be less eager to eat or otherwise be repulsed by foods which are unfamiliar in sight, smell, and consistency. In these instances, the PEAR may need to be modified to be more palatable to the animal, perhaps by compounding with more familiar foods. For instance, Lum et al. (2013) reports that the PEAR from *Staurospira* sp. can replace 7.5 % of corn and soybean meal without affecting the growth performance or health status of broiler chickens, and the growth of chicks has shown no adverse response when fed 20 % sewage-grown *Chlorella* and *Scenedesmus* sp when compared to a corn-soybean meal based diet. However, 20 % blue-green algae feeds were found to lower growth rates as opposed to lower level algae contents in 3 week-old broiler chickens. Pigs were able to be fed by replacing 33 % of the soybean meal with a mixture of *S. maxima*, *A. platensis*, or *Chlorella* sp without affecting their weight gain or feed efficiency without any apparent toxicity. However, pigs could not tolerate a 15 % replacement of their corn and soybean meal with de-fatted *Staurospira* sp. The use of microalgae and PEAR is an active area of investigation, but there is no over-arching solution, since what is palatable and nutritious is specific to the type of animal, the type of algae, the components extracted from the biomass, and how the biomass was processed.

One obvious application of PEAR is in aquaculture, especially for the feeding of aquatic animals which naturally consume algae. There is a large market for foods derived from aquaculture. In 2010, the worldwide aquaculture production of food (excluding plants) was 60 million tons valued at $119 billion. (FAO 2012) It should be noted again that different aquatic species have different nutritional needs and dietary preferences (Makkar 2012). For instance, microalgae which are rich in polyunsaturated fatty acids serve as good feed for mollusks, crustacean larvae, and zooplanktons for crustacean, and fish larvae. Microalgae have been studied as a fish meal that has the ability to improve the rate of weight gain as well as the muscle protein composition and quality. The color, texture, and taste of the flesh can also be improved by using microalgae-based feeds. The use of microalgal additives has also noted to cause efficacious assimilation of dietary protein, improvement in physiological activity, stress response, starvation tolerance, and disease resistance (Hasan and Chakrabarti 2009).

Waste algae may also be composted to create soil conditioner or fertilizer (Han et al. 2014). Depending on the water source, the composted algae may contain high amounts of salt or heavy metals, and some species create toxins. However, some toxins such as microcystins can be degraded in the composting process. Algae may also have a low carbon-to-nitrogen ratio, which may allow for a faster composting process due to the nitrogen available to the microbes, but this ratio is highly variable across different biomasses. If the ratio is too low, nitrogen may be lost through ammonia volatilization, but this can be compensated for by adding co-composting materials with a high C/N ratio such as timber byproducts, straw, or animal manure. Co-composting materials can also be used to dilute the concentration of biological toxins and heavy metals that may be present in an algal biomass. The carbon dioxide released in composting was originally absorbed by the algae from the environment, so it can be considered to be carbon-neutral for composting's "carbon footprint." However, the release of methane should be considered from the composting operation, since it has a higher impact than carbon dioxide as a greenhouse gas.

From a study of algal sludge from Taihu Lake by (Zhang et al. 2014a), it was found that the algal sludge, which acts as an environmental pollutant, can be converted to bio-organic fertilizer through solid-state fermentation that had more nutrients than other commonly used composts. These composts were also found to host large amounts of plant growth-promoting rhizobacteria which are associated with plant roots and synergistically encourage plant growth by competing with pathogens for the root surface area and by acting as biopesticides towards pathogens including bacteria, fungi, and nematodes. These rhizobacteria also serve to fertilize their host plants. One challenge in utilizing these rhizobacteria is that they need to survive and reproduce in the compost until the plants have established a root system to host these highly beneficial microbes. It was found that the compost made from this waste algal sludge through solid-state fermentation successfully hosted stable cultures of the rhizobacteria, while rapidly degrading the harmful and toxic microcystins. The ability to degrade these microcystins biologically in this process is particularly useful, because they are relatively thermally stable and non-volatile, making them difficult to remove otherwise. While this study from Zhang et al. (2014a) was from a naturally-occurring, refloated algal sludge, these results are very encouraging for other algal product streams exiting a biorefinery.

6 Conversion of PEAR into Glues and Composite Materials

One common strategy for utilizing waste proteins from both plant and animal sources is to convert the proteins into glue. Similarly, waste algal proteins and biomass may be transformed into marketable, useful materials by converting them into glues and composite materials. Converting proteinaceous biomasses into glues is a relatively simple process, but the conversion of PEAR into glues for direct use or for incorporation into composite materials has not been widely investigated.

Glues and composite materials from natural proteins have been made in a primitive fashion for millennia, and significant developments in soybean protein glues have been made in the early 1900s (Laucks and Davidson 1932). However, the research and development of glues and composites from natural sources has accelerated recently within the past decade including textbooks with chapters on this subject (Wool and Sun 2005). The general strategy is to utilize the broad base of knowledge of synthesizing glues, polymers, and composites from other plant proteins that has been published to create analogous polymeric products from algae cake. Useful glues might be made by simply denaturing the protein, but further modification, including the implementation of crosslinking agents, is likely to improve the mechanical properties, water resistance, and range of potential applications. It should be noted that the complex protein chemistries discussed typically affect the primary, secondary, and tertiary structures of the proteins, and it is not possible to sketch all of the specific chemical mechanisms which occur in a protein denaturization process. However, robust mechanisms for converting various proteins from a variety of plant and animal sources into adhesives by denaturing them through non-specific processes have been well-established.

There are a variety of mechanisms for denaturing proteins into adhesive states. Proteins exist with highly complex primary, secondary, tertiary, and quaternary structures. The goal when denaturing the proteins into glues is to maintain the primary structure and the molecular weight of the protein while breaking up the higher order structures, although it may be useful to retain some of the secondary structures. The higher order structures are held together primarily by hydrogen bonds and chemical cross-links. One strategy for converting proteins into glues is to use a surfactant, such as sodium dodecyl sulfate, which interferes with the hydrogen bonding thereby causing the denaturization of the protein. Another strategy would be to use enzymes to denature the protein. Enzymes could be used to target specific types of cross-links or bonds to selectively degrade the proteins, but this option would be difficult to make profitable on a large scale since enzymes tend to be expensive reagents, and in addition, enzymatic reactions often require specific, controlled conditions. A third strategy is to denature the proteins by modifying the pH of the algal slurry. This has been historically proven to be a simple, yet robust, method for converting waste proteins into glues for a variety of plant and animal proteins, although there has been very little work done on waste algae proteins. While proteins will denature in both acid and alkaline solutions, most glues are typically made under alkaline conditions, which is the method used in the following line of work.

One of the major drawbacks of denaturing the PEAR in an alkaline solution is that the primary structure will eventually hydrolyze into smaller fragments, causing a reduction in the molecular weight and the mechanical properties. Therefore, it is very important to control the reaction kinetics using this approach, as the goal is to provide enough time to allow for the protein to denature and expose the hydrophilic, adhesive functional groups, but not enough time to cause excessive hydrolysis. The optimal reaction time will depend on the temperature and the strength of the base added. Stronger bases, such as NaOH, are known to have a higher initial adhesive strength, but may hydrolyze the protein rapidly, while glues made from weaker bases such as CaOH will be able to have a longer pot life than stronger bases and will hydrolyze the proteins to a far lesser extent over time. Some glue formulations may use multiple hydroxides in the formulation to achieve the optimal combination of properties for an application.

6.1 Kinetics of Denaturization

Some preliminary tests to examine how the time and temperature of denaturation affected the adhesion of the proteins have been conducted. In two series of experiments, 9 g of proteinaceous algal byproduct was denatured in 90 mL of 1 M NaOH. One series was conducted at 25 °C and the other at 50 °C. Five shear test samples were made at 30, 60, and 90 min by applying aliquots of the mixture to wooden sticks. Both the time and temperature had a significant effect on the resultant shear strength. As expected, the shear strength increased with reaction for a period of time as the proteins become denatured and disentangled. However, a drop in shear strength was also observed at longer times as expected, since excessive reaction time results cause the protein chains to break down through hydrolysis under alkaline conditions as shown in Fig. 1. The molecular weight degradation of

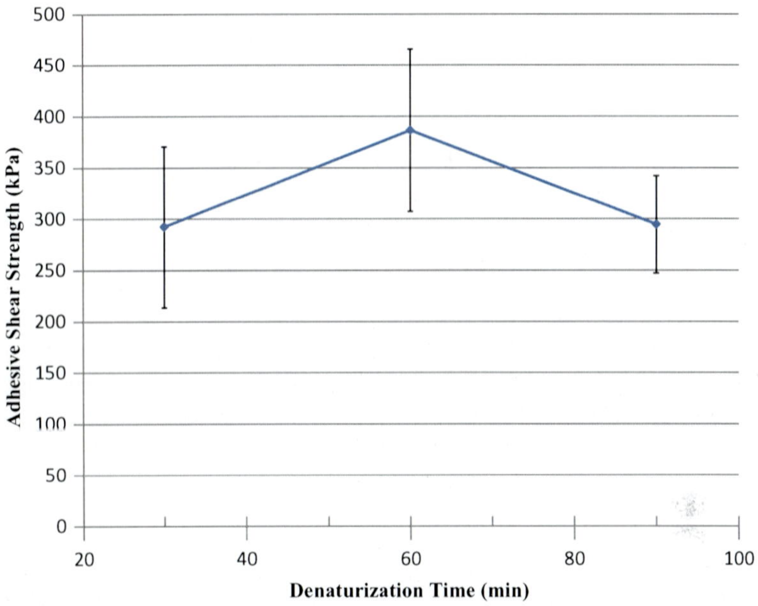

Fig. 1 Shear test results for algae glue from hydroxide reaction on wooden sticks showing the glue's shear strength for various reaction times

the protein chains from this hydrolysis is hypothesized to cause a loss of cohesive strength within the protein glue resulting in a reduction of the effective strength of the glue as determined by shear strength tests.

6.2 *Effects of Extraction Process Conditions*

Among the many variables that affect the properties of PEARs and their developed coproducts are those relating to lipid extraction. The solvents and temperatures used as well as how many times the extraction is repeated will affect the resultant PEAR. Higher extraction temperatures are known to typically yield higher amounts of lipid, but what is less understood is how the higher extraction temperature affects the PEAR and its potential to make useful coproducts. To begin investigating these considerations, lipids were extracted from an algal biomass at various extraction temperatures ranging from 40 to 80 °C. The resultant PEAR was made into glue by reacting with NaOH and then tested by conducting shear strength tests on wooden sticks with the results shown in Fig. 2.

While higher extraction temperatures have been known to yield larger amounts of lipids without significantly reducing the mass of the protein yield, these results show that using higher temperatures results in glues that have lower shear strength. The mechanism of the deterioration of the shear strength at elevated temperatures

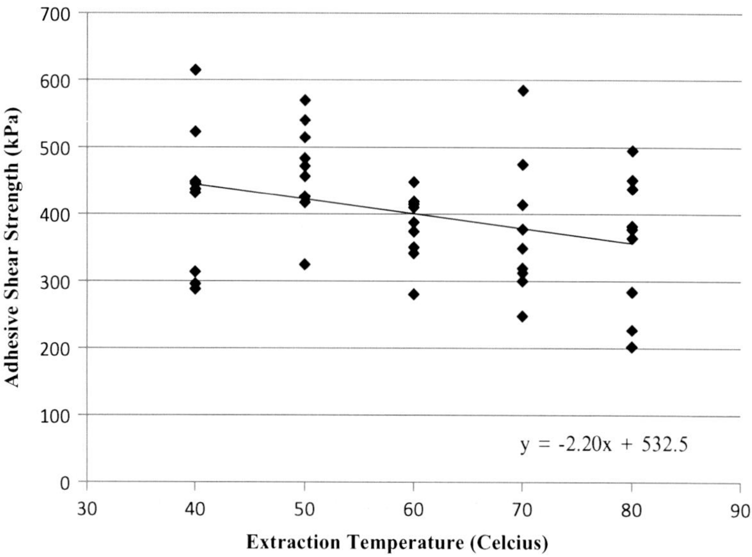

Fig. 2 Shear test results for algae glues that have been made from algal residues from lipid extractions at various temperatures showing a reduction in adhesive strength with increasing extraction temperature ($p = 0.03$)

cannot be determined from this mechanical test, but it is hypothesized that the elevated temperatures result in hydrolysis and molecular weight degradation that causes a loss of cohesive strength within the glue.

This finding may have a transformative effect on how some technologists approach the economics of algal bioprocessing industries. If the algal residue can be converted into valuable glues, it would be unwise to maximize the lipid yield by increasing the extraction temperature, using more aggressive solvents, and performing multiple extractions because at some point the use of overly-aggressive extraction techniques will damage the quality and value of the adhesive co-products. Therefore, instead of unilaterally maximizing the lipid yield, the value of the additional lipid extracted must be balanced against the lost value from the deterioration of the protein properties in order to maximize the profitability of biorefinery.

6.3 *Algal Binder for Paper, Wood, and Agricultural Byproducts*

Preliminary studies by the author on the conversion of PEAR into various glues and binders have been promising thus far with the development of various prototypes. A simple, low-cost glue was made by mixing the proteins with urea in an aqueous solution at 50 °C for 3 h. This resulted in a glue that is strong enough for paper and

Fig. 3 Paper pad glued together using denatured algae proteins

labeling applications as shown in Fig. 3. The glue was used to assemble the paper pad that is shown in Fig. 3. Undiluted, the glue was too strong and caused the paper to rip. However, with the proper concentration, the glue will dry with sufficient strength to hold the paper pad together, but not too strong so as to cause the paper to rip. Upon removing a sheet the glue does not feel sticky. The glue does appear faintly yellow on the paper, but this a common aspect of many natural glues.

Among the most promising materials that have been made from this technology are composite materials which use the PEAR glue as a binder for fibrous plant products. These composite materials reap the collateral benefit of utilizing other natural waste sources from agriculture and forest products including sugarcane bagasse, rice husks, wood chips, sawdust, post-consumer paper products, and similar materials.

One composite material that has been developed includes formaldehyde-free bagasse/sawdust/algae protein composites with esthetic and mechanical properties similar to conventional particleboards. Coasters made from this type of material are shown in Fig. 4. After sealing, these coasters have been able to be used for multiple years without any noticeable degradation of properties. It is proposed that these sustainable materials may be used in furniture, cabinetry, or other similar particleboard applications.

Some preliminary mechanical tests have already been completed on the algae-bagasse particleboard composites. The effect of the clamping pressure on the mechanical properties of the composites has been investigated with encouraging results. The two most important mechanical properties of particleboard composites are the flexural modulus and flexural strength. In the particleboard industry, these are known as the modulus of elasticity and modulus of rupture, respectively. As shown in Figs. 5 and 6, under optimal processing conditions, the algae/bagasse/sawdust composites have moduli of elasticity (MOE) ranging from 1.3 to 2.4 GPa, which exceed the MOE values of particleboard composites made from conventional urea-formaldehyde resins that have been found to range from 0.6 to 2 GPa (Rathke

Fig. 4 Coasters made with algal glue binder. Coaster on *left* uses glue with sugarcane bagasse. Coaster on *right* also contains sawdust, resulting in a smoother surface

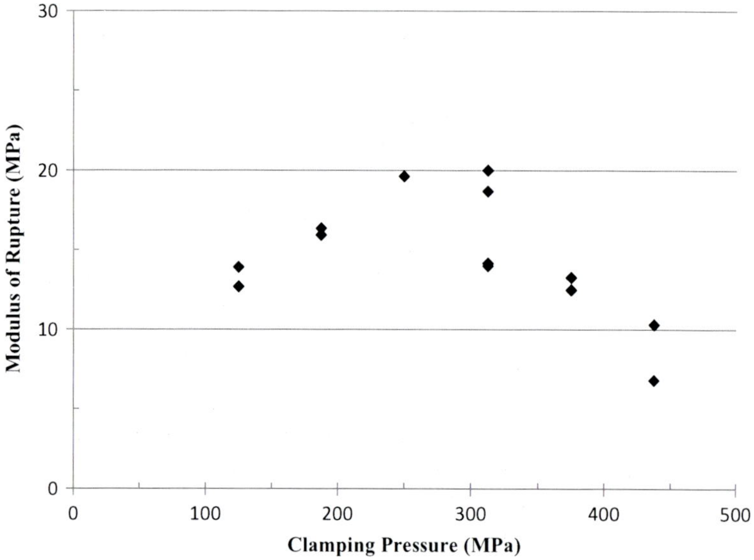

Fig. 5 Modulus of rupture of composites of algal protein, sugarcane bagasse, and sawdust

et al. 2012). These composites have moduli of rupture (MOR) ranging from 12 to 20 MPa, which exceed the MOR values of particleboard composites made from urea-formaldehyde resins that have been found to range from 2 to 14 MPa (Rathke et al. 2012) .

Another potential application of these composites is shown in Fig. 7 as a set of Mardi Gras beads made from algal glue and sawdust. Louisiana residents who have cleaned up their yards or neighborhoods after a Mardi Gras parade would be able to appreciate disposable toys, beads, and party favors that could biodegrade over time. Of course, this technology could transfer to a variety of low-cost, disposable

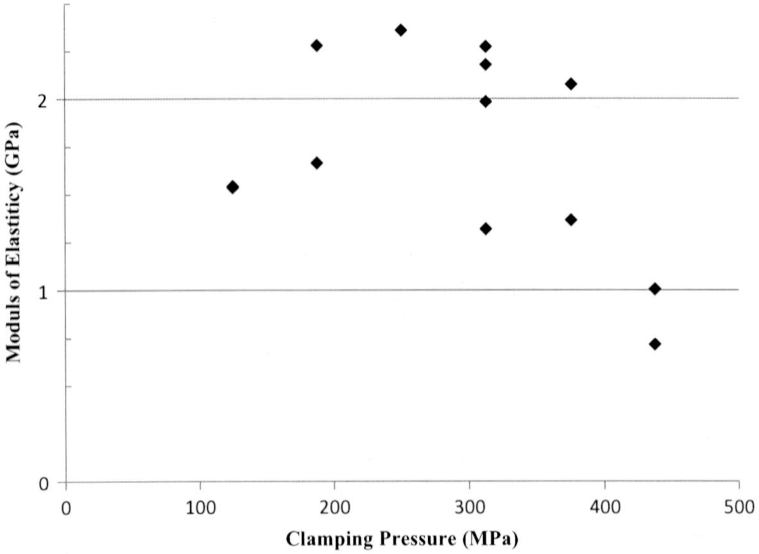

Fig. 6 Modulus of elasticity of composites of algal protein, sugarcane bagasse, and sawdust

Fig. 7 Mardi Gras beads made from algal protein and sawdust and then painted with the algae protein

materials which could be used by consumers with a clear conscience as a substitute for non-biodegradable plastics that are used in disposable or short-life applications.

The algal glue has been found to be especially useful for binding fibrous plant products. In addition to the bagasse filled composites, composites from rice husks have also been made as shown in Fig. 8. Light-weight composites similar to drywall or ceiling tiles have been made using the algae glue and post-consumer paper products. The results thus far provide encouragement for the development of

Fig. 8 Composite made with rice husks similar to conventional particleboard

additional composites that utilize other types of natural fibrous fillers in addition to the ones already tested, perhaps including grass clippings or other yard waste.

One characteristic, which is common to protein-based glues and composites, is that they are innately biodegradable. However, the composites can be conferred with resistance to moisture, mold, and insects by coating the materials with a thin layer of sealant. Future work will include the testing of natural sealants, such as waxes from sustainable sources, to create products that are durable with a longevity that can be "programmed" by the diffusion barrier provided by the coating.

6.4 Composites from Algal Protein and Limestone

In the refining of petroleum, the higher molecular weight components, such as bitumen or asphalt, are removed from the lighter components which are more suitable for fuels. Since these high molecular weight materials tend to be viscoelastic, semisolid materials, they cannot be directly used as a liquid fuel and have found greater application in sand and/or gravel composites, and most of the asphalt currently generated is used in roadways. This has inspired some investigation into whether the PEAR from biofuels production could be converted into structural composites suitable for infrastructure purposes. Furthermore, this sort of technology may become necessary, because if biofuels are to significantly displace petroleum fuels, replacements for these petroleum coproducts must also be developed. Unfortunately, attempts to convert PEAR into a binder for use in composites with properties similar to hot mix asphalt composites have not been successful thus far. However, some work along these lines has been conducted and the results and conclusions are discussed in this section because the composites from algal protein and limestone may still have some applications even if the strength and durability of conventional structural composites cannot be matched at present. It is also hoped that these initial steps in this young field will spur further innovation.

6.4.1 Mechanical Properties of Protein-Limestone Composites

Spirulina dried protein powder was mixed in an aqueous 1 N sodium hydroxide solution with a ratio of 1 g of powder per 2.7 mL of solution for 1 h before combining with limestone aggregates. Since water will need to be evaporated or boiled out of the composite, the minimum amount of water that will allow the dissolution of the protein and coating of the aggregates should be used. The gradation of the aggregates is also important when designing these composite materials. It is necessary to have some amount of interconnected porosity in the composite, because the composite hardens by drying.

These composites can only be made with a narrow range of algal binder (glue) content. It was discovered that samples with less than 5 % binder had insufficient strength for mechanical testing and would often fail with casual handling. Samples with over 10 % binder resulted in the bubbling within the sample, which resulted in strength loss or the destruction of the sample. Oftentimes when designing composite mixes, engineers seek to fill the porosity within the filler material with binder. However, it has been found to be necessary to design systems with enough residual porosity to allow for evaporation. Otherwise, the binder on the outside of the composite will be likely to dry to form an air-tight seal, and the entrapped water when heated may form pressurized bubbles resulting in strength loss and risk of explosion.

Compression Testing

Tests were conducted similar to the ASTM C39 test method for compressive strength of molded concrete cylinders in strain-controlled tests on 2″×4″ cylinders. The compressive strengths for composites with a binder-to-aggregate ratios ranging from 5 to 10 wt% are shown in Fig. 9. The trends in compressive strengths as shown in Fig. 9 show that the composites have a maximum compressive strength at 6.25 wt%. There is an abrupt increase in compressive strength from 5 to 6.25 wt% and a more slight, but significant decrease from 6.25 to 10 wt%.

The elastic modulus found in compression in shown in Fig. 10. Again, the increase from 5 to 6.25 wt% is highly significant, but there is no statistically significant change from 6.25 to 10 wt%.

Tensile Testing

The tensile strengths of composites as determined using the test method described in AASHTO T132 are displayed in Fig. 11 for various ratios of algal binder. The 5 wt% composites do not have sufficient strength to be mounted and tested, and only one of the five 6.25 wt% samples made was strong enough to be tested.

These composite materials have been found to be very weak in tension, especially with binder contents below 7.5 %. Also, the tensile strengths have a different

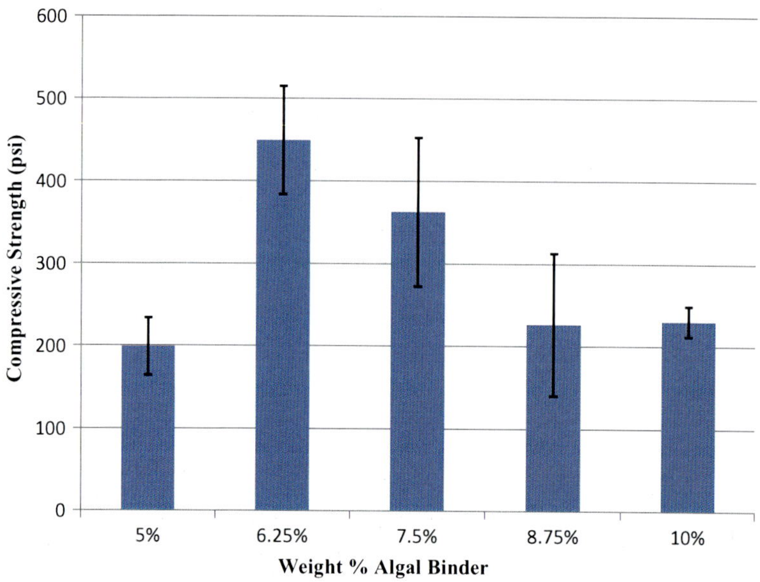

Fig. 9 Compressive strength averages for different weight percentages of algae in the composite materials. *Error bars* indicate standard deviation of five samples

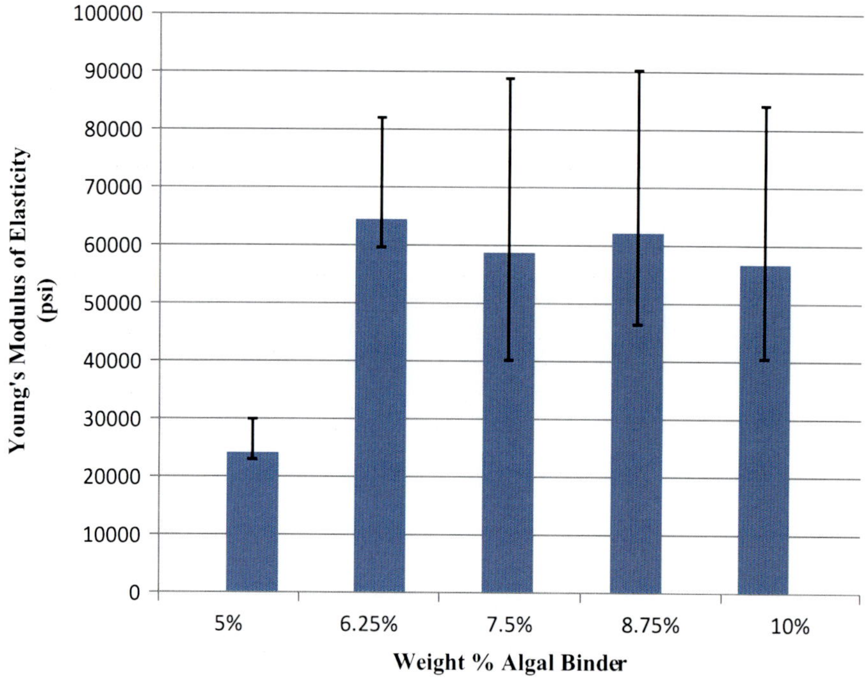

Fig. 10 Bar graph representing Young's Modulus of Elasticity for different weight percentages of algae in the composite materials from compression testing. *Bars* indicate standard deviation

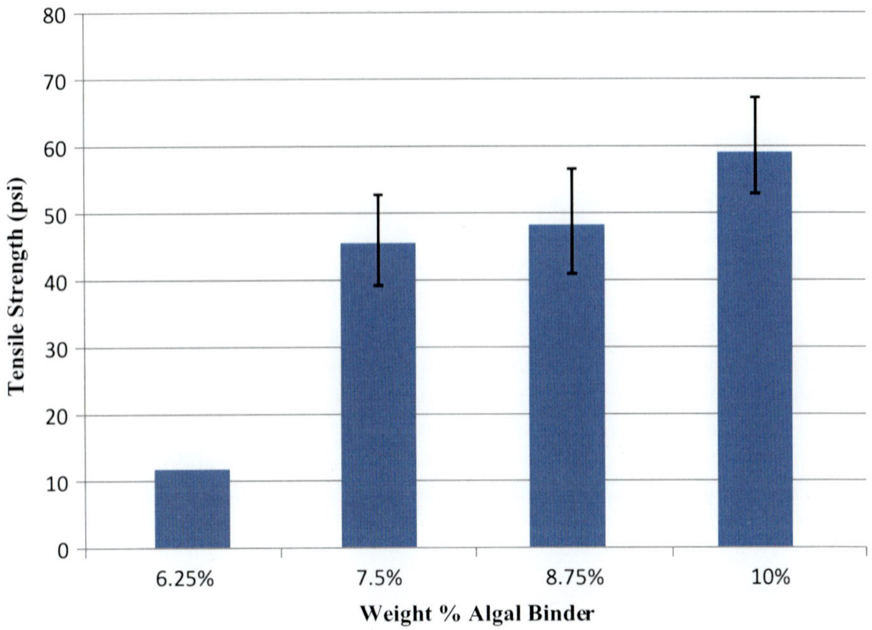

Fig. 11 Tensile strength averages for different weight percentages of algae. *Error bars* indicate standard deviation

trend than the compressive strength with respect to binder content. The tensile strength monotonically increases with increasing binder content with a dramatic increase between 6.25 and 7.5 wt%.

6.4.2 Potential Applications

Since algae-based composites represent a relatively new technology that is still under development, there has not been any application testing completed to date. However, potential applications are suggested here based on the observation of the properties of the composites. It can be clearly seen from the mechanical properties that the algae-limestone composites made from this formulation should not be used in structural, load-bearing applications due to a lack of strength. The other major limitation is that the composites are protein-based and remain biodegradable. While it is likely that the chemistry and formulation can be modified to improve the mechanical properties and anti-microbial agents can be added to mitigate microbial attack, it is highly unlikely that these composites will be useful for permanent, structural applications without an extreme chemical modification of the proteins.

Even with these limitations, there may still be a variety of niche applications for which these composites may still be considered, especially in applications where low strength and biodegradability are assets, not drawbacks. For instance, these composites may be useful for excavatable fill applications. That is, there may be applications where strengths greater than compacted soil are needed, but a low strength is desired so that the material can be mechanically excavated easily. While the biodegradable nature of the composites is a major drawback for permanent structural applications, it may be an ideal solution for temporary infrastructure. The construction of temporary roads and pathways for lumber and oilfield industries has a significant environmental impact. Algae-based composites may have applications in making temporary roads or providing soil stabilization for heavy machinery or equipment. Since the binder is biodegradable, the infrastructure around a worksite could be more quickly reclaimed by the environment after a project is completed. Also, since the binder hardens by evaporation, it may be able to be placed more rapidly by applying heat and pressure using hot rolling or similar machinery. This would be one advantage over concrete, which requires time to hydrate for hours or days to achieve adequate strength. If the soil around a work site has been excavated and needs to be returned to its natural state, the algal binder could provide temporary stabilization in the short term and then be biodegraded and replaced by the action of local microbes, insects, and plants, potentially providing both stabilization and fertilization.

6.5 Advantages of Using Glue as a Coproduct from PEAR

There are significant advantages to converting PEAR into glues and composite materials. While the price of these glues cannot be accurately foreseen until they are brought to the market, by using similar plant-based adhesives as a reference, these glues should provide significantly more value per unit of biomass than existing feed or fertilizer applications. While not as valuable as more refined chemicals, it also has the ability to be sold in a large volume market which has the potential to consume tons of PEAR. The glue can also be used with other forest or agricultural solid waste products to make composite materials.

Glues made through this process also do not have any volatile organic compounds or similar toxic chemicals added, making them safer for manufacturers and consumers which adds to their value. These materials are also biodegradable, and the rate of degradation can be controlled using sealants or additives. The process described would also have two major advantages if commercialized. First, since the chemistry occurs in an aqueous slurry, there is no need to de-water the PEAR, which eliminates an expensive industrial process. However, the PEAR may still be dewatered and dried to stabilize it if it needs to be stored for a period of time or shipped. Second, all of the PEAR goes into the product with no additional waste streams. This avoids the need for consideration of any additional waste streams. Although the PEAR may contain a large number and variety of impurities, they have been found to only have a minor impact on the mechanical properties of the resultant composite. This also means that any contaminants, such as heavy metals, would be incorporated into the composite. This should especially be considered if the composites are used in consumer products or in environmentally sensitive applications. Conversely, this process could be used intentionally in contaminated protein-rich streams to sequester contaminants into a solid material which may serve as a safer and more concentrated way to contain, store, transport, or treat contaminants rather than keeping them in an aqueous slurry.

6.6 Challenges of Using Glue as a Coproduct from PEAR

A variety of challenges persist for converting PEAR into glues and composite materials. One challenge is that the glue hardens with drying, preferably by the application of heat to cause boiling or evaporation. In order for the center of the composite to dry, the composite must be designed with some amount of porosity to allow for the water vapor to vent through the composite in the drying process. This limits the ratio of binder to filler that can be used, which limits the mechanical properties attainable. This also presents an occupational hazard in that if the steam is not vented safely in a controlled manner, an explosion may result. The risk of explosion can be mitigated by proper composite formulation and process design. However, the explosion hazard should be addressed using appropriate protective gear and equipment at all times, although explosions rarely occur after the process and formulation have been properly designed.

Other disadvantages include the pot life of the glue. If strong bases, such as NaOH, are used, the glue will have a short pot life and should be used soon after mixing. Therefore, the glue cannot be made, stored, and sold as a single tank or tube of glue. Weaker bases, such as calcium hydroxide, may be used to partially or completely replace stronger bases, but they will have less initial adhesive strength.

With regard to the start-up of this technology on a commercial scale, a large amount of entrepreneurship will be required. A biofuels company will be unlikely to start-up without identifying a use for their PEAR. Making PEAR composites will most likely require a group with particleboard manufacturing experience and capability. These composites will require filler materials such as sawdust and bagasse which will need to be supplied. This illustrates the need for a comprehensive biorefinery approach towards making fuels and materials. Individually, it would be difficult to start up a biofuel production facility without a strategy for the large volumes of PEAR, and similarly, it would be unwise to start-up a PEAR composite manufacturing facility without a dependable source of both PEAR and filler materials. Hence, this sort of project would require great entrepreneurship, since multiple businesses must be started up simultaneously. Also, the facilities should be located nearby, or be connected by some inexpensive transportation mechanism, because the PEAR, fillers, and composite products have a relatively low value per ton, so the transportation costs of the PEAR and the filler to the composite manufacturing site should be low to optimize profitability. Ideally, a biorefinery would be located where the PEAR could be utilized in approximately the same location. That is, it would be advisable to start a biofuel production facility near a filler source, such as a sawmill or sugarcane refinery, and then build a particleboard manufacturing facility adjacent to or within the biorefinery. Locating the biorefinery near interstates, railways, and/or shipping ports would also reduce the price of bringing the products to market.

7 Other Uses for PEAR

While this chapter primarily addresses the conversion of PEAR in to solid products that can be sold in large volumes, other uses for PEAR are briefly discussed in the following subsections.

7.1 Burning or Cofiring

PEAR can be directly burnt to create energy, but to increase the heat release upon burning, the PEAR should be dry, which seems to make this option impractical, since it requires significant heat energy input to dry the algae to be used as a fuel to burn for heat energy. Even if the PEAR is dried by the sun, there is a significant energy cost in spreading out the PEAR and collecting after drying. Cofiring algal biomass with coal to generate power was considered by Kucukvar and Tatari (2011), who performed a life cycle assessment on this technology and raised several concerns, especially with regard to the comprehensive consideration of the culture of the algae and the drying of algae, and found that cofiring coal with algae dried with natural gas consumes relatively more non-renewable resources than 100 % coal firing. These issues, particularly the water content, with PEAR make burning it directly as a fuel a relatively poor option, especially since the value per ton as a poor-quality fuel would be significantly less than if it were to be used in most of the other applications discussed elsewhere in this chapter.

7.2 Biogas Production

PEAR can be used in subsequent bioprocesses to generate other fuels, including methane fermentation to produce biogas. Dębowski et al. (2013) have found there are a number of challenges with converting algal biomass into biogas, including the resistance of the cellulose in cell walls to degradation and the production of compounds that are toxic to anaerobic bacterial that occurs in some algal species. He also found that the carbon to nitrogen ratio found in the algal biomasses studied were typically below 10:1, while fermentation tanks should have ratios of at least 20:1. This issue can be addressed by co-fermenting with a biomass with a higher carbon to nitrogen ratio if available. It should be noted that this process creates a post-fermentation sludge as a byproduct which may be used as a fertilizer or recycled as a substrate into the biorefinery. The effectiveness of this conversion is also species-dependent, which is likely to be due to the proportions of the major components as well as variation in the cell wall structure of various species (Dębowski et al. 2013).

Algal biomass, particularly the carbohydrates, can also be converted into hydrogen using dark fermentation of the biomass using thermophilic bacteria. However, to ensure optimal conversion, it was found that pre-treatments were necessary including heating and sonication and adding hydrochloric acid and hydrogen peroxide to make the simple sugars readily available to the microbes. Roy et al. (2014) have conducted studies on this technology using a mixed thermophilic culture predominantly consisting of the *Thermoanaerobacterium* sp. genus which was fed with biomass from cultivated *Chlorella sorokiniana*, whereas Nguyen et al. (2010) conducted studies feeding *Thermotoga neapolitana* from cultures of *Chlamydomonas reinhardtii*. Subash and Mohan (2014) used a deoiled cake from a mixed microalgal culture with an acidogenic biocatalyst resulting in the production of hydrogen gas and volatile fatty acids with a high Chemical Oxygen Demand (COD) removal suggesting the technology may be useful for wastewater treatment.

7.3 Heavy Metal Adsorption

Algal biomass can also be used to remove heavy metal contamination. Lesmana et al. (2009) have reviewed the biosorbent potential of a variety of agricultural and algal biomasses, and has noted that the carboxyl, hydroxyl, sulfate, and amino groups in algal cell-wall polysaccharides serve as potential binding sites for metals with a strong dependence of the biosorption effectiveness on the chemistry between the contaminant and the biomass as well as the pH. Davis et al. (2003) have published a comprehensive review on the biochemistry and biosorbency of brown algal biomass and has found brown algal biomasses to be efficient and resilient, and specifically that that the orders of *Laminariales* and *Fucales* have an abundance of cell wall matrix polysaccharides and extracellular polymers that act as effective biosorbents. In addition to the option of using PEAR directly as a bioadsorbent, Zhang H et al. (2010) have shown that the biomass can be converted into activated charcoal, which is a powerful adsorbent for a number of established applications.

7.4 Pyrolysis and Hydrothermal Liquefaction

Pyrolysis uses high temperatures (typically over 400 °C) to thermally decompose dry biomass into gaseous, liquid, and solid phases. While pyrolysis can be catalyzed by a number of heavy metals which are often toxic, it was noted by Xu et al. (2014) that alkali and alkaline metal compounds can also be used to reduce the initial temperatures and activation energies for the pyrolysis of microalgal biomass. Watanabe et al. (2014) performed a study on the pyrolysis of *Botryococcus braunii* and found that in addition to the extracted oil, the biomass can be pyrolytically converted into fuels that are more valuable as fuels than biomasses from other species due to the relatively high content of C, H, and volatile matter with relatively lower ash contents. Although it is too early to confidently predict which technologies will be the most successful in biorefineries in the future, Zhang et al. (2014b) has noted that hydrothermal liquefaction (HTL) has been shown to yield more net energy and a lower economic cost for nutrient recovery than pyrolysis. Vardon et al. (2012) compared slow pyrolysis and HTL technologies for the conversion of algal biomasses, including PEAR, and found that HTL yields more bio-oil than slow pyrolysis.

In HTL, the biomass is heated in water to subcritical temperatures (below 374 °C). In addition to requiring lower temperatures, utilizing the biomass in a wet state allows for a major cost savings over pyrolysis by eliminating the high energy cost of drying. In HTL, the proteins and carbohydrates in a biomass can be converted into oils in addition to the lipids. For instance, Matsui et al. (1997) have been able to achieve a 78.3 % oil yield from HTL using *Spirulina* biomass which is low in lipid content using an iron catalyst. Algal biomass has been considered for use in jet fuels, and Fortier et al. (2014) has shown that this should be possible to accomplish with lower life cycle greenhouse gas emissions compared to conventional jet fuels. As with many processes for PEAR, the products of HTL can be improved through pre-treating, but ultimately a cost/benefit analysis should be considered before implementing additional pre-treatment processes. Chen et al. (2014) has found that centrifugation and ultrasonication pre-treatments were useful for significantly reducing ash contents, reducing the apparent activation energy of decomposition, and increasing the bio-crude oil yield from 30 to 55 %. Ruiz et al. (2013) have published a review on HTL technology as applied to lignocellulosic materials and aquatic biomass, and notes that the HTL treatment of the different biomasses are significantly different since microalgal biomasses are not lignocellulosic and typically will not contain any cellulose, and the HTL process will work primarily on the major constituents of proteins, carbohydrates, and lipids. Barreiro et al. (2013) have also reviewed the HTL processing of microalgal biomass and discusses some challenges. These challenges include the issue that the optimal process conditions depend on the biomass feedstock composition, including particle size, which may be variable. Beyond giving non-optimal products, variations in composition can cause precipitation which can cause clogging of lines or catalyst poisoning. Also, since the reaction pathways and kinetics are still poorly understood, HTL processes are difficult to design and optimize.

7.5 Recycle to Biorefinery

In some algal biorefinery applications, it may be advisable to recycle some or all of the PEAR into the algal culturing operations as a fertilizer and nutrient source, especially if other coproducts cannot be profitably produced from the PEAR, or if other fertilization options are expensive due to a high price of available fertilizers or a high nutrient demand in a given culture. Moon et al. (2014) has shown that combining the hydrolysate of PEAR could be used as a nutrient source that could be combined with sugar factory waste which would serve as a carbon source for the cultivation of *Ettlia* sp. Zhang et al. (2014b) considered using PEAR for energy and nutrient recycling through anaerobic digestion and hydrothermal liquefaction (HTL). Anaerobic digestion can be used to produce methane gas and the nutrient-rich effluent, especially the liquid phase, may be recycled to the algae culturing operation. When processing PEAR via HTL, the biomass is reacted in water at subcritical temperatures to create bio-crude oil in the liquid phase along with the creation of gaseous and solid (char) phases. The bio-crude oil and the char may be burnt for energy recovery, and the carbon-rich gas can be recycled into the biorefinery. Part of the liquid stream may be used to recycle nitrogen and phosphorous back to the algal culture. However, the amount of the liquid stream that can be recycled may be limited by compounds in the stream generated in the HTL process that may be toxic to the microalgae. The amount of liquid that can be recycled will be largely variable depending on the HTL process conditions and the sensitivity of the given culture species and conditions. Zhang et al. (2014b) also urges careful consideration for the use of catalysts in HTL, because while they may be useful in the HTL process, the elements found in catalysts are often toxic to microalgae as well as humans and can contaminate the liquid stream. With these considerations, Zhang et al. (2014b) concluded that with the state of technology at the time of publication, anaerobic digestion had a better potential for energy and nutrient recovery.

8 Conclusion

This chapter discusses a variety of options for material and energy coproducts that may be derived from PEAR. The best solution for PEAR a given biorefinery is likely to be variable and depend on the microalgal culture, the processes used (especially how the lipids are extracted), the capacity of the biorefinery, and the demand for products locally and globally. Also, as is the case in the petroleum industry, there may not be a single coproduct that can maximize the profitability of a biorefinery.

For a large biorefinery operation, it may be profitable to extract the high-value components first, such as pigments and pharmaceutical components. The remaining PEAR may be converted into a glue for particleboard composite materials. This may be a particularly good option if there are agricultural or forest byproducts nearby that may be used as a filler. It may be useful to separate the higher molecular weight proteins to be used for the glue, and use the remaining portion for animal feed or fertilizer. If only some components of the PEAR are useful for feed, the rest may be converted into low-grade fuels through the HTL process, or portions of the PEAR at any point may be recycled to the biorefinery as a nutrient source. However, all of the separations and processes add expenses that may only be justifiable for a large-scale operation. Whereas for a smaller scale, it may be preferable to just use a process which consumes all of the PEAR, such as the conversion to glue, which can utilize all of the PEAR without generating additional byproducts and yield a reasonable revenue per ton of biomass.

The optimal usage of PEAR will be necessary to create a viable and profitable biorefinery. Correspondingly, the creation of valuable coproducts from PEAR must be a part of the comprehensive design of these systems. With this having been stated, the processes used to derive products from the biomass should be carefully designed to balance the value derived from the primary products, presumably lipids for biofuels, and those of the coproducts. For instance, the use of multiple extractions at higher temperatures may yield more lipids for biofuels, but will reduce the quality of the proteins in the PEAR for use in glues. Other considerations might include the use of heavy metal catalysts in processes such as HTL, which may yield favorable amounts of products, but risk the contamination of products with the heavy metals, making them unfit for nutrient recycling in the biorefinery as well as for plant or animal feed. Thus, when planning a biorefinery operation, all of the processes should be designed with consideration of all of the products and coproducts to optimize the economic viability of the system.

List of Acronyms

PEAR	Post-extracted algal residue
MOE	Moduli of elasticity
MOR	Moduli of rupture
HTL	Hydrothermal liquefaction

References

Barreiro DL, Prins W, Ronsse F, Brilman W (2013) Hydrothermal liquefaction (HTL) of microalgae for biofuel production: state of the art review and future prospects. Biomass Bioenerg 53:113–127

Batista AP, Gouveia L, Bandarra NM, Franco JM, Raymundo A (2013) Comparison of microalgal biomass profiles as novel functional ingredient for food products. Algal Res 2:164–173

Bryant HL, Gogichaishvili I, Anderson D, Richardson JW, Sawyer J, Wickersham T, Drewery ML (2012) The value of post-extracted algae residue. Algal Res 1:185–193

Chen W, Ma J, Zhang Y, Gai C, Qian W (2014) Physical pretreatments of wastewater algae to reduce ash content and improve thermal decomposition characteristics. Bioresour Technol 169:816–820

Davis TA, Bolesky B, Mucci A (2003) A review of the biochemistry of heavy metal biosorption by brown algae. Water Res 37:4311–4330

Dębowski M, Zieliński M, Grala A, Dudek M (2013) Algae biomass as an alternative substrate in biogas production technologies—review. Renew Sust Ene Rev 27:596–604

Dufossé L, Galaup P, Yaron A, Arad SM, Blanc P, Murthy KNC, Ravishankar GA (2005) Microorganisms and microalgae as sources of pigments for food use: a scientific oddity or an industrial reality? Trends Food Sci Tech 16:389–406

FAO (2012) The state of world fisheries and aquaculture 2012. Food and Agriculture Organization, Rome, p 209. ISBN 1020-5489

Fortier MP, Roberts GW, Stagg-Williams SM, Sturm BSM (2014) Life cycle assessment of bio-jet fuel from hydrothermal liquefaction of microalgae. Appl Energy 122:73–82

Han W, Clarke W, Pratt S (2014) Composting of waste algae: a review. Waste Manage 34:1148–1155

Hasan MR, Chakrabarti R (2009) Use of algae and aquatic macrophytes as feed in small-scale aquaculture: a review. FOA Fisheries and Aquaculture Technical Paper, Rome, p 15

Ionescu M, Petrovic ZS (2009) Cationic polymerization of biological oils with superacid catalysts. US Patent 7501479

Kucukvar M, Tatari O (2011) A comprehensive life cycle analysis of cofiring algae in a coal power plant as a solution for achieving sustainable energy. Energy 36:6352–6357

Laucks I, Davidson G (1932) Vegetable protein glue and process of making. US Patent 310427, 19 Apr 1932

Lesmana SO, Febriana N, Soetaredjo FE, Sunarso J, Ismadji S (2009) Studies on potential applications of biomass for the separation of heavy metals from water and wastewater. Biochem Eng J 44:19–41

Lum KK, Kim J, Lei XG (2013) Dual potential of microalgae as a sustainable biofuel feedstock and animal feed. J Anim Sci Biotech 4:53

Makkar HPS (2012) Biofuel co-products as livestock feed: opportunities and challenges. Food and Agriculture Organization (FAO) of the United Nations, Rome. ISBN 978-92-5-107299-8

Matsui T, Nishihara A, Ueda C, Ohtsuki M, Ikenaga N, Suzuki T (1997) Liquefaction of microalgae with iron catalyst. Fuel 76(11):1043–1048

Moon M, Kim CW, Farooq W, Suh WI, Shrivastav A, Park MS, Mishra SK, Yang J (2014) Utilization of lipid extracted algal biomass and sugar factory wastewater for algal growth and lipid enhancement of *Ettlia* sp. Bioresour Tech 163:180–185

Nguyen TD, Kim K, Nguyen M, Kim MS, Kim D, Sim SJ (2010) Enhancement of fermentative hydrogen production from green algal biomass of Thermotoga neapolitana by various pretreatment methods. Int J Hydrogen Energy 35:13035–13040

Rathke J, Sinn G, Harm M, Teischinger A, Weigl M, Muller U (2012) Effects of alternative raw materials and varying resin content on mechanical and fracture mechanical properties of particle board. Bioresources 7(3):2970–2985

Richardson JW, Johnson MD, Lacey R, Oyler J, Capareda S (2014) Harvesting and extraction technology contributions to algae biofuels economic viability. Algal Res 5:70–78

Roy S, Kumar K, Ghosh S, Das D (2014) Thermophilic biohydrogen production using pre-treated algal biomass as substrate. Biomass Bioenerg 61:157–166

Ruiz H, Rodriguez-Jasso RM, Fernandes BD, Vicente AA, Teixeira JA (2013) Hydrothermal processing, as an alternative for upgrading agriculture residues and marine biomass according to the biorefinery concept: a review. Renew Sust Energ Rev 21:35–51

Subash GV, Mohan SV (2014) Deoiled algal cake as feedstock for dark fermentative biohydrogen production: an integrated biorefinery approach. Int J Hydrogen Energy 39:9573–9579

Vardon DR, Sharma BK, Blazina GV, Rajagopalan K, Strathmann TJ (2012) Thermochemical conversion of raw and defatted algal biomass via hydrothermal liquefaction and slow pyrolysis. Bioresour Tech 109:178–187

Watanabe H, Li D, Nakagawa Y, Tomishige K, Kaya K, Watanabe MM (2014) Characterization of oil-extracted residue biomass of *Botryococcus braunii* as a biofuel feedstock and its pyrolytic behavior. Appl Energy 132:475–484

Wool R, Sun X (2005) Bio-based polymers and composites, 1st edn. Elsevier Academic Press, Burlington

Xu Q, Ma X, Yu Z, Cai Z (2014) A kinetic study on the effects of alkaline earth and alkali metal compounds for catalytic pyrolysis of microalgae using thermogravimetry. App Therm Eng 73:355–359

Yaakob Z, Ali E, Zainal A, Mohamad M, Takrif MS (2014) An overview: biomolecules from microalgae for animal feed and aquaculture. J Bio Res 21:6

Zhang H, Tang Y, Cai D, Liu X, Wang X, Huang Q, Yu Z (2010) Hexavalent chromium removal from aqueous solution by algal bloom residue derived activated carbon: equilibrium and kinetic studies. J Hazard Mater 181:801–808

Zhang M, Li R, Cao L, Shi J, Liu H, Huang Y, Shen Q (2014a) Algal sludge from Taihu Lake can be utilized to create novel PGPR-containing bio-organic fertilizers. J Environ Manage 132:230–236

Zhang Y, Kendall A, Yuan J (2014b) A comparison of on-site nutrient and energy recycling technologies in algal oil production. Resour Conserv Recy 88:13–20

Utilization Alternatives of Algal Wastes for Solid Algal Products

Didem Özçimen, Benan İnan, Sevgi Akış, and Anıl Tevfik Koçer

Abstract Recently, there is a growing interest in utilization of algae to produce biofuels and valuable products. In order to use algal biomass effectively and decrease cost of the algal processes, researches on utilization of algal wastes for different purposes are carried out. Valuable algal products can be obtained from algal wastes and used in energy, food and environmental applications. Thermochemical processes are the most common methods for conversion of algal wastes to solid products such as algal biochar and algal activated carbon which can be used as energy source, adsorbent and soil improver. This chapter is especially focused on investigation of conversion potentials of algal wastes, which remain after various industrial processes and producing valuable solid products and also direct usage areas of algal wastes.

Keywords Microalgae • Macroalgae • Biomass • Biochar • Activated carbon • Biosorbent • Adsorbent • Animal feed • Carbonization • Adsorption • Biofuel • Pyrolysis • Biosorption • Algal wastes • Biofertilizer • Soil amendment • Energy source • Renewable source • Algal products • Seaweed

1 Introduction

Algae have come into prominence over the last decade as a commercial biofuel feedstock with their appropriate structural properties and high production capacity. Algae are photosynthetic organisms which can use solar energy to fix carbon dioxide in the atmosphere and produce glucose as a main product, then convert it to other important components such as lipids. Because of their flexible metabolic structures, algae can use three main carbon fixation metabolism (C_3, C_4 and crassulacean acid metabolism) and possess the enzyme pyruvate formate lyase which enables fermentation in the absence of oxygen (Anoop et al. 2011). Owing to these characteristics, they can utilize sun or artificial light for photosynthesis and so they can grow everywhere such

D. Özçimen (✉) • B. İnan • S. Akış • A.T. Koçer
Faculty of Chemical and Metallurgical Engineering, Bioengineering Department, Yıldız Technical University, 34220 Istanbul, Turkey
e-mail: ozcimen@yildiz.edu.tr

A. Prokop et al. (eds.), *Algal Biorefineries*, DOI 10.1007/978-3-319-20200-6_12

as lakes, seas and also deserts. Algae can be cultivated phototrophically or heterotrophically according to their sort of substrate. Phototrophic algae convert carbon dioxide in the atmosphere to nutrients such as carbohydrate. On the contrary, in addition to maintain the basic functional reactions, heterotrophic algae utilize organic carbon sources such as glucose to continue their existence (Wen and Chen 2003). Algae can be manipulated by applying different cultivation systems such as open ponds or closed systems (photobioreactors). Even though algae can survive in hard conditions, algal biomass yield is affected from some growth parameters like pH, light penetration, carbon dioxide and mixing. These parameters should be adjusted carefully in cultivation systems to produce algal biomass efficiently and with high productivity (Bitoga et al. 2011). Algae can be divided in two main group as macroalgae and microalgae. Microalgae are cellular microorganisms as their name implies and they can be found in unicellular form or simple colony structure. They can produce high amount of lipid under stress conditions to protect themselves and extend their life time and their lipid content may reach to 70 % of their dry weight under these conditions (Mata et al. 2010). Generally, they double their biomass in 3.5–24 h with high growth rate. Therefore, they have higher productivity than conventional foresty, agricultural products and aquatic plants. They need smaller places for cultivation in open ponds or photobioreactors compared to other raw materials requiring larger fields for cultivation. Also it is possible that they can be modified genetically according to their usage areas (Radakovits et al. 2010). Due to these outstanding features against traditional biodiesel feedstock, they gain importance on biodiesel production. In addition to biodiesel production, they can be used for producing bioethanol (Vergara-Fernandez et al. 2008), biogas (Yen and Brune 2007), biobutanol (Nakas et al. 1983) and other valuable products like omega-3 oil, eicosapentaenoic acid and pigments (Belarbi and Molina 2000; Cheng-Wu et al. 2001; Besada et al. 2009) which can be utilized in food, pharmaceutical and cosmetic industry (Spolaore et al. 2006; Olaizola 2003). In contrast to microalgae, macroalgae which are known as seaweeds, are plants that usually exist in coastal areas of the seas. They are classified within as green, red and brown macroalgae accordingly their pigments (Jung et al. 2013). Algae are important carbon sources for also utilization in biorefineries to obtain different products. Unlike their appearances, their morphologic and physiological features and chemical compositions of algae are different from terrestrial plants. They are different from lignocellulosic materials with having less or no lignin in their structures (Daroch et al. 2013). While microalgae are used for biodiesel production with their high lipid contents, macroalgae have high carbohydrate content which can be utilized for bioethanol production (Olaizola 2003; Özçimen et al. 2012). Despite of their great potential, obtaining algal biofuels and other products requires more energy and input and their production processes are quite expensive. In order to overcome production cost, researches are performed to utilize algal wastes which remain after various processes. The most common utilization choices of these wastes are anaerobic digestion, thermochemical conversion and direct usage (Rashida et al. 2013). While biogas production from algal wastes is mostly preferred utilization method, biochar and biosorbent production from solid algal wastes are underrated. Figure 1 shows the utilization ways of algal wastes.

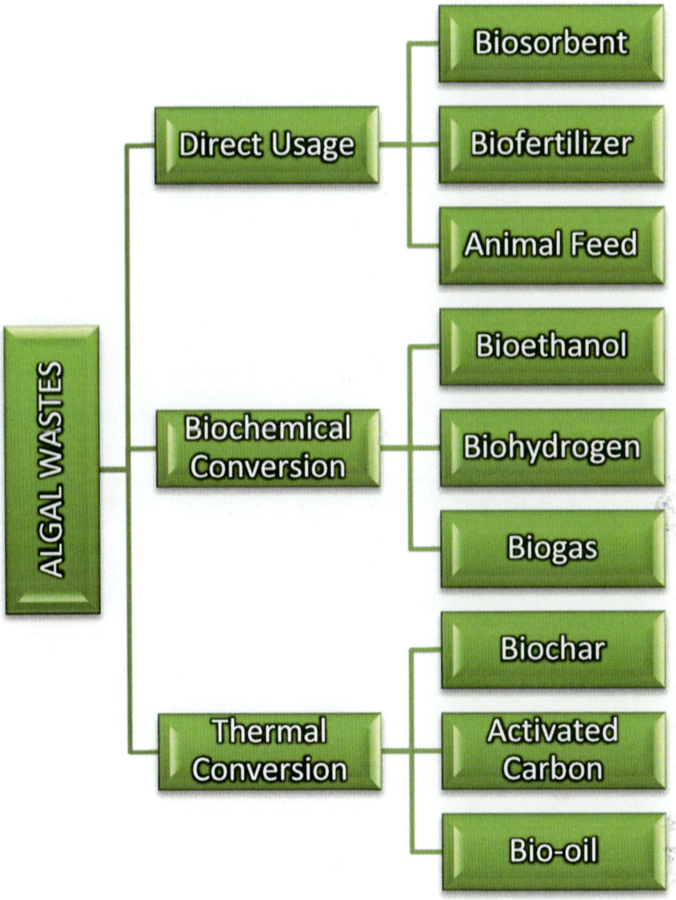

Fig. 1 The utilization ways of algal wastes

The aim of this chapter is investigation of conversion potentials of algal wastes which remain after various industrial processes, to obtain valuable solid products (biochar, biosorbent, animal feed etc.).

2 Utilization of Algal Wastes

2.1 The Usage of Algal Wastes for the Production of Biochar

Various types of wastes are utilized for many years via thermochemical processes which include carbonization, pyrolysis and gasification methods. The main advantage of this utilization type is to be able to use different type of biomass having

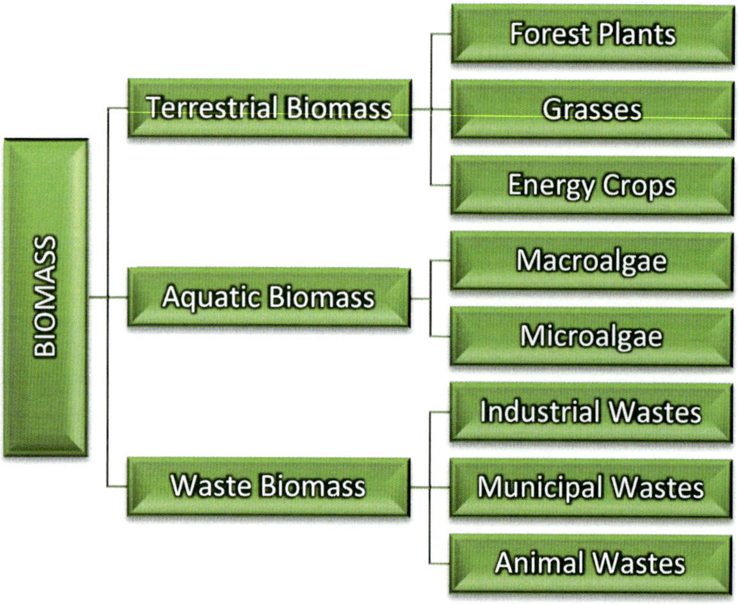

Fig. 2 The classifications of biomass

composition variety. Therefore, these processes also can be suggested for utilizing algal wastes remaining after different applications with varying composition. Carbonization or slow pyrolysis process which is one of the conversion methods and it degrades dry biomass in oxygen-free environment thermochemically to biochar (Murthy 2011).

Biochar is a solid product having high carbon content and obtained by thermal decomposition of organic material subjected to low temperature (<700 °C), in the absence of air (Lehmann and Joseph 2009). Various organic materials can be used for biochar production, such as woody material, algae, grasses, corn stover, straw, peanut shells, sorghum, olive pits, bark, and sewage wastes. Scientific researches on biochar are especially related with wood materials due to its consistency and relatively low ash content. The sector of forest/wood products is the primary source of biochar raw materials (Winsley 2007).

All plant and animal substances having carbohydrate content are called biomass energy sources. Although there is a wide variety of biomass sources, biomass can be usually classified as terrestrial, aquatic biomass sources and other type of wastes. Figure 2 shows the classifications of biomass (Özçimen D 2007).

Biochar is generally composed of aromatic structures and it is similar to the structure of graphite, but it shows the difference with uneven settlement of the aromatic ring (Lehmann and Joseph 2009). Carbon in the feedstock is converted to stable and unstable biochar structures as the final product. Because of having the aromatic ring structure, the stable biochar can remain in the soil for hundreds of years, whereas unstable compounds remain in the soil for weeks or years, depending on climate changes (Jirka and Tomlinson 2014). Besides the usage of biochar as an energy source, charcoal has been used for centuries for improving soils. Charcoal can be found in many places of the world due to forest fires and historical soil improvement practices. Use of charcoal in soils of Amazon is believed to date back at least 2000 years and caused the formation of nutrient-rich and dark *Terra Preta* soils ('black earth of the Indian' is called in Portuguese). These lands show differences from adjacent soils significantly with having high carbon content and efficiency. These *Terra Preta* soils, even today and even in harsh climatic conditions, is still fertile due to the presence of biochar as a resulting of retaining carbon in soils for long periods. Biochar usage to improve soil as traditional farming practices is not only applied in the Amazon, also applied in Japan, China, Africa, North America and Europe (Jirka and Tomlinson 2014).

2.1.1 Production of Biochar

Biomass resources can be used directly or after many conversion processes for producing energy. Thermochemical conversion processes are used for producing gas, liquid and solid products. These products are used for various applications (Fig. 3). Carbonization and pyrolysis are the thermochemical processes applied for producing liquid, solid and non-condensable gases products. In a low temperature and low heating rate conditions, high solid product yield is obtained and this process is called carbonization. Thermal decomposition of biomass under absence of air and inert atmosphere is called as the process of carbonization and a product that has high carbon content "biochar" is obtained at the end of this process. If low temperature and high heating rate conditions is applied for maximizing the yield of liquid product, this process is called pyrolysis (Bridgwater and Bridge 1991).

Biochar surface has hydrophilic, hydrophobic, acidic and basic properties. Surface of biochar has various functional groups, such as pyranone, phenolic, carboxylic, amine and lactone groups (Topsak 2011). Macroporous structure of biochar depends on the architectural structure of the raw material used (Sohi et al. 2010). The type of feedstock (organic material) and the production conditions such as temperature and heating rate affect the quality of biochar in soil amendment applications significantly. (McClellan et al. 2007; McLaughlin et al. 2009).

There are not so many studies on biochar production from algae yet, however the number of studies gradually increases. Studies that mentioned above, are summarized in the Table 1.

Effect of Carbonization Parameters

Factors that affect carbonization of biomass can be analyzed under two titles as characteristics of biomass and process parameters. Process parameters consist of temperature, heating rate, properties of gas atmosphere (inert gas, reactive gas, the

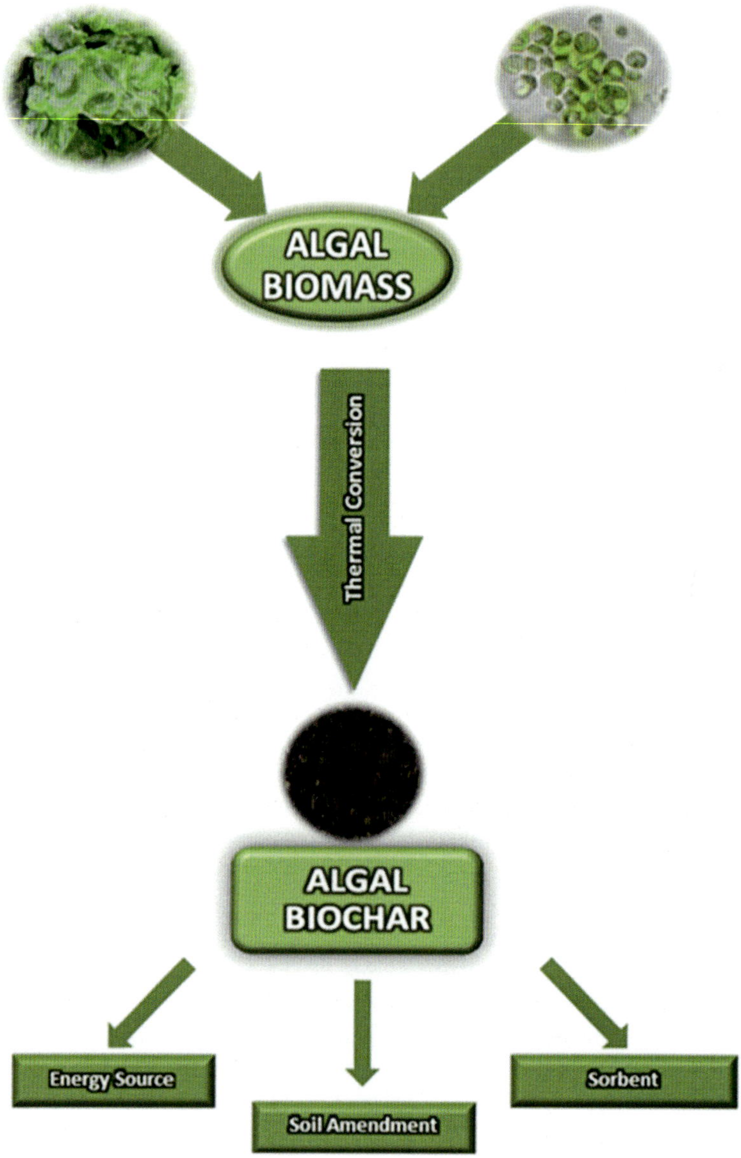

Fig. 3 Production and usage of algal biochar

pressure), catalyst, residence time in the reactor and the reactor geometry (Özçimen D 2007) There are various researches on biochar production from algae under different conditions and characterization of algal biochar.

The heating rate is important for the structure of biochar. For low heating rates, volatile components are released from the solid structure and no important change

Table 1 Conditions and aims of studies on algal biochar

Algae	Pre-treatment	Conditions	Aim	Reference
Chlorella sp.	washed twice with distilled water to remove salt and then dried at 80 °C for 12 h.	N_2 flow rate: 100 mL/min	Comparison of algal biochars that produce in two different reactors	Kirtania et al. 2014
		Temperature: 30–200 °C (heating rate of 5 K/min), 200–1000 °C (heating rate of 10 K/min)		
Chlorella vulgaris, Dunaliella salina	dried at 45 °C in a nitrogen atmosphere	N_2 flow rate: 400 mL/min	Investigation on pyrolysis products obtained from microalgae	Gong et al. 2014
		Temperature: 300–700 °C		
		Holding time: 20 min		
Saccharina japonica	dried at 105 °C for 24 h	N_2 flow rate: 600 mL/min	Research of yields of biochar and bio-oil obtained from algae	Choi et al. 2014
		Final temperature: 450 °C		
Oedogonium sp. *Cladophora vagabunda* *Tarong polyculture* *Ulva ohnoi*	Wash and dry	N_2 flow rate: 20 mL/min	Comparative assessment of the thermochemical conversion of freshwater and marine micro- and macroalgae	Kan et al. 2014
		Temperature: 1000 °C		
		Heating rate: 10 and 60 °C/min		
Gelidium sesquipedale		N_2 flow rate: 150 mL/min	Investigation of pyrolysis characteristics of macroalgae	Ferrera-Lorenzo et al. 2014a
		Temperature: 750 °C		
		Heating rate: 5 °C/min		
Laminaria digitata, Fucus serratus and mix macroalgae species	Washed in water and dried in oven at 60 °C	N_2 flow rate: 0.64–0.72 m^3/h	Comparison of bioproduct yields obtained algal biomass	Yanik et al. 2013
		Final temperature: 500 °C, Heating rate : 10 °C /min		
Undaria pinnatifida	Wash and dry	N_2 flow rate: 50 mL/min	Macroalgal biochar was for removal of copper	Cho et al. 2013
		Final temperature: 500 °C		
		Holding time: 1 h		
Sagarssum sp.	–	N_2 flow rate: 25 mL/min	Yields of biochar that produced under different conditions from macroalgae	Kim et al. 2013
		The heating rates: 5, 10, 15, and 20 °C / min		
		Final temperature: 800 °C.		

(continued)

Table 1 (continued)

Algae	Pre-treatment	Conditions	Aim	Reference
Prasiola crispa, Monostroma arcticum, Polysiphonia arctica, Devaleraea ramentacea, Odonthalia dentata, Phycodrys rubens, Sphacelaria plumosa Gigartina skottsbergii, Plocamium cartilagineum, Myriogramme manginii, Hymencladiop-sis crustigena, Kallymenia antarctica	Dried at 70 °C for 24 h	He flow rate: 50 mL/min	Investigation of thermo-chemical behaviour and chemical product formation	Kebelmann et al. 2013
		Temperature: 900 °C		
		Heating rate: 100 K/min		
Spirulina spp. *Spirogyra* spp. *Cladophora* spp.		N_2 flow rate: 30 mL/min	Analysis of yields of biochar that produced from different algae species	Chaiwong et al. 2012
		Temperatures: 50–135 °C		
		(10 °C/min.); constant at a 135 °C		
		(5 min.); 135–900 °C (100 °C/min)		
Dried algae	Washed with distilled water and air-dried	N_2 flow rate: 800 mL/min	Investigation of biochar yields obtained from different biomass feedstocks	Ronsse et al. 2011
		Heating rate: 17 °C/min		
		Residence times : 10 min and 60 min		
Lyngbya sp. *Cladophora* sp.	Washed with distilled water and then dried in an oven at 45 °C for 24 h	He flow rate: 200 cm^3/min	Investigation of the yields of algal biochar that produced under similar conditions	Maddi et al. 2011
		Ar flow rate: 20 cm^3/min		
		Heating rate: 30 °C /min		
		Final temperature: 600 °C		

Cladophora coelothrix, Cladophora Patentiramea, Chaetomorpha indica, Chaetomorpha linum, Cladophoropsis s, Ulva flexuosa, Cladophora vagabunda, Caulerpa taxifolia	Unwashed, pressed dry and oven dried at 65 °C for 48 h	N_2 flow rate: 3.5 L /min	Analysis of biochar characteristics from eight different macroalgae species	Bird et al. 2011
		Heating period: 30–40 min		
		Final hold temperature: 307 ± 5, 414 ± 5, 450 ± 5 and 512 ± 5 °C,		
Tetraselmis chui, Chlorella like, Chlorella vulgaris, Chaetoceros muelleri, Dunaliella tertiolecta and Synechococcus	Dried at 70 °C for 3 h	Heating rate: 10 °C/min	Thermal characterization of macroalgal and microalgal biochar	Grierson et al. 2009
		Maximum temperature: 750 °C		
Fucus vesiculosus, Chorda filum, Laminaria digitata, Fucus serratus, Laminaria hyperborea, and Macrocystis pyrifera	Washed with distilled water and air-dried	N_2 and air flow rate: 50 mL/min	Investigation of macroalgal bioproduct yields	Ross et al. 2008
		Temperature: 40 °C to 950 °C		
		Heating rate: 25 °C/min		
Chlorella protothecoides, Microcystis aeruginosa		Biomass-feeding rate: 4 g/min	Investigation of biofuels obtained from microalgae	Miao et al. 2004
		N_2 flow rate: 0.4 m^3/ h		
		Temperature: 500 °C		
		Heating rate: 600 °C/s		
		Vapour residence time: 2–3 s		

happens in the particle structure. However, the originality of cellular structure is disappeared at high heating rates (Kirtania et al. 2014). Kirtania et al. (2014) and Ross et al. (2008) found similar results in their studies. Kirtania et al. accentuated that the reactivity of biochar decreases with increased heating rate. Ross et al. indicated that interaction between biochar yield and heating rate is similar to interaction between the reactivity of biochar and heating rate.

In the literature, there are many studies about the relationship between biochar and process temperatures. These studies shown that increase of temperature affects biochar yields and reactivity of biochar negatively. In these studies, the maximum yields of algal biochar were found that obtained at temperature of 350–550 °C. (Ross et al. 2008; Grierson et al. 2009; Bird et al. 2011; Maddi et al. 2011). Yanik et al. (2013) calculated biochar yields which were produced at temperature of 500 °C as 29–36 %. Chaiwong et al. (2012) calculated biochar distribution as 28–31 % at same temperature. Kirtania et al. (2014) interpreted that reactivity of algal biochar produced at temperature of 800–950 °C is similar with reactivity of woody biochar produced under similar conditions. However, they found that the reactivity of biochar decreases with increase of temperature.

Ronsse et al. found that surface area in biochar increases with an increase in temperature and residence time. At high temperature, they observed that the relationship between surface area and residence time is changed. However, the researcher found that the interaction of surface area with temperature didn't changed.

The Effect of Algae Properties on Biochar

Algal cell walls are similar to lignocellulosic materials due to consist of mainly cellulose and hemicelluloses. The ratio of cellulose and hemicellulose content in algae is lower in comparison with that in terrestrial plants and trees, even so, these substances make algae much resistant. Different from lignocellulosic materials, lignin which is a material acting like cement in the layers of the cell walls of plants and protect the structure from degradation, is not found in the algae except for *Ulva* sp. which is only at concentrations of 3 % dry weight (Kraan 2012). This structural difference causes variety in the thermal behavior of algae in the thermochemical conversions.

One of the most important parameters of carbonization process is the particle size of the biomass. If the particle size of biomass samples used in carbonization process increases, solid yield increases. Because, to decompose the large diameter of the particles completely is difficult and thermal degradation from the surface to the center of particles take time. When the decrease of particle size occurs, there is an increase in the yield of solid product. The reason of reduction of mass transfer resistance and impacts is the more uniform heating of small particles. Thus, mass loss of small particles is greater than big particles by the heat effect in the process of carbonization carried out in the same conditions and the solid product yield is reduced. In contrast, temperature profile that is formed in the coarse particles by heat transfer resistance make an increase in the efficiency of the solid product while the yield of volatile substances decreases (White and Plaskett 1981; Goldstein 1983; Knight 1976).

For this reason, to increase the yield of the solid product in carbonization process big particles may be preferred instead of smaller. Based on this information; an increase in the solid yield value can be observed when macroalgae biomass samples is preferred instead of microalgae (Özçimen D 2007).

Mineral content of algae is also an important parameter for thermochemical conversion. It affects both the conversion process and the product yield. Mineral content of algae can change according to their species, their cultivation sites, physical effects and chemical effects like wave exposure and process type, and mineralization method. It is higher than many terrestrial plants and nutrients from animals (Ruperez 2002). For instance, ash content of many vegetables varies from 5 to 10 g/100 g dry weight such as 10.4 for potato, 7.1 for carrot and 7.1 for tomato. Despite these values, some vegetables can have low or high mineral content, like sweet corn which has low ash content like 2.6 % and spinach which has 20.4 % mineral content (USDA 2001). Unlike these materials, macroalgae species which are also known as seaweeds have quite high ash content in comparison with land vegetables. Seaweed ash contains sodium carbonate, potassium carbonate, and calcium carbonate, and among seaweed species, brown seaweeds have the highest ash content as 30.1–39.3 %. Beside ash, almost all of the seaweeds also contain sulphate which varies from 1.3 to 5.9 % and trace elements which are heavy metals (As, Cd, Cu, Hg, Pb, Zn) (USDA 2001). Contents of these heavy metals are usually less enough for toxic limits which allowed in some countries. In addition to these, amounts of copper and zinc are also below for the maximum amount allowed in seaweeds for consumption by human in Japan and France (Ruperez 2002). In a study of classification of macroalgae; *Fucus vesiculosus, Chorda filum, Laminaria digitata, Fucus serratus, Laminaria hyperborea,* and *Macrocystis pyrifera*, it was found that these species have high amount of macrominerals (Na, K, Ca, P, Si, and Mg and trace elements (Fe, Zn, Mn, Al, and Cu) than vegetable plants (Ross et al. 2008)

Thermochemical conversion processes have not been used for utilization of algae generally due to high moisture content and high alkali metals. In these processes, amount of these minerals gain importance due to cause some problems like slagging, fouling and other ash related problems and restrict its usage as fuel in combustion and gasification. However, availability of alkali metals in algal biomass improves the biochar yield. Therefore, algae shouldn't be disregarded as biochar feedstock and should be considered for utilization (Haykiri-Acma et al. 2013; Ross et al. 2009).

2.1.2 Potential Usage Areas of Biochar

Biochar, that is called as solid product of thermal carbonization of biomass, is used in many areas such as energy, metallurgical and chemical and also it is used to produce adsorbents and activated carbon (Özçimen 2013).

Biochar as Energy Source

Biochars can be used as alternative of conventional fuels depending on their high fixed carbon content and calorific value (Özçimen 2013).

Nowadays, 2.4 billion people use traditional biomass such as wood and agricultural wastes directly as a fuel in rural and poor urban areas for heating and cooking so indoor air pollution affects these people worldwide, especially housewives. And greenhouse gases arise as a result of biomass combustion directly.

However, using biochar that is obtained with carbonization process of biomass instead of direct usage of biomass for heating and cooking, reduces damage of local and global environment (Özçimen 2007).

Biochar as Soil Improver

As well as usage of algal biochar to obtain energy, it can be utilized for carbon sequestration and soil improvement. Biochar can change the physical, chemical and biological properties of soil and can improve nutrient and water holding capacity and plant growth (Duku et al. 2011). While soil acidity causes aluminum and manganese toxicity for plants, it reduces the availability of calcium, magnesium, phosphorus and molybdenum and affects plant growth negatively. Biochar has usually alkaline properties and when it is applied to soil, it neutralizes the soil acidity and increases the pH (Yuan et al. 2011). Biochar which has high amount of nitrogen, shows up alkaline properties. Thus, pH content of biochar can vary according to components of biomass and carbonization conditions. In addition to having resistance to degradation, biochar modifies long term water holding capacity of soil with its macropore structure which shows the cell structure of raw material (Sohi et al. 2010). Porous structure and holding soluble organic carbon capability of biochar provide a habitat which microbiota feeds from organic carbon and it is protected from grazers. Because of this, microbial biomass and activity increase in soils which contain biochar (Beesley et al. 2011). Biochar also reduces availability of heavy metals and other organic contaminants by binding them to its surface (Atkinson et al. 2010; Beesley et al. 2011). There are some researches on biochar for soil improvement. Gaskin et al., have investigated the cation exchange capacity of different biochar samples on soil which they were produced from peanut shell pellets, pine shavings, pine sawdust pellets, pine bark and oak shavings. It was reported that biochars from peanut shell pellets had the highest cation exchange capacity and increased water holding capacity of soil (Gaskin et al. 2007). Rondon et al. studied the effects of different amounts of biochar application (0, 30, 60 ve 90 g kg^{-1}) on biological nitrogen fixation of beans. Nitrogen amount which was obtained from biological nitrogen fixation, increased from 50 to 72 % by adding 90 g kg^{-1} biochar. It improved pH of the soil and increased K, Ca and P availability in the soil (Rondon et al. 2007). Kwapinski et al., researched the effect of biochar on microbial composition. Biochar samples of willow, pine and miscanthus were mixed with clay soil and corn was planted.

Microbiological analysis revealed that an increase of microorganism amount was occurred in plant roots which were cultivated in soil that contains biochar (Kwapinski et al. 2010). Algae, mostly macroalgae are low in carbon content, however high in nitrogen, phosphorus and other nutrients content in comparison with terrestrial plants (Ruperez 2002; Ruperez et al. 2002; Ross et al. 2008). For this reason, it is expected that algal biochar have high mineral content and this makes it beneficial for soil amendment. On the other hand, nutrition content of algae can change between species and according to cultivation types and places. Grierson et al. investigated characterization and carbonization effect of the biochar and bio-oil fractions of marine algae, *Tetraselmis chui*. It was seen that biochar which obtained from slow carbonization, had high cation exchange capacity, high amount of N content and low C:N ratio and it was concluded that biochar is a valuable product for agricultural and has a potential of a net reduction of atmospheric CO_2 (Grierson et al. 2011). Bird et al. also investigated biochar of eight green algae species (*Cladophora coelothrix Kützing, Cladophora patentiramea (Montagne) Kützing, Chaetomorpha indica (Kützing) Kützing, Chaetomorpha linum (O.F. Müller) Kützing, Cladophoropsis* sp., *Ulva flexuosa Wulfen, Cladophora vagabunda (Linnaeus) Hoek*). It was found that biochar samples had low carbon and high nitrogen content and minerals. Their study has showed that algal biochar is suitable for both use in soil amendment and long-term carbon sequestration (Bird et al. 2011).

2.2 *Utilization of Algal Wastes for Other Applications*

2.2.1 Production of Activated Carbon

Activated carbon, known as activated charcoal, is a form of carbon that has been processed with oxygen to create pores. Activated carbon has a large surface area and it is a porous material that removes organic compounds from liquids and gases (Özçimen 2007). Since 3750 BC activated carbon is used in various fields by humans. In the early years, activated carbon was used in production of bronze and the elimination of the unpleasant odor. Nowadays, activated carbon that being used in the pharmaceutical industry, is used in a variety of industries including gas adsorption, fat and oil removal, dry cleaning and much more. Likewise, activated carbons are used in water treatment to remove organic compounds that produce carcinogens during the disinfection of water (Çeçen 2011).

Properties and quality of activated carbon depends on the intended use. Generally, activated carbon must have a large surface area and porous structure. In addition to, some properties such as density, particle size distribution, mesh size and ash content, are important for use. Pore structure of activated carbon determines the adsorption capacity; chemical structure determines the interaction with polar or non-polar chemicals; active regions determine the viability of the chemical adsorption (Özçimen 2007).

Production Methods of Activated Carbon

For the production of activated carbon, all the carbon-rich substances can be used by various activation procedures. Therefore, the raw materials activated with various chemicals under different conditions (Özçimen 2007). Activated carbon is produced through two different processes: physical activation and chemical activation. The physical activation process known as steam activation is known for yielding higher quality activated carbon. However, it is more expensive and requires industry manufacturing due to the amount of heat necessary. Physical activation and pyrolysis processes at a high temperature (usually 600–900 ° C) create char product. Then, the raw material is exposed to different gases such as argon and nitrogen. As a last step, the char is oxidized or activated at temperatures above 600–1200 ° C, blasted by steam. (Azner 2011)

Chemical activation is the preferred method because of its shorter production time and lower temperatures required. During chemical activation, the source material is impregnated with certain chemicals, typically an acid, a strong base or a salt such as phosphoric acid, potassium hydroxide, calcium chloride and zinc chloride. The raw material is carbonized at a low temperature, usually 450–900 ° C. It is believed that the activation steps proceed simultaneously in this process (Aygün 2002).

Usage Areas of Activated Carbon

Activated carbon is utilized in a number of industries because of its purification properties. The biggest application of activated carbon is in the purification of water and potable water treatment. It is used in a variety of water treatment industries, from municipal water supply treatment, wastewater treatment, swimming pools, aquariums and even home filtration systems. The second application of activated carbon is air purification. Activated carbon is also used to control potentially harmful, environmentally damaging, unpleasant odors in a number of environments, including homes and manufacturing facilities. The food industry uses activated carbon in as part of various processes, such as the decolorization of sugar, purify organic compounds, chlorine removal, decaffeination and many other practices. In medical industry, activated carbon can be found in almost every hospital or clinic in the world. It is used for odor control, filtration, respiration masks and also wound dressing. The vapor and liquid phase applications of activated carbon are shown in Table 2 (Özçimen 2007).

Biochars which are produced from algal wastes, can be used as activated carbon later by applying chemical activation. In literature, there are just a few studies on activated carbon which is produced from algal biochars. Ferrera-Lorenzo et al. (2014b) produced activated carbon from *Gelidium sesquipedale* solid residue by chemical activation with KOH. For chemical activation, two different heating methods were used: traditional chemical activation and microwave chemical activation. The best activated carbons from algae were obtained with traditional chemical activation. Under optimum activation conditions, S_{BET} and pore volume were seen as

Table 2 Applications of activated carbon

		Applications of activated carbon
Vapor Phase Applications	Solvent recovery	Solvent recovery for process economy and control of gas: acetate fibers (acetone), pharmaceuticals, film coating and paint industry applications.
	Carbon dioxide production	Purification of carbon dioxide: amines, mercaptans and alcohols adsorption.
	Industrial ventilation	Adsorption of organic vapors.
	Disposal of hazardous waste	Removal of heavy metals from flue gases.
	Cigarette production	Removing some of the harmful elements in cigarette smoke.
	Composite fibers	Impregnation of powdered activated carbon in foam/fiber components
	Odor neutralizer	Elimination of unpleasant odors: filter units.
Liquid Phase Applications	Potable water procedures	The removal of organic compounds and for removal of malodors.
	Non Alcoholic beverages	Obtaining process water, sterilization with chlorine, removal of organic compounds and removal of malodors.
	Recovery of Gold	Leaching process: gold recovery in sodium cyanide
	Petrochemistry	The removal of hydrocarbon and oils.
	Underground Water	Removal of unwanted substances in underground water
	Industrial Waste Water	Released water during the process to be suitable for the environment.
	Swimming Pools	Control of Chloramine level
	Semiconductors	Production of high purity water
	Alcohols	Production of process water: removal of phenol and trihalomethanes.

2118 m^2/g and 1.14 cm^3/g, respectively. Salima et al. (2013) investigated utilization of activated carbons were produced from *Ulva lactuca* and *Systoceira stricta* for the removal of hazardous cationic dyes. In addition to, they measured adsorption parameters such as pH, temperature, contact time and ionic strength. As a result of this study, algal activated carbons are an alternative to commercially available adsorbent used for the removal of hazardous cationic dyes. Altenor et al. (2012) investigated to physiochemical properties and adsorption efficiencies of *Turbinaria turbinata* and its activated carbon. Results of this study showed that algal activated carbons are more useful than raw algal materials for the removal of methylene blue. El-Sikaily et al. (2011) researched copper adsorption onto dried *Pterocladia capillacea* and its activated carbon in this study. Besides this they investigated to effect of adsorption conditions such as contact time, activated carbon concentration and copper concentration. El Nemr et al. (2011) researched adsorption of toxic Cr(VI) ion onto dried *Pterocladia capillacea* and its activated carbon in this study. For *Pterocladia capillacea* and its activated carbon, the maximum adsorption capacities were calculated as 12 mg/g and 66 mg/g, respectively. Aravindhan et al. (2009) produced activated carbons from *Sargassum longifolium* ve *Hypnea valentiae* macroalgae species by zinc chloride activation and investigated to removal of phenol from aqueous solutions using activated carbons. Surface areas of obtained activated carbon were seen as 802 m^2/g and 783 m^2/g, respectively. For the removal of phenol from aqueous solutions, obtained activated carbons could be used. Ncibi et al. (2009) produced activated carbons from marine *Posidonia oceanica* fibres and investigated characteristics of these activated carbons. The results showed that the porous structure develops due to the physical activation. El-Sikaily et al. (2007) investigated removal of Crom ion from saltwater and wastewater onto *Ulva lactuca* and its activated carbon. Yields of removal of Crom ions were calculated as 92 % and 98 %, respectively. For the removal of Crom ions from saltwater and wastewater, *Ulva lactuca* and its activated carbon could be used. Studies that mentioned above, are summarized in the Table 3.

2.2.2 Usage of Algal Wastes as Biosorbent

Algal wastes can be also utilized by using directly, without performing any conversion method. One of these alternative utilization methods is its usage as biosorbent. The most basic definition of biosorption is removal of various contaminants depend on ambient conditions by microorganisms or biological origin materials (biomass). Biosorption process can occur with alive or dead biosorbents. The advantages of biosorption by dead biosorbents are quite high. When working with living biosorbent, high concentrations of pollutants resulting from the accumulation of excessive pollutants stop the cell growth by toxic effects. There is no such a problem in dead biosorbents. In addition, alive biosorbents need nutrient continuously (Aksu et al. 2010).

The most important advantages of biosorption can be listed as:

- Biosorbents can be easily obtained. If there is a suitable laboratory, it is quite easy to produce biomass.
- Biosorbents are unlimited and very cheap. Used biomass is utilized by regenerating again.
- The remaining wastes from the process are harmless and can be easily destroyed.
- Due to the biosorbents are organic material, biosorbents do not damage to nature.
- Metal removal capacity of biosorbents is very high. Biosorbents can be used even in very dilute solutions (Şahan 2008)

Biosorption Mechanism

Biosorption mechanism is not fully understood yet, but with from different perspectives assessment, a classification was made as follows (Veglio and Beolchini 1997) (Fig. 4).

For biosorption of metals in the living cells, intracellular accumulation (dependent on metabolism) and metal compounds binding to the cell surface (independent

Table 3 Conditions and aims of studies on algal activated carbon

Algae	Impregnation ratio	Conditions	Aim	References
Gelidium sesquipedale	KOH and K_2CO_3 0.5:1	Conventional and microwave chemical activation Temperature : 750 °C	Investigation of the effect of chemical material type used for activation	Ferrera-Lorenzo et al. 2014b
Gelidium sesquipedale	KOH 0.5:1 and 1:1	Conventional and microwave chemical activation Temperature : 750 °C	Investigation of the influence of the activation temperature and KOH weight ratio	Ferrera-Lorenzo et al. 2014c
Ulva fasciata	10 % calcium chloride 10 % sodium carbonate 10 % sodium sulphate	Carbonization process at 400 °C thermally activated at 800 °C	Adsorption of lead (II) ions	Jeyakumar and Chandrasekaran 2014
Ulva lactuca and Systoceira stricta	phosphoric acid (20 %)	443 K for 90 min, followed by pyrolysis at 873 K during 180 min	Removal of hazardous cationic dyes	Salima et al. 2013
Pterocladia capillacea	98 % H_2SO_4	1 h at room temperature	Removal of toxic chromium	El Nemr et al. 2011
Pterocladia capillacea	98 % H_2SO_4	24 h at room temperature	Sorption of copper	El-Sikaily et al. 2011
Gracilaria and Sargassum sp.	97 % H_2SO_4	24 h at room temperature	Removal of hexavalent chromium	Esmaeili and Ghasemi 2009; Esmaeili et al. 2010
Gracilaria	97 % H_2SO_4	24 h at room temperature	Biosorption of Ni (II)	Esmaeili and Ghasemi 2009

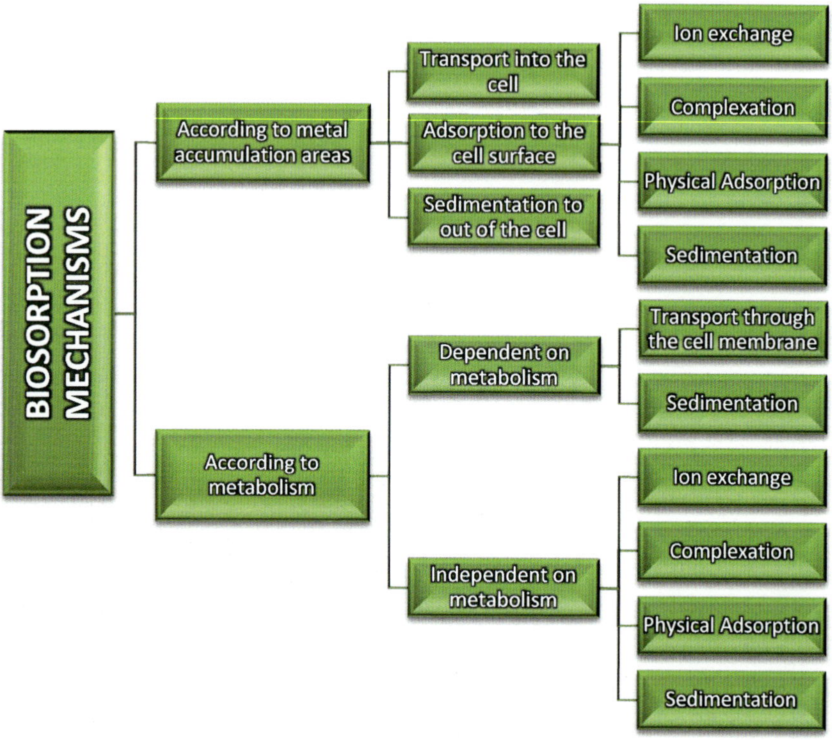

Fig. 4 Classification of biosorption mechanisms

on metabolism) are known as two basic mechanisms. Intracellular accumulation or metabolic process is the result of relatively large amounts of metals in the environments. The rate of these processes is slow and generally depended on the environmental conditions and nutrients. Binding to cell surface process is a type of passive process. This process can occur on dead or alive biomass. Non-metabolic surface binding is very fast and generally, it occurs less than a few minutes (Kılıç 2004).

Intracellular transport phenomena is related to cell metabolism and this mechanism may be same as the important transport mechanisms for cells such as potassium, sodium and magnesium ions transport mechanisms. In many instances in the literature, biosorption with living organism includes two basic steps: binding to cell membrane and transport into the cell through the cell membrane (Gourdon et al. 1990; Huang et al. 1990).

Physical adsorption occurs with the presence of van der Waals forces and other weak intermolecular forces such as dipole-dipole forces. In addition, electrostatic interactions are effective in the biosorption with bacteria and algae (Şahan T 2008).

Biosorption via complexation can occur depending on complex structure on the cell surface after the interaction between metal ions and active groups . Metal ions can bind to ligands on the cell wall. The compatibility between metal ions and the nitrogen in the cell wall leads to biosorption (Veglio and Beolchini 1997).

Sedimentation mechanism can depend on or not depend on the metabolism. Generally, this mechanism may be result of chemical interaction between cell surface and metal ions. The mechanism of ion exchange is due to exchange properties of carbohydrates in the organism (Veglio and Beolchini 1997).

As can be seen from the literature, the mechanism of biosorption is not only kind. It can consist in more than one mechanism simultaneously (Şahan 2008).

Dittert et al. (2012, 2014) investigated the usage of *Laminaria digitata* after pretreatment with acid for biosorption of crom ions and effects of some conditions such as pH, temperature, contact time, biomass and concentration on biosorption. The effect of pH on biosorption was found to be greater than the effect of temperature. Also biosorption capacity of crom ions in algae was found to increase with pH.

Xiong et al. (2013) researched the usage of *Laminaria japonica* that modified with sulfuric acid for biosorption of Re (VII) from aqueous solution. In addition, Liu et al. (2009) investigated the biosorption of Cd^{2+}, Cu^{2+}, Ni^{2+} and Zn^{2+} ions onto *Laminaria japonica* that has exposed to various pretreatments. These pretreatments were cross-linking with epichlorohydrin, oxidation with potassium permanganate, cross-linking with glutaraldehyde and washing with distilled water.

Vilar et al. (2008a) studied the biosorption from Pb^{2+}/Cu^{2+}, Pb^{2+}/Cd^{2+}, Pb^{2+}/Zn^{2+} and Cd^{2+}/Zn^{2+} solutions onto algal waste, composite material and algae *Gelidium.* In a different study, Vilar et al. (2008b) investigated continuous biosorption of Pb/Cu and Pb/Cd using algae *Gelidium* and agar extraction algal waste. In 2005, Vilar et al. investigated impact of pH, ionic strength and temperature on biosorption by *Gelidium* and agar extraction algal waste for Pb^{2+} removal from aqueous solutions. These studies showed that biosorption capacity of *Gelidium* is higher than biosorption capacity of algal waste.

Sarı and Tuzen (2008a, b researched the biosorption of Pb(II) and Cd(II) onto the *Ceramium virgatum* and *Ulva lactuca* from aqueous solution and investigated the effects of experimental parameters such as pH, contact time, algae dosage and temperature on the biosorption process. In 2010, Tuzen and Sarı investigated selenium (IV) biosorption from aqueous solution onto *Cladophora hutchinsiae* and determined the optimum biosorption conditions. These studies showed that algal biomass is very useful for metal removal from aqueous solution.

Vijayaraghavan and Yun (2008) carried out biosorption of reactive black 5 by using *Laminaria* that treated with acid. Also in these experiments, the effect of temperature and pH on biosorption were investigated.

Nakiboğlu and Sevindir (2006) studied the biosorption of crom ions on *Chlorella* sp. and *Scedenesmus Obliquus.* For the biosorption, optimum temperature, pH and algae concentration were determined. Optimum conditions for both of algal species were similar. Besides, Elmacı et al. (2005) researched the biosorption of Remazol Turkish Blue-G onto the dried *Chara* sp., *Cladophora* sp. and *Chlorella* sp. Each of three species of algae are reported as effective for removing heavy metals but the best removing is achieved by *Cladophora* sp.

2.2.3 Usage of Algal Wastes as Fertilizer

Algal wastes can be utilized for growth of plants as bio-fertilizer. Nowadays bio-fertilizers are preferred rather than chemical fertilizers due to being environmental friendly and cost-effective. Bio-fertilizers contain microorganisms which can fix nitrogen, solubilize phosphate, and promote plant growth. Algae can be helpful in agriculture with these functions. Seaweeds and microalgae -especially blue-green algae- are used as nutritional supplements and biofertilizers to improve growth of plant and production yield (Guedes et al. 2014). Structure of algae includes regulatory macro- and micronutrients like cytokinins, auxins, gibberellins, and betaines which can increase plant growth by inducing (Valente et al. 2006). Although there are some fertilizers that are produced from marine algae, used in agriculture, researches for utilizing algae as bio-fertilizer continue to develop this application. Thorsen et al., have investigated effects of utilizing *Laminaria digitata* on plant growth and it is reported that *Laminaria digitata* increase seed germination and improves rooting in terrestrial plants (Thorsen et al. 2010). Algae are found beneficial in cultivating plants and improving productivity with a number of substances like vitamins, amino acids, polypeptides, and antibacterial and antifungal matter which are exist in composition of algae (de Mule, et al. 1999). Schwartz and Krienitz also implied that different in-direct growth-promotion effects may have influence such as enhancing the water-holding capacity of soils or substrates and producing antifungal and antibacterial compounds (Schwartz and Krienitz 2005). In addition to these, algae have a potential as fertilizer by providing high N:P ratio to plants.

2.2.4 Usage of Algal Wastes as Animal Feed

Usage as animal feed is another option for utilizing algal biomass which remains after different processes. As described previously, the content of algae comprise important nutritional elements which can meet the requirement for animal feed (Zubia et al. 2008). Protein content of plants is low for fish diets and demand of protein source for aquatic feed cannot be met with these materials. In order to overcome this demand, microalgae have been used as protein source traditionally. Besides protein content, algae contain other important nutrients which can be participate in food chain later, such as vitamins, essential PUFAs, pigments, and sterols, fish gain resistance to bacterial contamination (Guedes and Malcata 2012). Microalgal fatty acids which have longer than ten carbon atoms can stimulate lysis of bacterial protoplasts and such bacterial infections can be avoided (Guedes et al. 2011). In addition to that, carotenoid content of algae can provide natural pigments to some organisms like salmon which have a characteristic red color in their muscles (Guedes and Malcata 2012).

Investigations which have been performed to evaluate nutritional and toxicological values of algal biomass showed that it is a convenient feed source for animals. (Gendy and El-Temtamy 2013) The most used algal species for animal feed are

Spirulina, *Chlorella*, and *Scenesdesmus* species. Positive results on health have been obtained from usage of algal biomass to feed some animals such as cows, horses, pigs, poultry, cats and dogs (Spolaore et al. 2006). Besides, it has beneficial effects in livestock raising such as in a research which *Laminaria digitata* used as animal feed for pigs resulted with 10 % weight increases on a daily basis (Harun et al. 2010). However, in some cases like poultry feed, utilization of algal biomass in higher concentrations can cause reduction in growth rate and color and flavor changes can be seen in chicken eggs (Hudek et al. 2014).

Although this chapter focuses on utilization of algal wastes from different industrial processes to obtain solid products, spent biomass of algae also can be in liquid or gaseous form and may be utilized in this way. Biogas production from algal wastes which remains after bioethanol production can be given as an example for such cases.

3 Conclusion

Algal investment is not economically feasible due to operational and capital cost which occur from mostly harvesting and dewatering of algae. Although many innovations are seen in production of algal biomass day by day, utilization of algal wastes is the most appropriate option for energy recovery and cost reduction. Solid products of algal wastes offer great potential for usage in different application areas. In this chapter, utilization methods of algal wastes are explained and disregarded potential of solid products from algal wastes is highlighted. Along with the usage as energy source, utilization of these solid products in various fields such as environmental, food and agriculture is also mentioned. Although there are a few researches on using algal solid products, it can be said that these products can be useful for various water treatments and soil amendment as biosorbent and biochar, and can be also utilized to meet the demand of the animal feed in the present and future.

References

Aksu Z, Ertuğrul S, Dönmez G (2010) Methylene Blue biosorption by Rhizopus arrhizus: effect of SDS (sodium dodecylsulfate) surfactant on biosorption properties. Chem Eng J 158:474–481

Altenor S, Ncibi MC, Emmanuel E et al (2012) Textural characteristics, physiochemical properties and adsorption efficiencies of Caribbean alga Turbinaria turbinata and its derived carbonaceous materials for water treatment application. Biochem Eng J 67:35–44

Anoop Singh A, Nigamb P, Murphy J (2011) Mechanism and challenges in commercialisation of algal biofuels. Bioresour Technol 102:26–34

Aravindhan R, Rao JR, Nair BU (2009) Preparation and characterization of activated carbon from marine macro-algal biomass. J Hazard Mater 162:688–694

Atkinson CJ, Fitzgerald JD, Hipss NA (2010) Potential mechanisms for achieving agricultural benefits from biochar application to temperate soils: a review. Plant Soil 337:1–18

Aygün A (2002) Production of activated carbon from domestic natural raw materials and investigation of adsorption properties. Master thesis, İTÜ, İstanbul

Azner JZ (2011) Characterization of activated carbon produced from coffee residues by chemical and physical activation. Master Thesis, Chemical Engineering, Sweden, Stockholm
Beesley L, Moreno-Jiménez E, Gomez-Eyles JL et al (2011) A Review of biochar's potential role in the remediation, revegetation and restoration of contaminated soils. Environ Pollut 159:3269–3282
Belarbi EH, Molina EY (2000) A process for high yield and scaleable recovery of high purity eicosapentaenoic acid esters from microalgae and fish oil. Process Biochem 35:951–969
Besada V, Andrade JM, Schultze F et al (2009) Heavy metals in edible seaweeds commercialised for human consumption. J Mar Syst 75:305–313
Bird MI, Wurster CM, de Paula Silva PH et al (2011) Algal biochar – production and properties. Bioresour Technol 102:1886–1891
Bitoga JP, Lee IB, Lee CG et al (2011) Application of computational fluid dynamics for modeling and designing photobioreactors for microalgae production. Comput Electron Agric 76:131–147
Bridgwater AV, Bridge SA (1991) Biomass pyrolysis liquids upgrading and utilization. Elsevier Applied Science, New York
Çeçen F (2011) Water and wastewater treatment: historical perspective of activated carbon adsorption and its integration with biological processes. In: Çeçen F, Aktaş Ö (eds) Activated carbon for water and wastewater treatment: integration of adsorption and biological treatment, 1st edn. Wiley-VCH, Weinheim
Chaiwong K, Kiatsiriroat T, Vorayos N et al (2012) Biochar production from freshwater algae by slow pyrolysis. Maejo Int J Sci Technol 6(02):186–195
Cheng-Wu Z, Zmora O, Kopel R et al (2001) An industrial-size flat plate glass reactor for mass production of Nannochloropsis sp. (Eustigmatophyceae). Aquaculture 195:35–49
Cho HJ, Baek K, Jeon JK et al (2013) Removal characteristics of copper by marine macro-algae-derived chars. Chem Eng J 217:205–211
Choi J, Choi JW, Suh DJ et al (2014) Production of brown algae pyrolysis oils for liquid biofuels depending on the chemical pretreatment methods. Energ Conversv Manage 86:371–378
Daroch M, Geng S, Wang G (2013) Recent advances in liquid biofuel production from algal feedstocks. Appl Energ 102:1371–1381
De Mule MCZ, de Caire GZ et al (1999) Effect of cyanobacterial inoculation and fertilizers on rice seedlings and postharvest soil structure. Commun Soil Sci Plant Anal 30(1–2):97–107
Dittert IM, Vilar VJP, da Silva EAB et al (2012) Adding value to marine macro-algae Laminaria digitata through its use in the separation and recovery of trivalent chromium ions from aqueous solution. Chem Eng J 193–194:348–357
Dittert IM, Brandão HL, Pina F et al (2014) Integrated reduction/oxidation reactions and sorption processes for Cr(VI) removal from aqueous solutions using Laminaria digitata macro-algae. Chem Eng J 237:443–454
Duku MH, Gua S, Haganb EB (2011) Biochar production potential in Ghana—a review. Renew Sust Energ Rev 15:3539–3551
El Nemr A, El-Sikaily A, Khaled A et al (2011) Removal of toxic chromium from aqueous solution, wastewater and saline water by marine red alga Pterocladia capillacea and its activated carbon. Arab J Chem. doi:10.1016/j.arabjc.2011.01.016
Elmacı A, Yonar T, Özengin N et al (2005) The Investigation of biosorption characteristics of dried *Chara* sp., *Cladophora* sp. and *Chlorella sp.* for the removal of zn(II), cd(II), co(II) and remazol turkish blue-g in aqueous solutions. Ekoloji 55:24–31
El-Sikaily A, El Nemr A, Khaled A et al (2007) Removal of toxic chromium from wastewater using green alga *Ulva lactuca* and its activated carbon. J Hazard Mater 148:216–228
El-Sikaily A, El Nemr A, Khaled A (2011) Copper sorption onto dried red alga Pterocladia capillacea and its activated carbon. Chem Eng J 168:707–714
Esmaeili A, Ghasemi S (2009) Evaluation of the activated carbon prepared of algae marine *gracilaria* for the biosorption of ni (II) from aqueous solutions. World Appl Sci J 6(4):515–518

Esmaeili A, Ghasemi S, Rustaiyan A (2010) Removal of hexavalent chromium using activated carbons derived from marine algae *gracilaria* and *sargassum* sp. J Mar Sci Technol 18(4):587–592

Ferrera-Lorenzo N, Fuente E, Bermudez JM et al (2014a) Conventional and microwave pyrolysis of a macroalgae waste from the Agar–Agar industry. Prospects for bio-fuel production. Bioresour Technol 151:199–206

Ferrera-Lorenzo N, Fuente E, Suárez-Ruiz I et al (2014b) KOH activated carbon from conventional and microwave heating system of a macroalgae waste from the Agar–Agar industry. Fuel Process Technol 121:25–31

Ferrera-Lorenzo N, Fuente E, Suárez-Ruiz I et al (2014c) Sustainable activated carbons of macroalgae waste from the Agar–Agar industry. Prospects as adsorbent for gas storage at high pressures. Chem Eng J 250:128–136

Gaskin JW et al (2007) Potential for pyrolysis char to affect soil moisture and nutrient status of a loamy sand soil. Proceedings of the 2007 Georgia water resources conference, Ürdün, 27–29 Mar 2007

Gendy TS, El-Temtamy SA (2013) Commercialization potential aspects of microalgae for biofuel production: an overview. Egypt J Pet 22:43–51

Goldstein IS (1983) Organic chemicals from biomass. CRC, Florida (Second Printing)

Gong X, Zhang B, Zhang Y et al (2014) Investigation on pyrolysis of low lipid microalgae chlorella vulgaris and dunaliella salina. Energy Fuels 28:95–103

Gourdon R, Bhende S, Rus E et al (1990) Comparison cadmium biosorption by gram-positive and gram-negative bacteria from activated sludge. Biotechnol Lett 12(11):839–842

Grierson S, Strezov V, Ellem G et al (2009) Thermal characterisation of microalgae under slow pyrolysis conditions. J Anal Appl Pyrol 85:118–123

Grierson S, Strezov V, Shah P (2011) Properties of oil and char derived from slow pyrolysis of tetraselmis chui. Bioresour Technol 102:8232–8240

Guedes AC, Malcata FX (2012) Nutritional value anduses of microalgae in aquaculture. In: Muchlisin ZA (ed) Aquaculture. InTech, Crotia. ISBN 978-953-307-974-5

Guedes AC, Amaro HM et al (2011) Microalgae as sources of high added-value compounds—a brief review of recent work. Biotechnol Prog 27(3):597–613

Guedes AC, Amaro HM, Sousa-Pint I et al (2014) Applications of spent biomass. In: Pandey A, Lee DJ, Chisti Y, Rsoccol C (eds) Biofuels from algae. Elsevier. ISBN: 978-0-444-59558-4

Harun R, Singh M, Forde GM et al (2010) Bioprocess engineering of microalgae to produce a variety of consumer products. Renew Sust Energ Rev 14:1037–1047

Haykiri-Acma H, Yaman S, Kucukbayrak S (2013) Production of biobriquettes from carbonized brown seaweed. Fuel Process Technol 106:33–40

Huang CP, Huang CP, Morehart AL (1990) The removal of Cu(II) from dilute aqueous solution by *Saccharomyces cerevisiae*. Wat Res 24:433–439

Hudek K, Davis LC, Ibbini J (2014) Commercial products from algae. In: Bajpai R, Prokop A, Zappi M (eds) Algal biorefineries volume 1: cultivation of cells and products. Springer, Dordrecht. ISBN 978-94-007-7493-3

Jeyakumar RPS, Chandrasekaran V (2014) Adsorption of lead(II) ions by activated carbons prepared from marine green algae: equilibrium and kinetics studies. Int J Ind Chem 5:2

Jirka S, Tomlinson T (2014) 2013 State of the biochar industry, a survey of commercial activity in the biochar field. A report by the International Biochar Initiative (IBI)

Jung KA, Lim SR, Kim Y et al (2013) Potentials of macroalgae as feedstocks for biorefinery. Bioresour Technol 135:182–190

Kan T, Grierson S, de Nys R et al (2014) Comparative assessment of the thermochemical conversion of freshwater and marine micro- and macroalgae. Energy Fuels 28:104–114

Kebelmann K, Hornung A, Karsten U et al (2013) Thermo-chemical behaviour and chemical product formation from Polar seaweeds during intermediate pyrolysis. J Anal Appl Pyrol 104:131–138

Kılıç M (2004) Examination of lead (II) and mercury (II) ions adsorption on biomass. phd thesis, Süleyman Demirel University, Isparta
Kim SS, Ly HV, Kim J et al (2013) Thermogravimetric characteristics and pyrolysis kinetics of alga Sagarssum sp. biomass. Bioresour Technol 139:242–248
Kirtania K, Joshua J, Kassim MA et al (2014) Comparison of CO2 and steam gasification reactivity of algal and woody biomass chars. Fuel Process Technol 117:44–52
Knight JA (1976) Pyrolysis of fine sawdust. 172nd American Chemical Society national meeting, San Francisco
Kraan S (2012) Algal polysaccharides, novel applications and Outlook. In: Chang C (ed) Carbohydrates–comprehensive studies on glycobiology and glycotechnology. InTech, Crotia. ISBN 978-953-51-0864-1
Kwapinski W, Byrne CMP, Kryachko E et al (2010) Biochar from biomass and waste. Waste Biomass Valor 1:177–189
Lehmann J, Joseph S (2009) Biochar for environmental management, 1st edn. Earthscan, Londra, pp 1–2
Liu Y, Cao Q, Luo F et al (2009) Biosorption of Cd2+, Cu2+, Ni2+ and Zn2+ ions from aqueous solutions by pretreated biomass of brown algae. J Hazard Mater 163:931–938
Maddi B, Viamajala S, Varanasi S (2011) Comparative study of pyrolysis of algal biomass from natural lake blooms with lignocellulosic biomass. Bioresour Technol 102:11018–11026
Mata TM, Martins AA, Caetano NS (2010) Microalgae for biodiesel production and other applications: a review. J Renew Sust Ener 14(1):217–232
McClellan T, Deenik J, Uehara G et al (2007) Effects of flashed carbonized macadamia nutshell charcoal on plant growth and soil chemical properties. ASA-CSSA-SSA ınternational annual meetings, New Orleans, LA
McLaughlin H, Anderson PS, Shields FE et al (2009) All biochars are not created equal, and how to tell them apart. Proceedings North American Biochar Conference, Boulder, CO
Miao X, Wu Q, Yang C (2004) Fast pyrolysis of microalgae to produce renewable fuels. J Anal Appl Pyrol 71:855–863
Murthy GS (2011) Overview and assessment of algal biofuels production technologies biofuels: alternative feedstocks and conversion processes. In: Pandey A, Larroche C, Ricke RS, Dussap CG, Gnansounou E (eds). Elsevier. ISBN: 978-0-12-385099-7
Nakas JPS, Parkinson CM, Coonley CE et al (1983) System development for linked-fermentation production of solvents from algal biomass. Appl Environ Microb 46:1017–1023
Nakiboğlu A, Sevindir HC (2006) Biosorption of Crom ions with various algae from leather industry wastewater. SDÜ Fen Bilimleri Enstitüsü Dergisi 10(2)
Ncibi MC, Jeanne-Rose V, Mahjoub B et al (2009) Preparation and characterisation of raw chars and physically activated carbons derived from marine Posidonia oceanica (L.) fibres. J Hazard Mater 165:240–249
Olaizola M (2003) Commercial development of microalgal biotechnology: from the test tube to the marketplace. Biomol Eng 20:459–466
Özçimen D (2007) Evaluation of various vegetable residues by carbonization. PhD thesis, Istanbul Technical University
Özçimen D (2013) An approach to the characterization of biochar and bio-oil. Renewable Energy for Sustainable Future, iConcept Press, pp 41–58
Özçimen D, Gülyurt MÖ, İnan B (2012) Algal biorefinery for biodiesel production. In: Fang Z (ed) Biodiesel – feedstocks, production and applications. InTech, Crotia. ISBN 978-953-51-0910-5
Radakovits R, Jinkerson RE, Darzins A et al (2010) Genetic engineering of algae for enhanced biofuel production. Eukaryot Cell 9:486–501
Rashida N, Rehmana MSU, Han JI (2013) Recycling and reuse of spent microalgal biomass for sustainable biofuels. Biochem Eng J 75:101–107
Rondon MA, Lehmann J, Ramirez J et al (2007) Biological nitrogen fixation by common beans (phaseolus vulgaris l.) increases with bio-char additions. Biol Fertil Soils 43:699–708
Ronsse F, Van Hecke S, Nachenius R et al (2011) Production and characterisation of slow pyrolysis biochar. In Proceedings of the 19th European biomass conference and exhibition, Berlin, Germany, pp 1681–1685

Ross AB, Jones JM, Kubacki ML et al (2008) Classification of macroalgae as fuel and its thermochemical behavior. Bioresour Technol 99:6494–6504

Ross AB, Anastasakis K, Kubacki ML et al (2009) Investigation of the pyrolysis behaviour of brown algae before and after pre-treatment using PY-GC/MS and TGA. J Anal Appl Pyrolysis 85:3–10

Ruperez P (2002) Mineral content of edible marine seaweeds. Food Chem 79:23–26

Ruperez P, Ahrazem O et al (2002) Potential antioxidant capacity of sulfated polysaccharides from the edible marine brown seaweed Fucus vesiculosus. J Agric Food Chem 50(4):840–845

Şahan T (2008) Removal of some heavy metals from wastewaters via biosorption and optimization of biosorption conditions. Phd thesis, Yüzüncü Yıl University, Van

Salima A, Benaouda B, Noureddine B et al (2013) Application of Ulva lactuca and Systoceira stricta algae-based activated carbons to hazardous cationic dyes removal from industrial effluents. Wat Res 47:3375–3388

Sarı A, Tuzen M (2008a) Biosorption of cadmium(II) from aqueous solution by red algae (*Ceramium virgatum*): equilibrium, kinetic and thermodynamic studies. J Hazard Mater 157:448–454

Sarı A, Tuzen M (2008b) Biosorption of Pb(II) and Cd(II) from aqueous solution using green alga (*Ulva lactuca*) biomass. J Hazard Mater 152:302–308

Schwartz D, Krienitz L (2005) Do algae cause growth-promoting effects on vegetables grown hydroponically? In: Price MR (ed) Fertigation: optimizing the utilization of water and nutrients. International Potash Institute, Beijing, pp 161–170

Sohi SP, Krull E, Lopez-Capel E et al (2010) Advances in agronomy, Ith edn. Elsevier/Academic, Burlington

Spolaore P, Joannis-Cassan C, Duran E et al (2006) Commercial applications of microalgae. J Biosci Bioeng 101:87–96

Thorsen M, Woodward S et al (2010) Kelp (Laminaria digitata) increases germination and affects rooting and plant vigour in crops and native plants from an arable grassland in the Outer Hebrides. Scotland J Coast Conservat 14(3):239–247

Topsak E (2011) Pyrolysis process effects on the structural properties of biomass and its functional group distribution. Master thesis, ITU, İstanbul

Tuzen M, Sarı A (2010) Biosorption of selenium from aqueous solution by green algae (Cladophora hutchinsiae) biomass: equilibrium, thermodynamic and kinetic studies. Chem Eng J 158:200–206

USDA (2001) Agricultural research service. Nutrient Database for Standard Reference, Release 14

Valente LMP, Gouveia A et al (2006) Evaluation of three seaweeds Gracilaria bursa-pastoris, Ulva rigida and Gracilaria cornea as dietary ingredients in European sea bass (Dicentrarchus labrax) juveniles. Aquaculture 252(1):85–91

Veglio F, Beolchini F (1997) Removal of metals by biosorption: a review. Hydrometallurgy 44:301–316

Vergara-Fernandez A, Vargas G, Alarcon N et al (2008) Evaluation of marine algae as a source of biogas in a two-stage anaerobic reactor system. Biomass Bioenerg 32:338–344

Vijayaraghavan K, Yun YS (2008) Biosorption of C.I. Reactive Black 5 from aqueous solution using acid-treated biomass of brown seaweed Laminaria sp. Dyes Pigments 76:726–732

Vilar VJP, Botelho CMS, Boaventura RAR (2005) Influence of pH, ionic strength and temperature on lead biosorption by Gelidium and agar extraction algal waste. Process Biochem 40:3267–3275

Vilar VJP, Loureiro JM, Botelho CMS et al (2008a) Continuous biosorption of Pb/Cu and Pb/Cd in fixed-bed column using algae *Gelidium* and granulated agar extraction algal waste. J Hazard Mater 154:1173–1182

Vilar VJP, Botelho CMS, Boaventura RAR (2008b) Kinetics modelling of biosorption by algal biomass from binary metal solutions using batch contactors. Biochem Eng J 38:319–325

Wen Z, Chen F (2003) Heterotrophic production of eicosapentaenoic acid by microalgae. Biotechnol Adv 21:273–294

White LP, Plaskett LG (1981) Biomass as fuel. Academic, London
Winsley P (2007) Biochar and bioenergy production for climate change mitigation. New Zeal Sci Rev 64(1):5–10
Xiong Y, Xu J, Shan W et al (2013) A new approach for rhenium(VII) recovery by using modified brown algae Laminaria japonica adsorbent. Bioresour Technol 127:464–472
Yanik J, Stahl R, Troeger N et al (2013) Pyrolysis of algal biomass. J Anal Appl Pyrol 103:134–141
Yen HW, Brune DE (2007) Anaerobic co-digestion of algal sludge and waste paper to produce methane. Bioresour Technol 98:130–134
Yuan JH, Xu RK, Zhang H (2011) The forms of alkalis in the biochar produced from crop residues at different temperatures. Bioresour Technol 102:3488–3497
Zubia M, Payri C et al (2008) Alginate, mannitol, phenolic compounds and biological activities of two range-extending brown algae, Sargassum mangarevense and Turbinaria ornata (Phaeophyta: Fucales), from Tahiti (French Polynesia). J Appl Phycol 20(6):1033–1043

Algal Cell Disruption and Lipid Extraction: A Review on Current Technologies and Limitations

Chandra S. Theegala

Abstract Although numerous laboratory-based analytical techniques were developed and tested over the last five or six decades, industrial-scale algal oil extraction can be considered to be at its infancy.

Cost-effective, industrial-scale algal lipid extraction has been considered only after the advent of the algal biofuel industry. Presently, there is clearly a dearth of literature or reported results from commercial algal extraction technologies. When compared to land-based oil-seed crops, several fundamental differences exist for algal lipid extraction. Starting with the need for cost-effective harvesting and dewatering of dilute algal cultures (with 0.015–0.03 % solids) to differences in cell wall chemistry, and from the unsuitability of standard oil-seed pressing techniques to the need for cell disruption before drying, create unique challenges for microalgal lipid extraction. The present chapter discusses the limitations, challenges, and findings from numerous laboratory-based cell disruption and lipid extraction experiments and analytical techniques developed specifically to characterize or quantify algal lipids for nutraceutical, aquacultural, fine-chemical, or other value-added applications. Some potential industrial-scale, lipid extraction technologies are also discussed.

Keywords Biofuel • Biodiesel • Omega-3 • PUFAs • Algal oils • Solvent extraction • Pressurized lipid extraction • Microwave lipid extraction • Soxhlet • Co-solvent extraction • Sonication • Transesterification

1 Introduction

Commercial algal industry initially focused on harvesting and utilizing the entire cell contents as nutritive supplements and aquacultural feeds (e.g. *Spirulina* as human/animal nutritive supplement and *Chlorella* for aquacultural live feed). In the early

C.S. Theegala (✉)
Biological and Agricultural Engineering, Louisiana State University & LSU AgCenter, 161 EB Doran Bldg, Baton Rouge, LA 70803, USA
e-mail: Ctheegala@agcenter.lsu.edu

A. Prokop et al. (eds.), *Algal Biorefineries*, DOI 10.1007/978-3-319-20200-6_13

1950s, the projected world population figures and insufficiency in protein supply triggered a search for unconventional protein (Spolaore et al. 2006). Algal biomass appeared at that time as a good candidate (Becker 2004). Aquaculture industry relies heavily on microalgae, which as a group represent the third-largest aquacultured crop in the world today (after freshwater fishes and mollusks) (Wijkstrom et al. 2000; Wikfors and Ohno 2001). Most aquacultural applications utilize small unicellular algal strains with easy digestibility and with appropriate proteins, fatty acids, and nutrients. In the 1980s commercial production of *Dunaliella salina*, as a source of β-carotene, became the third major microalgal industry (Spolaore et al. 2006). In the following years, the potential for numerous extracts and biomolecules (such as Astaxanthin, Lutein, Polyunsaturated Fatty Acids (PUFAs), etc.) has been identified. As a consequence of the directions of the algal industry to date, commercial lipid extraction did not receive significant attention. Although numerous laboratory-based analytical techniques were developed and tested over the last six decades, current industrial-scale algal oil extraction technology can be considered to be at its infancy. The true industrial-scale algal lipid extraction has been considered only after the advent of the algal biofuel industry. Presently, there is clearly a dearth of literature or reported results from industrial-scale algal extraction technologies. The present chapter discusses the limitations, challenges, and findings from numerous laboratory-based cell disruption and lipid extraction experiments and analytical techniques developed specifically to characterize or quantify algal lipids for nutraceutical, aquacultural, or other fine-chemical applications. Some potential industrial-scale, lipid extraction technologies are also discussed.

2 Background Information

Before discussing the various lipid extraction techniques, it is very important to understand the underlying facts and challenges pertinent to algal cultures, types of algal lipids, cell wall chemistry, differences between oil-seeds and algae, and difficulties in drying and employing oil-pressing techniques. These underlying facts and challenges are critical for understanding and overcoming the limitations of industrial-scale algal lipid extraction.

2.1 Algal Culture Densities and Moisture Content

Unlike land-based oil seeds which are relatively dry at the harvesting stage, algal biofuels start with extremely dilute liquid cultures. High rate open algal ponds typically have algal densities of 150–300 mg-dry/L, which relates to 0.015–0.03 % solids content. Although higher biomass densities are attainable, denser cultures often result in reduced productivity due to light limitations. Presently, Stage-1 harvesting (also known as dewatering) concentrates the dilute cultures to approximately 1–2 %

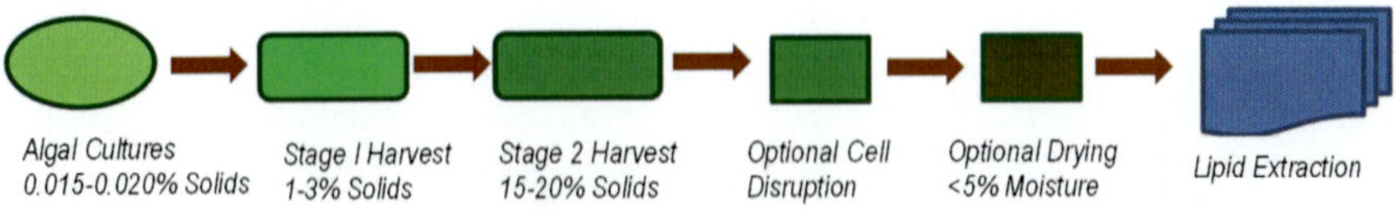

Fig. 1 Unlike oilseeds, significant energy is expended on preparation of algae for lipid extraction and usually involves: (1) stage-1 harvesting, (2) stage-2 dewatering, and an optional (3) stage-3 drying (to <5 % moisture)

solids (Dassey et al. 2014; Cooney et al. 2009). A high-powered centrifuge is usually employed as a Stage-2 system to concentrate the algae to about 15 20 % solids (approximately the consistency of peanut butter) (Fig. 1). The wet paste (with 80–85 % moisture) can be further dried to yield dry algal cake (<5 % moisture). Oil extraction can be performed either on the wet paste or dry algal cake. It is very important to note that significant amount of energy has to be expended to bring dilute algal cultures to ~20 % solids content. Therefore, drying beyond the wet paste consistency is energy intensive (Halim et al. 2011) may not be viable or cost-effective for biofuel or other low-value applications. In fact, this last drying step is known to create negative energy balances.

2.2 *Types of Lipids in Algae*

Algal lipids can broadly be defined based on the polarity of the molecular head group (Kates 1986a) as: (1) neutral (NL) or non-polar lipids, which comprise of acylglycerols and free fatty acids (FFAs) and (2) polar lipids which can be further sub-categorized into phospholipids (PL) and glycolipids (GL) (Halim et al. 2012a). Acylglycerol consists of fatty acids with ester-bonds to a glycerol backbone. Based on the number of fatty acids, they can be classified as triacylglycerols (TAG), diacylglycerols (DAG), and monoacylglycerols (MAG). FFAs on the other hand are fatty acids bonded to a hydrogen atom. Algal fatty acids range from 12 to 22 carbons in length and can be either saturated or unsaturated (Halim et al. 2012a). Neutral lipids are produced by microalgae for energy storage, therefore are also known as storage lipids. Neutral lipids are bound by relatively weak non-covalent forces (Van der Waals or hydrophobic associations) through their hydrocarbon chains to other lipids and to hydrophobic regions of proteins in the microalgae (Enssani 1990). Due to the weak bonding, neutral lipids are relatively easy to extract. Polar lipids, on the other hand, are part of the molecular building blocks of cell membranes. These polar lipids are harder to extract as they are capable of forming covalent and hydrogen bonds with adjacent molecules. Figure 2 shows the differences in composition of neutral and polar lipids in *Phaeodactylum tricornutum* obtained from two subsequent extractions with 1:1 chloroform: methanol co-solvent mixture (Ryckebosch et al. 2012). Apart from these two main classes of lipids,

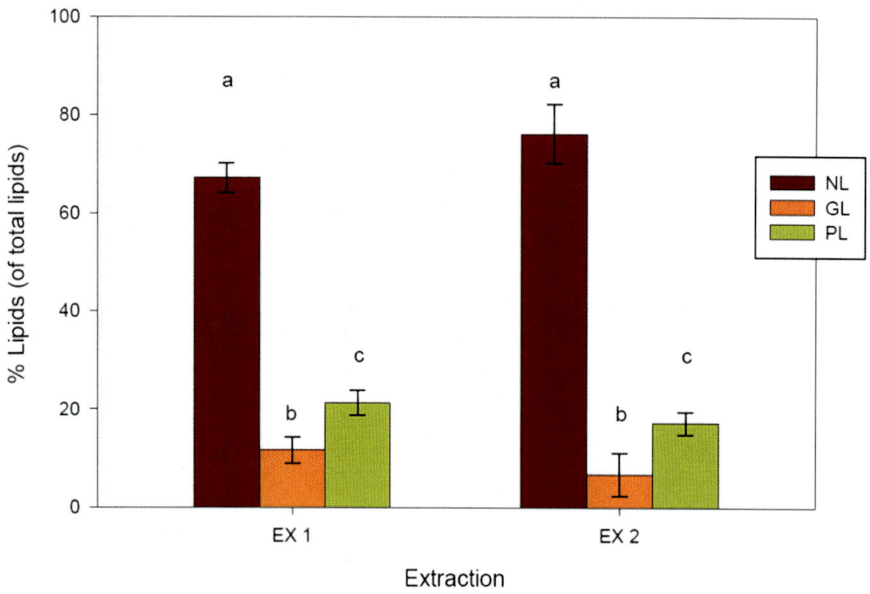

Fig. 2 Compositional variations in neutral (NL) and polar lipids in two consecutive solvent extractions using chloroform and methanol, mixed in 1: 1 (v/v) ratios (Reprinted with permission from Springer (Ryckebosch et al. 2012))

microalgae also contain neutral lipids without fatty acids, such as hydrocarbons, sterols, ketones, and pigments (carotenes and chlorophylls), which cannot be converted to biodiesels (Halim et al. 2012a). Furthermore, photosynthetically grown microalgae produce high levels of fat-soluble pigments, which hinder the lipid extraction and subsequent biodiesel production (Cooney et al. 2009).

2.3 Variations in Lipid Contents and Compositions

With an estimated 300,000 or more identified species of microalgae, it is impossible to assign a generic lipid composition profile to microalgal lipids. The lipid contents and compositions vary drastically between species and the amounts can range from 15 to 77 % (Chisti 2007). Even for the same species, the lipid content and composition is heavily influenced by culture conditions (light, light/dark ratio, temperature, nutrients, culture densities, etc.). Several researchers have demonstrated that selective nutrient starvation favors metabolic pathways that increase production of storage lipids (Halim et al. 2012a). In general, the oil content (percent) in most microalgal strains selected for biodiesel applications is comparable to or marginally higher than the best land-based oilseeds like rapeseed. However, the real benefit of employing microalgae lies with its potential to produce 10–20 times more lipids than oilseed crops on an aerial productivity basis (L/ha/y).

2.4 *Difficulties with Algal Lipid Extraction*

The algal cell walls have major variations in their structures, compositions, thickness and chemistry, all of which have a major influence on the choice of the lipid extraction process. For example, several unicellular algal cells with rigid cell walls will not be crushed but will rather flow with water through the thousands of micro-channels that exist in pressing equipment (Cooney et al. 2009). Apart from the cellulosic cell walls that are present for most green and brown algae, diatoms have hard silica frustules that are difficult to break. On the other hand, cells with weak cell walls (like *Dunaliella salina*) can be cracked easily by passing them through a homogenizer or through a pressure expansion valve. "Cell milking", which is a new concept (discussed later), appears to be viable only for cells with weak or porous cell walls. The well-established oil pressing techniques that was perfected for land-based oil seeds may not be practical in all cases for extracting algal lipids as the process requires relatively dry algal biomass with low moisture content (<5 %). Numerous researchers looking into the energy balances have clearly indicated that drying algae to less than 5 % moisture levels, although attempted at an analytical-scale, is impractical as input energy exceeds the energy content of the produced oil. A good and proven alternative to oilseed pressing is the solvent extraction process. However, the extremely low algal cell density and high moisture content in the harvested algal paste (80–85 % moisture after Stage-2 harvesting), coupled with the need for extraction of two different types of lipids (neutral and polar) adds additional complexity to the algal lipid extraction process.

3 Algal Lipid Extraction

Until recently, most algal lipid extraction techniques were based on lipid extraction from wet algal paste or dry algal cake, with or without pre-treatment or cell disruption. Several solvent extraction methods, starting from (Folch et al. 1957; Bligh and Dyer 1959) co-solvent mixture-based extraction to supercritical CO_2 extraction to pressurized lipid extraction have been developed and tested, mostly at laboratory-scale. Various combinations of cell disruptions and lipid extraction techniques have been employed. Some of the commonly used cell-disruption, extraction-augmenting methods, and extraction techniques are presented below.

3.1 *Cell Disruption and Extraction-Augmenting Methods*

Numerous researchers have employed cell-disruption prior to lipid extraction. Cell disruption techniques shatter the cell wall and facilitate better lipid extraction. Some of the most commonly employed cell-disruption/pre-treatment techniques include: bead beating (or bead milling), sonication, high pressure homogenization, heat

Table 1 Cell disruption and/or extraction augmenting techniques

Bead beating/Bead mill/Dyno mill	Bead beating uses grinding balls and high speed spinning (or agitation) to disrupt the cell walls. This is a well-established and widely used method, which has been used both on a laboratory as well as an industrial scale.
High pressure homogenization	This method was originally employed in the dairy industry, therefore is a well-established technology, both at the laboratory and industrial scale. High pressure pumps (positive displacement pumps) are used to force cells through a valve seat, which can be tightened for a narrower orifice and extreme pressures (up to 2500 bar). The cells are sheared as they are forced through a narrow orifice. As the pressurized fluids (with sheared cells, intact cells, liquids) exits the orifice, they are subjected to sudden expansion, causing an explosion that causes additional dell disruption.
Sonication	Ultrasound in the frequencies of 20 kHz and higher are employed for cell disintegration. As the high intensity waves propagate through the liquid, it creates alternating high-pressure and low-pressure cycles. These cycles create micro-bubbles that collapse violently in a process called as cavitation. These implosions cause very high localized temperatures (5000 °K) and pressures (2000 atm.), which facilitates cell-disruption. Although this technology is not as established as the earlier two methods or was proven economically viable for algae at the commercial scale, the technology lends itself ideally to a continuous-flow pre-treatment process, therefore, was grouped with other established methods.

disruption (including autoclaving, boiling, microwave heating), osmotic shocking, lyophilization (freeze drying), liquid nitrogen, lipolysis, alkaline/chemical pre-treatment, enzyme pre-treatment, and anti-oxidant addition. It is important to note that some of these methods are not always employed at the pre-treatment stage, but also in conjunction with solvent-extraction process to augment the extraction efficiency (e.g. simultaneous sonication during solvent extraction). A brief description of two mature and one promising industrial-scale cell-disruption methods, along with a short description for each are listed in Table 1. One important but neglected step in algal cell disruption and lipid extraction is the critical assessment of industrial viability of cell-disruption/extraction-augmenting technologies. Improvements in these areas can lead to significant cost-savings in the overall lipid extraction from microalgae.

The true benefit of cell-disruption is heavily dependent on the employed extraction method and/or algal strain. For example, if effective co-solvent mixtures (such as chloroform and methanol) are employed in sufficient volumes or if cells with thin or weak cells walls (ex. *Dunaliella salina*) are used, pre-treatment or cell disruption may not be necessary. Ryckebosch and co-workers (Ryckebosch et al. 2012) demonstrated that cell-disruption techniques (fresh algae-control, lycophilization, lycophilization and sonication, lycophilization and liquid nitrogen, lycophilization and bead beating) had no significant effect on the amount of total lipids extracted using 1:1 chloroform: methanol mixture, when compared to total lipids extracted from fresh algae. However, when ethyl ether alone was employed, lycophilization with bead beating performed significantly better than all other methods and extracted 92 % of non-polar lipids in the very first extraction. They indicated that petroleum ether could not sufficiently penetrate the cell wall or dissolved the components in the cell wall of intact cells.

In contrast to the findings of Ryckebosch and coworkers, Lee at al. (1998) reported almost twice the crude lipid yield by employing mechanical disruption of *B. braunii* with their chloroform/methanol (2/1 v/v) co-solvent extraction. More recently, Lee and co-workers (Lee et al. 2010) worked with aliquots of 0.5 g dry cell biomass blended with 100 ml distilled water, which was subjected to cell disruption using five different methods (autoclaving, bead beating, microwave, sonication, and osmotic shock). Their results indicated marked differences between the different cell-disruption methods, with bead beating and microwave treatments delivering consistently high lipid yields. However, it is very important to note that the reported results for various methods were not standardized against the energy consumed during the cell-disruption (e.g. per kWh). Such standardized comparisons are crucial for assessing the viability of an industrial-scale cell-disruption process.

Another recent comparative microalgal cell-disruption study was undertaken by Prabakaran and Ravindran (2011) who tested the efficacy of sonication, osmotic shock, microwave, autoclave, and bead beating on lipid extraction efficiency of three microalgal species (*Chlorella* sp., *Nostoc* sp., *and Tolypothrix* sp.). Their results indicated that all pre-treatment methods for all three species had a marked improvement over lipid extraction from the controls (no disruption). Among the tested methods, sonication, microwaves, and bead beating had the best extractions (Fig. 3).

If one were to base their decision on the well-established Folch or Bligh and Dyer methods, which do not require cell disruption and drying for achieving extremely high lipid extraction efficiency from most algal cells, it appears as though cell disruption can be avoided with a careful selection of the choice of solvents, co-solvent ratios, and solvent: sample proportions. Increasing the duration of the solvent interaction, agitation, or solvent temperature may also be of crucial importance

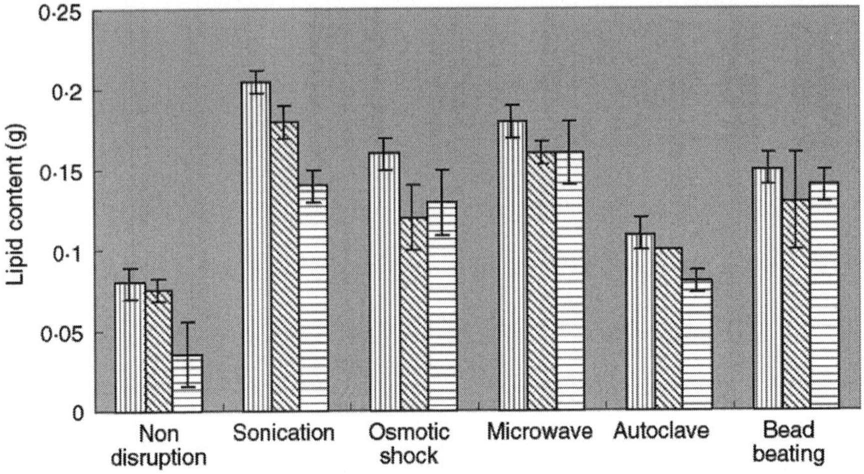

Fig. 3 Among the tested cell-disruption methods, sonication, microwave, and bead beating resulted in consistently high lipid contents. The three bars in each method represent *Chlorella* sp., *Nostoc* sp., and Tolypothrix sp., respectively (Reprinted with permission from Wiley (Prabakaran and Ravindran 2011))

for effective lipid extraction without cell-disruption. However, if one were to consider algal lipid extraction on process and economic viability, introducing a low-energy demanding cell disruption technique that will lower the solvent usage and solvent recovery costs sounds logical. Due to these mixed and contradicting results and numerous unanswered questions, better clarity on the role and need for cell-disruption or pre-treatment techniques is needed. New experiments that will not only quantify the improvements in lipid yields for each pre-treatment, but also account the energy/economic burden of each pre-treatment is very critical. Future research should also quantify lipid yield improvements for each cell-disruption method and standardize it against energy consumption.

3.2 Physical Extraction Techniques

3.2.1 Expeller or Mechanical Pressing

Oil presses or expellers are the most common method employed for extracting oils from oilseeds and nuts. The oily materials are mechanically squeezed under high pressures, which causes the material to heat up due to friction. The higher temperatures facilitate better oil recovery. Despite the simplicity of the unit and suitability for continuous operation, the extraction efficiency for commercially viable expellers is usually around 75 %. The same technology can also be used for microalgae if algae can be subjected to cell-disruption and cost-effectively dried to <5 % moisture levels. As mentioned earlier, drying algae to 5 % moisture content (or less) can induce negative energy balances (Halim et al. 2011; Scott et al. 2010), therefore, may not be a viable alternative for biofuel or other low-value applications. Another hurdle to overcome with the drying process is linked to the need for cell-disruption under moist conditions (Halim et al. 2012a; Cooney et al. 2009). In other words, the wet algal paste has to be subjected to cell-disruption prior to drying.

3.3 Solvent Extraction Techniques

3.3.1 Single Solvent Extraction

Organic solvents such as benzene, cyclo-hexane, hexane, acetone, and chloroform have been used for extracting lipids from microalgae. Solvent destroys the algal cell wall, and extracts oil from aqueous medium because of their higher solubility in organic solvents than water (Singh and Gu 2010). The oils may be extracted after subsequent distillation or solvent evaporation. Although hexane is reported to be one of the best solvents for lipid extraction, the overall lipid efficiency is relatively low as a portion of the neutral lipids are held in the cytoplasm as a complex with polar lipids. Releasing lipids from this complex is not easy as this complex is

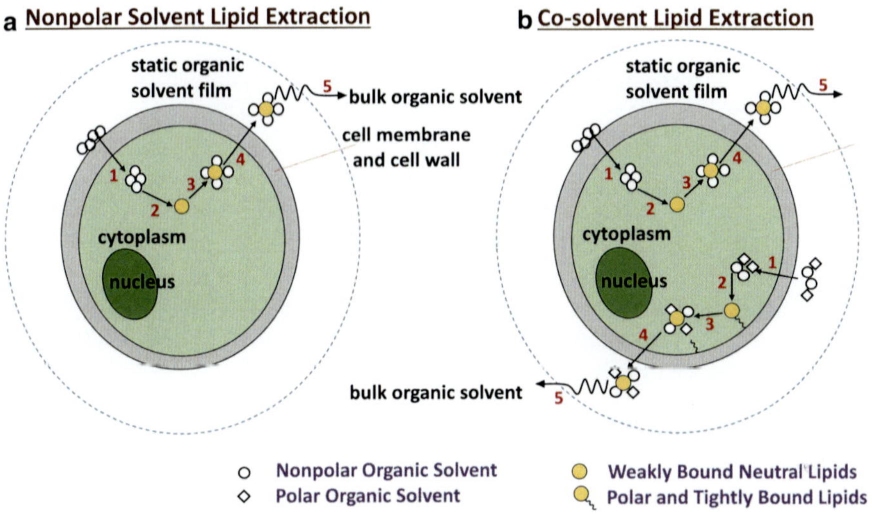

Fig. 4 Conceptual mechanisms of a single non-polar solvent (**a**) and co-solvent (**b**) based lipid extractions. Both mechanisms can be described in five steps. Step 1: penetration of organic solvent through the cell membrane. Step 2: interaction of organic solvent with the lipids. Step 3: formation of organic solvent–lipids complex. Step 4: diffusion of organic solvent–lipids complex across the cell membrane. Step 5: diffusion of organic solvent–lipids complex across the static organic solvent film into the bulk organic solvent (Recreated and modified with permission from Elsevier (Halim et al. 2012a))

strongly linked via hydrogen bonds to proteins in the cell membrane (Halim et al. 2012a). Due to this limitation, only a portion of the neutral lipids are extracted. The polar lipids in biomembranes, on the other hand are in intimate contact with aqueous phase of the electrolytes (Enssani 1990), therefore require the presence of membrane wetting medium such as polar solvents for effective extraction. These limitations led to the development of co-solvent based extraction procedures. Figure 4 depicts the conceptual mechanisms behind the single and co-solvent based extraction techniques.

3.3.2 Co-solvent Based Extraction

The co-solvent extraction method relies on the concept of "like dissolves like" and employs two-solvents for effective extraction. Lipids that are largely hydrophobic (neutral lipids) will favorably interact with relatively non-polar solvents (such as chloroform, ethyl ether, benzene), while membrane-associated polar lipids will require polar solvents (such as ethanol, methanol, isopropanol) to disrupt the hydrogen bonding and electrostatic forces between the lipids and proteins (Kates 1986b; Cooney et al. 2009).

Folch et al. (1957) were the first researchers to report a chloroform/methanol/water phase system for extraction of lipids from biological materials. This method is still considered as a classic and most reliable method for quantitative extraction of lipids (Iverson et al. 2001). This method uses 1 part of sample to 20 parts of co-solvent (2:1, chloroform/methanol) for the initial extraction into a single phase solution (Folch et al. 1957; Iqbal 2012). After the initial extraction in a single phase liquid, the mixture is subjected to several washings with water, which induces biphasic separation. Neutral and polar lipids will partition to the organic phase containing both the solvents. The non-lipid contaminants (dissolved proteins and carbohydrates) will partition to the aqueous phase.

Bligh and Dyer's co-solvent extraction is the most cited reference method in literature for the extraction of lipids from biological materials (Burja et al. 2007). Although, both Folch and Bligh & Dyer methods are reported comparable (Iverson et al. 2001), the later method uses reduced volumes of solvents. In short, Bligh and Dyer method involves mixing 1 part sample with 3 parts co-solvent (1:2, chloroform/methanol) and conducting the initial extraction in a single phase. The mixture is later converted to biphasic solution by adding metered quantities of chloroform and water. The lipids partition to the heavier chloroform layer, while the non-lipids remain in the upper methanolic layer (Iqbal 2012).

Despite very successful and reliable extractions at the lab-scale, oftentimes without any prior cell-disruption, the co-solvent based system is not fully tested for algal lipid extraction at the industrial-scale. The complicated and delicate steps of monophasic extraction, followed by conversion to bi-phasic solutions and water addition/washings is not very conducive to a continuous-flow, industrial-scale lipid extraction systems. Due to these complexities, improved methods of extraction with single solvent were explored (e.g. PLE, discussed below).

3.4 Augmented or Modified Solvent Extraction Methods

3.4.1 Soxhlet Extraction

The solubility of an analyte in solvent (single and co-solvent) is governed by the Gibbs free energy of the dissolution process, which is directly related to the equilibrium constant governing the concentration of the analyte in either phase (Mead et al. 1986). A batch extraction with a limited solvent volume will reach a saturation point as the system will be limited by the lipid mass transfer equilibrium. One way to address this problem is to add continuous fresh solvent, which allows additional solubilization of the analyte in the solvent. However, continuous addition of fresh solvent adds costs and complexity due to recover of lipids from large volumes of solvent. Soxhlet extraction process uses a Soxhlet extraction apparatus, which employs a series of ingenious cycles of solvent evaporation and condensation to provide a continuous supply the fresh solvent to the analyte held in a special thimble. This apparatus overcomes two primary limitations as the solvent is reused multiple times and the extracted lipids are held in a concentrated form within a limited volume of solvent. However, the repeated evaporation and condensation cycles add an additional economic burden on lipid extraction (Wang and Weller 2006). Halim et al. (2011) found Soxhlet operation of hexane extraction to be significantly more (280 %) efficient than a batch system when used for extracting lipids from *Chlorococcum* sp. However, the elevated temperatures potentially caused lipid degradation (Halim et al. 2012a). Therefore, despite the technical merits, the viability and suitability of an industrial-scale soxhlet extraction system for algal lipids is not clear from the reported literature.

3.4.2 Pressurized Lipid Extraction

In a pressurized lipid extraction (PLE) system, the extractions are carried out at elevated temperatures and pressures. With PLE systems (also known as accelerated solvent extraction or pressurized solvent extraction), the solubility of the analyte is greatly enhanced and the extraction process is completed in a shorter time as the desorption kinetics are greatly accelerated. Higher temperature increases molecular motion of the molecules and thereby decreasing the molecular interactions of hydrogen bonds, van der Waals forces, and dipole interactions (Cooney et al. 2009). Higher pressures increase the penetration power of the solvent through the cell wall and improve the transport of solvent to hard-to-reach areas of the cells (Cooney et al. 2009; Richter et al. 1996). The elevated pressures can reduce the dielectric constant of an otherwise immiscible solvent to values that better match the polarity of the lipids (Cooney et al. 2009; Richter et al. 1996; Herrero et al. 2006). Due to the improved penetration power and lipid extraction efficiency, many researchers are considering PLE systems that employ a single solvent, which is a technically viable alternative to the complicated co-solvent systems.

Accelerated solvent extraction was first reported by Richter et al. (1996) for extraction of chemicals from environmental samples. Numerous researchers have worked on laboratory-scale batch PLE systems with different solvents and different algal strains. Denery et al. (2004) extracted carotenoids and kavalactones from *Haemotococcus pluvialis* and *Dunaliella salina*, respectively. They found the optimum temperature and pressure to be 60 °C and 2000 psi. Presently, this technique is well known for its efficiency, shorter extraction times, and reduced solvent needs (Denery et al. 2004). Apart from the well documented literature on extraction of bioactive compounds from microalgae, the PLE method was not reported for extraction of microalgal lipids for biofuel applications (Iqbal 2012). Due to process advantages of a PLE system (e.g. use of single solvent), this system has potential for adoption at the industrial scale. As demonstrated by our research at LSU, PLE-based systems can be modified to Continuous Flow Lipid Extraction System (CFLES, Fig. 5) (Iqbal and Theegala 2013a). Our results indicated that CLFES achieved significant improvements in total glycerides at moderate temperatures and pressures (100 °C and 50 psi) when compared to Soxhlet extraction (Fig. 5). Despite potential advantages, more information is needed on the economic feasibility of algal lipid extraction using an industrial-scale PLE system.

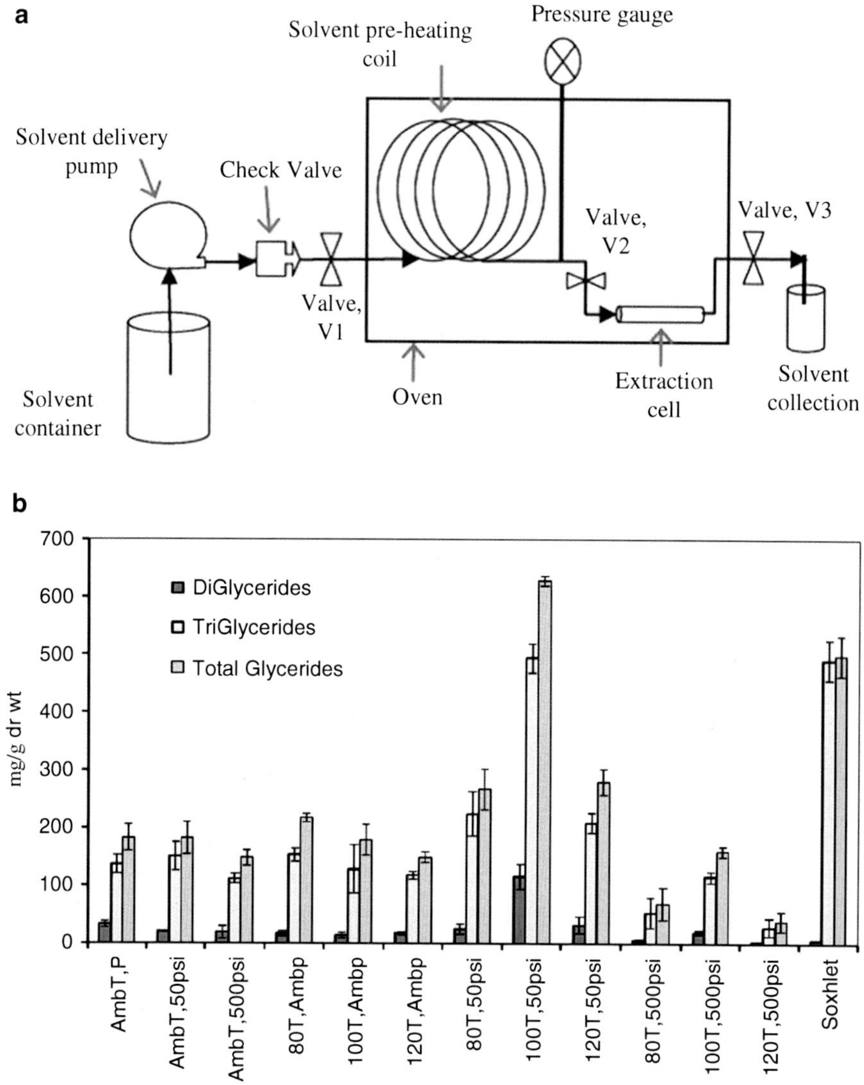

Fig. 5 (**a**) A schematic of the developed and tested Continuous Flow Lipid Extraction System (CFLES), and (**b**) comparison of total bound glycerides (mono-, di-, and triglycerides) extracted from *Nannochloropsis* sp. under different temperature and pressure combinations in CFLES and Soxhlet extraction (Reprinted with permission from Wiley (Iqbal and Theegala 2013a))

3.4.3 Microwave Assisted Solvent Lipid Extraction

Microwave radiation can be employed for assisting solvent extraction. When cells receive this radiation, localized superheating occurs which leads to instantaneous increases in temperatures and pressure within the cell matrices (Halim et al. 2012a).

The weak hydrogen bonds are disrupted by promoting the rotation of molecular dipoles, an effect opposed by the viscosity of the medium and strongly dependent upon the solvent and matrix (Cravotto et al. 2008). Our research at LSU has indicated that microwave assisted solvent extraction can have better yield than soxhlet extraction at 120 °C (Fig. 6a) (Iqbal and Theegala 2013b). As indicated earlier, microwaves have been effectively used for cell-disruption. Lee et al. (2010) have indicated that microwaves and bead-beating were the best cell-disruptors on their experiments with 3 different species of microalgae. Our experiments on extracting lipids from microwave assisted solvent extraction system indicated that cells are heavily disrupted (Fig. 6b). However, it is not clear if the microwave radiation merely causes cell-disruption or it directly aids in the solvent extraction process apart from cell disruption. This question can be answered by conducting controlled microwave assisted solvent extraction experiments on pre-disrupted algal cells.

The energy consumption for the microwave assistance is another important parameter that needs to be quantified. If one were to look at the well-established

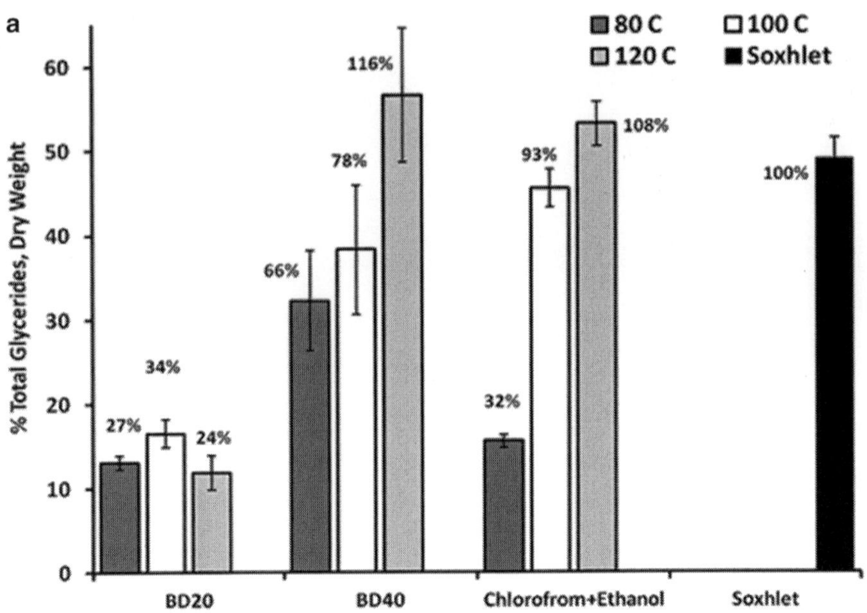

Fig. 6 (**a**) Total saturated fatty acid methyl esters (FAMEs) produced from oil extracted with microwave assisted extraction using BD20 (20 % ethanol in biodiesel), BD40 (40 % ethanol), and chloroform with ethanol as compared to conventional Soxhlet extraction. (**b**) Scanning electron microscope (SEM) images of *Nannochloropsis* sp. showing that microwave energy efficiently disrupted the microalgal cell structures. The top image shows intact cells, while the bottom image shows the cells exposed to BD40 in microwave assisted extraction at 100 °C (Reprinted with permission from Elsevier (Iqbal and Theegala 2013b))

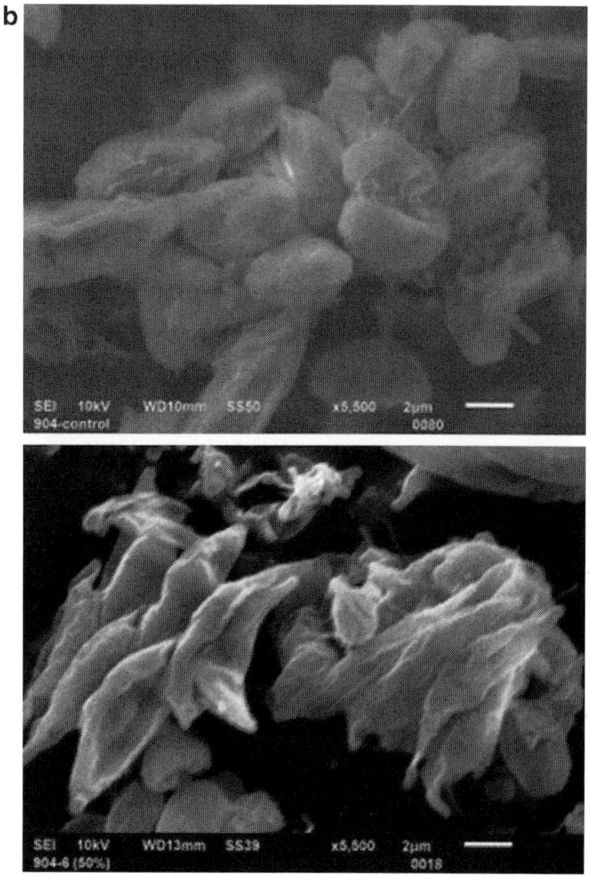

Fig. 6 (continued)

chemical industry, it can be clearly seen that electricity is the least preferred source for heating. If the efficiency factor for conversion of electricity to microwave energy is incorporated into the computations, the economic viability of microwave technology may be further impeded. Apart from these possible limitations, microwave assistance has several undisputed benefits for algal lipid extraction. When compared to traditional heating, microwaves can impart the energy in a very short time period. Secondly, as microwave heating is done without any direct liquid contact, development of an industrial-scale, continuous-flow microwave assisted solvent extraction system appears to be technically feasible. Thirdly, microwave assistance improves the lipid yields significantly (better than Soxhlet extraction). However, as indicated earlier, the improvements from microwave based methods have to be compared with other methods and standardized against energy consumption (e.g. energy consumed per unit increase in lipid yield).

3.4.4 Ultrasound Assisted Solvent Lipid Extraction

Apart from the use of ultrasound for cell-disruptions (discussed earlier), ultrasound has been employed during solvent extraction or along with a solvent. Wiltshire et al. (2000) have reported more than 90 % extraction of fatty acids and pigments from *Scenedesmus obliquus* with ultrasound (Wiltshire et al. 2000). Complete extraction of lipids from *Chaetoceros gracilis* by using ultrasound was subsequently evaluated by Pernet and Tremblay (2003). It was concluded that ultrasonic method increased the extraction rate, which directly influences the overall lipid recovery. However, it is not clear if ultrasound employed during/along-side the solvent extraction mainly assists in the cell-disruption or if it enhances extraction kinetics due to other mechanisms. Experimental results that quantify the improvements in lipid yield (per unit input energy) from employing ultrasound techniques, either at cell-disruption stage or during solvent extraction, are critically needed. In contrast to positive results, ultrasound was also reported as ineffective by researchers focusing on cell-disruption. Halim et al. (2012b) indicated that ultrasonication for 25 min of both low-density and high-density cultures, even at the highest power level (130 W), failed to effectively rupture the cells (*Chlorococcum* with thick cell walls), indicating that cell walls play a critical role. Apart from several unanswered technical questions, sufficient information on the feasibility or economics for a commercial-scale algal ultrasonication system is still not reported in literature (Singh and Gu 2010). Despite the drawback and limited/conflicting results, the ultrasound technology lends itself perfectly to a continuous-flow, industrial-scale process (be it cell-disruption or lipid extraction), therefore, merits further exploration and viability assessment.

3.5 *Other Extraction Methods*

3.5.1 Supercritical Fluids Extraction

Supercritical fluid extraction (SFE) is an emerging green technology that has the potential to replace the traditional organic solvent extraction (Halim et al. 2012a). In simple terms, when the pressure and temperature of a fluid are raised above their critical values (Tc and Pc), the fluid enters a supercritical region. In supercritical state, the fluid attains gas-like mass transfer properties and liquid-like solvating properties with diffusion coefficients greater than those of a liquid (Luque de Castro et al. 1994; Romanik et al. 2007; Leonard et al. 2008). While most of the applications employed CO_2 and water, several other fluids such as methanol, ethanol, and pentane have been reported in literature. Carbon dioxide has gained utmost importance as it has moderate critical properties (Fig. 7, 31.1 °C and 73.8 bar), low toxicity, low flammability, and chemical inertness (Cooney et al. 2009; Halim et al. 2012a). Several researchers have reported the use of SFE for effective algal lipid, fatty acid, and pigment extraction (Mendes et al. 1995, 2003, 2006; Taylor 1996).

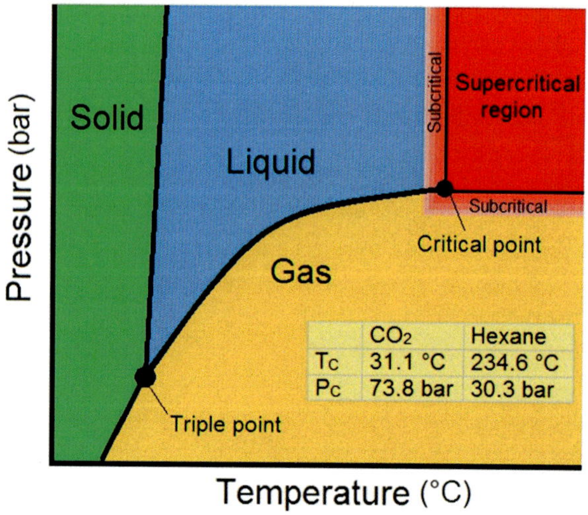

Fig. 7 A typical P-T phase diagram showing the solid, liquid, and gas phases and supercritical point. The supercritical temperature (Tc) and Supercritical pressure (Pc) for CO_2 and hexane are shown

Canela et al. (2002) looked into supercritical extraction of fatty acids and carotenoids from *Spirulina maxima*. Their experiments indicated that the temperature and pressure had little effect (above Tc and Pc), but the extraction rates were significantly different. Andrich et al. (2005) explored extraction of bioactive lipids from *Nannochloropsis* sp. using supercritical fluid extraction. They reported SFE to have comparable extraction efficiency to that of solvent extraction using hexane. Researchers have also looked into the variations to the traditional CO_2-based SFE. Subcritical solvent extraction was explored and found to have certain advantages over supercritical solvent extraction. The less intensive subcritical conditions retain certain features of the supercritical solvent extraction at a lower operational cost (Herrero et al. 2005). Chen et al. (2011) used ethanol at subcritical conditions and extracted lipids from wet past of *Nannochloropsis sp.* with a maximum efficiency of 90.21 % of total lipids.

Despite the lack of demonstrated deployment of the SLE technologies at an industrial-scale for algal lipids, the SLE process warrants further consideration as it has several potential merits, some of which include: (1) tunable solvating power, (2) improved lipid yield due to higher penetration power of supercritical fluids, (3) shorter extraction time as the fluids have liquid–gas properties, (4) production of solvent-free crude lipids, (5) inherent safety of an industrial-scale SFE system, and (6) process suitability for a continuous-flow, industrial-scale SFE unit (Halim et al. 2012a; Halim et al. 2011). Presently, the biggest drawback is its intensive energy needs (Cooney et al. 2009). Future research may perhaps offer a lower energy demanding process (or sub-critical process) with improved energy reuse.

3.5.2 Direct Transesterification

Historically, biodiesel oils are first extracted and purified and later subjected to transesterification to produce biodiesels. Direct transesterification is a "single-step" process that converts saponifiable lipids directly to fatty acid methyl esters (FAME) using an in-situ extraction/transesterification process. This method is gaining a lot of attention, at least at the laboratory-scale for analytical applications. In short, the process involves adding alcohol (such as methanol) and catalyst (base or acid catalyst) to algal paste (or dried algal biomass) and produce fatty acid methyl ester at elevated temperatures. Several researchers have employed reaction enhancing techniques such as microwaves and ultrasonication to achieve better extraction efficiencies. Positive results were reported from both assisted and un-assisted direct transesterification techniques. Griffiths et al. (2010) indicate that the most commonly used method of Bligh and Dyer was the least effective method for fatty acid production from three different microalgae, when compared to Smedes and Askland, Folch, and direct transesterification. They employed wet algal paste (19.6–27.4 % dry weight) and indicated that up to 10 % of water of the total reaction volume had no detrimental effects on transesterification. Koberg et al. (2011) demonstrated that both microwave and sonification- assisted direct transesterification process had better yields than the 2-stage processes. In another study, wet algal biomass (*Nannochloropsis* sp.) with 90 % water was subjected to a simultaneous lipid extraction/transesterification process using supercritical methanol (Patil et al. 2011). They indicated the single-step process has favorable energy balance as the drying and extraction needs are eliminated. Despite promising results at the laboratory scale, it remains to be seen whether direct transesterification can be proven to be viable and cost-effective for industrial-scale conversion of algal lipids to biodiesel.

3.5.3 Cell Milking

Cell milking for microalgal lipids is a relatively new concept. In concept, just like "milking cows", a portion of the lipids inside the cells are extracted without affecting the cell viability. The "milked cells" are returned for continued growth and repeatedly milking. Hejazi et al. (2002) reported successful extraction of β-carotene from *Dunaliella salina*. According to their observations, solvents with higher hydrophobicity (decane and deodecane) are gentle on the cell walls and can extract triglycerides from microalgal cells without loss of cell viability. However, Cooney et al. (2009) stated that the effectiveness of cell milking is limited to cells that are "porous" or have "open pores" such as *Dunaliella*. They also indicated that long term testing of cell viability remains to be done.

3.5.4 Genetic Engineering

Although not fully documented or demonstrated, one emerging concept of lipid extraction involves genetic modification of photosynthetic organisms to secrete lipids through their cell membrane into the culture media. REG Life Sciences is one company that claims to have produced bacteria-based biological catalysts, which have been engineered to selectively convert various sugars using a single-step fermentation process to drop-in and differentiated products (REG Life Sciences 2015). Although conceptually appealing, such processes are reliant on a sugar source supporting fermentation (Cooney et al. 2009). Therefore, genetic engineering on autotrophic microalgae is perhaps the most logical direction from a sustainability point-of-view.

3.5.5 Hydrothermal Liquefaction

Hydrothermal liquefaction (HTL) is one technology that has caught the attention of numerous researchers and industries. However, the HTL process does not produce saponifiable lipids that can be converted to biodiesels. The HTL process produces bio-oils (that resemble fossil crude) from wet biomass slurries at elevated temperatures and pressures. Our prior research has demonstrated that oil production from the HTL process using pine sawdust and switchgrass is particularly promising (Midgett et al. 2012). However using oily biomass as feedstock for the HTL process does not appear to be justifiable. Producing a concoction of oxygenated products from an oily feedstock that is rich in triglycerides does not sound like a logical approach (Fig. 8). Our research has indicated that oily feedstocks (such as tallow seeds, peanuts, and pure vegetable oil), although may result in higher oil production and energy density (MJ/kg), they do not offer major advantages (energy content or compositional) over low value cellulosic or waste feedstocks (such as dairy manure, poultry litter, pine sawdust, and switchgrass) (Midgett et al. 2012). Looking from a practicality and economic perspective, it makes sense to use a slightly larger volume of low-value cellulosis feedstock as opposed to oily feedstocks. For example, approximately 1.4 tons of switchgrass produces the same energy (in MJ) as 1 ton of crushed peanuts in a HTL system. Therefore, using algal biomass in a HTL system may not be justifiable due to difficulties in production and drying of large quantities of algal biomass.

4 Potential Industrial Scale Algal Lipid Extraction Technologies

Despite decades of results from laboratory-based experiments, there is no settlement on the most effective algal lipid extraction method (Lee et al. 2010). Identification of a universally effective and economically viable algal lipid extraction technology is difficult given the diversity of algal lipids, variations in cell wall composition, end use of the extracted lipids, and limitations of various cell

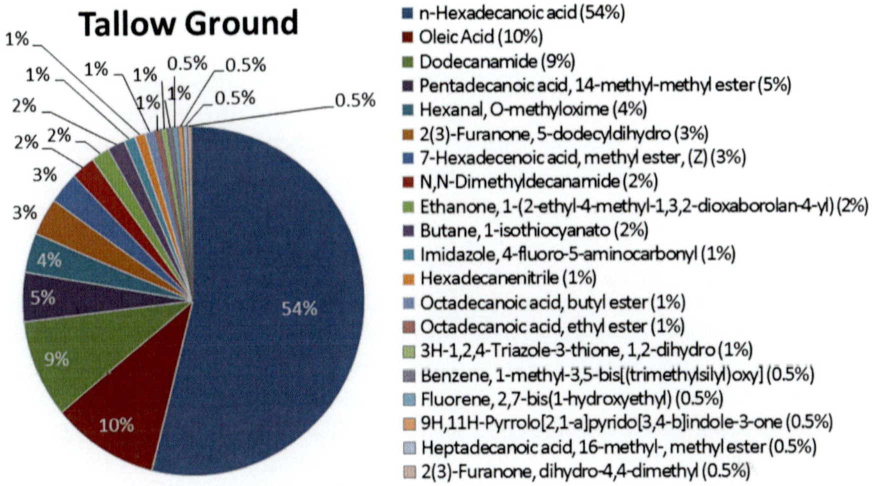

Fig. 8 HTL process converts homogenous triglycerides in oily seeds (including microalgae) to a concoction of low-value oxygenated compounds. The top 20 most abundant compounds in the acetone soluble fraction produced from tallow ground accounted for 100 % of the area of the feedstock. The top 20 selection was based on the largest peak areas in the GC-MS chromatogram. All values are reported as weight percent (Reprinted with permission from Springer (Midgett et al. 2012))

disruption and extraction methods. There is clearly a dearth of information on pilot or commercial-scale algal cell disruption and lipid extraction (Mercer and Armenta 2011; Halim et al. 2012b). Therefore any recommendations and suggestions for potential industrial-scale lipid extraction processes are based on the reported (at laboratory scale) process advantages, limitations, economics, logical reasoning, and perceived process viability at an industrial-scale.

Irrespective of the solvent or solvents used for lipid extraction, cell disruption appears to be a logical choice due to relatively lower energy needs. Bead beating or sonication process, although not proven for industrial algal applications, appear to be the most suitable options for continuous flow industrial operations. As drying algal paste beyond the 20 % solids content may introduce negative energy balance, lowering the moisture below 80 % does not make economic sense. However, availability of reject heat or alternative-energy-based drying options (e.g. solar or geothermal energy) may justify further drying. If cells are subjected to further drying (below 80 % moisture) it is important to disrupt the algal cells first before drying as drying is known to irreversibly close the pores in the cell wall and retard solvent access (Roder and Sixta 2004).

Disrupted cells can then be subjected to lipid extraction. Selection of the best lipid extraction method is more complicated than cell disruption as several technically viable options exist. The final choice of the process will have to be based on process efficiency, amenability to continuous flow mode of operation, commercial scalability, operational costs, solvent cost, chemical stability of the desired end product (which is affected by the choice of extraction method), environmental impact of the method, and suitability to a biorefinery model. Direct transesterification or use of biodiesel as a co-solvent showed promise at the laboratory level and may be an option if biodiesel is the primary target product. However, this method seriously hampers the biorefinery approach as exposure to toxic chemicals limit the usability of the leftover proteins, carbohydrates, and other value-added products for human or animal uses. Cell milking is likely to have very limited applicability as only selective species with weak cell walls can be milked. Out of the remaining methods, pressurized solvent extraction, ultrasound assisted solvent extraction, microwave assisted solvent extraction, and supercritical fluid extraction processes appear to be conducive to continuous-flow, industrial-scale algal lipid extraction systems. Microwave technology, despite its merits, may be limited by the high operational expenses. Microwave technology starts with electricity (which is an expensive option for industrial heat) and adds an additional efficiency factor (conversion of electricity to microwave energy), which further adds to the economic burden and therefore may be limited to high-value applications (other than biodiesel). Exclusion of the MW technology points to pressurized solvent extraction, ultrasound solvent extraction, and supercritical fluid (CO_2) extraction as the three finalist algal lipid extraction technologies. Each of the finalist processes have their own merits and limitations and need validation at a pilot-plant or industrial-scale. From an environmental perspective, supercritical CO_2 is perhaps the most benign technology as it uses naturally abundant and non-toxic CO_2. The residual biomass and byproducts from supercritical CO_2 extraction process will likely have the widest applications and therefore is the most suitable option for algal biorefineries.

List of Abbreviations

BD20	Solvent mixture with 20 % ethanol and 80 % biodiesel
BD40	Solvent mixture with 40 % ethanol and 60 % biodiesel
CFLES	Continuous flow lipid extraction system
DAG	Diacylglycerols
FAME	Fatty acid methyl esters
FFA	Free fatty acids
GL	Glycolipids
HTL	Hydrothermal Liquefaction
MAG	Monoacylglycerols
NL	Neutral lipids
PL	Phospholipids
PLE	Pressurized lipid extraction
TAG	Triacylglycerols
SEM	Scanning electron microscope
SFE	Supercritical fluid extraction

References

Andrich G, Nesti U, Venturi F, Zinnai A, Fiorentini R (2005) Supercritical fluid extraction of bioactive lipids from the microalga *Nannochloropsis* sp. Eur J Lipid Sci Technol 107(6):381–386

Becker W (2004) Microalgae in human and animal nutrition. In: Richmond A (ed) Handbook of microalgal culture. Blackwell Publishers, Oxford, pp 312–351

Bligh EG, Dyer WJ (1959) A rapid method of total lipid extraction and purification. Can J Biochem Physiol 37(8):911–917

Burja AM, Armenta RE, Radianingtyas H, Barrow CJ (2007) Evaluation of fatty acid extraction methods for *Thraustochytrium* sp. J Agric Food Chem 55(12):4795–4801

Canela APRF, Rosa PTV, Marques MOM, Meireles MAA (2002) Supercritical fluid extraction of fatty acids and carotenoids from the microalgae *Spirulina maxima*. Ind Eng Chem Res 41(12):3012–3018

Chen M, Chen X, Liu T, Zhang W (2011) Subcritical ethanol extraction of lipid from wet microalgae paste of *Nannochloropsis* sp. J Biobased Mater Bioenerg 5(3):385–389

Chisti Y (2007) Biodiesel from microalgae. Biotechnol Adv 25:294–306

Cooney M, Young G, Nagle N (2009) Extraction of bio-oils from microalgae. Sep Purif Rev 38(4):291–325

Cravotto G, Boffa L, Mantegna S, Perego P, Avogadro M, Cintas P (2008) Improved extraction of vegetable oils under high-intensity ultrasound and/or microwaves. Ultrason Sonochem 15(5):898–902

Dassey AJ, Hall SG, Theegala CS (2014) An analysis of energy consumption for algal biodiesel production: comparing the literature with current estimates. Algal Res 4:89–95

Denery JR, Dragull K, Tang CS, Li QX (2004) Pressurized fluid extraction of carotenoids from *Haematococcus pluvialis* and *Dunaliella salina* and kavalactones from *Piper methysticum*. Anal Chim Acta 501(2):175–181

Enssani E (1990) A Method for the extraction of liquid hydrocarbons from microalgal biomass. Energy conversion engineering conference, IECEC-90, Proceedings of the 25th Intersociety, Piscataway, NJ, 6:250–255

Folch J, Lees M, Sloane-Stanley GH (1957) A simple method for the isolation and purification of total lipids from animal tissues. J Biol Chem 226(1):497–509

Griffiths MJ, Van Hille RP, Harrison STL (2010) Selection of direct transesterification as the preferred method for assay of fatty acid content of microalgae. Lipids 45(11):1053–1060

Halim R, Gladman B, Danquah MK, Webley PA (2011) Oil extraction from microalgae for biodiesel production. Bioresour Technol 102(1):178–185

Halim R, Danquah MK, Webley PA (2012a) Extraction of oil from microalgae for biodiesel production: a review. Biotechnol Adv 30(3):709–732

Halim R, Harun R, Danquah MK, Webley PA (2012b) Microalgal cell disruption for biofuel development. Appl Energy 91(1):116–121

Hejazi MA, de Lamarlie C, Rocha JMS, Vermue M, Tramper J, Wijffels RH (2002) Selective extraction of carotenoids from the microalga *Dunaliella salina* with retention of viability. Biotechnol Bioeng 79(1):29–36

Herrero M, Senorans FJ, Cifuentes A, Ibanez E (2005) Optimization of accelerated solvent extraction of antioxidants from *Spirulina platensis* microalga. Food Chem 93(3):417–423

Herrero M, Cifuentes A, Ibanez E (2006) Sub- and supercritical fluid extraction of functional ingredients from different natural sources: plants, food-by-products, algae and microalgae. A review. Food Chem 98(1):136–148

Iqbal J (2012) Development of cost-effective and benign lipid extraction system for microalgae. Doctoral Dissertation, Louisiana State University

Iqbal J, Theegala CS (2013a) Optimizing a continuous flow lipid extraction system (CFLES) used for extracting microalgal lipids. GCB Bioenergy 5(3):327–337

Iqbal J, Theegala CS (2013b) Microwave assisted lipid extraction from microalgae using biodiesel as co-solvent. Algal Res 2(1):34–42

Iverson SJ, Lang SLC, Cooper MH (2001) Comparison of the Bligh and Dyer and Folch methods for total lipid determination in a broad range of marine tissue. Lipids 36(11):1283–1287

Kates M (1986a) Definition and classification of lipids. Techniques of lipidology: isolation, analysis, and identification of lipids. Elsevier Science Publisher, Amsterdam

Kates M (1986b) Techniques of lipidology: isolation, analysis, and identification of lipids. In: Bruden RH, van Knippenberg PH (eds) Laboratory techniques in biochemistry and molecular biology, 2nd edn, Vol 3, Part 2. Elsevier, New York, pp 100–111

Koberg M, Cohen M, Ben-Amotz A, Gedanken A (2011) Bio-diesel production directly from the microalgae biomass of *Nannochloropsis* by microwave and ultrasound radiation. Bioresour Technol 102(5):4265–4269

Lee SJ, Yoon BD, Oh HM (1998) Rapid method for the determination of lipid from the green algae *Botryococcus braunii*. Biotechnol Tech 12(7):553–556

Lee JY, Yoo C, Jun SY, Ahn CY, Oh HM (2010) Comparison of several methods for effective lipid extraction from microalgae. Bioresour Technol 101(1):S75–S77

Leonard E, Nielsen D, Solomon K, Prather KJ (2008) Engineering microbes with synthetic biology frameworks. Trends Biotechnol 26(12):674–681

Luque de Castro MD, Valcarcel M, Tena MT (1994) Analytical supercritical fluid extraction. Springer, Heidelberg

Mead JF, Alfin-Slater RB, Howton DR, Popjak G (1986) Lipids: chemistry, biochemistry, and nutrition. Plenum Press, New York

Mendes RL, Coelho JP, Fernandes HL, Marrucho IJ, Cabral JM, Novais JM, Palavra AF (1995) Applications of supercritical CO_2 extraction to microalgae and plants. J Chem Technol Biotechnol 62(1):53–59

Mendes RL, Nobre BP, Cardoso MT, Pereira AP, Palavra AF (2003) Supercritical carbon dioxide extraction of compounds with pharmaceutical importance from microalgae. Inorg Chim Acta 356:328–334

Mendes RL, Reis AD, Palavra AF (2006) Supercritical CO_2 extraction of gamma-linolenic acid and other lipids from *Arthrospira (Spirulina) maxima*: comparison with organic solvent extraction. Food Chem 99(1):57–63

Mercer P, Armenta RE (2011) Developments in oil extraction from microalgae. Eur J Lipid Sci Technol 113(5):539–547

Midgett JS, Stevens BE, Dassey AJ, Spivey JJ, Theegala CS (2012) Assessing feedstocks and catalysts for production of bio-oils from hydrothermal liquefaction. Waste Biomass Valorization 3(3):259–268

Patil PD, Gude VG, Mannarswamy A, Deng S, Cooke P, Munson-McGee S, Rhodes I, Lammers P, Nirmalakhandan N (2011) Optimization of direct conversion of wet algae to biodiesel under supercritical methanol conditions. Bioresour Technol 102(1):118–122

Pernet F, Tremblay R (2003) Effect of ultrasonication and grinding on the determination of lipid class content of microalgae harvested on filters. Lipids 38(11):1191–1195

Prabakaran P, Ravindran AD (2011) A comparative study on effective cell disruption methods for lipid extraction from microalgae. Lett Appl Microbiol 53(2):150–154

REG Life Sciences (2015) Technology Overview. http://www.reglifesciences.com/technology/technology-overview. Accessed 2 Feb 2015

Richter BE, Jones BA, Ezzell JL, Porter NL, Avdalovic N, Pohl C (1996) Accelerated solvent extraction: a technique for sample preparation. Anal Chem 68(6):1033–1039

Roder T, Sixta H (2004) Thermal treatment of cellulose pulps and its influence to cellulose reactivity. Lenzinger Berichte 83:79–83

Romanik G, Gilgenast E, Przyjanzy A, Kaminski M (2007) Techniques of preparing plant material for chromatographic separation and analysis. J Biochem Biophys Methods 70(2):253–261

Ryckebosch E, Muylaert K, Foubert I (2012) Optimization of an analytical procedure for extraction of lipids from microalgae. J Am Oil Chem Soc 89(2):189–198

Scott SA, Davey MP, Dennis JS, Horst I, Howe CJ, Lea-Smith DJ, Smith AG (2010) Biodiesel from algae: challenges and prospects. Curr Opin Biotechnol 21(3):277–286
Singh J, Gu S (2010) Commercialization potential of microalgae for biofuels production. Renew Sust Energ Rev 14(9):2596–2610
Spolaore P, Joannis-Cassan C, Duran E, Isambert A (2006) Commercial applications of microalgae. J Biosci Bioeng 101(2):87–96
Taylor LT (1996) Supercritical fluid extraction. Wiley, New York
Wang L, Weller CL (2006) Recent advances in extraction of nutraceuticals from plants. Trends Food Sci Technol 17(6):300–312
Wijkstrom U, Gumy A, Grainger R (2000) The state of world fisheries and aquaculture. Food and Agriculture Organization of the United Nations, Rome
Wikfors GH, Ohno M (2001) Impact of algal research in aquaculture. J Phycol 37(6):968–974
Wiltshire KH, Boersma M, Moller A, Buhtz H (2000) Extraction of pigments and fatty acids from the green alga *Scenedesmus obliquus* (Chlorophyceae). Aquat Ecol 34(2):119–126

Microalgal-Derived Biomethanization and Biohydrogen Production – A Review of Modeling Approaches

Pascal Kosse, Marc Wichern, and Manfred Lübken

Abstract Microalgae represent an excellent example of the pressing need of our society for sustainable energy as they have experienced increased momentum of interest as a promising feedstock for biomethane and biohydrogen production. While biomethane can be obtained from various microalgal species, biohydrogen production mostly involves the single-cell green alga *Chlamydomonas reinhardtii*. In this context, it is of general agreement in science that the development of mathematical models supports the understanding of the biochemical processes involved besides helping to optimize the process engineering. This chapter reviews the approaches to mathematically model processes of microalgal-derived biomethanization and biohydrogen production. Regarding biomethanization a standardized model framework exists in the form of the Anaerobic Digestion Model No. 1 (ADM1), which has been developed on an ongoing basis. In the context of anaerobic digestion of microalgae, studies available regarding the application of the ADM1 focus entirely on the improvement of kinetic description (Contois) and on a closer consideration of the effect of salinity for marine cultures. Future attention will certainly be given to a detailed determination of stoichiometric model parameters for microalgae as has already been done for other substrates. Besides the ADM1, there are hardly any other modeling approaches published to date for modeling microalgal digestion. Though biohydrogen production, on the contrary, is subject to a wider variety of modeling approaches, a standardized model framework has not yet been distilled from the majority of available models. However, a uniform trend is clearly seen in the application of the S-system modeling framework with further modifications.

Keywords Anaerobic digestion • Microalgae • Biomethane • Biomethanization • Biohydrogen • Mathematical simulation • Anaerobic Digestion Model No. 1 (ADM1) • S-system • *Chlamydomonas reinhardtii* • Modeling • Substrate characterization • Sulfur-deprived photosynthesis • Direct biophotolysis • Indirect biophotolysis • Alga protein determination • Stoichiometric coefficients • Photosynthesis

P. Kosse (✉) • M. Wichern • M. Lübken
Institute of Urban Water Management and Environmental Engineering, Ruhr-Universität Bochum, Bochum, Germany
e-mail: pascal.kosse@ruhr-uni-bochum.de

A. Prokop et al. (eds.), *Algal Biorefineries*, DOI 10.1007/978-3-319-20200-6_14

1 Introduction

Microalgae represent our world's "green lung" as they generate about 50 % of our planet's oxygen, while converting atmospheric carbon dioxide to organic compounds in return. Their true significance today goes beyond this ancient biochemical process known as photosynthesis. Nowadays, our society has to face the upcoming energy shortage of conventional fossil fuels more and more, and this demands creative sustainable solutions. Microalgae fulfill the sustainability criterion and have, thus, become an important basis for biorefinery concepts regarding the production of biofuels, such as biohydrogen, biodiesel and bioethanol. Among these products, biohydrogen is of great interest, since it possesses the highest energy density known for any fuel of 142 $MJ \cdot kg^{-1}$ (for comparative purposes: kerosene 46.3 $MJ \cdot kg^{-1}$, diesel 45.3 $MJ \cdot kg^{-1}$ and gasoline 45.8 $MJ \cdot kg^{-1}$), while it does not possess a carbon footprint and emits water on combustion (Kamat and Bisquert 2013). Nevertheless, the production of biomethane in the perspective of anaerobic microalgal digestion plays an equivalently important role as it is used to generate onsite electrical power or thermal heat to offset biomass processing and extraction processes (Ward et al. 2014).

However, even if the processes of anaerobic biomethane and biohydrogen production are environmentally sound, it is important to consider the complexity of their interrelated chemical and microbiological mechanisms, in order to understand and optimize them. In this context, the development and validation of mathematical models can assist the understanding of the process dynamics and reveals optimization opportunities (Lauwers et al. 2013).

This chapter will provide an overview of state-of-the-art modeling approaches towards anaerobic microalgae digestion for both biomethane and biohydrogen production based upon the latest scientific publications. The approaches will be comparatively analyzed to show their individual limits of applicability.

2 Modeling of Biohydrogen Production in Microalgae

The majority of models found in literature dealing with microalgal biorefinery products are devoted to biohydrogen production. This is hardly surprising, due to its reputation as a promising alternative to circumvent the upcoming energy shortage of conventional fuels.

The majority of modeling approaches orient themselves around the facultative photoautotrophic and photoheterotrophic microalga *Chlamydomonas reinhardtii* (Fig. 1) (Melis 2007; Melis et al. 2000), while other species, such as *Chlorella vulgaris* (Guan et al. 2004b), *Chlorella pyrenoidosa* (Kojima and Lin 2004), *Platymonas subcordiformis* (Guan et al. 2004a, b) or *Spirulina platensis* (Aoyama et al. 1997), are in the focus of research for hydrogen production. Despite the employed microalgal species, the individual photosynthetic hydrogen production rate depends on various factors, such as strain type, nutrient composition, temperature, pH and light

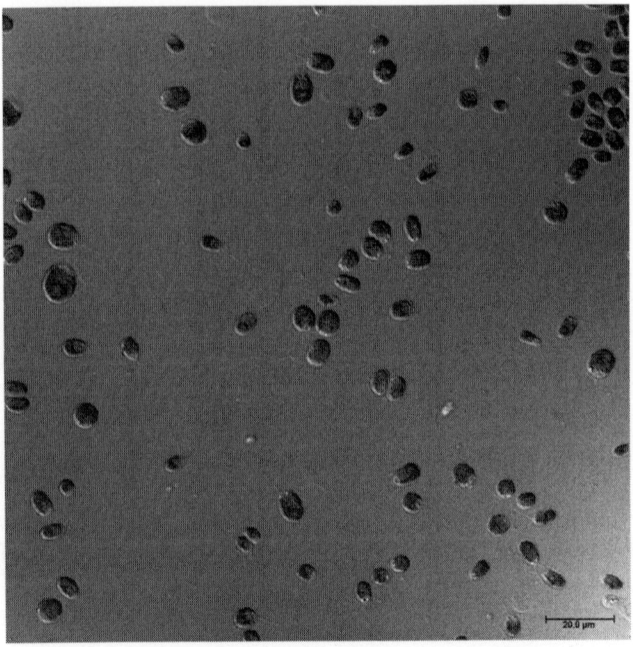

Fig. 1 Transmission image of *Chlamydomonas reinhardtii* at 630× magnification (200 Hz). Image taken with confocal laser scanning microscope TSC SP8-cLSM (Leica Microsystems GmbH, Wetzlar, Germany)

intensity (Jo et al. 2006). In order to define modeling structures, it is indispensable to have fundamental knowledge of the interactions between the process variables and the process behavior beforehand, however, these are often discounted.

Microalgae carry out plant-like oxygenic photosynthesis using chlorophyll a and other pigments to capture energy from sunlight using the photosynthetic systems II and I (PSII and PSI). The process starts, under aerobic conditions, with the splitting of water into electrons (e^-), oxygen (O_2) and hydrogen (H^+) ions. An electron from water is donated to the oxidized photosystem II molecule following the absorption of a quantum of light near 680 nm. The light energy converts the PSII into a moderately strong reductant capable of raising an electron from a lower to a higher energy state. As a result, the photosystem II loses its electron to an electron acceptor that belongs to the electron transport system (ETS) embedded within the thylakoid membrane. On the ETS, the electron travels through several membrane carriers known as pheophytin, quinones, cytochromes and a copper-containing protein called plastocyanin. As the electron travels from one electron carrier to another, it releases its energy, which is used to pump proton ions (H^+) ions into the thylakoid membrane. In that way, the ETS creates a proton concentration gradient inside the thylakoid membrane that is used to synthesize adenosine triphosphate (ATP). The process of ATP production is called photophosphorylation. ATP is produced out of adenosine diphosphate (ADP) and inorganic phosphorus (P_i) when H^+ ions leaving the thylakoid membrane through the enzyme ATP synthase. Therefore, the whole process has two sources of H^+ ions, one out of the splitting of water, and a second one through the pumping action of the ETS. Finally, the electron is accepted by the chlorophyll of photosystem I, which has previously absorbed light quanta. As a result, PSI mediates the electron through plastocyanin to another electron acceptor from a second ETS, known as ferredoxin (Fd). The latter one passes the electron to $NADP^+$, which becomes reduced to NADPH by Fd-$NADP^+$ reductase (FNR) (Batie and Kamin 1984). Under anaerobic conditions, electrons (e^-) from the reduced ferredoxin (Fd) can also be used by the hydrogenase or nitrogenase to reduce protons evolving molecular hydrogen (Vargas et al. 2014; Ghysels et al. 2013) (Eq. 1).

$$2H^+ + 2Fd^- \rightarrow H_2 + 2Fd \tag{1}$$

This reaction pathway is included under the terms *direct* (Eq. 2) and *indirect photolysis* (Eq. 3), which constitute the main relevant processes for hydrogen production.

$$\begin{aligned} &2H_2O \xrightarrow{h\cdot v} 2H_2 + O_2 \\ &12H_2O + 6CO_2 \xrightarrow{h\cdot v} C_6H_{12}O_6 + 6CO_2 \end{aligned} \tag{2}$$

$$C_6H_{12}O_6 + 12H_2O \rightarrow 12H_2 + 6CO_2 \tag{3}$$

The main difference between direct and indirect photolysis is that indirect photolysis involves a temporal separation between the light and dark reaction of photosynthesis. In a fist stage, photoautotrophically grown cultures accumulate carbohydrates, such as starch in microalgae or glycogen in cyanobacteria, while producing molecular oxygen in parallel. In a second stage, these carbohydrates can be decomposed during fermentation under anaerobic conditions producing molecular hydrogen with CO_2 evolution (Huesemann et al. 2010; Kennes 2013). In that way indirect biophotolysis overcomes two major drawbacks from direct photolysis, which are oxygen inhibition of the hydrogenase and the generation of potentially explosive H_2-O_2 mixtures (Huesemann et al. 2010).

Keeping to the example of *Chlamydomonas reinhardtii*, one approach to artificially induce the necessary anoxic conditions is to place *C. reinhardtii* cells in a sulfur-deprived medium (Fouchard et al. 2005). The oxygenic photosystem II activity is, consequently, reduced, while maintaining the cell's respiration, resulting in a decrease in O_2 concentration (Wykoff et al. 1998). When the photosynthetic oxygen generation falls below the rate of O_2-uptake by respiration, the cells' culture gradually becomes anaerobic, inducing the activity of reversible Fe-hydrogenase, which occurs after 1 or 2 days (Melis et al. 2000). This metabolic process constitutes the basis for the majority of models dealing with microalgal hydrogen production (Horner 2002; Jorquera et al. 2008).

For the purpose of completeness, other relevant microalgal-based hydrogen processes will be mentioned here briefly, such as photo-fermentation and dark fermentative hydrogen production during the acidogenic phase of anaerobic digestion of organic matter (Levin 2004; Kapdan and Kargi 2006). The emphasis in dark fermentation is rather on the production of a mixed biogas containing primarily H_2, CO_2, methane (CH_4), CO and hydrogen sulfide (H_2S). The preferred substrates for dark fermentation are glucose or other polymers, such as starch or cellulose that are produced by microalgae beforehand. Nonetheless, these processes have not been addressed in modeling attempts so far. A combination of dark and photobiological processes has, however, been shown to be the most efficient approach and is considered for setting up models (Chandra and Venkata Mohan 2011), which are mostly limited to S-system formulations (Horner 2002; Park and Moon 2007; Jorquera et al. 2008; Zhang 2011, 2012, 2013). Additionally, there are only two further models worth mentioning: the Volume Element Model proposed by Vargas et al. (2014) and a mechanistic model proposed by Williams and Bees (2014). Apart from these independent model formulations, there are certain techniques available that result in model formulations, such as the Response Surface Methodology, which is a statistical approach, or the flux balance analysis (Jo et al. 2006; Yun et al. 2012), which leads to a stoichiometric model (Boyle and Morgan 2009).

S-system modeling was most frequently applied and further improved for modeling microalgal biohydrogen production. S-system modeling dates back to the 1960s, when it was first proposed by Savageau (1969a, b, 1976) as a modeling approach based on nonlinear differential equations. The actual first attempt to model biohydrogen production in microalgae started in 2002 with an S-system model introduced by Horner (2002). The model consists of 15 ordinary differential equations with 75 parameters. The letter "S" refers to the main property of this model, that is, to describe the essential *saturable* and *synergistic* characteristics of complex biological systems or cellular networks, where microalgae can be added as well. These nonlinear differential equations always describe the change in concentration of an element of the system, X_i, with time. Thus, two terms are yielded: one represents the net production or influx, and the other represents the net degradation or efflux. Elements may be a quantifiable component of a system or specifically a microalgal culture. From a mathematical point of view, net production and degradation terms are expressed by power-law functions.

$$\frac{dX_i}{d_t} = \alpha_i \prod_j X_j^{gi\_j} - \beta_i \prod_j X_j^{hi\_j} \tag{4}$$

The left-hand side of the equation shown above is the first derivative of X_i with respect to time, with $i=1, 2, 3 \ldots N$. The right hand side X_j represents independent and dependent variables that have an action for any given X_i, with $j=1, 2, 3 \ldots N$. The first term on the right-hand side corresponds to only those entities that increase or inhibit the production of X_i, while the factors in the second term correspond to only those entities that contribute to, or inhibit, the consumption of X_i, with α_i, $\beta_i > 0$.

S-system models contain relatively few parameters and can, thus, be easily applied to describe the electron transport chain in PSII and PSI throughout the

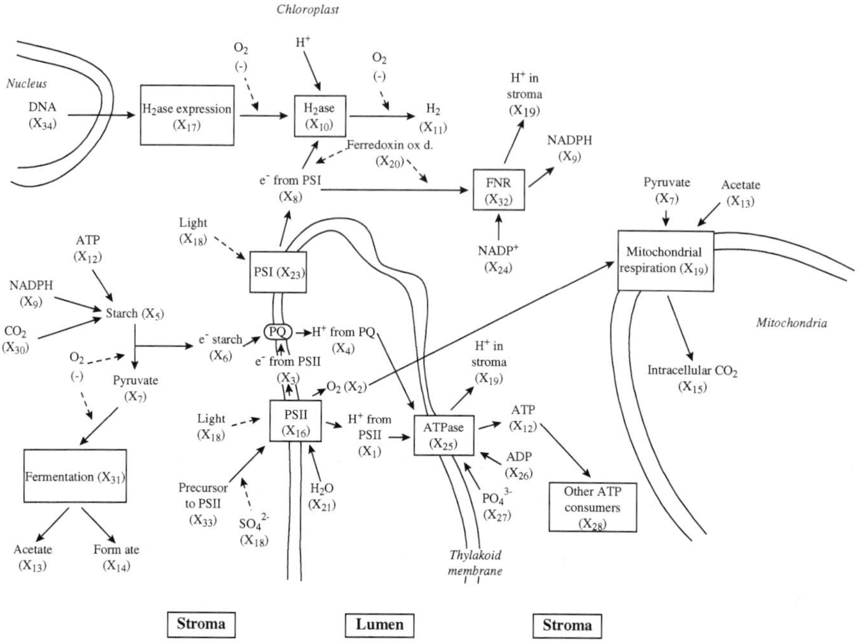

Fig. 2 Scheme of the metabolic map indicating the process of biohydrogen production by sulfur-deprived *Chlamydomonas reinhardtii*. The scheme is the basis for the majority of S-system modeling approaches found in literature (Horner 2002; Jorquera et al. 2008). (Reproduced with permission from the International Journal of Hydrogen Energy (IJHE) (2008), Elsevier)

thylakoid membrane. For that purpose, a metabolic map shows the main processes for hydrogen production by sulfur-deprived *Chlamydomonas reinhardtii* (Fig. 2).

The metabolic map includes dependent and independent variables that can further be subdivided into photosynthetic components, fermentation-related species and generalized cellular functions (Jorquera et al. 2008). Dependent variables are protons from PSII (X_1), oxygen (X_2), electrons from PSII (X_3), protons from plastoquinone (PQ) (X_4), starch (X_5), electrons from starch oxidation (X_6), pyruvate (X_7), electrons from PSI (X_8), NADPH (X_9), hydrogenase (X_{10}), hydrogen (X_{11}), ATP (X_{12}), acetate (X_{13}), formate (X_{14}), intracellular CO_2 (X_{15}), PSII (X_{16}) and H_2ase expression (X_{17}). Dependent variables comprise light (X_{18}), protons in the stroma (X_{19}), ferredoxin oxidized (X_{20}), water (X_{21}), sulfate (X_{22}), PSI (X_{23}), NADP$^+$ (X_{24}), ATPase (X_{25}), ADP (X_{26}), phosphate (X_{27}), other ATP consumers (X_{28}), mitochondrial respiration (X_{29}), extracellular CO_2 (X_{30}), fermentation (X_{31}), FNR (X_{32}), precursor to PSII (X_{33}) and DNA (X_{34}).

The S-system approach by Horner (2002) comprised 15 ordinary differential equations in the general form of Eq. 4 capturing oxygen, hydrogen, protons from photolysis of water, electrons from PSII, lower potential energy electrons from the electron transport chain between PSII and PSI, electrons from PSI, ATP, ADP, water, inorganic phosphate (P_i), reduced Fd, oxidized Fd, consumers of O_2 and ATP. It has been pointed out by Horner (2002) that adequate kinetic modeling of biohydrogen production by sulfur-deprived *Chlamydomonas reinhardtii* has to predict the precise start and the end time of gas production, as well as the rates of H_2 and O_2 production. This model was the basis for further upcoming modeling approaches. The S-system model by Park and Moon (2007) specifically models eight primary metabolites of the biochemical photosynthetic process and is the only model that considers the biohydrogen production by microalgae under sulfur-free conditions. The eight primary metabolites comprise oxygen, protons from the photolysis of water, electrons from PSII, ATP, ADP, lower potential energy electrons from electron-transport chain between PSII and PSI, electrons from PSI, and hydrogen. However, when taking a critical look at the model, it explicitly models the release of hydrogen gas and the effects of illumination, but omits the role of endogenous substrates, such as protein and starch, that are known to increase significantly in the initial stages of sulfur-deprivation before hydrogen is produced (Fouchard et al. 2005; Kosourov et al. 2002; Posewitz 2004; Melis et al. 2000). As a drawback from this, the model results in an important number of parameters that are difficult to identify from the experimental data available (Williams and Bees 2014; Fouchard et al. 2009). Finally, the model appears to be a discrete, multi-state model rather than a continuous formulation (Williams and Bees 2014). Jorquera et al. (2008) developed an S-system model that expands the model from Horner (2002) and from Park and Moon (2007), but includes further reaction steps and pathways. The model comprises a total of 17 ordinary differential equations, 17 state variables, 34 rate constants and 77 kinetic orders. Its ordinary differential equations comprised protons from PSII, oxygen, electrons from PSII, protons from PQ, starch, electrons from starch, pyruvate, and electrons from PSI, NADPH, hydrogenase, hydrogen, ATP, acetate, formate, intracellular CO_2, active PSII and H_2ase expression. The models by Park and Moon (2007) and Jorquera et al. (2008) assume a homogenous microalgal culture, which is actually not the case in a real photobioreactor. This critical aspect forms the basis for the model modification by Zhang (2011). Zhang has developed a model that considers the heterogeneity in a photobioreactor, which is induced by light attenuation and boundary conditions. The model is based on an Advective-Diffusive Reaction Equation (ADRE) (Zhang 2011, 2012, 2013). Fouchard et al. (2009) developed a model that describes the kinetics of extra- and intracellular sulfur, total biomass, and intracellular starch concentrations as a function of environmental conditions. The model has been thoroughly described, together with model parameter identification, and ties up to the model of Park and Moon (2007). Formulating mathematical models for biohydrogen production in microalgae is always challenging concerning parameter estimation. The object of Williams and Bees (2014) was to develop a model that does not try to refine parameter values arbitrarily to obtain quantitative agreement, but rather produces a robust mechanistic model that exhibits the same quality as other modeling approaches. Their modeling approach results in a set of coupled ordinary differential equations that comprehend the key features sulfur-dependent photosynthesis, growth, changes in endogenous substrate and hydrogen gas release. Nevertheless, its construction is based upon previous studies by Melis et al. (2000), Kosourov et al. (2002) and Zhang and Melis (2002). In the interest of clarity, the model structure is shown below in words (for explicit details of nomenclature, consult Williams and Bees 2014):

$$\frac{dS}{dt} = \textit{sulfur system input} - \textit{sulfur uptake by cell} \tag{5}$$

$$\frac{ds}{dt} = \textit{sulfur uptake by cell} - \textit{PSII repair} + \textit{protein breakdown} \\ - \textit{protein production} - \textit{growth / decay} \tag{6}$$

$$\frac{dp}{dt} = -\textit{protein breakdown} + \textit{protein production} - \textit{growth / decay} \tag{7}$$

$$\frac{d\omega}{dt} = \textit{photosynthesis} - \text{respiration} - \textit{supersaturation loss} \tag{8}$$

$$\frac{dh}{dt} = \Lambda \bullet O_2\,\text{sensitivity} \bullet (\textit{PSII dependent} + \textit{PSII independent}) \bullet e^{-}\,\textit{pathway} \tag{9}$$

The model formulated turned out to agree qualitatively with published experimental results from biohydrogen production. It is, therefore, a powerful tool, which can be used to improve on microalgal biohydrogen production in a two-stage reactor with sulfur-deprived photosynthesis.

Independently from previous modeling approaches, which mostly follow on from one another, Vargas et al. (2014) developed a mathematical model that predicts the H_2 production through microalgal cultivation during indirect biophotolysis. The model considers the reactants mass fractions, the availability of enzymes for catalysis and temperature as the variable capable of affecting the processes. The transient model provided good quantitative and qualitative results in agreement with direct measurements from a photobioreactor, and thus, the model is applicable for predicting microalgal growth in photobioreactors.

Summing up, no clear standardized model framework has emerged for modeling biohydrogen production so far, which is hardly surprising, as various ways exist that end in the production of hydrogen gas such as sulfur-deprivation, anaerobic conditions during photosynthesis or due to operational procedures (direct and indirect photolysis). Therefore, different models are needed to address the different processes. However, a uniform trend is clearly seen in the application of S-system models that have been further modified and well-established since 2002. However, the applicability of S-system models is limited to hydrogen production by sulfur-deprived microalgal cultures in a one-stage reactor and in terms of direct photolysis. In the case of a two-stage reactor system with sulfur-deprived photosynthesis, the mechanistic model formulation presented by Williams and Bees (2014) is an appropriate approach that is already very mature, but still bears opportunities for improvement, such as by performing more independent measurements of parameters instead of fitting the data (Williams and Bees 2014). Vargas et al. (2014) made progress in the field of indirect biophotolysis. We can, therefore, expect more approaches here in the near future as the modeling of indirect biophotolysis has unfortunately had less priority in recent studies.

3 Available Models for Simulating Biomethane Production Through Anaerobic Digestion of Microalgae

Anaerobic digestion is a well-known microbiological process that converts organic substances into biogas mainly composed by methane (50–75 % by vol. CH_4), carbon dioxide (25–50 % by vol. CO_2), nitrogen (0–10 % by vol. N_2), oxygen (<1 % by vol. O_2), and hydrogen (<1 % by vol. H_2). Microalgae, as they are rich in proteins, lipids and carbohydrates, represent an excellent feedstock for anaerobic digestion. Within these molecules, lipids are the most energy-rich compound in microalgae with 37.6 $kJ \cdot g^{-1}$, followed by proteins with 16.7 $kJ \cdot g^{-1}$ and carbohydrates with 15.7 $kJ \cdot g^{-1}$ (Wilhelm and Jakob 2011).

Though carbohydrates possess the lowest energy content, they represent, nonetheless, the most important biomass compounds for biofuel production, since carbohydrates in microalgae mainly comprise cellulose in cell walls and starch in plastids without lignin and low hemicelluloses contents making them readily convertible into fermentable sugars (Chen et al. 2013). In case of *Chlamydomonas reinhardtii* the carbohydrate contents in biomass is around 17 % of dry matter (Becker 2007). Therefore, microalgae are theoretically capable of yielding high biogas rates as for instance reported by González-Fernández et al. (2012) with 0.1–0.5 $L_{CH4} \cdot g_{VS}^{-1}$, with 60–80 % methane in biogas, depending on the process temperature (15–52 °C) and hydraulic retention time (HRT) (3–64 days). By comparison, Koch et al. (2009) reported a biogas production rate for the mono fermentation of grass silage of 0.5 $L_{CH4} \cdot g_{VS}^{-1}$, with a methane concentration of 52 %. Methane production of other agricultural crops such as maize, wheat, rice and sugarcane wastes range between 0.28 $L_{CH4} \cdot g_{VS}^{-1}$ and 0.34 $L_{CH4} \cdot g_{VS}^{-1}$ (Chandra et al. 2012). However, this technology faces several hurdles regarding microalgal digestion for which simulation and modeling could play an effective tool to find a strategic solution that helps to modify the applied physical system. Such hurdles are following enlisted:

- Inhibition of methanogenic archaea in the case of high ammonia concentration resulting from high nitrogen content of microalgae (Chen et al. 2008; Koster and Lettinga 1984). Inhibiting concentrations reported vary in a wide range from 1.7 to 14 $g_{NH4\text{-}N} \cdot L^{-1}$ (Sialve et al. 2009).
- Low biodegradability of biomass due to the nature of the cell wall of some microalgal species (Sialve et al. 2009).
- Reduction of digester performance due to the potential toxicity of sodium for marine microalgae digestion (Sialve et al. 2009). A strong inhibition is reported at sodium levels exceeding 10 $g \cdot L^{-1}$ (Lefebvre and Moletta 2006).

A wide variety of models for modeling purposes of anaerobic digestion exist that range from simple models to more mechanistic models, such as the IWA Anaerobic Digestion Model No. 1 – ADM1 (Batstone et al. 2002). A scheme of the processes included in ADM1 and their interrelation is presented in Fig. 3.

ADM1 describes the main processes in anaerobic digestion: hydrolysis, acidogenesis from sugars and amino acids, acetogenesis from long chain fatty acids, propionate, butyrate and valerate, and acetoclastic and hydrogenotrophic methanogenesis.

Moreover, it has already been applied to model substrates, such as activated and primary sludge, similarly agricultural substrates (Lübken et al. 2007, 2010; Wichern et al. 2009), to which microalgae may also be added as co-digestion feedstock (Schwede et al. 2013).

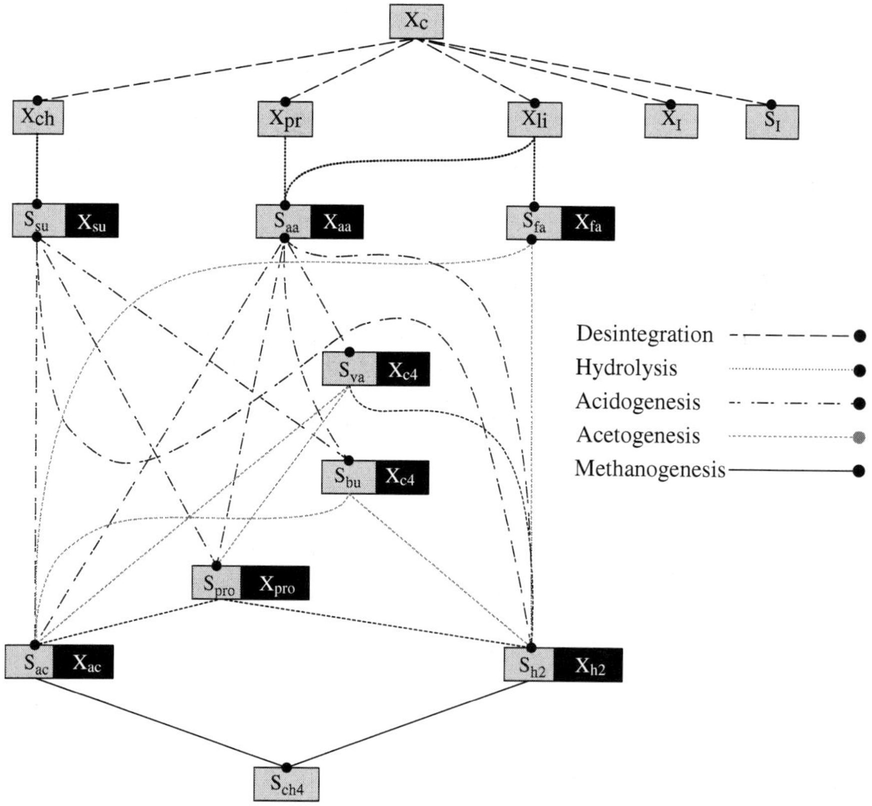

Fig. 3 Basic process scheme of the anaerobic digestion model No. 1 (ADM1) for organic components. X_c = composite material, X_{ch} = particulate carbohydrates, X_{pr} = particulate proteins, X_{li} = lipids, X_I = particulate inerts, S_I = soluble inerts, S_{su} = monosaccharides, S_{aa} = amino acids, S_{fa} = long chain fatty acids, S_{va} = valerate, S_{bu} = butyrate, S_{pro} = propionate, S_{ac} = acetate, S_{h2} = hydrogen, S_{ch4} = methane, X_{su} = sugar degrader, X_{aa} = amino acids degrader, X_{fa} = long chain fatty acids degrader, X_{c4} = valerate and butyrate degrader, X_{pro} = propionate degrader, X_{ac} = acetate degrader, X_{h2} = hydrogen consumer. Basic unit of organic components is $kg_{COD} \cdot m^{-3}$

Within the ADM1 framework (Fig. 3), composite material (X_c) with lumped characteristics is split up into a non-biodegradable part that is composed of particulate inerts (X_I) and soluble inerts (S_I). The biodegradable part divides across its principal constituents: carbohydrates (X_{ch}), proteins (X_{pr}) and lipids (X_{li}). Carbohydrates and proteins are enzymatically hydrolyzed to soluble substrates, such as sugars (S_{su}) and amino acids (S_{aa}). Lipids are hydrolyzed into amino acids (S_{aa}) and long chain fatty acids (S_{fa}). Thereafter, acidogens degrade sugars and amino acids into a mixture of organic acids comprised of acetic (S_{ac}), propionic (S_{pro}), butyric (S_{bu}) and valeric acid (S_{va}). Afterwards, valeric, butyric and propionic acid are converted into acetic acid (S_{ac}), which is, along with hydrogen (S_{h2}) from fatty acids and propionic acid, one precursor for the final product – methane (S_{ch4}).

The first application of ADM1 to model anaerobic digestion of microalgal biomass dates back to the study by Mairet et al. in 2011 using the green alga *Chlorella vulgaris*. The original version of ADM1 showed a good ability to describe microalgae digestion, except for a low biomass-to-substrate ratio that resulted in an overestimation of inorganic nitrogen, which was attributed to the predetermined first-order kinetics in ADM1. Thereby, it was demonstrated that hydrolysis is rate-limiting in anaerobic digestion of microalgae, as was also shown in following studies, such as the example of anaerobic co-digestion of the marine microalga *Nannochloropsis salina* (Schwede et al. 2013). In this context, it is quite challenging to model the rate of organic degradation at the hydrolysis step as several factors are influential, such as the particle size, the access of the microorganisms to the particular components of the substrate, the mixing of the reactor or the concentration of the microorganisms. Mairet et al. (2011) used a Contois model in their approach to better reflect hydrolysis characteristics, which has already proven to be sound in former studies describing biodegradation, ranging from sewage sludge to municipal solid waste (Sötemann et al. 2006; Vavilin et al. 2008; Nopharatana et al. 2007). Batstone (2002) also suggested that Contois kinetics might be used if the biomass-to-substrate ratio is low enough to be rate-limiting. Regarding microalgal digestion, Mairet et al. (2011) succeeded in reproducing a 140-day experiment of *Chlorella vulgaris* using data from Ras et al. (2011). The underlying idea was to overcome the disadvantage of first-order models in predicting transient or dynamic behavior. While Contois models maintain the simplicity of a first-order model, they can be seen as an improvement on Monod's microbial growth model, thus keeping its accuracy. In this context, the general mathematical expression of the Contois model equation (Eq. 10) is comprised of p as the process rate [$kg_{COD} \cdot m^{-3} \cdot d^{-1}$], p_m as the maximum specific hydrolysis rate [d^{-1}], K_X [$kg_{COD} \cdot m^{-3}$] as the half-saturation coefficient for the ratio S/X, S as the particulate compound concentration [$kg_{COD} \cdot m^{-3}$], and X as the hydrolytic (disintegration) biomass concentration [$kg_{COD} \cdot m^{-3}$].

$$p = p_m \cdot X \cdot \frac{S}{K_X \cdot X + S} = p_m \cdot X \cdot \frac{S/X}{K_X + S/X} \tag{10}$$

In the equation shown above, the microbial specific growth rate is determined by the ratio S/X that could be expressed in other words as the substrate-to-biomass density, hence defining the difference to Monod's model, where the microbial growth rate is defined by an absolute substrate concentration. Moreover, it could be argued that Monod's model is ideal for slow growing pure strains of low density, while the Contois model is widely applied to real-life "dirty" situations, characterized by high densities, mixed communities and heterogeneous substrates, and is, therefore, ideal for microalgae (Arditi and Ginzburg 2012). When the Contois kinetics are used to describe the hydrolysis step in ADM1, three terms are gathered for carbohydrates, proteins and lipids, as shown below:

$$k_{hyd,ch} \cdot X_{ch} \overset{modified\ to}{\rightarrow} k_{hyd,ch} \frac{X_{ch}}{K_{s,ch} \cdot X_{su} + X_{ch}} \cdot X_{su} \tag{11}$$

$$k_{hyd,pr} \cdot X_{pr} \overset{modified\ to}{\rightarrow} k_{hyd,pr} \frac{X_{pr}}{K_{s,pr} \cdot X_{aa} + X_{pr}} \cdot X_{aa} \tag{12}$$

$$k_{hyd,li} \cdot X_{li} \overset{modified\ to}{\rightarrow} k_{hyd,li} \frac{X_{li}}{K_{s,li} \cdot X_{fa} + X_{li}} \cdot X_{fa} \tag{13}$$

As the proposed modification step in ADM1 led to a model comprised of 30 state variables, Mairet et al. (2012) introduced a simpler three reaction model called MAD (Microalgae Anaerobic Digestion model), which also achieved a good fit with the data obtained by Ras et al. (2011) (Fig. 4).

The idea of the MAD is to model (1) the mass balance in the headspace of a continuous-stirred-tank anaerobic digester for the dynamics of the partial pressures of CO_2 and CH_4, (2) the dynamics of molecular concentrations in the liquid phase,

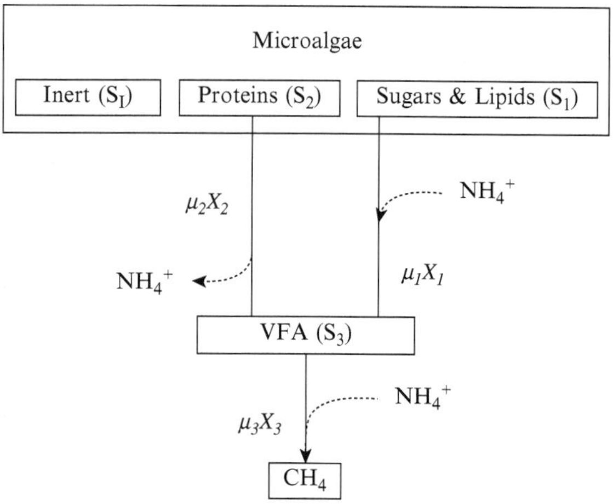

Fig. 4 Basic scheme of the MAD model according to Mairet et al. (2012). (Reproduced with permission from Biotechnology & Bioengineering, John Wiley and Sons)

such as sugars, lipids, proteins and inerts, (3) the charge balance, (4) the specific growth rates for the hydrolysis-acidogenesis reactions as Contois function and (5) the liquid–gas transfer rate of CO_2 [$mol \cdot L^{-1} \cdot d^{-1}$]. All these processes are expressed through the following three equations:

$$\alpha_1 \cdot S_1 + \alpha_2 \cdot NH_4^+ \overset{\mu_1(\cdot)X_1}{\rightarrow} X_1 + \alpha_3 \cdot S_3 + \alpha_4 \cdot CO_2 \tag{14}$$

$$\alpha_5 \cdot S_2 \overset{\mu_2(\cdot)X_2}{\rightarrow} X_2 + \alpha_6 \cdot S_3 + \alpha_7 \cdot NH_4^+ + \alpha_8 \cdot CO_2 \tag{15}$$

$$\alpha_9 \cdot S_3 + \alpha_{10} \cdot NH_4^+ \overset{\mu_3(\cdot)X_3}{\rightarrow} X_3 + \alpha_{11} \cdot CH_4^+ + \alpha_{12} \cdot CO_2 \tag{16}$$

Equations 14 and 15 represent the hydrolysis-acidogenesis reactions, while Eq. 16 models the specific growth rate of the methanogenesis. Moreover, S_1 is the sugar-lipid concentration ($g_{COD} \cdot L^{-1}$), S_2 the protein concentration ($g_{COD} \cdot L^{-1}$) and S_3 the VFA concentration ($g_{COD} \cdot L^{-1}$). X_i represents the microbial population concentration associated with the reaction ($g_{COD} \cdot L^{-1}$), while α_i shows the stoichiometric parameters.

Mairet et al. (2012) emphasized that the MAD model in its reduced complexity towards the modified ADM1 model allows for the development of advanced, model-based control and monitoring strategies. However, a limitation is expected when it is applied to marine species that are sensitive to sodium toxicity (Sialve et al. 2009), as MAD does not capture this phenomenon. This issue was taken up by Hierholtzer and Akunna (2014) by modifying the original ADM1 process inhibition for acetate uptake, which was originally compiled of a pH ($I_{pH,ac}$), inorganic nitrogen ($I_{IN,lim}$) and free ammonia inhibition term (I_{NH3}). Consequently, the uptake of acetate is inhibited within a finite range, meaning $I_{acetate}$ works as an inhibition function when $I_{acetate} \rightarrow 0$, while no effect is observed when $I_{acetate} \rightarrow 1$. The modification adds an extra inhibition factor $I_{cations}$ that takes account of all cations in saline water that have an inhibitory effect on acetoclastic methanogens: sodium, magnesium, calcium and potassium.

$$I_{acetate} = I_{pH,ac} \cdot I_{IN,\lim} \cdot I_{NH_3} \cdot I_{cations} \tag{17}$$

$$I_{cations} = \frac{1}{1 + \left(\frac{S_{Na^+}}{K_{I,Na^+}}\right) + \left(\frac{S_{Mg^{2+}}}{K_{I,Mg^{2+}}}\right) + \left(\frac{S_{Ca^{2+}}}{K_{I,Ca^{2+}}}\right) + \left(\frac{S_{K^+}}{K_{I,K^+}}\right)} \tag{18}$$

In the function shown above, $I_{cations}$, the concentrations of the individual cations are expressed as $S_{Na}{}^+$, $S_{Mg}{}^{2+}$, $S_{Ca}{}^{2+}$ and $S_K{}^+$, while $K_{I,Na}{}^+$, $K_{I,Mg}{}^{2+}$, $K_{I,Ca}{}^{2+}$ and $K_{I,K}{}^+$ represent their inhibitory concentrations.

Nonetheless, the proposed model is based upon first-order parameters corresponding to the hydrolysis kinetics of carbohydrates, lipids and proteins and does not take up the established modification using the Contois model, as established by Mairet et al. (2011).

4 Microalgal Feed Characterization in ADM1

Substrate characterization is a critical step in ADM1 that has been discussed extensively for various feeds and similarly for microalgae. In the original framework of ADM1, all embedded conversion processes are structured on a fixed-stoichiometry, while the degradation of composite particulate material (X_c) is described with lumped characteristics and in terms of the chemical oxygen demand (COD). Nonetheless, since ADM1 has an affinity to models derived for sewage sludge digestion (Siegrist et al. 2002), the default X_c composition in terms of carbohydrates, proteins, lipids and the inert fraction is, consequently, not in agreement with the composition of microalgae. Future modeling attempts certainly have to address this issue, which will demand an increase of measurements of process parameters. The following paragraphs will, therefore, briefly introduce two approaches that suit microalgal modeling with ADM1.

In 2006, Kleerebezem and Van Loosdrecht (2006a, b) proposed an algorithm where the total COD of the substrate is first split into each input state variable, dedicating the remaining COD to the X_c fraction (Kleerebezem and van Loosdrecht 2006a; Girault et al. 2012). The substrate characterization starts with some basic

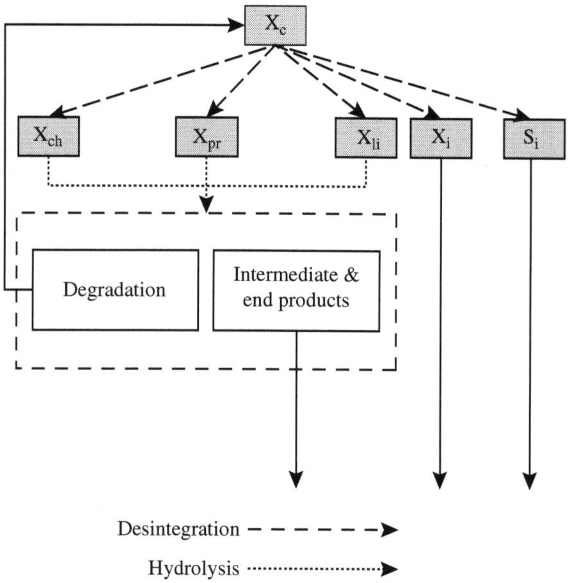

Fig. 5 Disintegration step in ADM1. Composite particulate matter (X_c) is described with lumped characteristics

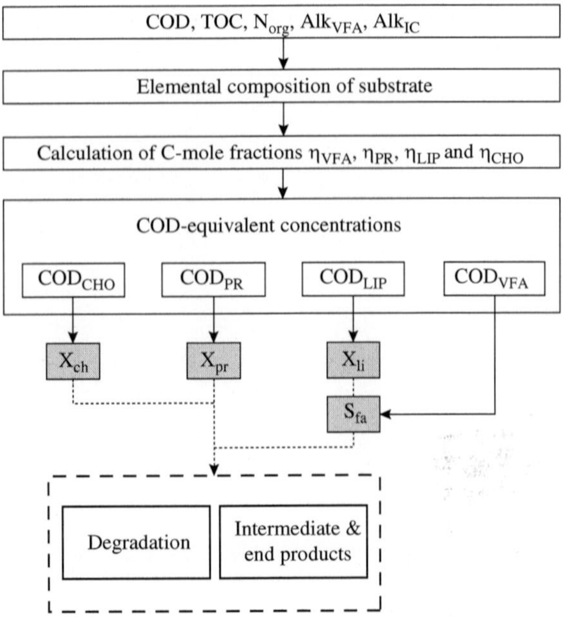

Fig. 6 Derivation of the substrate influent characteristic, according to Kleerebezem and Van Loosdrecht (2006b)

analytical measurements comprising the COD, total organic carbon (TOC), organic nitrogen (N_{org}) and alkalinity (Alk). However, criticism was expressed as many assumptions about the carbon and nitrogen inert content were made, similarly concerning the biochemical fractionation and the non-biodegradable variables (Jimenez et al. 2014).

The microalgal biomass characterization according to Kleerebezem and Van Loosdrecht (2006b) starts with basic analytical measurements that are used to calculate the elemental composition (Eq. 19) as follows:

$$C_xH_yO_zN_v^{-III u} \tag{19}$$

$$y = \frac{2 \cdot COD + Alk_{VFA} - 2 \cdot N_{org}}{TOC} \tag{20}$$

$$z = 2 - \frac{COD + 0.5 \cdot N_{org}}{TOC} \tag{21}$$

$$v = \frac{N_{org}}{TOC} \tag{22}$$

$$u = \frac{-Alk_{VFA}}{TOC} \tag{23}$$

The carbon mole fractions η_{CHO}, η_{VFA}, η_{LIP} and η_{PR} are calculated from the elemental composition, as outlined in Eqs. 24, 25, 26, and 27. Kleerebezem and Van Loosdrecht (Kleerebezem and van Loosdrecht 2006b) used standard molecules to obtain the values for the variables charge (Ch_{VFA}), nitrogen content of protein (N_{PR}) and oxidation state of lipids (γ_{LIP}). In this way, there are possibilities to be more accurate, for example, by determining the individual protein composition of the microalga.

$$\eta_{VFA} = \frac{u}{Ch_{VFA}} \tag{24}$$

$$\eta_{PR} = \frac{v}{N_{PR}} \tag{25}$$

$$\eta_{LIP} = \frac{y - 2 \cdot z - 3 \cdot v - u}{4 \cdot \left(\gamma_{LIP}^{-1}\right)} \tag{26}$$

$$\eta_{CHO} = 1 - \eta_{LIP} - \eta_{VFA} - \eta_{PR} \tag{27}$$

Finally, the actual COD concentrations can be calculated using the TOC value measured:

$$COD_{CHO}\left[\frac{g_{COD}}{L}\right] = TOC \cdot \eta_{CHO} \cdot \frac{\gamma_{CHO}}{4} \cdot MW_{O_2} \tag{28}$$

$$COD_{PR}\left[\frac{g_{COD}}{L}\right] = TOC \cdot \eta_{PR} \cdot \frac{\gamma_{PR}}{4} \cdot MW_{O_2} \tag{29}$$

$$COD_{LIP}\left[\frac{g_{COD}}{L}\right] = TOC \cdot \eta_{LIP} \cdot \frac{\gamma_{LIP}}{4} \cdot MW_{O_2} \tag{30}$$

$$COD_{VFA}\left[\frac{g_{COD}}{L}\right] = TOC \cdot \eta_{VFA} \cdot \frac{\gamma_{VFA}}{4} \cdot MW_{O_2} \tag{31}$$

Though Mairet et al. (2011) emphasized the importance of influent substrate characterization in their study, they used an approximate elemental composition for *Chlorella vulgaris* as proposed by Geider and La Roche (2002), which was $CH_{1.5801}O_{0.3251}N_{0.2619}S_{0.0049}$. Another way of characterizing the input substrate in a more detailed way is the algorithm proposed by Ramsay and Pullammanappallil (2001). This algorithm is based upon Stickland reactions and the amino acid composition, as exemplarily shown in Table 1 for *Chlorella vulgaris*.

As a result of Table 1, the individual chemical composition of the microalgal protein is yielded, as shown exemplarily for selected species in Table 2. The Ramsay

Table 1 Exemplary calculation of stoichiometric coefficients for protein degradation of a 5-day-old culture of *Chlorella vulgaris* according to Ramsay and Pullammanappallil (2001) and Becker (2007)

Amino acid (AA)	Content (mole AA/c-mole protein)	C_2 acid (mole/mole AA)	C_3 acid (mole/mole AA)	C_4 acid (mole/mole AA)	C_5 acid (mole/mole AA)
Alanine	0.0157	1.0			
Arginine	0.0127	0.5	0.5		0.5
Aspartic acid	0.0179	1.0			
Cysteine	0.0028	1.0			
Glutamic acid	0.0231	1.0		0.5	
Glycine	0.0115	1.0			
Histidine	0.0040	1.0		0.5	
Isoleucine	0.0076				1.0
Leucine	0.0175				1.0
Lysine	0.0167	1.0		1.0	
Methionine	0.0044		1.0		
Phenylalanine	0.0099				
Proline	0.0096	0.5	0.5		0.5
Serine	0.0082	1.0			
Threonine	0.0096	1.0		0.5	
Tryptophan	0.0042				
Tyrosine	0.0068	1.0			
Valine	0.0109			1.0	
TOTAL (α)	Mole/c--mole	0.301	0.064	0.271	0.278

and Pullammanappallil algorithm is of particular interest for modeling microalgae, as many complete amino acid charts for various microalgae are available in literature (Becker 2007; Khairy 2011; Fowden 1952; Brown 1991; Yuan et al. 2014). Nevertheless, it has to be considered that the amino acid composition varies according to the age of the microalgae species (Fowden 1952) and according to the kind of growth – autotrophic or heterotrophic (Khairy 2011).

Table 2 gives a short overview of the stoichiometric coefficients calculated as a result of applying the algorithm from Ramsay and Pullammanappallil (2001). In comparison to the standard ADM1 coefficients, it becomes obvious that this kind of substrate characterization for microalgae is meaningful due to the substantial deviations towards the default ADM1.

To sum up, modeling approaches regarding biomethanization to date mostly involve the standardized model framework of the ADM1, which is based on con-

Table 2 The *upper table* shows exemplary calculations of chemical protein compositions for selected microalgal species. The *lower table* gives an overview of the corresponding calculated stoichiometric coefficients compared to the default coefficients used in ADM1

Microalga	Chemical composition of protein
Chlorella vulgaris	$CH_{1.9230}O_{0.4966}N_{0.2559}S_{0.0072}$
Scenedesmus obliquus	$CH_{1.9573}O_{0.5153}N_{0.2702}S_{0.0047}$
Blue algae	$CH_{1.9947}O_{0.5557}N_{0.2558}S_{0.0033}$
Spirulina platensis	$CH_{1.9376}O_{0.5078}N_{0.2549}S_{0.0064}$
Dunaliella bardawil	$CH_{1.9406}O_{0.5060}N_{0.2573}S_{0.0068}$
Arthrospira maxima	$CH_{1.9123}O_{0.5021}N_{0.2499}S_{0.0038}$
Aphanizomenon sp.	$CH_{1.9916}O_{0.5244}N_{0.2651}S_{0.0034}$

Fermentation product	Acetic acid ($f_{ac,aa}$)	Propionic acid ($f_{pro,aa}$)	Butyric acid ($f_{bu,aa}$)	Valeric acid ($f_{va,aa}$)
ADM1[a]	0.400	0.050	0.260	0.230
Blue algae	0.325	0.041	0.242	0.293
Chlorella vulgaris	0.301	0.064	0.271	0.278
Scenedesmus obliquus	0.315	0.063	0.266	0.274
Spirulina platensis	0.293	0.063	0.231	0.314
Dunaliella bardawil	0.293	0.060	0.257	0.300
Arthrospira maxima	0.285	0.057	0.256	0.310
Aphanizomenon sp.	0.286	0.059	0.265	0.310

Calculations are based on amino acid compositions as given in literature (Becker 2007; Yuan et al. 2014)

[a]Suggested as standard values

stant stoichiometry. However, as the ADM1 was designed to be freely adjustable, it has also been adopted to model the anaerobic digestion of microalgae. Recent studies regarding the application of the ADM1 addressing microalgal digestion focus entirely on a closer consideration of the effect of salinity for marine microalgal cultures and on the improvement of kinetics. Regarding the latter, hydrolysis rates in ADM1 are predetermined by first-order kinetics that have been successfully substituted by a Contois model. Here, the kinetics do not depend on the substrate concentration, but on the amount of substrate per biomass unit. Nonetheless, future ADM1 modeling attempts have to focus on a refinement of the substrate feed characterization by methods such as those exemplarily outlined above.

List of Acronyms

ADM1	Anaerobic Digestion Model No. 1
ADRE	Advective-Diffusive Reaction Equation
Alk	Alkalinity

Alk_{IC}	Bicarbonate alkalinity
Alk_{VFA}	Alkalinity of neutralized fatty acids
ATP	Adenosine triphosphate
Ch_{VFA}	Charge of volatile fatty acids
COD	Chemical Oxygen Demand
COD_{CHO}	COD-equivalent carbohydrate concentration
COD_{LIP}	COD-equivalent lipid concentration
COD_{PR}	COD-equivalent protein concentration
COD_{VFA}	COD-equivalent volatile fatty acid concentration
ETS	Electron Transport System
Fd	Ferredoxin
FNR	Fd-$NADP^+$ reductase
HRT	Hydraulic retention time
MAD	Microalgae Anaerobic Digestion model
MW_{O2}	Molecular weight of oxygen
N_{org}	Organic nitrogen
N_{PR}	Nitrogen content of protein
PQ	Plastoquinone
PSI	photosynthetic system I
PSII	photosynthetic system II
TOC	Total Organic Carbon
γ_{LIP}	Oxidation state of lipids
ηCHO	Carbon mole fraction of carbohydrates
ηLIP	Carbon mole fraction of lipids
ηPR	Carbon mole fraction of proteins
ηVFA	Carbon mole fraction of volatile fatty acids

References

Aoyama K, Uemura I, Miyake J, Asada Y (1997) Fermentative metabolism to produce hydrogen gas and organic compounds in a cyanobacterium, *Spirulina platensis*. J Ferment Bioeng 83(1):17–20

Arditi R, Ginzburg LR (2012) How species interact. Altering the standard view on trophic ecology. Oxford University Press, Oxford

Batie CJ, Kamin H (1984) Electron transfer by ferredoxin: $NADP^+$ reductase. Rapid-reaction evidence for participation of a ternary complex. J Biol Chem 259(19):11976–11985

Batstone DJ (2002) Anaerobic digestion model no. 1 (ADM1). Scientific & technical report, vol 13. IWA Publishing, London

Batstone DJ, Keller J, Angelidaki I, Kalyuzhnyi SV, Pavlostathis SG, Rozzi A, Sanders WTM, Siegrist H, Vavilin VA (2002) The IWA Anaerobic Digestion Model No 1 (ADM1). Water Sci Technol 45(10):65–73

Becker EW (2007) Micro-algae as a source of protein. Biotechnol Adv 25(2):207–210

Boyle NR, Morgan JA (2009) Flux balance analysis of primary metabolism in *Chlamydomonas reinhardtii*. BMC Syst Biol 3(1):4

Brown MR (1991) The amino-acid and sugar composition of 16 species of microalgae used in mariculture. J Exp Mar Biol Ecol 145(1):79–99

Chandra R, Venkata Mohan S (2011) Microalgal community and their growth conditions influence biohydrogen production during integration of dark-fermentation and photo-fermentation processes. Int J Hydrog Energy 36(19):12211–12219

Chandra R, Takeuchi H, Hasegawa T (2012) Methane production from lignocellulosic agricultural crop wastes: a review in context to second generation of biofuel production. Renew Sust Energ Rev 16(3):1462–1476

Chen Y, Cheng JJ, Creamer KS (2008) Inhibition of anaerobic digestion process: a review. Bioresour Technol 99(10):4044–4064

Chen C, Zhao X, Yen H, Ho S, Cheng C, Lee D, Bai F, Chang J (2013) Microalgae-based carbohydrates for biofuel production. Biochem Eng J 78:1–10

Fouchard S, Hemschemeier A, Caruana A, Pruvost J, Legrand J, Happe T, Peltier G, Cournac L (2005) Autotrophic and mixotrophic hydrogen photoproduction in sulfur-deprived Chlamydomonas cells. Appl Environ Microbiol 71(10):6199–6205

Fouchard S, Pruvost J, Degrenne B, Titica M, Legrand J (2009) Kinetic modeling of light limitation and sulfur deprivation effects in the induction of hydrogen production with *Chlamydomonas reinhardtii* Part I. Model development and parameter identification. Biotechnol Bioeng 102(1):232–245

Fowden L (1952) The effect of age on the bulk protein composition of *Chlorella vulgaris*. Biochem J 52(2):310–314

Geider R, La Roche J (2002) Redfield revisited: variability of C:N:P in marine microalgae and its biochemical basis. Eur J Phycol 37(1):1–17

Ghysels B, Godaux D, Matagne RF, Cardol P, Franck F, Subramanyam R (2013) Function of the chloroplast hydrogenase in the microalga Chlamydomonas: the role of hydrogenase and state transitions during photosynthetic activation in anaerobiosis. PLoS ONE 8(5), e64161

Girault R, Bridoux G, Nauleau F, Poullain C, Buffet J, Steyer J, Sadowski AG, Béline F (2012) A waste characterisation procedure for ADM1 implementation based on degradation kinetics. Water Res 46(13):4099–4110

González-Fernández C, Sialve B, Bernet N, Steyer J (2012) Impact of microalgae characteristics on their conversion to biofuel. Part II: Focus on biomethane production. Biofuels Bioprod Biorefin 6(2):205–218

Guan Y, Deng M, Yu X, Zhang W (2004a) Two-stage photo-biological production of hydrogen by marine green alga *Platymonas subcordiformis*. Biochem Eng J 19(1):69–73

Guan Y, Zhang W, Deng M, Jin M, Yu X (2004b) Significant enhancement of photobiological H_2 evolution by carbonylcyanide m-chlorophenylhydrazone in the marine green alga *Platymonas subcordiformis*. Biotechnol Lett 26(13):1031–1035

Hierholtzer A, Akunna JC (2014) Modelling start-up performance of anaerobic digestion of saline-rich macro-algae. Water Sci Technol 69(10):2059

Horner J (2002) A power-law sensitivity analysis of the hydrogen-producing metabolic pathway in *Chlamydomonas reinhardtii*. Int J Hydrog Energy 27(11–12):1251–1255

Huesemann MH, Hausmann TS, Carter BM, Gerschler JJ, Benemann JR (2010) Hydrogen generation through indirect biophotolysis in batch cultures of the nonheterocystous nitrogen-fixing cyanobacterium *Plectonema boryanum*. Appl Biochem Biotechnol 162(1):208–220

Jimenez J, Gonidec E, Rivero C, Andrés J, Latrille E, Vedrenne F, Steyer J (2014) Prediction of anaerobic biodegradability and bioaccessibility of municipal sludge by coupling sequential extractions with fluorescence spectroscopy: towards ADM1 variables characterization. Water Res 50:359–372

Jo JH, Lee DS, Park JM (2006) Modeling and optimization of photosynthetic hydrogen gas production by green alga *Chlamydomonas reinhardtii* in sulfur-deprived circumstance. Biotechnol Prog 22(2):431–437

Jorquera O, Kiperstok A, Sales E, Embirucu M, Ghirardi M (2008) S-systems sensitivity analysis of the factors that may influence hydrogen production by sulfur-deprived *Chlamydomonas reinhardtii*. Int J Hydrog Energy 33(9):2167–2177

Kamat PV, Bisquert J (2013) Solar fuels. Photocatalytic hydrogen generation. J Phys Chem C 117(29):14873–14875

Kapdan IK, Kargi F (2006) Bio-hydrogen production from waste materials. Enzym Microb Technol 38(5):569–582

Kennes C (ed) (2013) Air pollution prevention and control. Wiley, Chichester

Khairy HM (2011) Comparative effects of autotrophic and heterotrophic growth on some vitamins, 2,2-diphenyl-1-picrylhydrazyl (DPPH) free radical scavenging activity, amino acids and protein profile of *Chlorella vulgaris* Beijerinck. Afr J Biotechnol 10(62):13514–13519

Kleerebezem R, van Loosdrecht M (2006a) Critical analysis of some concepts proposed in ADM1. Water Sci Technol 54(4):51

Kleerebezem R, van Loosdrecht M (2006b) Waste characterization for implementation in ADM1. Water Sci Technol 54(4):167

Koch K, Wichern M, Lübken M, Horn H (2009) Mono fermentation of grass silage by means of loop reactors. Bioresour Technol 100(23):5934–5940

Kojima E, Lin B (2004) Effect of partial shading on photoproduction of hydrogen by Chlorella. J Biosci Bioeng 97(5):317–321

Kosourov S, Tsygankov A, Seibert M, Ghirardi ML (2002) Sustained hydrogen photoproduction by *Chlamydomonas reinhardtii*: effects of culture parameters. Biotechnol Bioeng 78(7):731–740

Koster IW, Lettinga G (1984) The influence of ammonium-nitrogen on the specific activity of pelletized methanogenic sludge. Agric Wastes 9(3):205–216

Lauwers J, Appels L, Thompson IP, Degrève J, Impe V, Jan F, Dewil R (2013) Mathematical modelling of anaerobic digestion of biomass and waste: power and limitations. Prog Energy Combust Sci 39(4):383–402

Lefebvre O, Moletta R (2006) Treatment of organic pollution in industrial saline wastewater: a literature review. Water Res 40(20):3671–3682

Levin D (2004) Biohydrogen production: prospects and limitations to practical application. Int J Hydrog Energy 29(2):173–185

Lübken M, Wichern M, Schlattmann M, Gronauer A, Horn H (2007) Modelling the energy balance of an anaerobic digester fed with cattle manure and renewable energy crops. Water Res 41(18):4085–4096

Lübken M, Gehring T, Wichern M (2010) Microbiological fermentation of lignocellulosic biomass: current state and prospects of mathematical modeling. Appl Microbiol Biotechnol 85(6):1643–1652

Mairet F, Bernard O, Ras M, Lardon L, Steyer J (2011) Modeling anaerobic digestion of microalgae using ADM1. Bioresour Technol 102(13):6823–6829

Mairet F, Bernard O, Cameron E, Ras M, Lardon L, Steyer J, Chachuat B (2012) Three-reaction model for the anaerobic digestion of microalgae. Biotechnol Bioeng 109(2):415–425

Melis A (2007) Photosynthetic H_2 metabolism in *Chlamydomonas reinhardtii* (unicellular green algae). Planta 226(5):1075–1086

Melis A, Zhang L, Forestier M, Ghirardi ML, Seibert M (2000) Sustained photobiological hydrogen gas production upon reversible inactivation of oxygen evolution in the green alga *Chlamydomonas reinhardtii*. Plant Physiol 122(1):127–136

Nopharatana A, Pullammanappallil PC, Clarke WP (2007) Kinetics and dynamic modelling of batch anaerobic digestion of municipal solid waste in a stirred reactor. Waste Manag 27(5):595–603

Park W, Moon I (2007) A discrete multi states model for the biological production of hydrogen by phototrophic microalga. Biochem Eng J 36(1):19–27
Posewitz MC (2004) Hydrogen photoproduction is attenuated by disruption of an isoamylase gene in *Chlamydomonas reinhardtii*. Plant Cell 16(8):2151–2163
Ramsay IR, Pullammanappallil PC (2001) Protein degradation during anaerobic wastewater treatment: derivation of stoichiometry. Biodegradation 12(4):247–257
Ras M, Lardon L, Bruno S, Bernet N, Steyer J (2011) Experimental study on a coupled process of production and anaerobic digestion of *Chlorella vulgaris*. Bioresour Technol 102(1):200–206
Savageau MA (1969a) Biochemical systems analysis. I. Some mathematical properties of the rate law for the component enzymatic reactions. J Theor Biol 25(3):365–369
Savageau MA (1969b) Biochemical systems analysis. II. The steady-state solutions for an n-pool system using a power-law approximation. J Theor Biol 25(3):370–379
Savageau MA (1976) Biochemical systems analysis. A study of function and design in molecular biology. Advanced book program. Addison-Wesley, London
Schwede S, Kowalczyk A, Gerber M, Span R (2013) Anaerobic co-digestion of the marine microalga *Nannochloropsis salina* with energy crops. Bioresour Technol 148:428–435
Sialve B, Bernet N, Bernard O (2009) Anaerobic digestion of microalgae as a necessary step to make microalgal biodiesel sustainable. Biotechnol Adv 27(4):409–416
Siegrist H, Vogt D, Garcia-Heras JL, Gujer W (2002) Mathematical model for meso- and thermophilic anaerobic sewage sludge digestion. Environ Sci Technol 36(5):1113–1123
Sötemann SW, van Rensburg P, Ristow NE, Wentzel MC, Loewenthal RE, Ekama GA (2006) Integrated chemical, physical and biological processes modelling of anaerobic digestion of sewage sludge. Water Sci Technol 54(5):109–117
Vargas J, Mariano AB, Corrêa DO, Ordonez JC (2014) The microalgae derived hydrogen process in compact photobioreactors. Int J Hydrog Energy 39(18):9588–9598
Vavilin VA, Fernandez B, Palatsi J, Flotats X (2008) Hydrolysis kinetics in anaerobic degradation of particulate organic material: an overview. Waste Manag 28(6):939–951
Ward AJ, Lewis DM, Green FB (2014) Anaerobic digestion of algae biomass: a review. Algal Res 5(10):204–214
Wichern M, Gehring T, Fischer K, Andrade D, Lübken M, Koch K, Gronauer A, Horn H (2009) Monofermentation of grass silage under mesophilic conditions: measurements and mathematical modeling with ADM 1. Bioresour Technol 100(4):1675–1681
Wilhelm C, Jakob T (2011) From photons to biomass and biofuels: evaluation of different strategies for the improvement of algal biotechnology based on comparative energy balances. Appl Microbiol Biotechnol 92(5):909–919
Williams CR, Bees MA (2014) Mechanistic modeling of sulfur-deprived photosynthesis and hydrogen production in suspensions of *Chlamydomonas reinhardtii*. Biotechnol Bioeng 111(2):320–335
Wykoff DD, Davies JP, Melis A, Grossman AR (1998) The regulation of photosynthetic electron transport during nutrient deprivation in *Chlamydomonas reinhardtii*. Plant Physiol 117(1):129–139
Yuan X, Shi X, Yuan C, Wang Y, Qiu Y, Guo R, Wang L (2014) Modeling anaerobic digestion of blue algae: stoichiometric coefficients of amino acids acidogenesis and thermodynamics analysis. Water Res 49:113–123
Yun Y, Jung K, Kim D, Oh Y, Shin H (2012) Microalgal biomass as a feedstock for bio-hydrogen production. Int J Hydrog Energy 37(20):15533–15539
Zhang T (2011) Dynamics modeling of hydrogen production by sulfur-deprived *Chlamydomonas reinhardtii* culture in tubular photobioreactor. Int J Hydrog Energy 36(19):12177–12185

Zhang T (2012) Advection–diffusion constrained metabolic reaction in bioreactor analysis for hydrogen production. Biomass Bioenergy 47:324–333

Zhang T (2013) Effects of fluid and light dynamics on H_2 production in a mechanically stirred photobioreactor. Bioresour Technol 146:70–81

Zhang L, Melis A (2002) Probing green algal hydrogen production. Philos Trans Royal Soc B: Biol Sci 357(1426):1499–1509

Hydrothermal Pretreatments of Macroalgal Biomass for Biorefineries

Héctor A. Ruiz, Rosa M. Rodríguez-Jasso, Mario Aguedo, and Zsófia Kádár

Abstract Recently, macroalgal biomass is gaining wide attention as an alternative in the production of biofuels (as bioetanol and biogas) and compounds with high added value with specific properties (antioxidants, anticoagulants, anti-inflammatories) for applications in food, medical and energy industries in accordance with the integrated biorefineries. Furthermore, biorefinery concept requires processes that allow efficient utilization of all components of the biomass. The pretreatment step in a biorefinery is often based on hydrothermal principles of high temperatures in aqueous solution. Therefore, in this chapter, a review on the application of hydrothermal pretreatment on macroalgal biomass is presented.

Keywords Biorefinery • Biomass valorization • Hydrothermal pretreatment • Macroalgae • Biofuels • High value-added products • Antioxidants • Bioethanol • Severity factor • Brown macroalgae • Reactors • Fucoidan • Hydrothermal liquefaction • Methane • Bio-oil

1 Introduction

The term biorefinery is similar to the classical petroleum refinery but includes the conversion of biomass into fuels, power and chemicals with high added value (Ruiz et al. 2013a; http://www.nrel.gov/biomass/biorefinery.html). Biorefinery requires the need to identify different bio-products, chemicals and bioenergy carriers,

H.A. Ruiz (✉) • R.M. Rodríguez-Jasso
Biorefinery Group, Food Research Department, School of Chemistry,
Autonomous University of Coahuila, Saltillo 25280, Coahuila, Mexico
e-mail: hector_ruiz_leza@uadec.edu.mx

M. Aguedo
Lab of Biological and Industrial Chemistry, Gembloux Agro-Bio Tech – University of Liège,
Passage des Déportés 2, 5030 Gembloux, Belgium

Z. Kádár
Center for BioProcess Engineering, Department of Chemical and Biochemical Engineering,
Technical University of Denmark, 2800 Kgs Lyngby, Denmark

A. Prokop et al. (eds.), *Algal Biorefineries*, DOI 10.1007/978-3-319-20200-6_15

considering both environmental and economic aspects (Wei et al. 2013; Jung et al. 2013). Furthermore, this concept can be also integrated into already existing biofuel production processes (Hughes et al. 2013; Moncada et al. 2014). Recently, Kumar et al. (2013) showed that it is possible to develop a macroalgal biorefinery from *Gracilaria verrucosa*, which can be commercially viable for the production of agar (27–33 % of extraction) and bioethanol (0.43 g/g sugar). Furthermore, considering other important aspects in the integrated approach of macroalgal biorefineries design, Golberg et al. (2014) proposed the design of these biorefineries taking into account the thermodynamics modeling of biorefineries size and distribution, bioconversion technology and sustainability implications for developing economies. Therefore, they concluded that the combination of these factors, should promote the implementation of sustainable distributed biorefineries in developing countries. Also, the generation of co-products is required in a biorefinery to be economically feasible. Baghel et al. (2015) reported that 1 ton of fresh red algal biomass produced high value-added compounds as lipids (1.2–4.8 kg), agar (28.4–94.4 kg), cellulose (4.4–41.9 kg), mineral solution (3.1–3.6 kg), R-phycoerythrin (0.3–0.7 kg) and R-phycocyanin (0.1–0.3 kg) and concluded that the macroalgal biomass is manpower intensive and thus creates new additional employment and sustainable income sources improving the livelihoods and socio-economic status of economically underprivileged coastal communities. However to our knowledge, there are currently no commercial biorefining plants in operation using macroalgal biomass as raw material and limited studies have been conducted on the economics of industrial-scale biorefineries, so there is an excellent opportunity to implement new biorefineries, but it is important to consider all the aspects in the implementation of macroalgal biorefineries. According to Azapagic (2014), the sustainability of an integrated biorefinery depends on technological, economic (feedstock costs, capital costs, and biofuel costs), environmental (greenhouse gas emissions, land-use change, biodiversity, water use) and social factors (jobs and regional development, health issues, human and labour rights, land availability and food prices).

Recently macroalgae (seaweed) gained more and more interest as a promising raw material for biorefinery processing (Bozell 2008; Marquez et al. 2014; Suutari et al. 2015). Since the chemical structure of macroalgae is complex, it enables to produce high added value compounds and develop new processes in a biorefinery concept (Clarens et al. 2010; Kraan 2013). The fractionation of the biomass into different components is performed via pretreatment, which remains one of the most energy intensive and expensive process step in a biorefinery. Therefore the selection of an efficient pretreatment is essential also in environmental and economic points of view (Ruiz et al. 2013b; Fang 2013).

Since hydrothermal pretreatment has been reported as a potential clean technology to convert biomass according to biorefinery concept, its application to macroalgae is discussed in this chapter.

2 Macroalgal Biomass as Raw Material

Macroalgae as a biomass has several advantages over terrestrial biomass. Their use for second-generation biofuel production it does not compete with food supply, and they don not require: land and freshwater for their cultivation. It produces significantly higher yields per unit or per year compared to terrestrial biomass. Most macroalgae lack of lignin, which obstruct the enzymatic hydrolysis and further conversion of polysaccharides (Jung et al. 2013; Wei et al. 2013; Fasahati et al. 2015). The energy potential has been reported (Montingelli et al. 2015) approximately five times higher compared to the 22 EJ of terrestrial biomass. In 2012 the world annual macroalgae production was reported (FAO 2012) approximately 21 million tons wet weight. Considering that amount, macroalgae can be also considered as a source of raw material for biorefinery processes and for new industrial applications (Hayashi et al. 2014) along with other biomass. Asia (China, Korea, Japan, Indonesia, Philippines) is the main macroalgae producing area worldwide, however, USA, Chile, France, Portugal and Norway are also significant producers (Seaweed Site: http://www.seaweed.ie; Jung et al. 2013). Also, the macroalgae cultivation, harvest process and transporting has been successfully established in some Asian and Northern European countries and have been grown at commercial scale for food products for decades (Wei et al. 2013), offering the possibility of sustainable cultivation of a vast potential biomass feedstock.

According to Titlyanov and Titlyanova (2010), the macroalgae species for cultivation is chosen regarding to the location of a farm and cultivation facilities, for example: in the open sea, on the land, in the cold waters of a temperature zone, in warm waters od the tropics, on the productivity and adaptability of a species, factors as irradiance, pollution, water movement, degree of wave action and also due to its cost effectiveness and to the application of macroalgal biomass.

Macroalgae can be divided into three general types: red (*Rhodophyta*), brown (*Phaeophyta*), and green (*Chlorophyta*). The chemical composition of macroalgae significantly varies depending on the type, the species and on cultivation conditions (Mohamed et al. 2012; Haykiri-Acma et al. 2013; Hong et al. 2014). Another important aspect of macroalgae is the carbohydrate content which varies between 30 and 70 % (Cho et al. 2013); however the polysaccharide composition of macroalgae differs across the major macroalgal taxonomic groups: (Navarro and Stortz 2005; Siddhanta et al. 2009; Roesijadi et al. 2010; Ge et al. 2011; Delattre et al. 2011; Usov and Zelinsky 2013). Table 1 summarizes the composition of different macroalgae. Jung et al. (2013)) reported that *Laminaria japonica* (brown macroalgae), *Eucheuma* spp. (red macroalgae), *Kappaphycus alvarezii* (red macroalgae), *Undaria pinnatifida* (brown macroalgae) and *Gracilaria verrucosa* (red macroalgae) are the most promising macroalgae species for biorefinery feedstock. According to these authors, current mass-cultivation technology and market for macroalgae-based refinery technology needs to be focused on utilizing brown and red macroalgae rather than green macroalgae.

Table 1 Chemical composition of brown, red and green macroalgae (dry basis, % w/w)

Macroalgae	Carbohydrates	Fiber	Protein	Lipids	Minerals	Reference
***Chlorophyta*: Green Algae**						
Caulerpa lentillifera	38.66	32.99	10.41	1.11	37.15	Matanjun et al. (2009)
Caulerpa racemosa	83.2	dns	18.3	19.10	dns	Rameshkumar et al. (2013)
Chaetomorpha sp.	43.9	dns	11.1	3.3	36.6	Neveux et al. (2014a)
Cladophora rupestris	dns	24.70[b]	29.80	1.00	16.80	Marsham et al. (2007)
Derbesia sp.	26.9	dns	21.6	10.4	34.7	Neveux et al. (2014a)
Rhizoclonium riparium	29.53	11.93[b]	12.77	0.28	37.62	Chirapart et al. (2014)
Ulva sp.	43.9	dns	16.3	1.9	30.7	Neveux et al. (2014a)
Ulva clathrata	dns	40.60[a]	20.01	2.20	27.50	Peña-Rodríguez et al. (2011)
Ulva intestinalis	48.97	5.87[b]	10.59	0.62	20.65	Chirapart et al. (2014)
Ulva lactuca	61.50	60.05[a]	27.2	0.30	11.0	Ortiz et al. (2006)
Ulva linza	37.4	2.40[b]	31.6	1.8	29.2	Jang et al. (2012a)
Ulva clathrata	dns	40.60[a]	20.01	2.20	27.50	Peña-Rodríguez et al. (2011)
Ulva clathrata	43.00	dns	17.8	1.9	37.3	Pham et al. (2013)
Ulva pertusa	59.07	56.84	6.30	2.39	22.86	Jang et al. (2012b)
***Phaeophyceae*: Brown Algae**						
Ascophyllum nodosum	69.60	dns	1.20	1.20	22.50	Rioux et al. (2007)
Bifurcaria bifurcate	dns	37.42[a]	10.92	5.67	34.31	Gómez-Ordóñez et al. (2010)
Durvillaea Antarctica	64.65	63.9[a]	11.00	2.55	21.80	Ortiz et al. (2006)
Durvillaea Antarctica	dns	50.00[b]	8.20	1.80	19.20	Astorga and Mansilla (2013)
Fucus serratus	dns	16.00[b]	17.40	1.80	18.60	Marsham et al. (2007)
Fucus vesiculosus	65.70	dns	1.40	1.40	24.8	Rioux et al. (2007)
Himanthalia elongate	dns	37.14[a]	14.08	0.94	36.41	Gómez-Ordóñez et al. (2010)
Hydroclathrus clathratus	82.26	2.70	6.39	2.18	6.47	Zafar and Chowdhury (2009)
Laminaria sp.	60	dns	12.00	2.00	26.00	Fasahati et al. (2015)
Laminaria japónica	60.9–17.0	dns	6.8–10.3	7.2–11.5	13.8–21.1	Xu et al. (2014)

Laminaria japónica	50.60	dns	10.50	37.3	1.60	Pham et al. (2013)
Laminaria japonica	54.50	28.91	7.40	1.37	28.33	Jang et al. (2012b)
Laminaria saccharina	dns	30.23[a]	25.70	0.79	34.78	Gómez-Ordóñez et al. (2010)
Macrocystis pyrifera	dns	37.10[b]	10.09	1.80	34.10	Astorga and Mansilla (2013)
Mastocarpus stellatus	dns	31.70[a]	21.30	0.39	24.99	Gómez-Ordóñez et al. (2010)
Sargassum cristaefoliumb	46.78	16.10[b]	5.09	0.17	24.90	Borines et al. (2011)
Sargassum fulvellum	44.5	3.5[b]	19.9	0.5	35.1	Jang et al. (2012a)
Sargassum fusiforme	dns	62.30[a]	10.90	1.40	dns	Dawczynski et al. (2007)
Sargassum fusiforme	59.0	4.2[b]	13.9	0.4	26.6	Jang et al. (2012a)
Sargassum kushimontea	42.89	11.53[b]	6.37	0.17	26.67	Borines et al. (2011)
Saccharina longicruris	57.80	dns	12.40	2.10	27.70	Rioux et al. (2007)
Sargassum naozhouense	47.73	4.83	11.20	1.06	35.18	Peng et al. (2013)
Sargassum polycystum	33.49	39.67	5.40	0.29	42.40	Matanjun et al. (2009)
Sargassum thunbergii	40.3	11.3[b]	14.5	0.2	26.3	Li et al. (2012)
Sargassum vulgare	61.60	7.70[b]	13.60	0.04	19.40	Marinho-Soriano et al. (2006)
Undaria pinnatifida	dns	45.90[a]	18.90	4.50	dns	Dawczynski et al. (2007)
Undaria pinnatifida	48.5	3.5[b]	18.2	1.8	28.0	Cho et al. (2013)
***Rhodophyta*: Red algae**						
Eucheuma cottonii	26.49	25.05[a]	9.76	1.10	46.19	Matanjun et al. (2009)
Gelidium amansii	74.7	11.6[b]	18.3	0.0	7.4	Jang et al. (2012a)
Gelidium amansii	52.84 [c]	dns	10.19	7.44	2.52	Malihan et al. (2014)
Gigartina pistillata	dns	29.31[a]	15.59	0.57	34.56	Gómez-Ordóñez et al. (2010)
Gracilaria cervicornis	63.10	5.60[b]	19.70	0.04	10.50	Marinho-Soriano et al. (2006)
Gracilaria salicornia	46.22	9.21[b]	16.28	1.69	13.49	Chirapart et al. (2014)
Gracilaria tenuistipitata	54.89	4.96[b]	6.11	0.26	22.91	Chirapart et al. (2014)

(continued)

Table 1 (continued)

Macroalgae	Carbohydrates	Fiber	Protein	Lipids	Minerals	Reference
Grateloupia elliptica	59.5	dns	13.4	1.1	26.0	Pham et al. (2013)
Grateloupia turuturu	dns	60.40[a]	22.90	2.60	18.50	Denis et al. (2010)
Ochtodes secundiramea	45.00	dns	10.10	3.50	dns	Gressler et al. (2011)
Plocamium brasiliense	52.00	dns	15.70	3.60	dns	Gressler et al. (2011)
Porphyra sp	dns	47.70[a]	26.30	1.90	dns	Dawczynski et al. (2007)
Porphyra umbilicalis	dns	35.46	39.01	0.34	11.72	Cofrades et al. (2008)

Adapted and modified from Ruiz et al. 2015
dns data not shown
[a]Total dietary fiber = soluble dietary fiber + insoluble dietary fiber
[b]Insoluble dietary fiber
[c]Only is considered of glucose and galactose quantified

Presently, macroalgae are applied in human foods (e.g. source of food fibre, food additives, yogurt, ice cream, pudding, stabilisers in processing fish, meat, dairy, and confectionary products), cosmetics applications (antiaging skin care, anticellulite treatment, photoprotective, moisturising) and fertilizers (Podkorytova et al. 2007; Wei et al. 2013; Bedoux et al. 2014). Furthermore, several short and long chain chemicals with medicinal and industrial applications can be also extracted (Andrade et al. 2013; de Quirós et al. 2010; Yazdani et al. 2015; Radulovich et al. 2015; Oliveira et al. 2015; Hoang et al. 2015; Kalimuthu and Kim 2015). Recently macroalgae have been intensively studied for biofuel production such as biobutanol, biodiesel, bioethanol and biogas by anaerobic digestion (Poots et al. 2012; Xu et al. 2014; Sun et al. 2014; Oliveira et al. 2014). Fasahati et al. (2015) have estimated the preliminary cost for industrial scale bioethanol production using dry brown macroalgae as raw material, based on a simple (hot water wash) and combined (acid thermal hydrolysis) processes, they simulated at the scales of 80,000 and 400,000 ton/year of dry brown macroalgae. The maximum dry seaweed price for simple and combined processes was calculated as 64.6 and 26 $/ton (80,000 ton/year) and 91.3 and 71.5 $/ton (400,000 ton/year), respectively. The minimum ethanol-selling price for both processes was calculated as 2.39 and 2.85 $/gal (80,000 ton/year) and 2.08 and 2.33 $/ton (400,000 ton/year) respectively. They concluded that the sensitivity analysis suggests that biomass price had the highest impact on minimum ethanol selling price and the acid thermal hydrolysis pretreatment of brown algae is not economically viable.

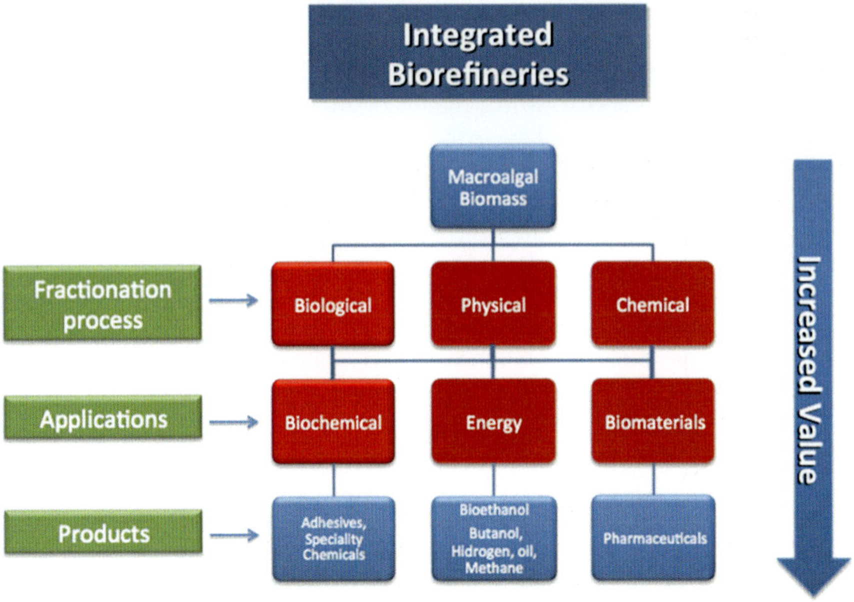

Fig. 1 Scheme of an integrated biorefinery using macroalgae as feedstock

Due to their composition and broad spectrum of applications, macroalgae can be considered as a promising feedstock for biorefineries. Figure 1 shows the scheme of an integrated biorefinery using macroalgae as feedstock.

3 Hydrothermal Pretreatment for Macroalgae Biomass

Numerous technologies have been applied for extraction of compounds from macroalgae biomass with different objectives (Rodriguez-Jasso et al. 2011; Schultz-Jensen et al. 2013; Wal et al. 2013; Pham et al. 2013; Tedesco et al. 2013; Malihan et al. 2014; Rodriguez-Jasso et al. 2014; Barbot et al. 2014. Vanegas et al. 2015). Similarly, the pretreatments for lignocellulosic materials also can be used and adapted on macroalgae biomass for many purposes as hydrolysis, extraction and structural modification of macroalgal biomass. These pretreatments for lignocellulosic material were reviewed previously (Alvira et al. 2010; Zheng et al. 2014) previously and they appeared to have operational and economical advantages and disadvantages.

Hydrothermal pretreatments (HTT) are an alternative for the fractionation of macroalgae biomass and the excellent solvent properties of water as a reaction medium and the high moisture content of macroalgal biomass make a promising processing technology for the direct use of this biomass in the production of biofuels and high added-value compounds. In general terms, the HTT can be considered cost-effective and ecofriendly processes. For this pretreatment, hot water is used at temperatures from 100 to 374 °C, under high pressure, corresponding to conditions below the water critical point (Meillisa et al. 2015). Moreover, this pretreatment can also applied on macroalgae as already demonstrated with (Vanegas et al. 2015) or without (Schultz-Jensen et al. 2013) addition of catalyst.

As a complement to HTT, there is the possibility of applying hydrothermal liquefaction (HL) (Tekin et al. 2014). In this process, water at conditions approaching its critical point, the wet biomass is decomposed to a liquid biocrude. Elevated temperatures and pressures reduce the density, polarity and relative permittivity/dielectric constant of water, resulting in the hydrolysis and dissolution of solid biomass in HL (Neveux et al. 2014b). Therefore, HL for macroalgal biomass has an important advantage: it does not require drying, resulting in a huge cost saving in water removal operations; the bio-oil produced is not miscible with water and has a lower oxygen content, and therefore higher energy content (Chen et al. 2009). The hydrothermal liquefaction processing has been recently reviewed by Tian et al. (2014), showing its advantages as fast reactions in the use of wet feedstocks of macroalgal biomass. Tekin and Karagoz (2013) concluded that hydrothermal liquefaction is an important technology for biofuels production.

For macroalgae biomass and specifically for brown macroalgae sulphated polysaccharides (fucans), the term "autohydrolysis" can be referred to the acid polysaccharide hydrolysis under very mild conditions (Shevchenko et al. 2014). The sulphated groups are present in brown macroalgae, sulphated groups are associated to polysaccharides and their hydrolysis causes a pH drop that contributes to effectively self-catalyze the process. In recent works, Anastyuk et al. (2014) and Shevchenko et al. (2014) studied the depolymerization and extraction of fucoidan from brown macroalgae through autohydrolysis processing and using sulphated compounds as catalyst. Rodriguez-Jasso et al. (2011) applied the hydrothermal pretreatment on *Fucus vesiculosus* biomass and showed that the pH decreased in the liquid phase possibly due to the hydrolysis of polysaccharides by sulphated groups as catalyst.

Moreover, the carrageenan from macroalgae contain sulfated compounds that can carry out the autohydrolysis reactions (Ciancia et al. 2005; Barabanova et al. 2010; Aguilar et al. 2011; Prajapati et al. 2014). According to Hoffmann et al. (1996), the effect of autohydrolysis occurs at low pH, as carra-geenans in acid solution are cleaved at the 3,6-anhydrogalactose linkage.

3.1 Operational Conditions and Reactors

The most important operational variables affecting the effectiveness of HTT and HL include temperature, residence time, particle size, biomass feedstock, heating rate, pressure, substrate concentration and moisture content of the biomass and ratio liquid/solid may also greatly influence the process (Ruiz et al. 2013b; Akhtar and Amin 2011). According to Elliot et al. (2015), the wet feedstocks of macroalgal biomass are particularly suited for HL.

For hydrothermal pretreatment (HTT) in an operating range between 150 and 230 °C, the effect of temperature and time can be combined and represented by the severity factor Log (R_0) (Eq. 1) proposed by Overend and Chornet (1987).

$$Log(R_0) = \int_0^t \exp\left[\frac{T-100}{14.75}\right] dt \tag{1}$$

where t is the resident time (min), T is the process temperature (°C), 100 is the temperature of reference and 14.75 is an empirical parameter related with activation energy, assuming pseudo-first order kinetics. This severity factor has also been used in hydrothermal liquefaction (HL) for the bio-crude production and the correlation of operational conditions with temperature and time (Eboibi et al. 2014). Furthermore, the severity factor Log (R_0) in a non-isothermal hydrothermal pretreatment, which includes the combination of temperature and reaction time along heating and cooling phases as represented by the Eq. 2.

$$\log R_0 = \log\left[R_{0\ Heating} + R_{0\ Cooling}\right]$$
$$\log R_0 = \left[\int_0^{t_{MAX}} \frac{T(t)-100}{\omega} dt + \int_{t_{MAX}}^{t_F} \frac{T'(t)-100}{\omega} dt\right] \tag{2}$$

The heating (conduction, convection or radiation) in the reactors for HTT and HL can be performed by steam, fluidized sand baths, oil baths, electric heating jackets and microwave radiation to achieve fairly uniform heating as well as fast heat-up (Okuda et al. 2008; Shi et al. 2011).

HTT can be operated in isothermal and non-isothermal heating regimen (Fig. 2a, b). The heating rate affects directly the product yields and the final products. Zhang et al. (2008, 2011) reported that the heating rate, as governed by the mode of heat transfer in hydrothermal liquefaction, is an important factor that needs to be considered during scale up. Xu et al. (2015) reported the production of crude bio-oil using macroalgae – *Enteromorpha prolifera* as raw material and a fast-heating rate (75 °C/min) in a batch rector for hydrothermal pretreatment (250–390 °C, 10–120 min, algae/water mass ratio and K_2CO_3 loading) and mentioned that the variations in the yields can be attributed to the feedstock composition, oil separation process and the heating rate. Anastasakis and Ross (2015) produced biocrudes and biochars from four brown macroalgae using hydrothermal pretreatment (350 °C/15 min) with a heating rate of 25 °C/min and they reported that the HL has higher energy output than fermentation and analogue to that from anaerobic digestion. Okuda et al. (2008) studied the effect of hydrothermal pretreatment (99.8–299.8 °C, 30 min) at a heating rate of 50 °C/min on the subsequent enzymatic hydrolysis of cellulose from *Monostroma nitidum Wittrock* (green alga) and *Solieria pacifica* (red alga) and concluded that low temperatures of pretreatment on algae increased the rates of enzymatic hydrolysis.

$$logR_0 = log[R_{0\,Heating} + R_{0\,Cooling}]$$

$$logR_0 = \left[\int_0^{t_{MAX}} \frac{T(t)-100}{\omega}dt + \int_{t_{MAX}}^{t_F} \frac{T'(t)-100}{\omega}dt\right]$$

$$Log\ (R_0) = \int_0^t exp\left[\frac{T-100}{14.75}\right]dt$$

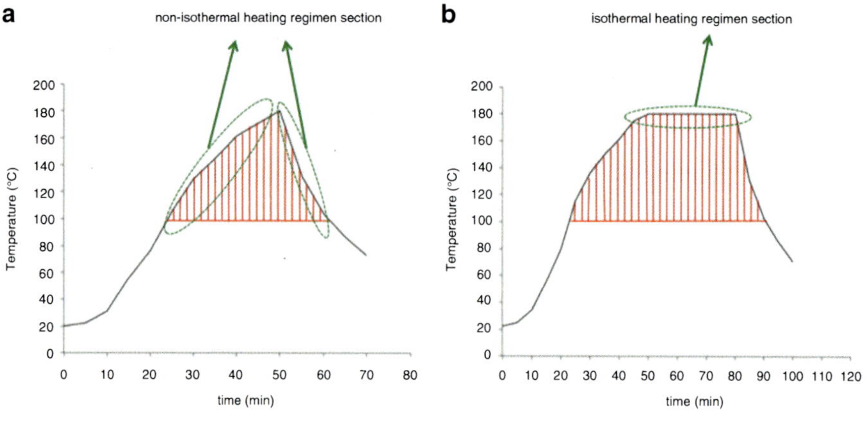

Fig. 2 Temperature profiles corresponding to: (**a**) non-isothermal and (**b**) isothermal regimens in hydrothermal processing (Adapted and modified from Ruiz et al. 2013b)

On the other hand, in the operation of HTT and HL on macroalgal biomass have been reported different types of reactor and configurations depending of the type of heating and the most used are: (1) Batch reactor (Fig. 3a): macroalgal biomass solid particles are mixed with water in the reactor and the residence time of the reacting solid is long. Rodriguez-Jasso et al. (2013) pretreated in a batch reactor *Fucus vesiculosus* (brown macroalgae) at 180 °C/20 min, the heating up and cooling were not taken into consideration. (2) Continuous flow reactor (Fig. 3b): macroalgal biomass is passing in one direction while water moves in the same or opposite direction and a continuous reactor system is typically required to operate at high temperatures and pressures to achieve a high conversion of the macroalgal biomass within a short residence time. Elliot et al. (2014) used a bench scale continuous flow reactor in hydrothermal pretreatment and wet macroalgal slurries biomass for liquid biocrude production and reported high conversion of this macroalgal biomass into fuel products. Additionally, continuous-flow reactors using macroalgal biomass have been recently revised by Elliot et al. (2015).

Other important point is the material for constructing reactors for HTT. Singh et al. (2015) reported that the reactor material in hydrothermal pretreatment consists mainly of stainless steel, this because the corrosion is more pronounced with increasing temperature.

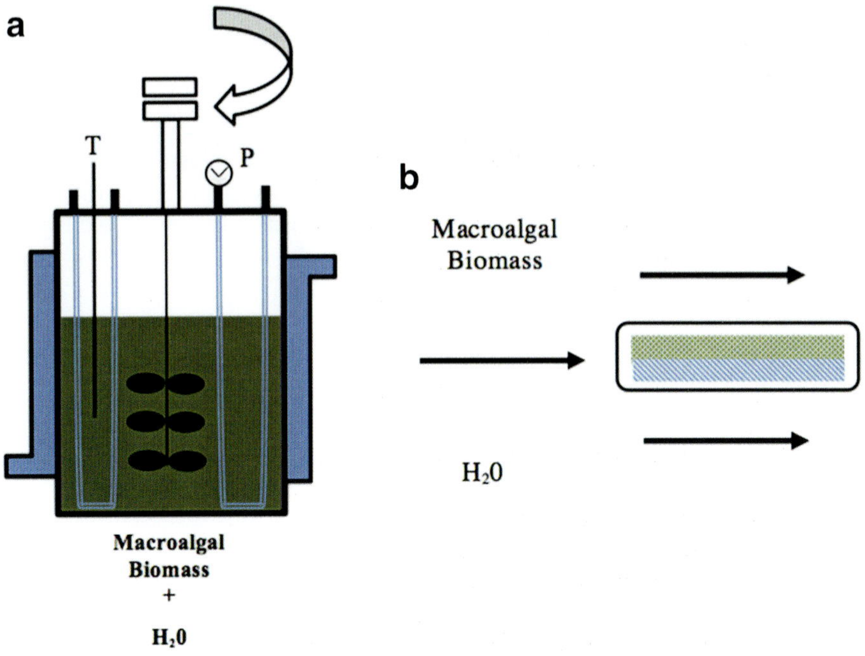

Fig. 3 Reactors for hydrothermal pretreatment using macroalgal biomass: (**a**) Batch; (**b**) Continuous reactor (Adapted and modified from Ruiz et al. 2013b)

3.2 *Applications of Macroalgal Biomass Using Hydrothermal Pretreatment*

To our knowledge, the studies and research hydrothermal pretreatment on macroalgae for the production of biofuels and high added-value compounds remain relatively limited. This section reports some products stemming from processes using hydrothermal pretreatment of macroalgal biomass as raw material.

3.2.1 Biofuels Production from Macroalgal Biomass Using Hydrothermal Pretreatment

Recent studies have been focused on the utilization of the sugars present in macroalgal biomass pretreated under hydrothermal processing for the production of biogas by anaerobic digestion, bioethanol, biobutanol, and hydrogen. Moreover, research on pretreatments of macroalgal biomass used as feedstock for the production of biofuels and high added value compounds are summarized in Table 2.

Methane

Anaerobic digestion can be applied to produce biogas, particularly methane, from various macroalgal biomasses using hydrothermal processing as pretreatment (Nielsen and Heiske 2011). Bruhn et al. (2011) used hydrothermal pretreatment at 130 °C for 20 min and *Ulva lactuca* as raw material, obtaining a methane yield of 187 ml g VS^{-1}. They concluded that the hydrothermal pretreatment had a minor positive impact on the methane yield. Jard et al. (2013) reported the methane potential (229 ml g VS^{1}) from hydrothermally pretreated *Palmaria palmate* at 160 °C for 30 min. They showed that at temperatures of pretreatment between 180 and 200 °C for 30 min, the methane potential decreased due to the formation inhibitory compounds in the liquid fraction.

Bioethanol and Butanol

Macroalgae are often rich in polysaccharides and sugars, and therefore well suited for bioethanol and biobutanol production using hydrothermal pretreatment (Yanagisawa et al. 2013; Suutari et al. 2015). In respect to the hydrothermal processing as pretreatment for the third generation of bioethanol, few analyses have been reported. This pretreatment is a necessary step to alter some structural characteristics of macroalgal biomass, thus the enzyme is favoured, increasing the potential of enzymatic hydrolysis to produce monomeric sugars and these fermentable sugars can be subsequently converted to bioethanol. Meillisa et al. (2015) produced fermentable sugars from *Saccharina japonica* using hydrothermal pretreatment in a

Table 2 Products from macroalgal biomass using hydrothermal pretreatment

Biomass Products	Macroalgae	Catalyst	Reactor	Conditions	References
Antioxidants	*Laminaria japonica* (South Korea)	Formic acid - 1 % (v/v) (1/25 algae/catalyst ratio)	Batch reactor. Subcritical water hydrolysis	180–260 °C/15–65 bar/30–75 min	Meillisa et al. (2015)
Antioxidants	*Laminaria saccharina*, *Laminaria digitata*, *Himanthalia elongate* (Ireland)	Deionised water (1/5 algae/catalyst ratio)	Batch reactor	121 °C/15 min/ Following by methanol extraction 60 %	Rajauria et al. (2010)
Glucose and xylose and Antioxidants	*Ulva pertusa* kjellmann (South Korea)	Distilled water (1/10 algae/catalyst ratio)	Batch reactor -microwave heating	100–200 °C/1–15 bar/2–12 min/1700 W	Choi et al. (2013)
Polysaccharides (Fucoidan)	*Sargassum muticum* (Spain)	Distilled water (1/30 algae/catalyst ratio)	Batch stainless steel reactor	150–240 °C/non-isothermal extraction	Balboa et al. (2013); Balboa et al. (2014)
Polysaccharides (Fucoidan)	*Fucus vesiculosus* (Portugal)	Distilled water (1/25 algae/catalyst ratio)	Batch reactor -microwave heating	30–120 psi/1–31 min	Rodriguez-Jasso et al. (2011)
Polysaccharides (Fucoidan)	*Fucus vesiculosus* (Portugal)	Distilled water (1/25 algae/ catalyst ratio)	Batch reactors submerged in a silicon oil bath	160–200 °C/10–30 min	Rodriguez-Jasso et al. (2013)
Cellulose fractions	*Monostroma nitidum* Wittrock, *Solieria pacifica* (Japan)	Distilled water (1/5 algae/catalyst ratio)	Micro reactor	100–300 °C/30 min	Okuda et al. (2008)
Fermentable sugars for Bioethanol	*Laminaria japonica* (South Korea)	H_2SO_4 – 40 and 94 mM (5–20 % w/v algae ratio)	Batch reactor	121 °C/15–60 min	Jang et al. (2012b)
Fermentable sugars for Bioethanol	*Ulva pertusa*, *Laminaria japonica*, *Gelidium amansii* (Japan)	H_2SO_4 – 0.5–5 % (v/v) (3–5 % w/v algae ratio)	Batch reactor	120 – 150 °C/15 – 120 min	Jang et al. (2012a)
Fermentable sugars for Bioethanol	*Chaetomorpha linum* (Denmark)	Distilled water (1/25 algae/catalyst ratio)	Loop batch reactor with continuous stirring	180–200 °C/10–15.5 bar/10 min	Schultz-Jensen et al. (2013)

(continued)

Table 2 (continued)

Biomass Products	Macroalgae	Catalyst	Reactor	Conditions	References
Fermentable sugars for Bioethanol	*Ulva fasciata* Delile (India)	Ammonia – 1, 2 and 4 % (v/v)	Batch reactor	120 °C/15 psi /30–90 min	Trivedi et al. (2013)
Fermentable sugars for Bioethanol	*Sargassum sagamianum* (South Korea)	Distilled water (1/10 algae/catalyst ratio)	High-temperature liquefying reactor.	200 °C /15 MPa/15 min.	Yeon et al. (2011)
Fermentable sugars for Butanol	*Ulva lactuca* (USA)	H_2SO_4 – 0.5 to 5 % (v/v) (1/10 algae/ catalyst ratio)	Steam batch reactor	125 °C/10–120 min	Poots et al. (2012)
Fermentable sugars for acetone, butanol and ethanol	*Ulva lactuca* (The Netherlands)	H_2SO_4 – 7.5 % w/w algae	Batch reactors submerged in a silicon oil bath	150 °C/10 min	van der Wal et al. (2013)
Fermentable substrate for enzymes	*Fucus vesiculosus* (Portugal)	Distilled water (1/25 algae/ catalyst ratio)	Batch reactor	180 °C/20 min	Rodriguez-Jasso et al. (2013)
Proteins, carbohydrates for Methane	Palmaria palmate (France)	NaOH – 0.04 % and HCl- 0.02 % (v/v) (50 g/L algae ratio)	Batch reactor with a stir-ring	120–200 °C/30 min	Jard et al. (2013)
Bio-oil	*Saccharina* spp. (USA)	Deionized water (22 % algae ratio)	Bench-scale continuous-flow reactor system.	350 °C/20 MPa / 30–60 min	Elliot et al. (2014)
Bio-oil	*Ulva fasciata*, *Enteromorpha* sp, *Sargassum tenerrimum* (India)	Distilled water biomass:water ratio of (1/6 algae/catalyst ratio)	Batch reactor	280 °C/15 min	Singh et al. (2015a)
Bio-oil	*Ulva fasciata* (India)	Water, organic solvents (CH_3OH, C_2H_5OH) (1/6 algae/catalyst ratio)	Batch Reactor with vertical agitation using a stirrer.	300 °C/1000 to 1300 psi/15 min	Singh et al. (2015b)

Bio-oil	*Enteromorpha prolifera* (China)	K_2CO_3 – 0 to 30 % (w/v) (0.5/8.4 – 5.5/8.4) algae/catalyst ratio)	Batch reactor	250–390 °C/10–120 min	Xu et al. (2015)
Bio-oil	*Enteromorpha prolifera* (China)	Na_2CO_3 – 0 to 5 % (w/v) (1/ 7.5 algae/catalyst ratio)	Batch reactor (GSH-0.25 zirconium)	220 – 320 °C / 15 min	Zhou et al. (2010)
Bio-oil and Bio-char	*Sargassum* spp. (Puerto Rico)	Deionized water (1/4 algae/catalyst ratio)	Batch reactor	350 °C/60 min/5–21 MPa/300 rpm	Díaz-Vázquez et al. (2015)
Biocrude	Algae biomass (USA)	Distilled water (35 % w/v algae ratio)	Continuous-flow reactor for obtaining engineering data for the catalytic hydrothermal gasification	305 °C /20 MPa	Elliott et al. (2013)
Biocrude	*Derbesia tenuissima*, *Ulva ohnoi*, *Oedogonium* sp. (Australia)	Water (1/14 algae/catalyst ratio)	Batch reactor tube fitted with a gasket and attached to a pressure-head, immersed in a fluidized sand bath	350 – 345 °C / 14–17 MPa / 8 min	Neveux et al. (2014b)
Bio-crude and bio-char	*Laminaria digitata*, *Laminaria hyperborea*, *Laminaria saccharina*, *Alaria esculenta* (Scotland)	Distilled water (1/3.75 algae/catalyst ratio).	Batch bomb type stainless steel reactor.	350 °C/15 min	Anastasakis and Ross (2015)
Hydrogen and methane-rich gas	*Fucus serratus*, *Laminaria digitata*, *Alaria esculenta*, *Bifurcaria bifurcate* (Ireland)	Distilled water (1/16.8 algae/catalyst ratio)	Batch reactor	500 °C /1 h	Schumacher et al. (2011)

range of temperatures of 180–260 °C. A temperature of 180 °C led to the highest production of total glucose (0.34 g/L). They concluded that high sugar recovery was achieved under experimental conditions without a catalyst, therefore these fermentable sugars are a renewable source of biofuel. Okuda et al. (2008) studied the enzymatic hydrolysis of hydrothermally pretreated *Monostroma nitidum Wittrock* (green alga) and *Solieria pacifica* (red alga) and the enzymatic hydrolysis yield were 79.9 and 87.8 %, respectively. They concluded that the solid pretreated was highly susceptible to enzymatic attack. Choi et al. (2013) pretreated the macroalgae *Ulva pertusa* Kjellmann in a high-pressure steam process at 180 °C for 8 min, after this process, the solid obtained was hydrolyzed with cellulase enzyme and the cellulose to glucose conversion yield was 77 %, with respect to the theoretical value. They concluded that the pressure steam process is an effective pretreatment for fermentable sugar production.

In a recent work, Yazdani et al. (2015) showed that the hydrothermal pretreatment at 121 °C for 60 min was a promising process to improve the digestibility of pretreated *Nizimuddinia zanardini*; increased the bioethanol yield, producing 34.6 g/kg of initial macroalgal biomass. Trivedi et al. (2013) produced bioethanol from hydrothermally pretreated *Ulva fasciata* at 120 °C for 1 h, obtaining 88.2 % of the theoretical ethanol yield. Ji-Hyeon et al. (2011) performed an operational strategy (repeated batch) for bioethanol production. They used as substrate for bioethanol production, the hydrolysates from hydrothermally pretreated *Sargassum sagamianum* at 200 °C for 15 min. They reported a bioethanol conversion of 84.3 % with respect to the theoretical value.

Regarding butanol, Wal et al. (2013) produced butanol, acetone and ethanol from *Ulva lactuca* hydrolysates using the microorganims *Clostridium acetobutylicum* and *Clostridium beijerinckii* and a hydrothermal pretreatment at 150 °C for 10 min. They concluded that the hydrolysates did not require supplementation with nutrients to support microbial growth, for butanol-acetone-ethanol production at high ugar concentration.

Hydrogen

In relation to hydrogen production, Jung et al. (2011) performed a hydrothermal pretreatment of *Laminaria japonica* under different conditions of temperature (150, 160, 170 and 180 °C), residence time (5–10 min). The maximum H_2 yield was 109.6 ml H_2/ g COM (chemical oxygen demand) at 170 °C for 20 min and it was concluded that as hydrothermal pretreatment increased, solubilization increased and the contenting cellulose and hemicellulose decreased. Schumacher et al. (2011) conducted a study on hydrothermal pretreatment using four species of macroalgal biomass (*Fucus serratus*, *Laminaria digitata*, *Alaria esculenta* and *Bifurcaria bifurcate*) at 500 °C, 1 h. The results showed that the production of hydrogen was obtained with a yield of 16 g H_2/kg of macroalgal biomass for *Bifurcaria bifurcate*.

Bio-oil

Anastasakis and Ross (2015) evaluated the hydrothermal liquefaction of four-macroalgal biomass (*L. digitata*, *L. hyperborea*, *L. saccharina* and *A. esculenta*) in order to maximize the bio-oil yield. A maximum bio-oil yield of 17.8 wt% was obtained for *A. esculenta* with a 1:3.75 macroalgal biomass: water ratio at 350 °C for 15 min without presence of the catalyst. The high heating value (HHV) for *A. esculenta* was 33.8 MJ/kg. Singh et al. (2015a) studied the effect of compositional changes of *Ulva fasciata*, *Enteromorpha* sp. and *Sargassum tenerrimum* macroalgal biomass on the production of bio-oil. These macroalgal biomasses were converted to bio-oil by hydrothermal liquefaction in a batch configuration reactor at 280 °C for 15 min with macroalgal biomass: water ratio of 1:6. The maximum bio-oil yield (12 wt%, dry basis) was observed for *Ulva fasciata*. They concluded that varying the feedstock composition affects the bio-oil yield.

3.2.2 Compounds with High Added-Value from Macroalgal Biomass Using Hydrothermal Pretreatment

Several polysaccharides and oligosaccharides derived from macroalgae are used for therapeutic and cosmetic applications due to their bioactivity capacity (Castro et al. 2014; Wang et al. 2014). At the present, the sulphated polysaccharides and saccharides present in the fucoidan from brown algae are recognized to have antithrombotic, antiviral, antiinflammatory, antilipidemic, and antioxidant activities (Fitton 2011; Senthilkumar et al. 2013). Therefore, the production of bioactive compounds with high added value is of great interest for biorefinery concept applying hydrothermal pretreatment.

On the other hand, microwaves as an alternative heating source for hydrothermal pretreatment have been successfully applied for the fucoidan extraction. Rodriguez-Jasso et al. (2011) applied the microwave-assisted hydrothermal pretreatment for the extraction of fucoidan from *Fucus vesiculosus* The highest fucoidan yield (18.22 g/100 g of macroalgal biomass) was obtained when the highest pressure (120 psi) and the lowest extraction time (1 min) and alga/water ratio (1 g/25 ml). They concluded that microwave-assisted hydrothermal pretreatment required short extraction time and use of non-corrosive solvents, resulting in reduced costs when compared to the acid–base extraction techniques.

In recent works, the fucoidan has been depolymerized and extracted using hydrothermal pretreatments. Rodriguez-Jasso et al. (2014) compared the fucoidan extraction from *Fucus vesiculosus* through two types of heating: microwave-assisted hydrothermal pretreatment (172 °C for 1 min) and conventional hydrothermal conduction (180 °C for 20 min), showing similar yields of extraction, 18.2 and 16.5 (g/100 g of macroalgal biomass). They mentioned that the chemical composition and antioxidant capacity of sulphated polysaccharides extracted vary according to the heating process. However, both fucans obtained by microwave-assisted pretreatment and conventional conduction have the potential for use as natural antioxidants in industrial applications. Balboa et al. (2013) extracted and depolymerized fucoidan from *Sargassum* sp. under hydrothermal pretreatment in the range of 150–240 °C. The fucoidan extracted is composed of fucose as the main sugar, and also galactose, xylose, glucose and mannose. They concluded that the hydrothermal pretreatment is an effective technology suitable for fractionating macroalgal biomass, allowing a simultaneous extraction and depolymerization of fucoidans in one single step.

Furthermore, the enzymes with known specificities that catalyze the degradation of fucoidans are important tools for studying the structural peculiarities and the biological role of this class of polysaccharide. Rodriguez-Jasso et al. (2013) used the hydrothermally pretreated *Fucus vesiculosus* as carbon source in the production of fucoidanase. *Fucus vesiculosus* was pretreated in non-isothermal regimen at 180 °C for 20 min. To our knowledge, this was the first study on the application of macroalgal biomass hydrothermally pretreated, as a substrate to produce enzymes.

Additionally, Balboa et al. (2014) produced compounds with antioxidant activity for application in cosmetic industry using non-isothermal hydrothermal pretreatment at 190 °C and *Sargassum muticum* as raw material. These extracts showed *in vitro* antioxidant properties comparable to commercial antioxidants. González-López et al. (2012) optimized the production of antioxidant compounds from *Sargassum muticum* after hydrothermal pretreatment. The pretreatment was performed under non-isothermal conditions (150–210 °C). Rajauria et al. (2010) evaluated the antioxidant capacity from *Laminaria saccharina*, *Laminaria digitata* and *Himanthalia elongata*. They found that the heat treatment causes improved effect in the activities of antioxidant compounds.

4 Conclusions

As illustrated in this chapter, the development of technologies and processes for the valorization of biomass according to the biorefinery concept is evidently essential for our society and a sustainable future. The marine biomass such as macroalgae can be a promising feedstock for biorefinery due of its wide geographic distribution without affecting the human food chain and environment. Moreover, the pretreatment stage plays an important role in the biorefinery concept, since the pretreatment allows the fractionation of the main components of the biomass. For this reason, the use of hydrothermal pretreatment and macroalgal biomass as feedstock are an important opportunity to develop sustainable and integrated biorefineries for the production of biofuels as bioethanol, methane, butanol, hydrogen, bio-oil and compounds with high added-value. Depending on the operational conditions and strategies, the hydrothermal pretreatment can be used for hydrolysis, extraction and structural modification of macroalgal biomass.

List of Abbreviations

EJ	Exajoule
HL	Hydrothermal liquefaction
HTT	Hydrothermal pretreatments
Log (R_0)	Severity factor
VS	Volatile Solids

References

Aguilar MJ, Batista AP, Nunes MC, Cordobés F, Raymundo A, Guerrero A (2011) From egg yolk/κ-Carrageenan dispersions to gel systems: linear viscoelasticity and texture analysis. Food Hydrocolloid 25:654–658

Akhtar J, Amin NAS (2011) A review on process conditions for optimum bio-oil yield in hydrothermal liquefaction of biomass. Renew Sustain Energ Rev 15:1615–1624

Alvira P, Tomás-Pejo E, Ballesteros M, Negro MJ (2010) Pretreatment technologies for an efficient bioethanol production process based on enzymatic hydrolysis: a review. Bioresour Technol 101:4851–4861

Anastasakis K, Ross AB (2015) Hydrothermal liquefaction of four brown macro-algae commonly found on the UK coasts: an energetic analysis of the process and comparison with bio-chemical conversion methods. Fuel 139:546–553

Anastyuk SD, Imbs IT, Dmitrnok PS, Zvyagintseva TN (2014) Rapid mass spectrometric analysis of a novel fucoidan, extracted from the brown alga *Coccophora langsdorfii*. Sci World J. doi:10.1155/2014/972450

Andrade PB, Barbosa M, Matos RP, Lopes G, Vinholes J, Mouga T, Valentão P (2013) Valuable compounds in macroalgae extracts. Food Chem 138:1819–1828

Astorga-España MS, Mansilla A (2013) Sub-Antarctic macroalgae: opportunities for gastronomic tourism and local fisheries in the Region of Magallanes and Chilean Antarctic Territory. J Appl Phycol 26:973–978

Azapagic A (2014) Sustainability considerations for integrated biorefineries. Trends Biotechnol 32:1–4

Baghel RS, Trivedi N, Gupta V, Neori A, Reddy CRK, Lali A, Jha B (2015) Biorefining of marine macroalgal biomass for production of biofuel and commodity chemicals. Green Chem. doi:10.1039/C4GC02532F (In press)

Balboa EM, Rivas S, Moure A, Domínguez H, Parajó JC (2013) Simultaneous extraction and depolymerization of fucoidan from *Sargassum muticum* in aqueous media. Mar Drugs 11:4612–4627

Balboa EM, Soto ML, Nogueira DR, González-López N, Conde E, Moure A, Vinardell MP, Mitjans M, Domínguez H (2014) Potential of antioxidant extracts produced by aqueous processing of renewable resources for the formulation of cosmetics. Ind Crops Prod 58:104–110

Barabanova AO, Tishchenko IP, Glazunov VP, Soloveva TF, Ermak IM (2010) Characteristics of polysaccharides and protein associated with them from dried and freshly collected red alga *Tichocarpus crinitus*. Chem Nat Compd 46:509–513

Barbot YN, Falk HM, Benz R (2014) Thermo-acidic pretreatment of marine brown algae *Fucus vesiculosus* to increase methane production—a disposal principle for macroalgae waste from beaches. J Appl Phycol. doi:10.1007/s10811-014-0339-x (In press)

Bedoux G, Hardouin K, Burlot AS, Nathalie Buorgougnon (2014) Bioactive components from seaweeds: cosmetics applications and future development. In: Nathalie Bourgougnon (ed)

Advances in Botanical Research focuses on sea plants, including algae, seaweed, and diatoms, Academic Press, ISBN:9780124080621
Borines MG, De Leon RL, McHenry MP (2011) Bioethanol production from farming non-food macroalgae in Pacific island nations: chemical constituents, bioethanol yields, and prospective species in the Philippines. Renew Sustain Energ Rev 15:4432–4435
Bozell JJ (2008) Feedstocks for the future – biorefinery production of chemicals from renewable carbon. Clean 36:641–647
Bruhn A, Dahl J, Nielsen HB, Nikolaisen L, Rasmussen MB, Markagr S, Olesen B, Arias C, Jensen PD (2011) Bioenergy potential of *Ulva lactuca*: biomass yield, methane production and combustion. Bioresour Technol 102:2595–2604
Castro LSEPW, Pinheiro TS, Castro AJG, Dore CMPG, Silva NB, Alves MGCF, Santos MSN, Leite EL (2014) Fucose-containing sulfated polysaccharides from brown macroalgae *Lobophora variegata* with antioxidant, anti-inflammatory, and antitumoral effects. J Appl Phycol 26:1783–1790
Chen P, Min M, Chen Y, Wang L, Li Y, Chen Q, Wang C, Wan Y, Wang X, Cheng Y, Deng S, Hennessy K, Lin X, Liu Y, Wang Y, Martinez B, Ruan R (2009) Review of the biological and engineering aspects of algae to fuels approach. Int J Agric Biol Eng 2:1–30
Chirapart A, Praiboon J, Puangsombat P, Pattanapon C, Nunraksa N (2014) Chemical composition and ethanol production potential of Thai seaweed species. J Appl Phycol 26:979–986
Cho Y, Kim H, Kim SK (2013) Bioethanol production from brown seaweed, *Undaria pinnatifida*, using NaCl acclimated yeast. Bioprocess Biosyst Eng 36:713–719
Choi W, Kang D, Lee H (2013) Enhancement of the saccharification yields of *Ulva pertusa* Kjellmann and rape stems by the high-pressure steam pretreatment process. Biotechnol Bioprocess Eng 18:728–735
Ciancia M, Sato Y, Nonami H, Cerezo AS, Erra-Balsells R, Matulewicz MC (2005) Autohydrolysis of a partially cyclized mu/nu-carrageenan and structural elucidation of the oligosaccharides by chemical analysis, NMR spectroscopy and UV-MALDI mass spectrometry. Arkivoc 12:319–331
Clarens AF, Resurreccion EP, White MA, Colosi LM (2010) Environmental life cycle comparison of algae to other bioenergy feedstocks. Environ Sci Technol 44:1813–1819
Cofrades S, López-López I, Solas MT, Bravo L, Jiménez-Colmenero F (2008) Influence of different types and proportions of added edible seaweeds on characteristics of low-salt gel/emulsion meat systems. Meat Sci 79:767–776
Dawczynski C, Schubert R, Jahreis G (2007) Amino acids, fatty acids, and dietary fibre in edible seaweed products. Food Chem 103:891–899
De Quirós AR, Lage-Yusty MA, López-Hernández J (2010) Determination of phenolic compounds in macroalgae for human consumption. Food Chem 121:634–638
Delattre C, Fenoradosoa TA, Michaud P (2011) Galactans: an overview of their most important sourcing and applications as natural polysaccharides. Braz Arch Biol Technol 54:1075–1092
Denis C, Morançais M, Li M, Deniaud E, Gaudin P, Wielgosz-Collin G, Barnathan G, Jaouen P, Fleurence J (2010) Study of the chemical composition of edible red macroalgae *Grateloupia turuturu* from Brittany (France). Food Chem 119:913–917
Díaz-Vázquez LM, Rojas-Pérez A, Fuentes-Caraballo M, Robles-Ramos IV, Jena U, Das K (2015) Demineralization of Sargassum spp. macroalgae biomass: selective thermochemical liquefaction process for bio-oil production. Frontiers Energy Res. doi:10.3389/fenrg.2015.00006
Eboibi BE, Lewis DM, Ashman PJ, Chinnasamy S (2014) Effect of operating conditions on yield and quality of biocrude during hydrothermal liquefaction of halophytic microalga *Tetraselmis* sp. Bioresour Technol 170:20–29
Elliot DC, Hart TR, Neuenschwander GG, Rotness LJ, Roessijadi G, Zacher AH, Magnuson JK (2014) Hydrothermal processing of macroalgal feedstocks in continuous-flow reactors. Sustain Chem Eng 2:207–215
Elliot DC, Biller P, Roos AB, Schmidt AJ, Jones SB (2015) Hydrothermal liquefaction of biomass: developments from batch to continuous process. Bioresour Technol 178:147–156

Elliott DC, Hart TR, Schmidt AJ, Neuenschwander GG, Rotness LJ, Olarte MV, Zacher AH, Albrecht KO, Hallen RT, Holladay JE (2013) Process development for hydrothermal liquefaction of algae feedstocks in a continuous-flow reactor. Algal Res 2:445–454

Fang Z (2013) Pretreatment techniques for biofuels and biorefineries. Springer, Berlin. ISBN 978-3-642-32735-3

FAO (2012) Global Aquaculture Production 1950–2012. Available from: http://www.fao.org/figis/servlet/TabSelector. Accessed 15 Nov 2014

Fasahati P, Woo HC, Liu JJ (2015) Industrial-scale bioethanol production from brown algae: effects of pretreatment processes on plant economics. Appl Energy 139:175–187

Fitton JH (2011) Therapies from fucoidan; multifunctional marine polymers. Mar Drugs 9:1731–1760

Ge L, Wang P, Mou H (2011) Study on saccharification techniques of seaweed wastes for the transformation of ethanol. Renew Energ 36:84–89

Golberg A, Vitkin E, Linshiz G, Khan SA, Hillson NJ, Yakhini Z, Yarmush ML (2014) Proposed design of distributed macroalgal biorefineries: thermodynamics, bioconversion technology, and sustainability implications for developing economies. Biofuels Bioprod Bioref 8:67–82

Gómez-Ordóñez E, Jiménez-Escrig A, Rupérez P (2010) Dietary fibre and physicochemical properties of several edible seaweeds from the northwestern Spanish coast. Food Res Int 43:2289–2294

González-López N, Moure A, Domínguez H (2012) Hydrothermal fractionation of *Sargassum muticum* biomass. J Appl Phycol 24:1569–1578

Gressler V, Fujii MT, Martins AP, Colepicolo P, Mancini-Filho J, Pinto E (2011) Biochemical composition of two red seaweed species grown on the Brazilian coast. J Sci Food Agric 91:1687–1692

Hayashi L, Bulboa C, Kradolfer P, Soriano G, Robledo D (2014) Cultivation of red seaweeds: a Latin American perspective. J Appl Phycol 26(22):719–727

Haykiri-Acma H, Yaman S, Kucukbayrak S (2013) Production of biobriquettes from carbonized brown seaweed. Fuel Process Technol 106:33–40

Hoang MH, Kim JY, Lee JH, You S, Lee SJ (2015) Antioxidative, hypolipidemic, and anti-inflammatory activities of sulfated polysaccharides from *Monostroma nitidum*. Food Sci Biotechnol 24:199–205

Hoffmann RA, Russell AR, Gidley MJ (1996) Molecular weight distribution of carrageenans. In: Philips GO, Williams PJ, Wedlock DJ (eds) Gums and stabilisers for the food industry. IRL Press at the Oxford University Press, Oxford, pp 137–148

Hong IK, Jeon H, Lee SB (2014) Comparison of red, brown and green seaweeds on enzymatic saccharification process. J Ind Eng Chem 20:2687–2691

Hughes SR, Gibbons WR, Moser BR, Rich JO (2013) Chapter 9: Sustainable multipurpose biorefineries for third-generation biofuels and value-added co-products(. In: Zhen Fang (ed) Biofuels – economy, environment and sustainability. InTech, Rijeka ISBN:978-953-51-0950-1

Jang SS, Shirai Y, Uchida M, Wakisaka M (2012a) Production of mono sugar from acid hydrolysis of seaweed. Afr J Biotechnol 11:1953–1963

Jang JS, Cho Y, Jeong GT, Kim SK (2012b) Optimization of saccharification and ethanol production by simultaneous saccharification and fermentation (SSF) from seaweed, *Saccharina japonica*. Bioprocess Biosyst Eng 35:11–18

Jard G, Dumas C, Delgenes JP, Marfaing H, Sialve B, Steyer JP, Carrere H (2013) Effect of thermochemical pretreatment on the solubilization and anaerobic biodegradability of the red macroalga *Palmaria palmate*. Biochem Eng J 79:253–258

Ji-Hyeon Y, Lee S, Choi WY, Kang DH, Lee HY, Jung KH (2011) Repeated-batch operation of surface-aerated fermentor for bioethanol production from the hydrolysate of seaweed *Sargassum sagamianum*. J Microbiol Biotechnol 21:323–331

Jung KW, Kim DH, Shin HS (2011) Fermentative hydrogen production from Laminaria japonica and optimization of thermal pretreatment conditions. Bioresour Technol 102:2745–2750

Jung KA, Lim SR, Kim Y, Park JM (2013) Potentials of macroalgae as feedstocks for biorefinery. Bioresour Technol 135:182–190

Kalimuthu S, Kim S (2015) Fucoidan, a sulfated polysaccharides from brown algae as therapeutic target for cancer (Chapter 7). In: Se-Kwon Kim (ed) Handbook of anticancer drugs from marine origin, Springer International Publishing, Switzerland, pp 145–164

Kraan S (2013) Mass-cultivation of carbohydrate rich macroalgae, a possible solution for sustainable biofuel production. Mitig Adapt Strateg Glob Change 18:27–46

Kumar S, Gupta R, Kumar G, Sahoo D, Kuhad RC (2013) Bioethanol production from *Gracilaria verrucosa*, a red alga, in a biorefinery approach. Bioresour Technol 135:150–156

Li D, Chen L, Chen S, Zhang X, Chen F, Ye N (2012) Comparative evaluation of the pyrolytic and kinetic characteristics of a macroalga (*Sargassum thunbergii*) and a freshwater plant (*Potamogeton crispus*). Fuel 96:185–191

Malihan LB, Nisola GM, Mittal N, Seo JG, Chung WJ (2014) Blended ionic liquid systems for macroalgae pretreatment. Renew Energ 66:596–604

Marinho-Soriano E, Fonseca PC, Carneiro MAA, Moreira WSC (2006) Seasonal variation in the chemical composition of two tropical seaweeds. Bioresour Technol 97:2402–2406

Marquez GPB, Santiañez JE, Trono GC, Montaño MNE, Araki H, Takeuchi H, Hasegawa T (2014) Seaweed biomass of the Philippines: sustainable feedstock for biogas production. Renew Sustain Energ Rev 38:1056–1068

Marsham S, Scott GW, Tobin ML (2007) Comparison of nutritive chemistry of a range of temperate seaweeds. Food Chem 100:1331–1336

Matanjun P, Mohamed S, Mustapha NM, Muhammad K (2009) Nutrient content of tropical edible seaweeds *Eucheuma cottonii*, *Caulerpa lentillifera* and *Sargassum polycystum*. J Appl Phycol 21:75–80

Meillisa A, Woo H, Chun B (2015) Production of monosaccharides and bio-active compounds derived from marine polysaccharides using subcritical water hydrolysis. Food Chem 171:70–77

Mohamed S, Hashim SN, Rahman HA (2012) Seaweeds: a sustainable functional food for complementary and alternative therapy. Trends Food Sci Technol 23:83–96

Moncada J, Tamayo JA, Cardona CA (2014) Integrating first, second, and third generation biorefineries: incorporating microalgae into the sugarcane biorefinery. Chem Eng Sci 118:126–140

Montingelli ME, Tedesco S, Olabi AG (2015) Biogas production from algal biomass. Renew Sustain Energ Rev 43:961–972

National Renewable Energy Laboratory (NREL) Available at: http://www.nrel.gov/biomass/biorefinery.html. Accessed 15 Sep 2014

Navarro DA, Stortz CA (2005) Microwave-assisted alkaline modification of red seaweed galactans. Carbohydr Polym 62:187–191

Neveux N, Magnusson M, Maschmeyer T, Nys R, Paul NA (2014a) Comparing the potential production and value of high-energy liquid fuels and protein from marine and freshwater macroalgae. GCB Bioenerg 7:673–689

Neveux N, Yuen AKL, Jazrawi C, Magnusson M, Haynes BS, Masters AF, Montoya A, Paul NA, Maschmeyer T, Nys R (2014b) Pre- and post-harvest treatment of macroalgae to improve the quality of feedstock for hydrothermal liquefaction. Algal Res 6:22–31

Nielsen HB, Heiske S (2011) Anaerobic digestion of macroalgae: methane potentials, pre-treatment, inhibition and co-digestion. Water Sci Technol 64:1723–1729

Okuda K, Oka K, Onda A, Kaijiyoshi K, Hiraoka M, Yanagisawa K (2008) Hydrothermal fractional pretreatment of sea algae and its enhanced enzymatic hydrolysis. J Chem Technol Biotechnol 83:863–841

Oliveira JV, Alves MM, Costa JC (2014) Design of experiments to assess pre-treatment and co-digestion strategies that optimize biogas production from macroalgae *Gracilaria vermiculophylla*. Bioresour Technol 162:323–330

Oliveira JV, Alves MM, Costa JC (2015) Optimization of biogas production from *Sargassum* sp. using a design of experiments to assess the co-digestion with glycerol and waste frying oil. Bioresour Technol 175:480–485

Ortiz J, Romero N, Robert P, Araya J, Lopez-Hernandez J, Bozzo C, Navarrete E, Osorio A, Rios A (2006) Dietary fiber, amino acid, fatty acid and tocopherol contents of the edible seaweeds *Ulva lactuca* and *Durvillaea Antarctica*. Food Chem 99:98–104

Overend RP, Chornet E (1987) Fractionation of lignocellulosic by steam-aqueous pretreatments. Philos Trans R Soc Lond 321:523–536

Peña-Rodríguez A, Mawhinney TP, Ricque-Marie D, Cruz-Suárez LE (2011) Chemical composition of cultivated seaweed *Ulva clathrata* (Roth) C. Agardh. Food Chem 129:491–498

Peng Y, Xie E, Zheng K, Fredimoses M, Yang X, Zhou X, Wang Y, Yang B, Lin X, Liu J, Liu Y (2013) Nutritional and chemical composition and antiviral activity of cultivated seaweed *Sargassum naozhouense* Tseng et Lu. Mar Drugs 11:20–32

Pham TH, Um Y, Yoon HH (2013) Pretreatment of macroalgae for volatile fatty acid production. Bioresour Technol 146:754–757

Podkorytova AV, Vafina LH, Kovaleva EA, Mikhailov VI (2007) Production of algal gels from the brown alga, *Laminaria japonica* Aresch., and their biotechnological applications. J Appl Phycol 19:827–830

Poots T, Du J, Paul M, May P, Beitle R, Hestekin J (2012) The production of butanol from Jamaica bay macro algae. Environ Prog Sustain Energ 31:29–36

Prajapati VD, Maheriya PM, Jani GH, Solanki HK (2014) Carrageenan: a natural seaweed polysaccharide and its applications. Carbohydr Polym 105:97–112

Radulovich R, Umanzor S, Cabrera R, Mata R (2015) Tropical seaweeds for human food, their cultivation and its effect on biodiversity enrichment. Aquaculture 436:40–46

Rajauria G, Jaiswal AK, Abu-Ghannam A, Gupta S (2010) Effect of hydrothermal processing on colour, antioxidant and free radical scavenging capacities of edible Irish brown seaweeds. Int J Food Sci Technol 45:2485–2493

Rameshkumar S, Ramakritinan CM, Yokeshbabu M (2013) Proximate composition of some selected seaweeds from Palk bay and Gulf of Mannar, Tamilnadu, India. Asian J Biomed Pharm Sci 3:1–5

Rioux LE, Turgeon SL, Beaulieu M (2007) Characterization of polysaccharides extracted from brown seaweeds. Carbohydr Polym 69:530–537

Rodriguez-Jasso RM, Mussatto SI, Pastrana L, Aguilar CN, Teixeira JA (2011) Microwave-assisted extraction of sulfated polysaccharides (fucoidan) from brown seaweed. Carbohydr Polym 86:1137–1144

Rodriguez-Jasso RM, Mussatto SI, Sepúlveda L, Agrasar AT, Pastrana L, Aguilar CN, Teixeira JA (2013) Fungal fucoidanase production by solid-state fermentation in a rotating drum bioreactor using algal biomass as substrate. Food Bioprod Process 91:587–594

Rodriguez-Jasso RM, Mussatto SI, Pastrana L, Aguilar CN, Teixeira JA (2014) Chemical composition and antioxidant activity of sulphated polysaccharides extracted from *Fucus vesiculosus* using different hydrothermal processes. Chem Pap 68:203–209, 23

Roesijadi G, Jones SB, Snowden-Swan, Zhu Y (2010) Macroalgae as a biomass feedstock: A preliminary analysis. Pacific Northwest Laboratory and United States Department of Energy. Available from:http://www.pnl.gov/main/publications/external/technical_reports/pnnl-19944.pdf. Accessed 20 Nov 2014

Ruiz HA, Cerqueira MA, Silva HD, Rodríguez-Jasso RM, Vicente AA, Teixeira JA (2013a) Biorefinery valorization of autohydrolysis wheat straw hemicellulose to be applied in a polymer-blend film. Carbohydr Polym 92:2154–2162

Ruiz HA, Rodríguez-Jasso RM, Fernandes BD, Vicente AA, Teixeira JA (2013b) Hydrothermal processing, as an alternative for upgrading agriculture residues and marine biomass according to the biorefinery concept: a review. Renew Sustain Energ Rev 21:35–51

Ruiz HA, Parajó JC, Teixeira JA (2015) Biorefinery strategies for macroalgae-based in bioethanol production. In: Energy science and technology. Studium Press LLC, Houston (in press)

Schultz-Jensen N, Thygesen A, Leipold F, Thomsen ST, Roslander C, Lilholt H, Bjerre AB (2013) Pretreatment of the macroalgae *Chaetomorpha linum* for the production of bioethanol – comparison of five pretreatment technologies. Bioresour Technol 140:36–42

Schumacher M, Yanik J, Sinag A, Kruse A (2011) Hydrothermal conversion of seaweeds in a batch autoclave. J Supercrit Fluids 58:131–135

Seaweed Site: information on marine algae. Available from: http://www.seaweed.ie. Accessed 20 Nov 2014

Senthilkumar K, Manivasagan P, Venkatesan J, Kim SK (2013) Brown seaweed fucoidan: biological activity and apoptosis, growth signaling mechanism in cancer. Int J Biol Macromol 60:366–374

Shevchenko NM, Anastyuk SD, Menshova RV, Vishchuk OS, Isakov VI, Zadorozhny PA, Sikorskaya TV, Zvyagintseva TN (2014) Further studies on structure of fucoidan from brown alga *Saccharina gurjanovae*. Carbohydr Polym. doi:10.1016/j.carbpol.2014.12.042 (In press)

Shi J, Pu Y, Yang B, Ragauskas A, Wyman CE (2011) Comparison of microwaves to fluidized sand baths for heating tubular reactors for hydrothermal and dilute acid batch pretreatment of corn stover. Bioresour Technol 102:5952–5961

Siddhanta AK, Prasad K, Meena R, Prasad G, Metha GK, Chhatbar MU, Oza MD, Kumar S, Sanandiya ND (2009) Profiling of cellulose content in Indian seaweed species. Bioresour Technol 100:6669–6673

Singh R, Balagurumurthy B, Bhaskar T (2015a) Hydrothermal liquefaction of macro algae: effect of feedstock composition. Fuel 146:60–74

Singh R, Bhaskar T, Balagurumurthy B (2015b) Effect of solvent on the hydrothermal liquefaction of macro algae *Ulva fasciata*. Process Saf Environ Prot 93:154–160

Sun C, Chen Y, Zhang X, Pan J, Cheng H, Wu M (2014) Draft genome sequence of *Microbulbifer elongatus* strain HZ11, a brown seaweed-degrading bacterium with potential ability to produce bioethanol from alginate. Mar Geonomics 18:83–85

Suutari M, Leskinen E, Fagerstedt K, Juparinen J, Kuupo Blomster J (2015) Macroalgae in biofuel production. Phycol Res 63:1–18

Tedesco S, Benyounis KY, Olabi AG (2013) Mechanical pretreatment effects on macroalgae-derived biogas production in co-digestion with sludge in Ireland. Energy 61:27–33

Tekin K, Karagoz S (2013) Non-catalytic and catalytic hydrothermal liquefaction of biomass. Res Chem Intermed 39:485–498

Tekin K, Karagoz S, Bektaş S (2014) A review of hydrothermal biomass processing. Renew Sustain Energ Rev 40:673–687

Tian C, Li B, Liu Z, Zhang Y, Lu H (2014) Hydrothermal liquefaction for algal biorefinery: a critical review. Renew Sustain Energ Rev 38:933–950

Titlyanov EA, Titlyanova TV (2010) Seaweed cultivation: methods and problems. Russ J Mar Biol 36:227–242

Trivedi N, Gupta V, Reddy CRK, Jha B (2013) Enzymatic hydrolysis and production of bioethanol from common macrophytic green alga *Ulva fasciata* Delile. Bioresour Technol 150:106–112

Usov AI, Zelinsky ND (2013) Chapter 2: Chemical structures of algal polysaccharides. In: Domínguez H (ed) Functional ingredients from algae for foods and nutraceuticals. Woodhead Publishing, Cambridge, pp 23–86

Vanegas CH, Hernon A, Bartlett J (2015) Enzymatic and organic acid pretreatment of seaweed: effect on reducing sugars production and on biogas inhibition. Int J Ambient Energ 36:2–7

Wal H, Sperber BLHM, Houweling-Tan B, Bakker RRC, Brandenburg W, López-Contreras AM (2013) Production of acetone, butanol, and ethanol from biomass of the green seaweed *Ulva lactuca*. Bioresour Technol 128:431–437

Wang HMD, Chen CC, Huynh P, Chang JS (2014) Exploring the potential of using algae in cosmetics. Bioresour Technol. doi:10.1016/j.biortech.2014.12.001 (In press)

Wei N, Quarterman J, Jin Y (2013) Marine macroalgae: an untapped resource for producing fuels and chemicals. Trends Biotechnol 31:70–77

Xu X, Kim JY, Oh YR, Park JM (2014) Production of biodiesel from carbon sources of macroalgae, *Laminaria japonica*. Bioresour Technol 169:455–461

Xu Y, Duan P, Wang F (2015) Hydrothermal processing of macroalgae for producing crude bio-oil. Fuel Process Technol 130:268–274

Yanagisawa M, Kawai S, Murata K (2013) Strategies for the production of high concentrations of bioethanol from seaweeds. Bioengineered 4:224–235

Yazdani P, Zamani A, Karimi K, Taherzadeh MJ (2015) Characterization of *Nizimuddinia zanardini* macroalgae biomass composition and its potential for biofuel production. Bioresour Technol 176:196–202

Yeon JH, Lee SE, Choi WY, Kang DH, Lee HY, Jung KH (2011) Repeated-batch operation of surface-aerated fermentor for bioethanol production from the hydrolysate of seaweed *Sargassum sagamianum*. J Microbiol Biotechnol 21:323–331

Zafar M, Chowdhury SMRA (2009) Water quality and biochemical components of *Hydroclathrus clathratus* in the Tidal shore area of St. Martin's Island, Bangladesh. Int J Phycol Phycochem 5:7–10

Zhang B, Keitz MV, Valentas K (2008) Thermal effects on hydrothermal biomass liquefaction. Appl Biochem Biotechnol 147:143–150

Zhang B, Shahbazi A, Wang L, Diallo O, Whitmore A (2011) Hot-water pretreatment of cattails for extraction of cellulose. J Ind Microbiol Biotechnol 38:819–824

Zheng Y, Zhao J, Xu F, Li Y (2014) Pretreatment of lignocellulosic biomass for enhanced biogas production. Prog Energy Combust Sci 42:35–53

Zhou D, Zhang L, Zhang S, Fu H, Chen J (2010) Hydrothermal liquefaction of macroalgae Enteromorpha prolifera to bio-oil. Energ Fuel 24:4054–4061

Conversion of Microalgae Bio-oil into Bio-diesel

Aimaro Sanna and Nur Adilah Abd Rahman

Abstract Microalgae are attracting worldwide attention as an alternative and renewable source for energy production. Microalgae, thanks to their high content in oil, can be used to produce many different kinds of biofuels such as biodiesel and jet fuels through various conversion technologies. This chapter aims to present an overview of the current thermochemical conversion technologies using microalgae with a specific emphasis on the conversion of algae into bio-oils by pyrolysis and their upgrading to biodiesel and jet fuel by catalytic pyrolysis and hydrotreating.

Keywords Microalgae • Microalgae bio-oil • Pyrolysis • Catalytic pyrolysis • Hydrothermal treating • Bio oil hydrotreating • Heterogeneous catalysis • Acid catalysts • Metal catalysts • Bio-oil characterisation • Renewable energy • Bio-refinery

1 Introduction

There is a strong worldwide interest in converting renewable sources into bio-fuels as a substitute to traditional fuels due to declining petroleum resources. Moreover, the increasing demand on fuel and energy, and consequences arising from petroleum fuel usage such as global warming, encourage governments to invest in renewable fuel.

Although lignocellulosic biomass is currently the best choice for bio-fuel production, microalgae's have very rapidly become a key player since they can be fed with flue gas rich in CO_2 and they are able to produce more bio-fuels per hectare compared to terrestrial plants (Luque 2010). Therefore, the utilization of microalgae as a bio-fuel source has attracted great interest globally. Microalgae are unicellular micro-organisms, which convert sunlight, water and carbon dioxide into algal biomass. Microalgae are a promising feedstock as their growing rate is fast and they

A. Sanna (✉) • N.A. Abd Rahman
Centre for Innovation in Carbon Capture and Storage (CICCS),
School of Engineering and Physical Sciences, Institute of Mechanical,
Process and Energy Engineering (IMPEE), Heriot-Watt University,
EH14 4AS, Edinburgh, UK
e-mail: a.sanna@hw.ac.uk

A. Prokop et al. (eds.), *Algal Biorefineries*, DOI 10.1007/978-3-319-20200-6_16

have the ability to grow either in fresh water, brackish water or even wastewater (Dillon 2009). They multiply quickly, in a period as short as 3.5 h during growth (Yusuf 2007) and can be harvested all year around.

However, despite the strong interest and investments in the area of bio-fuels from algae from oil companies, there are still technical and economic issues that may limit their widespread utilisation in the near future. For example, the optimistic figures given to estimate the potential of algae still need to be demonstrated, for both the average annual productivity of the algae and the available cellular level of oil content convertible to bio-fuels (Luque 2010). Moreover, even if the productivity of algae could be up to 50 times higher than that of fast growing terrestrial plants, the current estimated costs for algal cultivation are significantly higher (5–10 $/kg vs. 0.025–0.1 $/kg) (Garcia Alba et al. (2013)).

Compared to current leading oil crops such as palm, canola and jatropha, microalgae are more productive and have higher rates of biomass and oil production with an oil content ranging from 20 to 50 % by dry weight (Yusuf 2007). The main advantage is that microalgae are not in competition with food crops and can be grown in marginal lands and sea water (Demirbas 2011; Singh and Gu 2010). Other than that, microalgae use CO_2 as source of growth and help in reducing carbon dioxide in the atmosphere (Pragya et al. 2013). Moreover, algae are rich in nutrients such as proteins, vitamins, calcium, magnesium and trace minerals, which can be extracted before their conversion in bio-fuels. These multi-purpose functions are attracting investments worldwide, for the research and development of processes to convert microalgae into final products, in a biorefinery system.

Microalgae have different kinds of lipids, hydrocarbons and other complex oils depending on the species. Microalgae with high lipid contents are suitable for producing bio-oils (Yusuf 2007). Microalgae can be cultivated by different methods which are phototrophic (growth in light without additional nutrients), heterotrophic (growth in dark with additional nutrients) and mixotrophic (growth in either light or dark with additional nutrients) (Sawangkeaw and Ngamprasertsith 2013). Manipulating growing conditions such as light intensity, nutrient sources (organic carbon and nitrogen), temperature and pH influence the growth rate and lipid content of microalgae.

Researchers have focused on production of bio-diesel from microalgae by utilizing the algae lipid content. Lipids are extracted from microalgae cells and undergo conventional esterification and trans-esterification processes; while the remnant biomass (mainly proteins and cellulose) is considered as waste (Luque 2010; Miao and Wu 2004; Serrano-ruiz and Dumesic 2012). Table 1 summarizes the lipid content, biomass and oil productivity of selected microalgae. Table 1 shows that cultivation in autotrophic conditions produced microalgae with higher lipid content. Maximum oil productivity came from *Schizochytrium limacinum* species that produced about 525.1 kg/m^3/year, when cultivated in heterotrophic conditions.

It has to be noticed that many microalgae oils are generally rich in long-chain polyunsaturated acids including eicosapentaenoic (20:5 n-3, EPA) and

Table 1 Lipid content, biomass and oil productivity for different species of microalgae

Microalgae strain	Culture conditions	Lipid content (% w/w)	Productivity (kg/m^3/year)	
			Biomass	Lipid
Chlorella sp.	AT	22.4–33.9	158.4	53.7
Scenedesmus obliquus	AT, MT	12.6–58.3	153	54.2
Pseudochlorococcum sp.	AT	24.6–52.1	150	57.5
Chaetoceros muelleri	AT	11.7–25.3	585	108.2
Nannochloropsis sp.	AT	22.8–28	870	199.2
Chlorella protothecoides	HT	48.1–63.8	412.5	230.8
Schizochytrium mangrovei	HT	68	732	497.8
Schizochytrium limacinum	HT	50.3[a]	1044	525.1
Botryococcus braunii	AT	30		

Adapted from (Sawangkeaw and Ngamprasertsith 2013; De la Hoz Siegler et al. 2011; Araujo et al. 2011; Mandal and Mallick 2009)

[a]As total fatty acid content. AT, MT and HT are autotrophic, mixotrophic and heterotrophic cultivations

docosahexaenoic acids (22:6 u-3, C) which are generally undesirable in conventional biodiesel due to the negative impact of the polyunsaturated on oxidation stability (Luque 2010). In this chapter, the thermo-conversion of microalgae to bio-oils and the further upgrading of the latter to bio-crudes are reviewed. Also, the current technical limitations of algae conversion technologies are discussed.

2 Production of Bio-oil from Microalgae Biomass

There are a variety of processes available to convert microalgae biomass into bio-oil, such as thermochemical and biochemical processes. Many factors influence the conversion process selection, such as type and quantity of biomass feedstock, economic consideration, the desired form of energy and end-products (McKendry 2002). Thermochemical processes utilise the whole algae biomass for bio-fuel production, where the biomass is thermally decomposed into bio-oil under specific conditions. Thermochemical processes include hydrothermal liquefaction (HTL), pyrolysis and gasification while biochemical processes include alcoholic fermentation, anaerobic digestion and photo-biological hydrogen production. Figure 1 shows thermochemical process in converting microalgae into bio-oil.

The production of bio-oil from microalgae requires large quantities of biomass. The process takes place either in dry or wet conditions. However, microalgae contain high water content (80–90 %) and this is a major drawback for the direct implementation of thermo-chemical conversion of algae by dry processing (gasification or pyrolysis). Therefore, like the terrestrial biomasses, water content needs to be reduced before processing the raw algae, through energy intensive pre-drying.

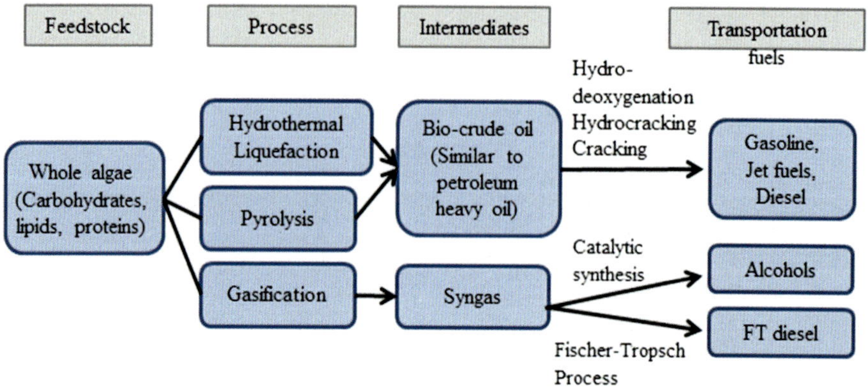

Fig. 1 Thermochemical technologies for energy production from microalgae (Adapted from (Tsukahara & Sawayama 2005; Shi et al. 2012))

2.1 *Hydrothermal Liquefaction*

Hydrothermal liquefaction is one of the alternative processes to convert wet algae biomass into liquid bio-crudes. Since, during HTL, water serves as the reaction medium, there is no need to dewater biomass, which represents a major energy input for thermochemical algal bio-fuel production (Vardon et al. 2011). Liquefaction occurs by processing algal biomass in hot compressed (sub-critical) water either with or without a catalyst. The temperature ranges between 200 and 350 °C and has a relatively high operating pressure (5–20 Mpa), where biomass breaks into smaller, unstable molecules and then, re-polymerize and re-combine forming long chain molecular products (Duan and Savage 2011a; Pragya et al. 2013; Sharma et al. 2011). Bio crude/bio-oil is the main product of HTL, while a solid residue and water containing soluble organic compounds represent the undesired by-products (Xiu and Shahbazi 2012). Self-separation of the bio-crude oil from water is facilitated as the reaction solution returns to standard conditions (Vardon et al. 2011).

Conversion of wet microalgae biomass through HTL is a promising algae conversion technology, where there is no net energy required to dry the wet microalgae biomass. Still, the complex system of HTL makes it very expensive (Pragya et al. 2013). The HTL process needs specialized materials that are able to resist high temperature and high pressure and robust catalysts that have to withstand the hydrothermal condition (Peterson et al. 2008)). Moreover, since a large amount of water exists in HTL, bio-oil is always emulsified within the water and large amount of organic solvent is needed for its extraction (Du 2013).

A number of studies have investigated the hydrothermal liquefaction of algae. Dote et al. performed HTL of *Botryococcus braunii* using Na_2CO_3 as a catalyst. A bio-crude in a yield of 57–64 wt% was obtained at 300 °C. The oil was equivalent in quality to petroleum oil (Dote et al. 1994). Biller and Ross treated *Chlorella vulgaris* and *Spirulina* at 350 °C, with a pressure of 200 bar in water and 1 M Na_2CO_3

and 1 M formic acid. The bio-crude from the process was 40 wt% for *Chlorella* sp. and *Nannochloropsis* sp., while 35 % for *Spirulina*. Aromatic hydrocarbons such as toluene, styrene, phenols, indoles along with fatty acids and alcohols were the main components. Their study also indicated that presence of Na_2CO_3 catalyst is favourable for HTL of algae with high carbohydrate fraction, while high protein or lipid containing algae are better liquefied without the use of a catalyst. The nitrogen fraction resulting in the bio-crude was considerably larger than the nitrogen containing models compounds (proteins and amino acids), whereas, the N content in bio-crude decreased when catalysts were used. The study also showed that N in bio-crude can be reduced by breaking down proteins to amino acids before HTL (Biller and Ross 2011). Chen et al. studied the conversion of mixed algae-culture from wastewater into bio-oil. The highest bio-oil yield (55 %) was obtained at 300 °C with 1 h retention time. While the highest heating value (HHV) was 33.3 MJ/kg produced at 320 °C with 1 h retention time. The bio-oil contained hydrocarbons and fatty acids, while the aqueous product contained organic acids and cyclic amines (Chen et al. 2014). According to Duan & Savage, bio-oils from wet biomass consisted of lower oxygen content and higher HHVs compared to bio-oil from pyrolysis of terrestrial biomass. However, heteroatoms such as O, N and S were still present in the algal bio-oil (Duan and Savage 2011a, b). Other works on algae HTL reported a bio-oil yield in the range of 10–50 % with a heating value of 35–40 MJ/kg, slightly lower than that of petroleum crude oil of 43 MJ/kg (Du 2013).

Overall, HTL bio-crudes contain a wide range of chemical compounds which include aliphatic compounds, aromatics and phenolic derivatives, carboxylic acids, esters, and nitrogenous ring structures. The starting algae composition and the HTL process conditions influence the functionalities present in the bio-crude and their abundance. The high heteroatom content (oxygenates and N-compounds) is the main factor distinguishing bio-crudes from petroleum oils leading to low thermal stability, high acidity, polymerization, high viscosity, and high-boiling distribution (Vardon et al. 2011).

2.2 *Gasification*

Gasification involves the partial oxidation of biomass into syngas at temperatures higher than 800 °C. Syngas is a mixture of hydrogen, carbon monoxide, carbon dioxide and some light hydrocarbons (Demirbas 2011). The produced H_2 is desirable as a clean energy carrier as well as in the chemicals industry. Syngas can be used in a gas engine or turbine as well as heat products or can be converted into bio-fuel products through the Fischer-Tropsch process, which produces long chain hydrocarbons (Suali and Sarbatly 2012). Algae gasification has not been widely investigated, with most of the work carried out in aqueous phase as hydrothermal gasification under 50–400 bar at 300–700 °C (Du 2013).

Chakinala et al. studied the supercritical water gasification of microalgae (*Chlorella vulgaris*) and glycerol. Algae gasification was more efficient at higher temperatures (600–700 °C) with low algae concentrations and longer residence time (15 min). With the addition of catalysts, gasification efficiency increased to a maximum conversion of 84 % at 600 °C in presence of Ni catalyst, with syngas mainly containing H_2 (Chakinala et al. 2010). A complete conversion of the reactants was instead obtained at 700 °C using excess Ru/TiO_2 catalyst.

2.3 *Pyrolysis*

Pyrolysis is thermochemical process that takes place in inert atmosphere, at mild temperatures. Pyrolysis requires dry feedstock to convert biomass into bio-oil. Pyrolysis decomposes biomass into compounds in vapour and gas phases, leaving a carbon-rich solid known as char (Bridgwater 2012). The condensable gas/vapours condensed into liquid which is known as bio-oil. Yields and properties of products are based on the conditions of pyrolysis system such as heating rate, temperature and gas flow rate (Bridgwater 2012; Dickerson and Soria 2013; French and Czernik 2010; Jahirul et al. 2012).

Many studies have been done on pyrolysis of microalgae. Demirbas et al., pyrolysed microalgae *Cladophora fracta* and *Chlorella protothecoides*, producing bio-oils with yields of 48.2 wt% and 55.3 wt%, respectively, at 500 °C (Demirbas 2011). Chaiwong et al. studied the bio-oil production from slow pyrolysis of *Spirulina* sp. The suitable temperature to obtain a high yield of bio-oil with composition similar to that of kerosene and diesel oil was between 450 and 600 °C (Chaiwong et al. 2013).

Miao et al. studied the fast pyrolysis of *Chlorella protothecoides* and *Microcystic aeruginosa* species. The bio-oil yield was 18–24 wt%. The experiment was done at 500 °C with a heating rate of 600 °C/s and N_2 as sweep gas. The bio-oil produced had a HHV of 29 MJ/kg and contained straight chain alkanes which were similar to diesel fuel (Miao et al. 2004). Peng et al. investigated the effect of temperature and holding time on pyrolysis of *Chlorella protothecoides*. It was found out that the maximum oil yield (52 wt%) can be obtained by pyrolysing the algae at 500 °C for 5 min (Peng et al. 2000).

Algae pyrolysis produces oils with yield comparable to that obtained by HTL but algae need to be in dry condition. Besides, pyrolysis technology is considered to be the cheapest amongst the thermo-chemical processes. However, as for bio-oils from lignocellulosic materials, the quality of algae bio-oils is poor due to the high oxygenates presence, high pH, low stability and variety of functionalities present. Therefore, upgrading of bio-oils is required to obtain final transportation fuels and chemicals comparable to those derived from crude oil. In particular, the bio-oil acidity has to be decreased and so that its energy density increased. The addition of a catalyst can improve the quality of bio-oil by favouring several O_2 removal reactions (decarbonylation and decarboxylation) during pyrolysis. Therefore, catalytic pyrolysis and hydro-deoxygenation have been proposed to upgrade pyrolysis bio-oils.

3 Microalgae Bio-oil

Microalgae composition is different from that of lignocellulose biomass, since algae are rich in proteins and minerals and poorer in lignin content. In consequence, the pyrolysis behaviour and bio-oil composition from algae is different from that of lignocellulose biomass (Demirbas 2011). Lipids are the main components to determine the yield of bio-oil from microalgae.

As a renewable liquid fuel, bio-oil can replace fuel oil or diesel in many power plant including boilers, furnaces, engines and turbines. Besides, bio-oil could serve as a raw material for production of adhesives, resins, flavours etc., after further processing and separation. Below are some industrial uses for bio-oil (Xiu and Shahbazi 2012):

1. Fuel combustion in boiler/furnace/ system for heat generation
2. Combustion in diesel engines/turbine for power generation
3. Transportation fuel after upgrading process
4. Production of anhydrosugars such as levoglucosan which are widely used in pharmaceuticals, surfactants and biopolymers industry
5. Production of chemicals and resins

Algal bio-oil has better qualities in many aspects than those produced from lignocellulosic biomass. For example, algal bio-oil has a higher heating value, lower oxygen content and a greater than 7 pH value. However, upgrading towards the removal of ash, nitrogen, phosphorus and oxygen in the bio-oil is still necessary before it can be used as drop-in fuels (Du 2013; Williams and Laurens 2010)). According to Bae et al., pre-treatment of microalgae by acid washing is able to remove most of ash content in algae (Bae et al. 2011).

Typically, bio-oil from the pyrolysis process consists of two phases: an aqueous phase which contains a variety of oxygenates compounds, and a non-aqueous phase which consists of water-insoluble organics (Demirbaş 2006). Table 2 shows the comparison between microalgae and other biomass elemental composition, energy content and proximate analysis. From the table, it can be seen that volatile matter and fixed carbon content of microalgae species (*Spirulina sp.*, *Chlorella vulgaris* and *Chlorella* sp.) are similar with those of terrestrial biomasses. However, microalgae species contain high ash content which is around 5 to 15 wt% that will inhibit the bio-oil production acting as in-situ catalyst. Besides, nitrogen content is much higher than terrestrial biomass species. This is because microalgae have high protein content.

Bio-oil from the pyrolysis process cannot be used directly as fuel due to high oxygen content, high viscosity, corrosiveness and thermal instability (Bridgwater 2012; French and Czernik 2010). Microalgae bio-oil has a lower oxygen contents compared to the terrestrial biomass as shown in Table 3, which lead to considerably high HHV.

Figure 2 shows the Van Krevelen, diagram comparing different biomasses to fossil fuels in terms of O:C and H:C ratios. The higher the respective H/C ratio and

Table 2 Proximate and ultimate analyses of different microalgae and terrestrial biomasses

Sample	*Nannochloropsis* sp.	*Spirulina* sp.	*C. vulgaris*	*Chlorella* sp.	Pine sawdust	Red oak
Proximate analysis (wt%)[a]						
Moisture		8.45	6.18	6.8	7.8	4.75
Volatile matter		65.48	66.56	72.19	75.38	83
Fixed carbon[b]		12.08	11.62	15.08	15.72	11.52
Ash		13.99	15.64	5.93	1.1	0.73
HHV (MJ/kg)	18.5	22.34	16.8	18.59	20.17	19.51
Ultimate analysis (wt%)[c]						
C	43	39.26	42.54	47.54	55.02	45.19
H	5.97	6.11	6.77	7.1	6.08	6.36
N	6.32	6.65	6.64	6.73	0.009	0.06
O	25.8	47.41	27.95	38.63	38.81	47.66
References	Duan and Savage (2011)	Chaiwong et al. (2013)	Wang and Brown (2013)	Phukan et al. (2011)	Zhang et al. (2013)	Wang and Brown (2013)

[a]As received basis
[b]By difference
[c]Dry and ash free basis

Table 3 Proximate and ultimate analyses of different microalgae bio-oils

	Spirulina sp.	*C. protothecoides (AT)*	*C. protothecoides (ET)*
HHV (MJ/kg)	33.2	30	41
C	68.9	62.07	76.22
H	8.9	8.76	11.61
N	6.5	9.74	0.93
O	14.9	19.43	11.24
References	[16]	[42]	[42]

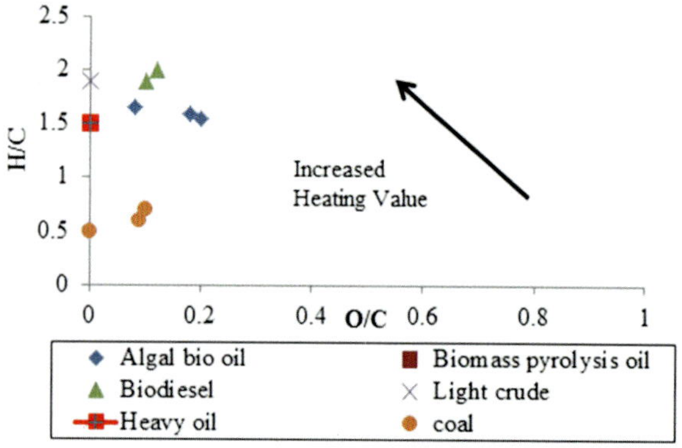

Fig. 2 Van Krevelen diagram for fuel and bio oils (Modified from (Thangalazhy-Gopakumar et al. 2012))

the lower the respective O/C ration the higher the energy content of the material. Algae bio-oil O:C and H:C ratios fall in the same region as biodiesel. Microalgae bio-oil has lower oxygen content than biomass bio-oil and closer in nature to heavy oil.

Nitrogenous compounds in algal bio-oil originate from the pyrolysis of protein and chlorophyll, for which nitrogen is an essential element. They are classified into amides, nitrile, and N-heterocycles. Generally, the relative percentage of the nitrogenous compounds that migrate into the bio-oil fraction is enhanced at higher temperatures, according to N distribution from algae pyrolysis carried out between 300 and 700 °C (Pragya et al. 2013). As typical pollutants, PAHs not only exist in fossil fuels but also can be produced from fuel combustion and pyrolysis. PAHs are not typically detected in microalgae bio-oil at 300–500 °C, but their percentage increases sharply when the pyrolysis temperature exceeds 600 °C (Pragya et al. 2013).Therefore, temperature is an important parameter to determine the quality of the bio-oil. Temperature also affects bio-oil yield together with other process conditions (e.g. retention time) and the composition of the pyrolysed microalgae, with 500 °C typically considered the best temperature to maximise bio-oil yield. An approach for increasing the yield of bio-oil production from fast pyrolysis of microalgae by manipulating the metabolic pathway in microalgae through heterotrophic growth was proposed by Miao and Wu. They used *Chlorella protothecoides*, which are microalgae that can be photo-autotrophically or heterotrophically grown under different culture conditions. However, heterotrophic growth leads to high production of biomass and accumulation of high lipid content in cells, whereas proteins are the main component of autotrophic *Chlorella protothecoides* algae (52.64 %). The yield of bio-oil (57.9 %) produced from heterotrophic *Chlorella protothecoides* cells was 3.4 times higher than from autotrophic cells by fast pyrolysis. Also, the bio-oil from heterotropic *Chlorella* sp. had lower oxygen content and consequently a higher heating value (41 MJ/kg). These properties were comparable to fossil oil (Miao and Wu 2004).

4 Catalytic Pyrolysis of Microalgae Biomass

Catalytic cracking has been indicated as a promising pathway to remove oxygenates in the bio-oil. In the presence of a catalyst, oxygenates can be removed through simultaneous dehydration, decarboxylation, and decarbonylation reactions (Babich et al. 2011; French and Czernik 2010; Zhao et al. 2013)). These catalysed reactions produce lighter bio-oil with improved properties and higher calorific content. The removal of oxygenated species in the oil transform the bio-fuel into a liquid fuel which is compatible with petroleum fuel (Brennan and Owende 2010). Consequently, researchers are focusing on upgrading bio-oil from microalgae through catalytic pyrolysis.

Catalytic pyrolysis, generally known as in-situ upgrading, is a system where both catalyst and biomass are incorporate and mixed, undergo pyrolysis reactions (e.g. using fixed, fluidised or circulating beds) at the same time or setting up an upgrading fixed bed at the outlet of the pyrolysis reactor to treat the pyrolysis vapours/gases (Du et al. 2013). The latter approach could eventually eliminate the costly condensation and re-evaporation procedures used in traditional upgrading of bio-oil (Du 2013).

Since catalytic upgrading is an attractive process to enhance the quality of algae bio-oil, it has been widely investigated using different catalysts. Most researchers are using *Chlorella vulgaris* as "algae model" because this specie is high in lipid content.

Table 4 shows the chemical composition of some algae bio-oils obtained by catalytic and non-catalytic pyrolysis; while Table 5 summarises the current research on catalytic pyrolysis of microalgae. A recent study by Gopakumar et al. compared catalytic and non-catalytic pyrolysis at 550 °C with temperature ramping rate of 2000 °C/min. HZSM-5 catalyst was used on pyrolysis of *Chlorella vulgaris*. Bio-oil yield from pyrolysis of microalgae was 52.7 wt% (with 60.7 wt% carbon yield). Aromatic hydrocarbons in bio-oil increased (from 0.9 to 25.8 wt%) as the catalyst loading was increased from zero to nine times the algae weight (Thangalazhy-Gopakumar et al. 2012).

Wang and Brown reported a promising microalgae biorefinery pathway, which uses catalytic pyrolysis with HZSM-5 catalyst to convert *Chlorella vulgaris* into aromatic hydrocarbons. The maximum carbon yield of aromatic hydrocarbons was 24 %, mostly as BTX molecules, with total aromatic selectivity of 75 %. Their results demonstrated that catalytic pyrolysis of *Chlorella vulgaris* produces better aromatic yields and distributions than catalytic pyrolysis of wood. Benzene, toluene, and xylene (BTX) were the most abundant products, with a combined carbon yield of 15 %. Naphthalene and alkyl-naphthalenes were the second major group of aromatic hydrocarbons, with carbon yields of 1.38 % and 2.23 %, respectively. Moreover, nitrogen was distributed among carbonaceous residue at low temperatures, but selectivity shifted to ammonia as temperature increased (Wang and Brown 2013). Zeng et al. studied the catalytic pyrolysis of algae from water bloom over nickel phosphide catalyst. Oxygen content was reduced from 41.7 wt% in the algae to 8.0 wt% in the bio-oil. Long chain alkanes (dodecane) became the primary components in the bio-oil followed by phenols and indoles. The obtained bio-oil had a high HHV of 37.2 MJ/kg, which matched the HHV of crude oil, 36.48 MJ/kg (Zeng et al. 2013).

Babich et al. also studied the catalytic pyrolysis of chlorella algae using Na_2CO_3 catalyst. In presence of Na_2CO_3 catalyst, the liquid yield decreased and gas yield increased. Also, the catalyst reduced the decomposition temperature to a lower value. The yield of the bio-oil produced was 28 wt% with a HHV of 33 MJ/kg. The bio-oil from catalytic pyrolysis also had a better quality because it contained lower acid and higher aromatics compounds compared to the non-catalytic pyrolysis (Babich et al. 2011).

Table 4 Chemical composition of microalgae bio-oils

Composition (%)	*Chlorella protothecoides* (AC)	*Chlorella protothecoides* (HC)	*Mycrocystis sp.* (400 °C)	*Mycrocystis sp.* (500 °C)	*Chlorella vulgaris* (no cat.)	*Chlorella vulgaris* (algae:ZSM-5 1:1)	*Chlorella vulgaris* (algae:ZSM-5 1:4)
Hydrocarbons	0.86$	1.02$	6.03	10.65	18.7*	16.3*	2.8*
Aromatics	0.75	2.21	3.95	10.07	19.8	23	44.9
Polars	33.82	89.83					
Asphaltenes	64.57	6.94					
Acids			26.05	19.05			
N compounds			14.34	20.88	16.7**	20.4**	5.3*
Sugars			26.43	15.28			
Reference	Miao and Wu (2004)	Miao and Wu (2004)	Pragya et al. (2013)	Pragya et al. (2013)	Thangalazhy-Gopakumar et al. (2012)	Thangalazhy-Gopakumar et al. (2012)	Thangalazhy-Gopakumar et al. (2012)

AC autotrophic, *EC* eterotrophic
* (alkanes and alkenes), ** (alkanes with N and aromatics with N), $saturated

Table 5 Summary on catalytic pyrolysis of microalgae for bio-oil production

Catalyst	Temp (°C)	Algae species	Catalyst effects	Ref
HZSM-5	800	*Chlorella vulgaris*	Bio-oil contains monocyclic aromatics components. In the presence of HZSM-5 catalyst, oxygen and nitrogen content were removed. The maximum carbon yield was 24 % compared to red oak which was 16.7 %	Wang and Brown (2013)
H-Y, H-Beta and H-SZM5	550	Algae from water bloom lake	H-SZM5 most effective with yield of 18.13 %, and maximum yield obtained with Si/Al ratio of 80. Aromatic increased with incorporation of gallium to HZSM-5	Du et al. (2013b)
HZSM-5	650	*Chlorella vulgaris*	Fixed bed pyrolysis give a yield of 52.7 wt% and as catalyst loading increased (0–9 times of biomass), the aromatic hydrocarbons increased	Thangalazhy-Gopakumar et al. (2012)
Na_2CO_3	300–450	*Chlorella*	Na_2CO_3 lowered the degradation temperature. Catalyst increased gas yield, decreased liquid yield. Resulting bio-oil had higher heating value, more aromatics and less acidic	I. V. Babich et al. (2011a)
Modified ZSM-5	650	*Chlorella pyrenoidosa*	The maximum carbon yield of light olefins (ethylene, propylene and butene) 31.9 % was obtained at temperature of 650 °C, with flow rate of 30 ml/h of steam atmosphere	Dong et al. (2013)
HZSM-5	400	*Nannochloropsis sp.*	The maximum yield obtained at 400 °C, with yield about 50 wt%. Bio-oil had lower oxygen content (19.5 wt%) compared to the non-catalytic pyrolysis, 30.1 wt%. Besides, HHV of catalytic bio yield also higher (32.7 MJ/kg) compared to non-catalytic bio-oil, with HHV of 24.6 MJ/kg	Pan et al. (2010)

Despite the fact that catalytic upgrading is a very promising route for the conversion of algae into bio-fuels and chemicals, catalyst deactivation represents a challenge. Catalyst deactivation occurs due to coke formation and strong adsorption of oxygenates components on surface of catalyst support (He and Wang 2012).

A way to enhance the quality of bio-oil is to hydrothermally pre-treat the wet algae at mild temperature (150–250 °C). Pre-treated bio-oils benefit from increased carbon content and HHV in the meantime, reduced nitrogen and ash contents compared with untreated algae. More than 70 % of the initial lipids can be retained in the pre-treated algae. Also, the pyrolytic bio-oil from pre-treated algae contains less N-containing compounds and is rich in fatty acids which can be readily converted to hydrocarbons in the presence of catalysts (Du 2013).

5 Algae Bio-oils Hydrotreating

Catalytic hydro-deoxygenation (HDO) is an important technique to decrease oxygen in fuels derived from biomass (Elliott 2007). HDO involves the reaction of oxygen with hydrogen to form water and saturated C–C bonds (Huber et al. 2006). Most of the HDO studies have been focused on lignocelluloses while HDO of algae, which is particularly challenging, due to its high nitrogen content that deactivate conventional catalysts, has received little attention (Sanna 2013).

Table 6 shows the elemental analysis and energy content of algae bio-oils after hydrotreatment under different conditions. Hydrocarbon oils extracted from a natural "bloom" of *Botryococcus bruunii* was hydrocracked to produce a bio-crude distillate comprising 67 % gasoline fraction, 15 % aviation turbine fuel fraction, 15 % diesel fuel fraction, and 3 % residual oil (Hillen et al. 1982). The oil and hydrogen gas at 3000 psi were passed through the reactor which was held at 400 °C. The liquid products were cooled and collected in knock-out pots with surplus hydrogen being vented. A total of 160 g of oil sample was fed to the hydrocracker at the rate of 230 mL/h and 127 g of product was obtained (Hillen et al. 1982). Duan and Savage investigated the hydrothermal conversion *Nannochloropsis sp.* in absence and presence of H_2 at 350 °C. The crude bio-oils produced from catalytic hydrothermal liquefaction contained near totality of the heating value of the algal feedstock, which was insensitive to the identity of the catalysts tested, and for liquefaction in H_2 also insensitive to whether a catalyst was present. The resulting bio-crude heating values were higher than those of petroleum heavy crudes. Water molecules have been found to be active participants in the liquefaction chemistry, with transfer of H atoms from the water of the bio-crudes. Also, this study indicated that high pressure H_2 gas did not improve the H/C ratios by much compared to the experiments carried out in absence of H_2. Presence of supported noble metal catalysts (Pt, Pd, Ru), even in the absence of added H_2, increased the H/C molar ratios even further, while the presence of Pt, Ni, and CoMo catalysts lead to bio-oils with lower O/C ratios than the oil produced in their absence, indicating catalytic hydrodeoxygenation during the hydrothermal liquefaction of the microalga (Duan and Savage 2011c).

As is well known, microalgae usually contain a high proportion of nitrogen due to its presence in chlorophyll and in protein. The hydrothermal liquefaction, in presence of Ru/C and Ni/SiO_2-Al_2O_3, produced oils with the lowest N content, both in the presence and in the absence of added H_2. Instead, Ru/C and $CoMo/Al_2O_3$ catalysts

Table 6 Major hydrocarbon components and elemental analysis of selected algal hydro-treated oils

Algae	Elements, wt% C	H	O	N	S	HHV (MJ/kg)	Yield, %	Reference
B. braunii	86.38	11.96	1.1	0.17	<0.1	/		Hillen et al. (1982)
***Nannochloropsis sp.*, No catalyst**	75.5	10.5	9.23	4.08	0.69	39	48	Duan and Savage (2011c)
***Nannochloropsis sp.*, PdC**	74.9	10.6	9.04	4.2	0.65	38.9	45	Duan and Savage (2011c)
***Nannochloropsis sp.*, Ni/$SiO_2Al_2O_3$**	76.2	10.7	9.01	3.64	<0.1	39.4	43	Duan and Savage (2011c)
***Nannochloropsis sp.*, zeolite**	74.2	10.5	8.92	4.05	0.88	38.5	43	Duan and Savage (2011c)
***Nannochloropsis sp.*, HZSM-5 (dry environment)**	84–87	10.16–11.0	0.4–2.7	2.3–2.4	<0.1	43.7	45	Li and Savage (2013)
***Nannochloropsis sp.*, Lipid Extracted Algae (LEA) fromSolix**	79.2	10	5.7	4.7	0.5		53.2	Elliott et al. (2013)
***Nannochloropsis sp.*, NB238 product**	78.6	10.4	5.3	4.2	0.5		38	Elliott et al. (2013)
***Nannochloropsis sp.*, Cellana, low lipid**	77	10.4	8	4.2	0.3		60.8	Elliott et al. (2013)
***Nannochloropsis sp.*, Cellana, high lipid**	77.6	10.6	7.2	4	0.3		63.6	Elliott et al. (2013)

had some sulphur removal activity (Duan and Savage 2011c). Crude bio-oil produced from hydrothermal liquefaction of *Nannochloropsis* sp. was hydrotreated over HZSM-5 catalyst (Li and Savage 2013). For a typical experiment, 0.4 g of crude bio-oil was charged into the reactor together with the desired amount of catalyst and then 43.5 bar H_2 (~6.3 mmol) was then charged to the reactor, where 4.0 mmol H_2 was calculated to be the theoretical amount of H_2 required to completely remove O, N, and S atoms in the bio-crude in the form of H_2O, NH_3, and H_2S. The N and O atom content were reduced to one-third of their starting values by treatment at 400 °C. A higher removal of O atom (2.81 vs. > 4 wt%) was achieved in presence of HZSM-5 compared to hydro-treated bio-oils obtained using Pt/C and Pd/C in aqueous environment. The authors did not clarify if the higher oxygen removal was due to higher catalyst activity or absence of water (in presence of zeolite).

Catalytic treatment over HZSM-5 can produce treated oil that retains about 80 % of the energy in the crude bio-oil, which is composed largely of hydrocarbon molecules and is >95 wt% C and H. Processing the bio crude at 500 °C produced a treated oil with 45 wt% yield that contains almost exclusively aromatic hydrocarbons (Li and Savage 2013). Elliott et al. converted wet algae slurries to bio crudes by hydrothermal liquefaction (HTL) and further hydro-treated the HTL bio-oils at 350 °C in a continuous-flow at 200 bar, recovering bio crude without the use of solvents (Elliott et al. 2013). A high slurry concentration of 35 solid% was used, which was larger than typically used in previous work (~10 solid %). Catalytic hydrotreating was effectively applied for hydro-deoxygenation, hydro-denitrogenation, and hydro-desulfurization of the bio-crude to form liquid hydrocarbon fuel in the presence of CoMoS (4 % Co, 15 % Mo) catalyst at 400 °C and 100–140 bar, with a hydrogen flow in great excess of the process requirement (Elliott et al. 2013).

Overall, there is growing number of work that have been dedicated to the hydroliquefaction/hydrotreating of algae to bio-crudes, which indicate that a wide range of microalgae can be processed by this route into a complex mixture oxygenated hydrocarbons that is liquid at or near room temperature at a high mass yield (Elliott et al. 2013).

Recently, three algal oils have been successfully converted to hydrocarbon fuels at commercial scale (Lupton 2012). All the algal oils were pre-treatment to remove metals and phosphorous and then converted in a two-step process to jet and diesel fuels with the required freeze and cold flow properties. In the first step, the oils were deoxygenated, de-nitrogenised and de-sulphureted producing straight-chain normal (n-) alkanes from the fatty acid component of the algal oil. Then, these straight-chain alkanes were cracked and isomerized into a mixture of highly branched (iso-) alkanes (Lupton 2012). The treated product was then fractionated by distillation into naphtha, synthetic paraffinic kerosene (SPK) jet fuel and diesel fractions. All three algal derived bio-fuels met ASTM specifications for bio-SPK jet fuel and ASTM D 975 No. 2-Ultra Low Sulfur Diesel (Lupton 2012).

6 Conclusion

Microalgae have the potential to replace fossil fuels biodiesel and jet fuel thanks to their high productivity, fast growing rate and absence of competition with food crops. Moreover, microalgae as fuels feedstock can reduce the emissions of carbon dioxide in the atmosphere, since they use CO_2 as source of carbon. Despite the fact that there are a variety of processes available to convert microalgae biomass into bio-fuels, the production of bio-oils by pyrolysis or liquefaction is very promising due to the low capital investments required. Despite this, bio-oils from pyrolysis or liquefaction of algae differ from petroleum oils for their high oxygenates and N-compounds content, which lead to low thermal stability, high acidity, polymerization, high viscosity and high-boiling distribution.

Catalytic pyrolysis is employed to maximise the yield of bio-oil and for their deoxygenation. *Chlorella* and *Nannochloropsis* species are the most studied algae strains, since they are able to produce bio-oil in high yield (48–58 wt%) at temperatures close to 500 °C. Typically, zeolite catalysts are employed for their capacity to generate aromatic compounds. Catalytic hydro-deoxygenation (HDO) is another technique employed to decrease O and N content in bio-oils by addition of hydrogen. HDO of algae bio-oils is the preferential technology to produce bio-diesel and jet-fuel with properties that match those of respective petroleum derived fuels.

The recent advances in bio-oil upgrading using catalytic cracking and HDO are very promising for the establishment of advanced biofuels in the coming years. Nevertheless, significant investments at R&D and demonstration scale are still required for their successful deployment.

List of Acronyms

BTX	Benzene, Toulene, Xylene
DHA	Docosahexaenoic acid
EPA	Eicosapentaenoic acid
HHV	Higher heating value
HDO	Hydro-deoxygenation
HTL	Hydrothermal liquefaction
PAH	Polyaromatic hydrocarbons

References

Araujo GS et al (2011) Bioprospecting for oil producing microalgal strains: evaluation of oil and biomass production for ten microalgal strains. Bioresour Technol 102(8):5248–5250

Babich IV, van der Hulst M, Lefferts L, Moulijn JA, O'Connor P, Seshan K (2011) Catalytic pyrolysis of microalgae to high-quality liquid bio-fuels. Biomass Bioenerg 35:3199–3207

Bae YJ, Ryu C, Jeon J-K, Park J, Suh DJ, Suh Y-W, Park Y-K (2011) The characteristics of bio-oil produced from the pyrolysis of three marine macroalgae. Bioresour Technol 102(3):3512–3520

Biller P, Ross AB (2011) Potential yields and properties of oil from the hydrothermal liquefaction of microalgae with different biochemical content. Bioresour Technol 102:215–225

Brennan L, Owende P (2010) Biofuels from microalgae—a review of technologies for production, processing, and extractions of biofuels and co-products. Renew Sust Energ Rev 14(2):557–577

Bridgwater AV (2012) Review of fast pyrolysis of biomass and product upgrading. Biomass Bioenerg 38:68–94

Chaiwong K, Kiatsiriroat T, Vorayos N, Thararax C (2013) Study of bio-oil and bio-char production from algae by slow pyrolysis. Biomass Bioenerg 56:600–606

Chakinala AG, Brilman DWF(W), van Swaaij WPM, Kersten SRA (2010) Catalytic and non-catalytic supercritical water gasification of microalgae and glycerol. Ind Eng Chem Res 49(3):1113–1122

Chen WT, Zhang Y, Zhang J, Yu G, Schideman LC, Zhang P, Minarick M (2014) Hydrothermal liquefaction of mixed-culture algal biomass from wastewater treatment system into bio-crude oil. Bioresour Technol 152:130–139

De la Hoz Siegler H et al (2011) The dynamics of heterotrophic algal cultures. Bioresour Technol 102(10):5764–5774

Demirbaş A (2006) Oily products from mosses and algae via pyrolysis. Energ Source Part A 28(10):933–940

Demirbas MF (2011) Biofuels from algae for sustainable development. Appl Energ 88(10):3473–3480

Dickerson T, Soria J (2013) Catalytic fast pyrolysis: a review. Energies 6(1):514–538

Dillon H (2009) Algae: fuel of the future? Environ Sci Technol 10:7160–7161

Dong X, Chen Z, Xue S, Zhang J, Zhou J, Liu Y, Liu Z (2013) Catalytic pyrolysis of microalga *Chlorella pyrenoidosa* for production of ethylene, propylene and butene. RSC Adv 3(48):25780

Dote Y, Sawayama S, Inoue S, Minowa T, Yokoyama S (1994) Recovery of liquid fuel from hydrocarbon-rich microalgae by thermochemical liquefaction. Fuel 73:1855–1857

Du Z (2013) Thermochemical conversion of microalgae for biofuel production. Published doctoral dissertation, University of Minnesota, Twin Cities

Du Z, Ma X, Li Y, Chen P, Liu Y, Lin X, Ruan R (2013) Production of aromatic hydrocarbons by catalytic pyrolysis of microalgae with zeolites: catalyst screening in a pyroprobe. Bioresour Technol 139:397–401

Duan P, Savage PE (2011a) Catalytic hydrotreatment of crude algal bio-oil in supercritical water. Appl Catal B-Environ 104:136–143

Duan P, Savage PE (2011b) Catalytic treatment of crude algal bio-oil in supercritical water: optimization studies. Sus World 4(4):1447

Duan P, Savage PE (2011c) Hydrothermal liquefaction of a microalga with heterogeneous catalysts. Ind Eng Chem Res 50(1):52–61

Elliott DC (2007) Historical developments in hydroprocessing bio-oils. Energ Fuel 21:1792–1815

Elliott DC, Hart TR, Schmidt AJ, Neuenschwander GG, Rotness LJ, Olarte MV, Holladay JE (2013) Process development for hydrothermal liquefaction of algae feedstocks in a continuous-flow reactor. Algal Res 2:445–454

French R, Czernik S (2010) Catalytic pyrolysis of biomass for biofuels production. Fuel Process Technol 91(1):25–32

Garcia Alba L, Torri C, Fabbri D, Kersten SRA, (Wim) Brilman DWF (2013) Microalgae growth on the aqueous phase from hydrothermal liquefaction of the same microalgae. Chem Eng J 228:214–223

He Z, Wang X (2012) Hydrodeoxygenation of model compounds and catalytic systems for pyrolysis bio-oils upgrading. Catal Sus Energ 1:28–52

Hillen LW, Pollard G, Wake LV, White N (1982) Hydrocracking of the oils of *Botryococcus braunii* to transport fuel. Biotechnol Bioeng XXIV:193–205

Huber GW, Iborra S, Corma A (2006) Synthesis of transportation fuels from biomass: chemistry, catalysts, and engineering. Chem Rev 106:4044–4098

Jahirul M, Rasul M, Chowdhury A, Ashwath N (2012) Biofuels production through biomass pyrolysis—A technological review. Energy 5(12):4952–5001

Li Z, Savage PE (2013) Feedstocks for fuels and chemicals from algae: treatment of crude bio-oil over HZSM-5. Algal Res 2(2):154–163

Lupton FS (2012) The refining of algal oil into fungible transportation fuels. Biomass Magazine, March 19. http://biomassmagazine.com/articles/7672/the-refining-of-algal-oils-into-fungible-transportation-fuels

Luque R (2010) Fuels of the future. Energy Environ Sci 3(3):253

Mandal S, Mallick N (2009) Microalga Scenedesmus obliquus as a potential source for biodiesel production. Appl Microbiol Biotechnol 84(2):281–291

McKendry P (2002) Energy production from biomass (part 2): conversion technologies. Bioresour Technol 83:47–54

Miao X, Wu Q (2004) High yield bio-oil production from fast pyrolysis by metabolic controlling of *Chlorella protothecoides*. J Biotechnol 110(1):85–93
Miao X, Wu Q, Yang C (2004) Fast pyrolysis of microalgae to produce renewable fuels. J Anal Appl Pyrol 71:855–863
Pan P, Hu C, Yang W, Li Y, Dong L, Zhu L, Fan Y (2010) The direct pyrolysis and catalytic pyrolysis of *Nannochloropsis* sp. residue for renewable bio-oils. Bioresour Technol 101:4593–4599
Peng W, Wu Q, Tu P (2000) Effects of temperature and holding time on production of renewable fuels from pyrolysis of *Chlorella protothecoides*. J Appl Phycol 12(2):147–152
Peterson AA, Vogel F, Lachance RP, Fröling M, Antal MJ Jr, Tester JW (2008) Thermochemical biofuel production in hydrothermal media: a review of sub- and supercritical water technologies. Energy Environ Sci 1(1):32
Phukan MM, Chutia RS, Konwar BK, Kataki R (2011) Microalgae *chlorella* as a potential bio-energy feedstock. Appl Energ 88(10):3307–3312
Pragya N, Pandey KK, Sahoo PK (2013) A review on harvesting, oil extraction and biofuels production technologies from microalgae. Renew Sust Energ Rev 24:159–171
Sanna A (2013) Advanced biofuels from thermochemical processing of sustainable biomass in Europe. Bioenerg Res 7(1):36–47
Sawangkeaw R, Ngamprasertsith S (2013) A review of lipid-based biomasses as feedstocks for biofuels production. Renew Sust Energ Rev 25:97–108
Serrano-ruiz JC, Dumesic JA (2012) Catalysis for alternative energy generation. Springer, New York, pp 29–57
Sharma YC, Singh B, Korstad J (2011) A critical review on recent methods used for economically viable and eco-friendly development of microalgae as a potential feedstock for synthesis of biodiesel. Green Chem 13(11):2993
Shi F, Wang P, Duan Y, Link D, Morreale B (2012) Recent developments in the production of liquid fuels via catalytic conversion of microalgae: experiments and simulations. RSC Adv 2(26):9727
Singh J, Gu S (2010) Commercialization potential of microalgae for biofuels production. Renew Sust Energ Rev 14(9):2596–2610
Suali E, Sarbatly R (2012) Conversion of microalgae to biofuel. Renew Sust Energ Rev 16(6):4316–4342
Thangalazhy-Gopakumar S, Adhikari S, Chattanathan SA, Gupta RB (2012) Catalytic pyrolysis of green algae for hydrocarbon production using H+ZSM-5 catalyst. Bioresour Technol 118:150–157
Tsukahara K, Sawayama S (2005) Liquid fuel production using microalgae. J Jpn Pet Inst 48(5):251–259
Vardon DR, Sharma BK, Scott J, Yu G, Wang Z, Schideman L, Strathmann TJ (2011) Chemical properties of biocrude oil from the hydrothermal liquefaction of *Spirulina* algae, swine manure, and digested anaerobic sludge. Bioresour Technol 102(17):8295–8303
Wang K, Brown RC (2013) Catalytic pyrolysis of microalgae for production of aromatics and ammonia. Green Chem 15(3):675
Williams PJLB, Laurens LML (2010) Microalgae as biodiesel & biomass feedstocks: review & analysis of the biochemistry, energetics & economics. Energ Environ Sci 3(5):554
Xiu S, Shahbazi A (2012) Bio-oil production and upgrading research: a review. Renew Sust Energ Rev 16(7):4406–4414
Yusuf C (2007) Biodiesel from microalgae. Biotechnol Adv 25:294–306
Zeng Y, Zhao B, Zhu L, Tong D, Hu C (2013) Catalytic pyrolysis of natural algae from water blooms over nickel phosphide for high quality bio-oil production. RSC Adv 3(27):10806
Zhang H, Xiao R, Jin B, Xiao G, Chen R (2013) Biomass catalytic pyrolysis to produce olefins and aromatics with a physically mixed catalyst. Bioresour Technol 140:256–262
Zhao C, Brück T, Lercher JA (2013) Catalytic deoxygenation of microalgae oil to green hydrocarbons. Green Chem 15(7):1720

A Framework for Sustainable Design of Algal Biorefineries: Economic Aspects and Life Cycle Analysis

Peam Cheali, Carina Gargalo, Krist V Gernaey, and Gürkan Sin

Abstract In this chapter, a framework for sustainable design of algal biorefineries with respect to economic and environmental objectives is presented. As part of the framework, a superstructure is formulated to represent the design space – describing technologies developed for processing various types of algae feedstock for the production of biodiesel and co-products. Relevant data and parameters for each process such as yield, conversion, operational cost is then collected using a standardized format (a generic model) and stored in a database. The sustainable design problem is then formulated mathematically as a mixed integer nonlinear programming problem, and is solved first to identify the optimal designs with respect to economic optimality. These optimal designs are then analyzed further in terms of environmental performance using life cycle analysis. For sustainability analysis, in total five impact categories are calculated including Photochemical oxidation potential (POP), global warming potential (GWP), aquatic ecotoxicity (EcotA), Carcinogenic emissions to urban air (EUA_C), and median lethal dose (LD_{50}). To add robustness to the analysis, the framework includes uncertainty analysis using Monte Carlo simulations as well. The application of the framework is highlighted on a case study focusing on feedstock microalgae cultivated in Raceway ponds to produce biodiesel. The framework with the database and superstructure provides an enabling tool to support systematic design and analysis of future and sustainable algal biorefinery concepts.

Keywords Algal biorefinery • Early-stage process design • Systematic framework • Process synthesis • Superstructure optimization • Generic process model • MINLP • Economic analysis • Sustainability analysis • Uncertainty analysis • Monte Carlo • Life-cycle assessment • GWP

P. Cheali • C. Gargalo • K.V. Gernaey • G. Sin (✉)
CAPEC-PROCESS, Department of Chemical and Biochemical Engineering,
Technical University of Denmark, Building 229, DK 2800 Lyngby, Denmark
e-mail: gsi@kt.dtu.dk

A. Prokop et al. (eds.), *Algal Biorefineries*, DOI 10.1007/978-3-319-20200-6_17

1 Introduction

The scarce availability of resources causes a serious challenge to economic growth of the chemical/ and biochemical industries. This motivates the development of more sustainable processes to process renewable feedstock for producing fuel, chemicals and materials.

Compared to other renewable feedstock (i.e. corn stover, wood, palm or soybean), algae give the highest oil yield per hectare per year (Demirbas and Demirbas 2011). Moreover, their high growth rate, their CO_2 consumption, the cleanliness of the processing technologies involved, and the potential to produce a variety of products (i.e. biofuels, bioenergy, animal feed, cosmetics, fertilizer, fibers, and intermediate proteins) further enhance the development of algae cultivation and conversion technologies (FAO 2010). The typical algal biorefinery consists of algae cultivation, harvesting, pretreatment, extraction, and conversion (i.e. transesterification). As regards the conversion step, thermochemical processes (i.e. hydrothermal liquefaction or pyrolysis) can also be used to convert algae to biofuels. Moreover, algae which contain a number of nutrients can also be converted into non-energy products (i.e. intermediate protein, animal feed or fertilizer). However, despite these advantages there are still two major challenges which need to be overcome: (i) a significant number of processing alternatives have been generated and developed due to the wide variety of products that can be produced (Petrick et al. 2013), and as a consequence it is difficult to select the most suitable processing alternative; and, (ii) the production cost and the yield of algae feedstock are high and uncertain (Demirbas and Demirbas 2011). The first challenge is addressed in this study and the second challenge will be addressed in future work.

A number of studies have been published on process synthesis, process design or process optimization identifying the optimal design concept of the algal biorefinery. Gong and You (2014) created a superstructure and performed a global optimization producing biodiesel from algae including life cycle and CO_2 mitigation optimization. Rizwan et al. (2013) also created a superstructure for a biorefinery converting algae to biodiesel. Martin and Grossmann (2013) identified the optimal algae composition to maximize the production of biodiesel and bioethanol. While these studies provided valuable information, however, they have been limited to only the production of biodiesel and a limited number of algae feedstocks. To overcome the design challenge, a systematic framework for synthesis and design of processing networks applied in an earlier study for lignocellulosic biorefinery (Cheali et al. 2014) is here adapted and used for creating the processing network of an algal biorefinery, including the use of a verified database, and to identify the optimal designs.

In this study, a systematic framework that uses superstructure-based optimization is used to identify the optimal algal biorefinery concept. The study starts with the generation of a database and a design space (superstructure, models). Consequently, the optimal designs are identified with respect to techno-economic criteria and life-cycle analysis. A superstructure representing the design space of the algal biorefinery is developed containing various types of microalgae and subsequent pre-treatment steps, reaction steps and separation technologies to produce biodiesel, and co-products (starch, intermediate proteins, fertilizer, animal feed, methane, and ethanol). Subsequently, the database (process specific parameters and data structured in a generic model) is integrated with the superstructure which is formulated by the combination of the alternatives available to produce biodiesel from algae (types of feedstock, technologies, and by-products). Finding the most optimal technology is then mathematically formulated as an optimization problem, which is solved to identify the optimal designs with respect to techno-economic criteria and life-cycle analysis related constraints.

This chapter is organized as follows: (i) the framework; (ii) techno-economic analysis, and (iii) life cycle assessment and (iv) conclusion.

2 Framework

This study uses an earlier developed framework (Cheali et al. 2014) presented in Fig. 1. We highlight the generation of a database, and in particular the data management and collection in the frame of a generic process modeling approach to collect and manage the complexity of the multi-disciplinary data related to algal biorefinery processes. The different steps which are part of the framework (Fig. 1) will be explained briefly.

Step 1: Problem definition

The first step includes the definition of the problem scope, the selection of suitable objective functions and optimization scenarios with respect to either techno-economic, sustainability and environmental impact criteria, or a combination of these objectives.

Step 2: Superstructure definition

A superstructure representing different concepts is formulated on the basis of a thorough literature review. A typical processing network consists of a number of processing steps converting or connecting feedstock to products. Each processing step is defined and represented in a generic way using the generic process model block (Fig. 2). Each process model block incorporates the generic model to represent various tasks carried out in the block such as mixing, reaction and separation. A corresponding superstructure is therefore defined as a processing network connecting the generic model blocks from the feedstock to the products. A detailed presentation of the generic models and the data required is provided in Step 3.

Step 3: Data collection and modeling

Once the superstructure is defined, the data are collected and modeling is performed. Generally, the models for each processing technology are rigorous, non-linear and complex models (e.g. kinetics, thermodynamics). In this study, however, a simple input–output type generic block model is used which is identified from the data

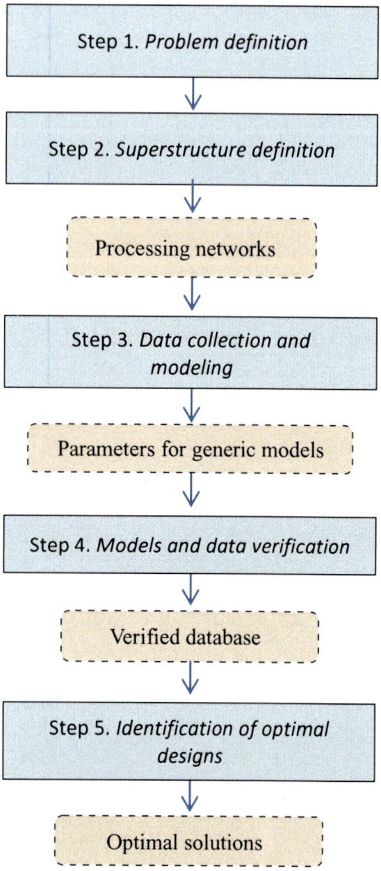

Fig. 1 The steps taken in the data management and identification of optimal design networks framework: the dashed boxes indicate the outcome of each step of the workflow

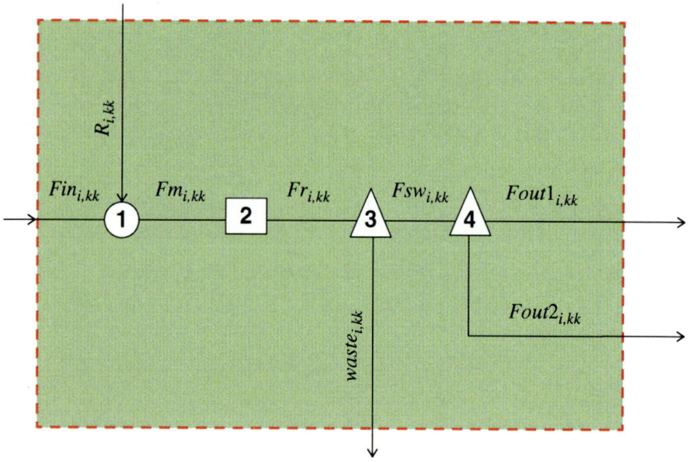

Fig. 2 The generic process model block representing the technology alternatives in a generic form (Cheali et al. 2014)

generated from the above-mentioned complex models. This generic process model block thus consists of four parts of the typical simple mass balance equations: (i) mixing; (ii) reaction; (iii) waste separation; and, (iv) product separation. The models and parameters used for the mass balance calculation are presented and explained in Table 1.

The Eqs. 1, 2, 3, 4, 5, 6, and 7 as presented in Table 1 form the simple mass balance equations used for the generic block to estimate the outlet mass flow $\left(F_{i,kk}^{out1},\ F_{i,kk}^{out2}\right)$ from the mass inlet flow rate $\left(F_{i,kk}^{in}\right)$. In the mixing section (Eqs. 1 and 2), the chemicals and utilities used $\left(R_{i,kk}\right)$ for each processing technology are calculated by using the ratio $\left(\mu_{i,j,kk}\right)$ to the inlet mass flow rate $\left(F_{i,kk}^{in}\right)$. The parameter $\alpha_{i,kk}$ represents the consumption of the utilities or chemicals: 0 corresponds to 100 % consumption; 1 represents no consumption. In the reaction section (Eq. 3), the reaction outlet mass stream $\left(F_{i,kk}^{R}\right)$ is calculated based on stoichiometry, $\gamma_{i,rr}$ and conversion fraction, $\theta_{react,rr}$. In the waste separation section (Eqs. 4 and 5), the waste stream $\left(waste_{i,kk}\right)$ and the remaining stream $\left(Fsw_{i,kk}\right)$ are calculated on the basis of the removal fraction, $SW_{i,kk}$. The product outlet streams are calculated in Eqs. 6 and 7 on the basis of a product separation fraction, $Split_{i,kk}$. The appropriate values for the above mentioned parameters can be collected in several ways including: (i) literature sources or technical reports; (ii) experimental data; (iii) simulation results; or, (iv) stream table or operating data of a designed flowsheet.

Moreover, in order to connect each generic process model block and thereby formulate the superstructure, the Eqs. 8, 9, 10, 11, 12, and 13 are used. The mass outlet flows mentioned earlier $\left(Fout1_{i,kk},\ Fout2_{i,kk}\right)$ are called primary and secondary mass outlet flow, respectively. The primary and secondary outlet flows are connected to the next generic blocks using binary variables $\left(S_p, S\right)$, respectively. The outlet flows between the generic blocks $\left(F1_{i,k,kk},\ F2_{i,k,kk}\right)$ of each stream (primary and secondary) are summed up as the input of the next generic process model block. The recycle flows can be considered by using these equations.

$$F_{i,k,kk}^{1} \le F_{i,kk}^{out1} \cdot S_p \tag{8}$$

$$F_{i,k,kk}^{2} \le F_{i,kk}^{out2} \cdot \left(S - S_p\right) \tag{9}$$

$$F_{i,kk}^{in} = \sum_k \left(F_{i,k,kk}^{1} + F_{i,k,kk}^{2}\right) \tag{10}$$

$$\sum_k F_{i,k,kk}^{1} = F_{i,kk}^{out1} \tag{11}$$

$$\sum_k F_{i,k,kk}^{2} = F_{i,kk}^{out2} \tag{12}$$

$$\sum_i F_{i,kk}^{out} \le F_{kk}^{MAX} \tag{13}$$

Table 1 Models and parameters used for the generic block

Equations	Parameters	Description	Min. value	Max value
Mixing section				
(1) $R_{i,kk} = \left(\mu_{i,j,kk} * F^{in}_{i,kk} * \alpha_{i,kk}\right)$	$\mu_{i,j,kk}$	Ratio of chemicals/utilities used to inlet flow	0	+inf
(2) $F^{M}_{i,kk} = F^{in}_{i,kk} + R_{i,kk}$	$\alpha_{i,kk}$	Amount of chemicals/utilities that is not consumed and mixed with process stream	0	1
Reaction section				
(3) $F^{R}_{i,kk} = F^{M}_{i,kk} + MW_i * \left(\gamma_{i,rr} * \theta_{react,rr} * F^{M}_{i,kk} / MW_{react}\right)$	$\gamma_{i,rr}$	Stoichiometric coefficient of component (*i*) in reaction (*rr*)	-inf	+inf
	$\theta_{react,rr}$	Conversion fraction of key reactant (*react*) in reaction (*rr*)	0	1
Waste separation section				
(4) $F^{out}_{i,kk} = \left(1 - SW_{i,kk}\right) * F^{R}_{i,kk}$	$SW_{i,kk}$	Removal fraction of waste	0	1
(5) $waste_{i,kk} = SW_{i,kk} * F^{R}_{i,kk}$				
Product separation section				
(6) $F^{out1}_{i,kk} = Split_{i,kk} * F^{out}_{i,kk}$	$Split_{i,kk}$	Separation fraction of component *i* in primary outlet stream	0	1
(7) $F^{out2}_{i,kk} = \left(1 - Split_{i,kk}\right) * F^{out}_{i,kk}$				

Equations 1, 2, 3, 4, 5, 6, 7, 8, 9, 10, 11, 12, and 13 which are used to estimate the inlet–outlet flow rate of each process model block are included in the mathematical formulation as process constraints presented in the Step 4.

Step 4: Models and data verification

After the superstructure is defined and the parameters are collected, a validation of the selected models and parameters needs to be performed for consistency check. The validation is performed in this step by performing a simulation (fixing the decision variables (y_k) of the structural constraints (Eqs. 14 and 15) in the MILP/MINLP problem formulation) for each processing technology or path, followed by comparison of the simulation results against the available data (experiment, technical reports, etc.).

Structural constraints,

$$\sum_k y_k \leq 1 \tag{14}$$

$$y \in \{0;1\}^n \tag{15}$$

The output of this step is a verified database representing the superstructure formulated in Step 2. All the necessary equations and constraints relevant to each processing technology are also formulated in this step prior to be solved as MILP/MINLP problems in GAMS in Step 5.

Step 5: Identification of optimal designs

An optimization problem (MILP/MINLP) is formulated and presented in this step. An objective function (e.g. max. Earnings Before Interest, Taxes, Depreciation and Amortization, Eq. 16) is formulated which is subjected to process constraints (Eqs. 1, 2, 3, 4, 5, 6, 7, 8, 9, 10, 11, 12, and 13); structural constraints (Eqs. 14 and 15) representing the superstructure which allows selection of only one process alternative in each step; and cost functions to calculate the operating costs (Eq. 17) using the cost parameters: raw material cost $\left(P1_{i,kk}\right)$; utility or chemicals cost $\left(P2_{i,kk}^{utilities/chemicals}\right)$; waste treatment cost $\left(P4_{i,kk}^{waste}\right)$. The capital cost is estimated using the six-tenth rules (Eq. 18) by collecting the available information from existing processes and the relative constant to each specific unit operation.

$$Max.\ EBITDA = Revenue - Expenses\left(excl.\ interest,\ taxes,\ depreciation\ and\ amortization\right) \tag{16}$$

Subject to:

Process models of the generic block (Eqs. 1, 2, 3, 4, 5, 6, 7)

$$h\left(\mu_{i,j,kk}, \alpha_{i,kk}, \gamma_{i,rr}, \theta_{react,rr}, MW_i, SW_{i,kk}, Split_{i,kk}\right)$$

Process constraints (Eqs. 8, 9, 10, 11, 12, and 13)

$$g\left(S_p^{\text{k,kk}}, S^{k,kk}\right) \le 0,$$

Cost constraints,

$$OPEX_{kk} = \left(P2_{i,kk}^{\frac{utilities}{chemicals}} \bullet R_{i,kk}\right) + \left(P4_{i,kk}^{waste} \bullet F_{i,kk}^{waste}\right) \quad (17)$$

$$CAPEX_{kk} = \sum_{kk}\left[a * F_{i,kk}^{in\ \text{n}}\right] \quad (18)$$

The problem is formulated and solved using GAMS. The generic model parameters and other data appearing in the constraints (e.g. $\alpha_{i,kk}, \gamma_{i,rr}, \theta_{react,rr}$, $P1_{i,kk}^{waste}$, $P2_{i,kk}^{utilities/chemicals}$, etc.) are accessed and called from the database. The data collection and verification (with help of a generic process model and its parameters stored in a database), and solution/ analysis of the problem result in the management of the complexity of formulating an MINLP-based optimization problem for algal processing networks.

3 Algal Biorefinery: Data Collection and Management, and Identification of Optimal Designs

3.1 Techno-economic Analysis and Screening

Step 1: Problem definition

The problem in this study is the identification of an optimal algal biorefinery network producing biodiesel from algae feedstock obtained from a raceway pond. The objective function is formulated using EBITDA as economic metric, which is then solved to identify solutions with respect to economic optimality.

Step 2: Superstructure definition

Algae conversion routes producing biodiesel were reviewed to formulate the superstructure (Fig. 3). The superstructure consists of four main processing steps to convert algae feedstock to products: harvesting, pretreatment, lipid extraction, and transesterification.

In this study, only algae feedstock (Raceway pond) is used, by applying the same basis (1300 tpd, 190 MM\$/a) as used by a Pacific Northwest National Laboratory (PNNL) study (Jones et al. 2014). The harvesting processing step consists of 12 alternatives (block no. 2–13). Different technologies use different types of utilities and methods resulting in a different operating and capital cost. The pretreatment

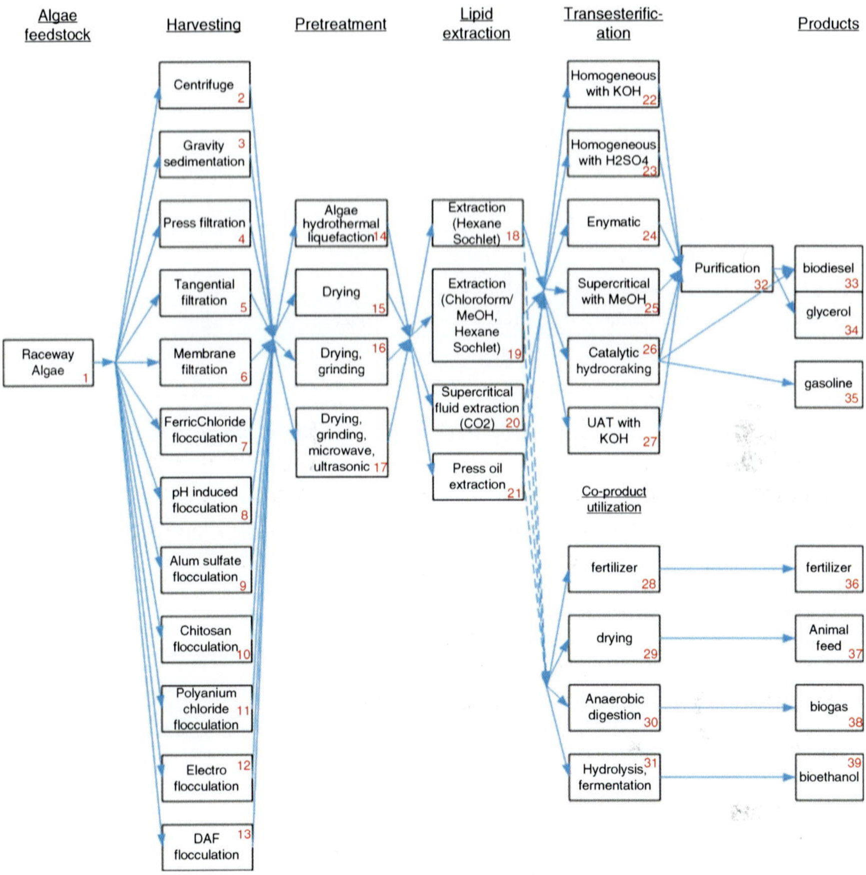

Fig. 3 The superstructure of algal biorefinery case study

processing step (block no. 14–17) consists of 4 alternatives, including a simple drying process for drying the influent, or combinations with other techniques (grinding, microwave, ultrasonic) and with the use of hydrothermal liquefaction for producing a higher yield of algae oil. Lipid extraction is one of the main processes for the algal biorefinery to extract algae oil (lipids) from algae-water feed slurry (20 %wt). There are four technologies presented (block no. 18–21) which are the extraction using different types of solvent, and mechanical extraction (oil press extraction). Algae oil (lipids), the primary product, is then processed further using a transesterification process thus producing fatty acid methyl ester (FAME or biodiesel) and glycerol. The secondary or co-products can be utilized as well for producing higher value products (fertilizer, animal feed, biogas or bio-methane, and bioethanol) by means of 4 process techniques (block no. 28–31). The defined superstructure represents a design space which includes a total of 1920 potential processing paths for conversion of algae to biodiesel.

Table 2 The parameters for the generic process block in harvesting processing step ($Split_{algae,kk} = 1$)

Process blocks		$\mu_{i,j,kk}$	$\alpha_{i,kk}$	*References*
(2) Centrifuge				Price et al. (1974)
(3) Gravity sedimentation				Sim et al. (1988)
(4) Press filtration				Sim et al. (1988)
(5) Tangential filtration				Petrusevski et al. (1995)
(6) Membrane filtration				Zhang et al. (2003)
(7) Ferric Chloride flocculation	Ferric Chloride	1.25	1	Granados et al. (2012)
(8) pH induced flocculation	NaOH	0.2	1	Wu et al. (2012)
(9) Alum sulfide flocculation	Alum	0.27	1	Sirin et al. (2012)
(10) Chitosan flocculation	Chitosan	0.18	1	Divakaran and Pillai (2002)
(11) Polyanium chloride flocculation	Polyanium chloride	0.27	1	Divakaran and Pillai (2002))
(12) Electro flocculation				Granados et al. (2012)
(13) Dissolved air floatation				Sim et al. (1988)

Table 3 The parameters for the generic process block in harvesting processing step

Process blocks		$\mu_{i,j,kk}$	$\alpha_{i,kk}$	$Split_{lipid,kk}$	References
(18) Extraction	Hexane Sochlet	0.18	0	0.26	Prommuak et al. (2012)
(19) Extraction	Chloroform and Hexane Sochlet	0.18	0	0.2	Long and Abdelkader (2011)
(20) Supercritical fluid extraction				0.12	Li et al. (2008)
(21) Press oil extraction				0.28	

Step 3: Data collection and estimation

The data and parameters required for the generic process model blocks that are used to define the superstructure (Fig. 3), are presented in this section and in Tables 2, 3, and 4. The alternative technologies presented and defined in this study are based on the available data from publications (i.e. literature, technical reports). Cost estimation was performed on the basis of the amount of utilities consumed (operating cost) and the available information of the existing plant/technologies with the six-tenth rule (capital cost). The sixth-tenth is an early-stage capital cost estimation which uses the capacity and capital cost of the existing plant and the capacity of the new plant to estimate the capital cost of the new plant (Cheali et al. 2015a).

Table 2 presents the parameters for the generic process model block in the harvesting processing step. The harvesting step is used to collect the algae diluted in the water. The following three parameters are identified and collected to describe the harvesting step: (i) the ratio of utilities or chemicals added $\left(\mu_{i,j,kk}\right)$; (ii) the con-

Table 4 The parameters for the generic process block in transesterification, co-product utilization, purification processing step

Transesterification		$\mu_{i,j,kk}$	$\alpha_{i,kk}$	$\theta_{react,rr}$	References
(22) Homogeneous	KOH, MeOH	0.05, 16	0, 1	0.92	Vicente et al. (2004)
(23) Homogeneous	H_2SO_4, MeOH	0.03, 9	0, 1	0.95	Miao and Wu (2006)
(24) Enzymatic	Enzyme, MeOH	0.0012, 6	0, 1	0.83	Levine (2013)
(25) Supercritical	MeOH	9	1	0.89	Levine (2013)
(26) Catalytic hydrocracking				0.82	Jones et al. (2014))
(27) Ultrasonic assisted transesterification (UAT)	KOH	9	0	0.925	Levine (2013)

sumption of added utilities or chemicals $(\alpha_{i,kk})$; (iii) separation or split of main- and by-product $(Split_{i,kk})$.

The pretreatment step is used to increase the concentration of the algae to a value of 20 wt% or higher. Therefore, the parameters here are (i) the heat and electricity required; and (ii) the concentration of algae at the outlet (20 wt%). In this step, there are four alternatives: (i) algae hydrothermal liquefaction (AHTL) with a yield of 0.52 wt% for the conversion of raw algae to algae oil (Jones et al. 2014); (ii) drying and grinding; (iii) drying, grinding and microwave; (iv) drying, grinding, microwave, and ultrasonic treatment.

Table 3 presents the parameters in the lipid extraction step which considers three types of technologies namely solvent based extraction, supercritical fluid extraction and press oil extraction. This step is used to extract algae oil (lipids) from the algae feed (20 wt%) after the drying processes. The primary product of this step is algae oil (lipid) and the secondary products which are separated are then processed to co-product utilization step.

Table 4 presents the parameters for the generic process model block in the transesterification processing step. Transesterification is used to convert algae oil (lipid) to biodiesel and glycerol. Catalysts (acids or base), enzymes and methanol are used in this step depending on the type of technology used. Information on the stoichiometric coefficient $(\gamma_{i,rr})$ following the typical transesterification reaction (Algae oil (lipid) + 3MeOH ↔ 3Fatty acid methyl ester (FAME) + glycerol) and the conversion fraction $(\theta_{react,rr})$ are further collected to describe this processing step.

In the co-product utilization step (block no. 28–31), the following alternatives are considered: There are four alternatives (fertilizer, dryer, anaerobic digestion, and fermentation) to utilize slurry cake of protein and starch to value added products (fertilizer, animal feed, methane, and ethanol). The data reported in Alabi et al. (2009) about the yields for these alternatives are used in this step. Fertilizer (block no. 28) in this study is used to produce potassium nitrate with the yield of 90 %wt. The amount of dry cake of protein and starch mixture produced by the dryer (block

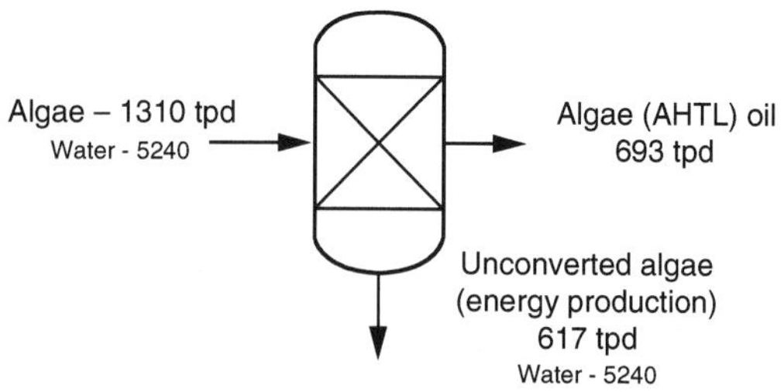

Fig. 4 The simplified process diagram showing mass inlet/outlet for hydrothermal liquefaction

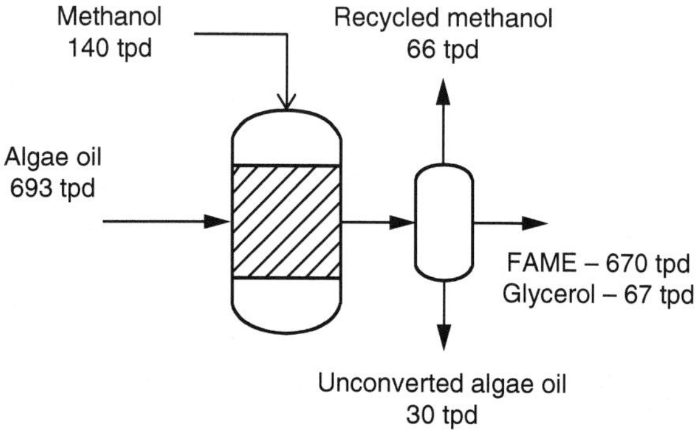

Fig. 5 The simplified process diagram showing mass inlet/outlet for homogeneous transesterification with H_2SO_4

no. 29) is converted into an animal feed product. In this alternative, 50 wt% of slurry cake (mostly water) is removed. Anaerobic digestion is used to produce bio-methane with a yield of 3 %wt (block no. 30). Hydrolysis and fermentation technologies (block no. 31) are used to produce bioethanol with a yield of 30 %wt.

Step 4: Models and data verification

In this step, models and data are verified by checking the conservation of mass for each process model block. The output of this step is the verified database of the algal biorefinery which will then be used as the input data for the optimization problem in the next step to identify the optimal processing paths. This step is highlighted for two processes below.

The first example is for the hydrothermal liquefaction process to produce algae oil (lipid) from raw algae. Heat is used as the main utility in this process. The mass balance (inlet stream(s) – outlet streams) for this process is closed by 100 % as shown in Fig. 4.

The second example is for homogeneous transesterification with H_2SO_4 to produce FAME (biodiesel) and glycerol from algae oil (lipid). Similarly the mass balance around this processing block is closed by 100 % as shown in Fig. 5.

Step 5: Identification of optimal designs

In this step, the formulated MILP/MINLP problem was solved; the optimal solutions were identified (max. EBITDA); and the results are presented in Table 5 illustrating the top-three ranking solutions. The production rate of biodiesel and glycerol,

Table 5 Top-three ranking of algae biorefinery optimal processing paths

Rank	Processing path	EBITDA (MM$/a)	Production (biodiesel/ glycerol) (tpd)	Capital cost (MM$)	Operating cost (MM$/a)
1	Algae, hydrothermal liquefaction, transesterification with H_2SO_4	316	670/67	252	198
2	Algae, hydrothermal liquefaction, transesterification with KOH	299	648/65	252	201
3	Algae, hydrothermal liquefaction, super/ subcritical transesterification with methanol	283	627/63	252	196

EBITDA, the total capital cost and operating cost as well as the optimal processing paths are presented.

This solution corresponded to the deterministic solution of the optimization problem, i.e. no uncertainties are considered. The formulation of the optimization problem consists of 99,437 equations and 97,319 variables and 40 decision variables. This problem was solved using the DICOPT solver using a Windows 7, Intel® Core™ i7 CPU@ 3.4GHz, 4GB RAM, and a solution resulted after about 10 s of calculation time.

As presented in Table 5, hydrothermal liquefaction was selected due to the highest yield of algae oil produced compared to other lipid extraction alternatives. The homogeneous transesterification using H_2SO_4 as catalyst was selected because this process reaches the highest conversion. The results are in agreement with the PNNL report (Jones et al. 2014) which used hydrothermal liquefaction and catalytic hydrotreating resulting in EBITDA of 280 MM$/a. Our study reported a relatively higher EBITDA compared to the PNNL study. The difference is due to the use of transesterification with H_2SO_4 instead of catalytic hydrotreating technology considered in PNNL study which has a lower yield and higher cost.

3.2 *Life Cycle Assessment of the Optimal Designs*

In this step, life cycle assessment of the two best designs selected from economic analysis performed in the previous step is performed. To add robustness to the life cycle analysis, uncertainty analysis using the Monte Carlo technique is also presented in this section.

The LCA steps described by the Environmental Protection Agency (EPA) were followed (EPA 2006), but extended by adding the uncertainty analysis steps as described by Gargalo and Sin (2015). For the uncertainty analysis, the Monte Carlo technique with Latin Hypercube sampling is used.

The methodology is divided into four parts, (A) goal and scope definition, (B) life cycle inventory, (C) life cycle impact assessment and, (D) life cycle interpretation and report generation. A brief description of the procedure and its application to the two selected algal biorefinery candidates is given below.

Part A: *Goal and scope definition*

This is the first stage of the analysis, where the user raises the questions to be answered in the last stage of the assessment, establishes the objectives and the categories of environmental impact that are relevant to the study.

Part B: *Life cycle inventory*

Part B of the extended LCA methodology is comprised of four steps.

Step 1: Definition of the system boundaries

The boundaries of the system to be analyzed were drawn around the manufacturing process and the utility scheme.

Step 2: Inventory of inputs and outputs

Mass and energy balances were collected from the two alternatives considered according to the system boundaries defined in the previous step.

Step 3: Identification and characterization of the source of uncertainty

This step aims to identify the source(s) of uncertainty and to characterize it (them) by means of a probability distribution function. The source of uncertainty here identified were the science-based characterization factors that are needed to convert the data inventory (from Step 2) into environmental impact categories.

In this study, the nominal values of the characterization factors were retrieved from an open-source database, IMPACT 2002+, whose models are reported in Humbert et al. (2002). Furthermore, due to the limited information, a uniform distribution was assumed to characterize the input uncertainties. The upper and lower bound of the uniform distribution are specified by performing an analysis of data variability reported across 6 different open-source LCA databases (CML-IA-2013, *Recipe*, *IChemE*, IMPACT 2002+, IMPACT World+).

Based on the extent of variability across different databases, low (5 %), medium (25 %) and high (50 %) levels of variation around the nominal values are assigned (Sin et al. 2009).

As a motivating example, the definitions of the uniform distribution of the characterization factors used to estimate the photochemical oxidation potential (POP) for four of the components involved in the processes are presented in Table 6. This uncertainty definition in the input database is performed for all the relevant components in the system.

Step 4: Latin hypercube sampling

The output from step 3 is a definition of the input uncertainty domain which is a *NxN* space where *N* refers to the total number of components (products, intermediates,

Table 6 Nominal value, upper and lower bound of the uniform distribution for the characterization factors used to estimate the POP

	Nominal/average value (IMPACT 2002+)	Lower bound	Upper bound
Methanol	0.132	0.066	0.198
Glycerol	0.861	0.431	1.292
Ethanol	0.392	0.196	0.588
Hexane	0.479	0.240	0.719
N_2O	0.700	0.350	1.050

raw materials, etc.) in the system. In this study, *N* was equal to 17. The Latin Hypercube Sampling technique was then used to generate random samples from the input uncertainty domain. Each sample is then used to calculate the LCA model output which is the environmental impact categories. It is noted that since the correlation matrix between input sources of uncertainties is not available, and therefore no correlation was assumed for the sampling.

More information on the LHS technique and correlation control can be obtained from (Helton and Davis 2003; Sin et al. 2009) and (Iman and Conover 2007), respectively.

In the present case study, 200 random samples were generated by LHS for each one of the characterization factors, collected for each one of the components in the system.

Part C & Part D: *Life cycle Impact Assessment & Life Cycle Interpretation and report generation*

In the previous section the data inventory is completed, and at this stage, the user is guided towards the estimation of relevant impact categories. The potential environmental impact (PEI) categories are calculated by Eq. 19 (Curran, 2006) as follows.

$$PEI_x = \sum_i CF_i \times F_i \tag{19}$$

Where, *x* represents a certain category of impact, *i* refers to a given component and *CF* represents the nominal value of the characterization factors that convert the component flow rates into potential environmental impacts used to analyze the environmental performance of the processes.

To select the categories of impact to be further analyzed under uncertainty, the expected values of the potential environmental impacts are calculated based on the nominal values of the characterization factors (CF).

The components in the system have contributions to the global warming potential (GWP), aquatic ecotoxicity (EcotA), carcinogenic emissions to urban air (log (EUA_C), median lethal dose (log (LD_{50})) and in the photochemical oxidation potential (POP). The Global Warming Potential (GWP) is used as a measure of the effect on radiation of a particular quantity of the substance over time relative to that of the same quantity of CO_2. The characterization factor for human toxicity impacts, here given by the carcinogenic emissions to urban air (log (EUA_C), is expressed in comparative toxic units regarding the estimated increase in mortality in the total human population, per unit mass of a chemical emitted. It is represented in disease cases per kg (of substance) emitted. The aquatic ecotoxicity potential (EcotA) is expressed in comparative toxic units (CTUe), an estimate of the potentially affected fraction of species (PAF) integrated over time and volume, per unit mass of a chemical emitted. The photochemical oxidation potential (POP) value of a particular hydrocarbon is a relative measure of how much the ozone concentration measured at a single location changes if emission of the hydrocarbon in question is altered by the same amount as that of a reference hydrocarbon, usually ethylene.

The expected values for the environmental impacts, corresponding to alternatives A and B, are presented in a bar chart form (Fig. 6).

Having selected the categories of impact *x*, for each LHS sample, Eq. 19 will be calculated resulting in 200 estimates for each impact category, *x*. The distribution of the impact category *x*, is then represented by building an empirical cumulative distribution function (CDF).

The cumulative distribution function of the model outputs, (Eq. 20) is obtained by rewriting Eq. 19, where *CF* represents the sample space *N* instead of nominal values.

$$CDF = F\left(Potential\ Environmental\ Impact, PEI\right) = \sum_{j=1}^{n} \Pr\left(PEI \leq PEI_j\right) \quad (20)$$

Where PEI_n is the largest possible value of PEI that is less than or equal to PEI .

The y-axis of the CDF plots shows the probability of *x* being less than or equal to a certain value *X*, $\Pr(x \leq X)$, while the x-axis refers to the actual values of the impact categories. The larger the range of the x-axis means the larger the uncertainty in the calculated value of PEI for a certain category. In this CDF plot, the mean value is indicated by the probability level at 0.5 (red arrows in the Figs. 7, 8,

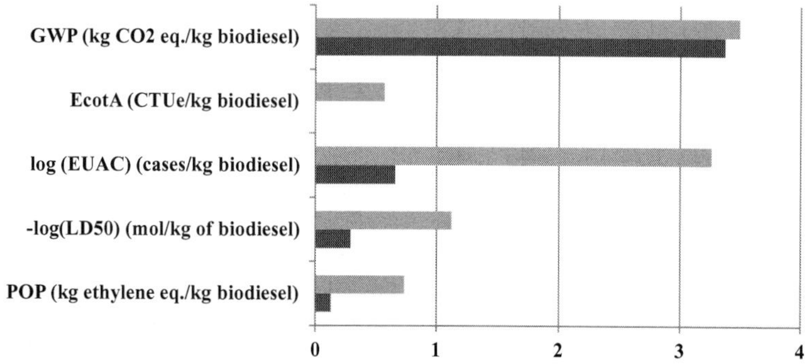

Fig. 6 Expected values for the potential environmental impacts for alternative A (*light grey*) and for alternative B (*dark grey*)

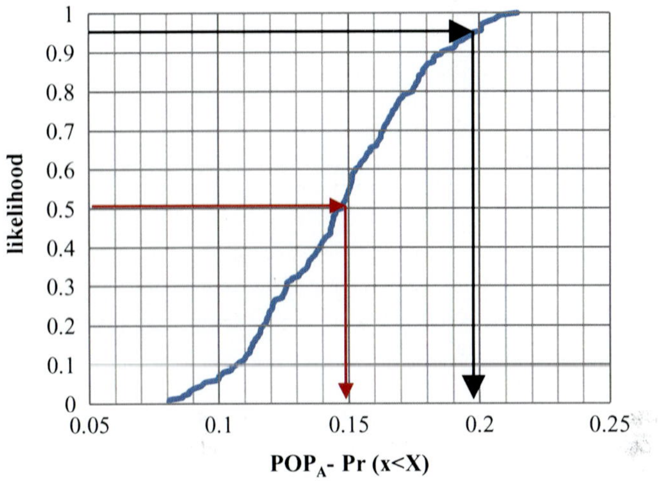

Fig. 7 *Alternative A* – Cumulative distribution function for the Photochemical Oxidation Potential (*POP*), upper bound identification for 95 % likelihood

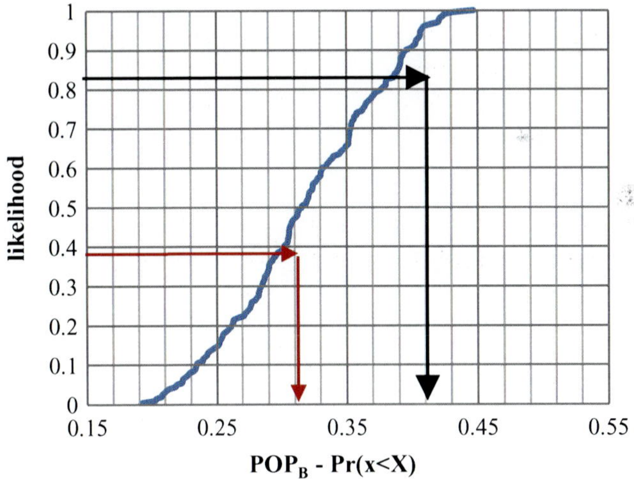

Fig. 8 *Alternative B* – Cumulative distribution function for the Photochemical Oxidation Potential (*POP*), upper bound identification for 95 % likelihood

9, 10, 11, 12, 13, 14, and 15). The mean values correspond to 'single point' analysis using average input (conversion) factors obtained in traditional LCA analysis.

As a summary, CDF plots are used to read out the impact category estimates that correspond to a probability of 0.95 (black arrows in the Figs. 7, 8, 9, 10, 11, 12, 13, 14, and 15 below). For example, the photochemical oxidation potential (Figs. 7 and 8) value at 0.95 probability is less than or equal to 0.20 and 0.41 kg ethylene eq./kg of biodiesel produced for alternative (A) and (B), respectively (Fig. 16).

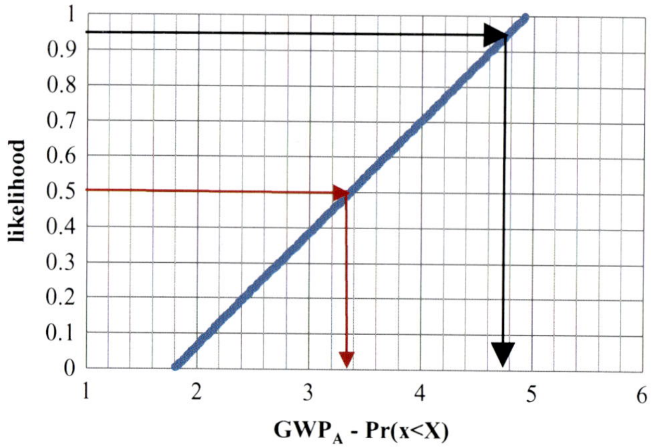

Fig. 9 *Alternative A* – Cumulative distribution function for the Global Warming Potential (*GWP*), upper bound identification for 95 % likelihood

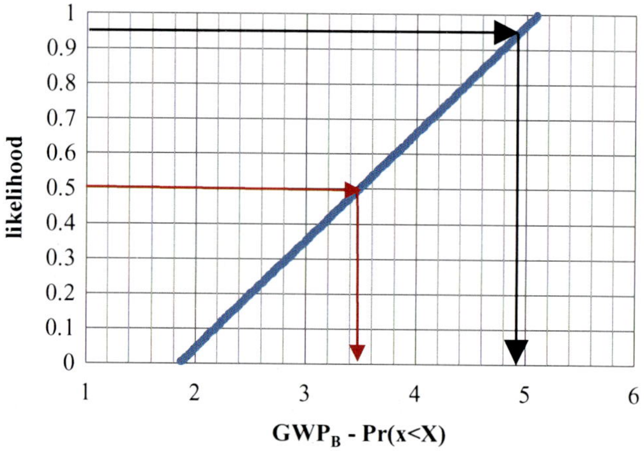

Fig. 10 *Alternative B* – Cumulative distribution function for the Global Warming Potential (*GWP*), upper bound identification for 95 % likelihood

Likewise, the global warming potential (Figs. 9 and 10) estimate at 0.95 probability is less than or equal to 4.77 and 4.93 kg CO_2 eq./ kg of biodiesel produced for alternative (A) and (B), respectively.

The comparison between the two alternatives can be based on the estimated PEI values at 0.95 probability level. In this case, alternative A seems to be the most promising considering all the input uncertainties that went into their calculations.

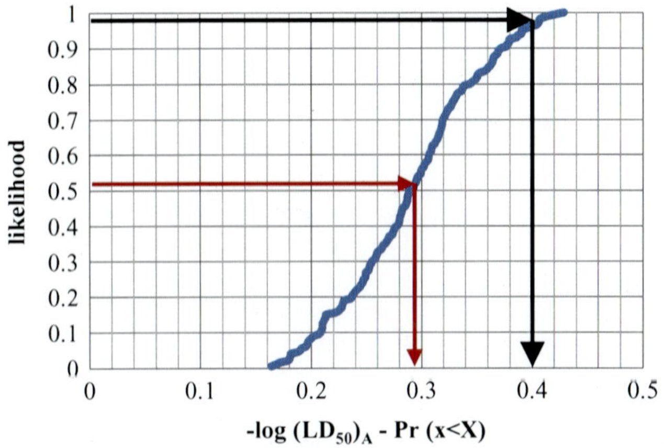

Fig. 11 *Alternative A* – Cumulative distribution function for the half population fatality (–log(LD_{50})), upper bound identification for 95 % likelihood

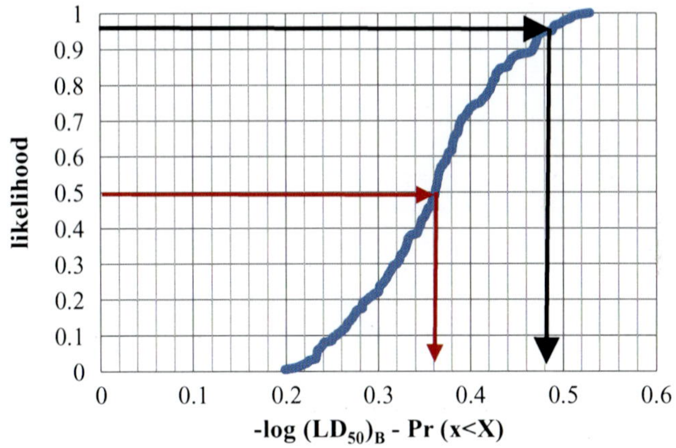

Fig. 12 *Alternative B* – Cumulative distribution function for thethe half population fatality (–log(LD_{50})), upper bound identification for 95 % likelihood

3.3 *Discussion*

The systematic framework for synthesis and design of processing networks followed in this study generated a large verified database resulting in a large design space which produced a number of scenarios prior to the identification of the optimal designs.

The input data in this study were collected from the literature and from a PNNL report. The resulting optimal design concept is algae cultivation in a raceway pond,

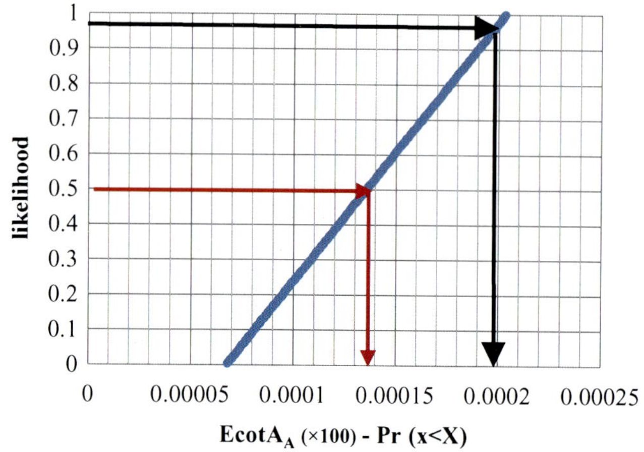

Fig. 13 *Alternative A* – Cumulative distribution function for the aquatic ecotoxicity (EcotA), upper bound identification for 95 % likelihood

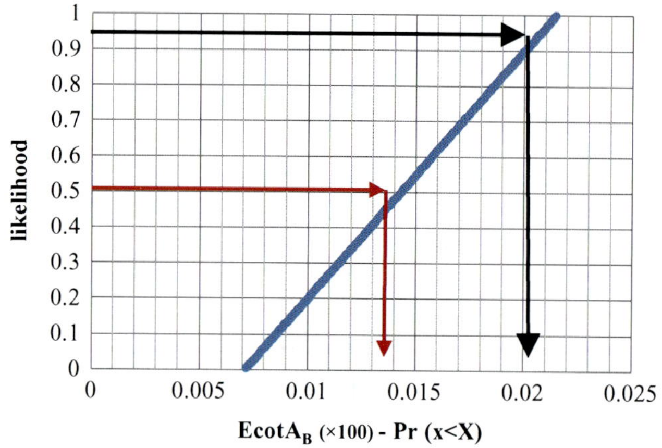

Fig. 14 *Alternative B* – Cumulative distribution function for the aquatic ecotoxicity (EcotA), upper bound identification for 95 % likelihood

hydrothermal liquefaction, and transesterification with H_2SO_4. The algae feedstock cost was estimated earlier by the U.S. Department of Energy (DOE 2013) as a fixed price of 430$/ton (340$/ton for cultivation, 90$/ton for dewatering) or 204 MM$/a for biodiesel production from lipid extraction. This cost was reduced to 300$/ton (or 190 MM$/a) due to the use of whole algae reported by PNNL (Jones et al. 2014). Therefore, an algae cost of 300$/ton was used in this study. Moreover, new optimal design concepts were found in this study resulting in a slightly higher EBITDA compared with the result in the PNNL report; 319 and 280 MM$/a, respectively. It is important to note that 90 % of the biodiesel production is related to the cost of the

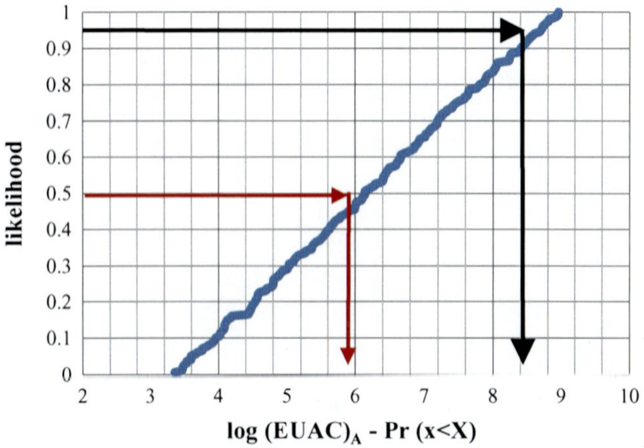

Fig. 15 *Alternative A* – Cumulative distribution function for the carcinogenic emissions to urban air (log (EUA$_C$), upper bound identification for 95 % likelihood

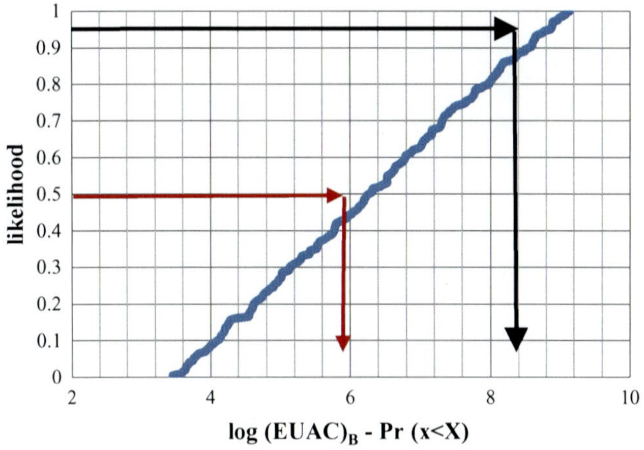

Fig. 16 *Alternative B* – Cumulative distribution function for the carcinogenic emissions to urban air (log (EUA$_C$), upper bound identification for 95 % likelihood

algae feedstock (190 MM$/a, 1300 tpd) which is much higher than lignocellulosic biomass (60 MM$/a, 2000 tpd). However, biodiesel yield for algal biorefinery (51 %) is much higher than lignocellulosic biorefinery (28 %) reported in the previous study (Cheali et al. 2014).

The results in this study are in agreement with the PNNL analysis, while different from the results in the report of British Columbia Council (Alabi et al. 2009). In their report, Alabi et al. performed economic analysis on algal biorefinery considering three different algae feedstocks. The report concluded that algae cultivated in a raceway pond and a photobioreactor cannot produce oil at competitive prices except for algae cultivated in a fermentation system due to the high capital cost associated with the cultivation. This difference in the analysis is due to the assumption of the oil content of algal biomass cultivated in raceway pond. In particular, the report of British Columbia Council assumed the oil content of algal biomass to be 20 %wt, while the PNNL study assumed the oil content to be 50 %wt (which is adopted in this study). This shows that oil content of algal biomass is a very important factor determining the viability of algal biorefinery.

In summary, the identification of optimal designs of the algal biorefinery remains still a challenging problem due to the quality of the data available and readiness and maturity of alternative technologies. The database of conversion technologies should be updated in line with further progress and developments in algae cultivation and processing technologies. In addition, to add robustness to the techno-economic analysis, uncertainty analysis need to be considered as well as discussed elsewhere in the study of Cheali et al. (2015b). Indeed the uncertainty analysis confirmed that oil content has a strong impact on the economic viability of algal biorefinery.

As regards the sustainability analysis, the extended LCA methodology used in this study provides a probabilistic interpretation of the calculated impact categories, while the traditional LCA methodology uses a "single-point" estimate for the impact categories which may lead to under- or over-estimation of impact categories. The uncertainty analysis of LCA provides a complete representation of uncertainties in the calculated impact categories by empirical CDFs. The decision making and comparison of process alternatives can then be made by using the percentile (e.g. 95 % probability) features of CDFs.

The environmental analysis on Alternative A and Alternative B was performed. The results show that Alternative A which has higher conversion, less methanol and unconverted algae oil recycled resulted in lower environmental impact. This study demonstrated that estimation of impact categories relevant for algal biorefinery is subject to significant uncertainties. These necessitate formal uncertainty analysis to enable robust and reliable decision making.

4 Conclusions and Future Work

In this study, a systematic framework for synthesis and design g optimal algal biorefinery with respect to techno-economic and sustainability analysis is presented.

The framework uses a superstructure to represent design space alternatives. The database of the algal biorefinery is developed by using a structured and generic model to represent process alternatives. The conservation of mass was used for data verification. A new optimal processing path was identified which includes the following processing scheme: hydrothermal liquefaction and transesterification with acid H_2SO_4 (alternative A).

It is also important to note that the microalgae cost is around 90 % of the biodiesel production cost which indicates that research and development efforts need to focus to bring down the raw material costs of the microalgal biorefinery.

As regards sustainability analysis, the LCA methodology of the EPA is extended with a comprehensive uncertainty analysis. The Monte Carlo technique with Latin Hypercube sampling was used for the uncertainty analysis. To interpret the results, a probabilistic framework that uses the 95 % probability level of the expected value of the impact category is used. Environmental sustainability analysis verified that the hydrothermal liquefaction and transesterification with acid H_2SO_4 (alternative A) have much better environmental performance (in particular lowest POP impact) and should be prioritized for further analysis.

Using the framework, many processing network alternatives are generated and evaluated at their optimality resulting in the identification of the optimal processing paths. The generated database and superstructure provides a versatile process synthesis toolbox to support designing future and sustainable algal biorefineries, in particular for the early stages of project development.

List of Acronyms

AHTL	Algae hydrothermal liquefaction
CAPEX	Capital Investment
CDF	Cumulative distribution function
CF	Characterization factor
CTUe	Comparative toxic units
EBITDA	Earnings Before Interest, Taxes, Depreciation and Amortization
EcotA	Aquatic ecotoxicity
EPA	Environmental protection agency
EUA_C	Carcinogenic emissions to urban air
EVPI	Expected value of perfect information
GAMS	General algebraic modeling system
GWP	Global warming potential
IRR	Internal rate of return
LD_{50}	Median lethal dose
MI(N)LP	Mixed integer (non)-linear programming
MM$/a	Million dollar per year
NPV	Net present value
PAF	Potential affected fraction
PEI	Potential environmental impact
POP	Photochemical oxidation potential

References

Alabi AO, Tampier M, Bibeau E (2009) Microalgae technologies & processes for biofuels/bioenergy production in british Columbia. The British Columbia Council, Vancouver

Cheali P, Quaglia A, Gernaey KV, Sin G (2014) Effect of market price uncertainties on the design of optimal biorefinery systems—a systematic approach. Ind Eng Chem Res 53(14): 6021–6032

Cheali P, Gernaey KV, Sin G (2015a) Uncertainties in early-stage capital cost estimation of process design – a case study on biorefinery design. Front Energ Res. doi:10.3389/fenrg.2015.00003

Cheali P, Vivion A, Gernaey KV, Sin G (2015b) Optimal design of algae biorefinery processing networks for the production of protein, ethanol and biodiesel. Comput Aid Chem Eng., 12th International symposium on process systems engineering and 25th European symposium on computer aided process engineering, 1151–1156

Curran MA (2006) Life cycle assessment: principles and practice. Scientific Applications International Corporation (SAIC). http://www.epa.gov/nrmrl/std/lca/pdfs/chapter1_frontmatter_lca101.pdf

Demirbas A, Demirbas MF (2011) Importance of algae oil as a source of biodiesel. Energy Convers Manag 52:163–170

Divakaran R, Pillai V (2002) Flocculation of algae using chitosan. J Appl Phycol 14:419–422

DOE (2013) MYPP, Biomass Multi-year program plan – May 2013. U.S.Do.Energy, Editor 2013, Bioenergy Technologies Office, USA

EPA (2006) Life cycle assessment: principles and practice, EPA/600/R-06/060. US Environmental Protection Agency/Office of research and development, Cincinnati

FAO Aquatic Biofuels Working Group (2010) Review paper – alage-based biofuels: applications and co-products. FAO Publications. ISBN 978-92-5-106623-2

Gargalo CL, Sin G (2015) Sustainable process design under uncertainty analysis: targeting environmental indicators. Copenhagen, Denmark: ESCAPE/PSE 25

Gong J, You F (2014) Global optimization for sustainable design and synthesis of algae processing network for CO2 mitigation and biofuel production using life cycle optimization. AICHE J 60(9):3195–3210

Granados MR, Ancien FG, Gomez C, Fernandez-Sevilla JM, Grima EM (2012) Evaluation of flocculants for the recovery of freshwater microalgae. Bioresour Technol 118:102–110

Helton JC, Davis FJ (2003) Latin hypercube sampling and the propagation of uncertainty in analyses of complex systems. Reliability Eng Syst Saf 81(1):23–69. doi:10.1016/S0951-8320(03)00058-9

Humbert S, Schryver A De, Margni M, Jolliet O (2002) IMPACT 2002 + : User Guide. Lausanne: Swiss Federal Institute of Technology Lausanne (EPFL). Retrieved from http://www.quantis-intl.com/pdf/IMPACT2002_UserGuide_for_vQ2.2.pdf

Iman RL, Conover WJ (2007) A distribution-free approach to inducing rank correlation among input variables. Commun Stat – Sim Comput 11(3):311–334. doi:10.1080/03610918208812265

Jones S, Zhu Y, Anderson D, Hallen R, Elliott D, Schmidt A, Albrecht K, Hart T, Butcher M, Drennan C, Snowden-Swan L (2014) Process design and economics for the conversion of algal biomass to hydrocarbons: whole algae hydrothermal liquefaction and upgrading. Pacifica Northwest National Laboratory, U.S. department of energy, PNNL-23227

Levine RB (2013) The production of algal biodiesel using hydrothermal carbonization and in situ transesterification. University of Michigan

Li Y, Horsman M, Wu N, Lan CQ, Dubois-Carlero N (2008) Biofuels from microalgae. Biotechnol Prog 24:815–820

Long RB, Abdelkader E (2011) Mixed-polarity azeotropic solvents for efficient extraction of lipids from nannochloropsis microalgae. Am J Biochem Biotechnol 7:70–73

Martin M, Grossmann IE (2013) Optimal engineered algae composition for the integrated simultaneous production of bioethanol and biodiesel. AICHE J 59(8):2872–2883

Miao X, Wu Q (2006) Biodiesel production from heterotrophic microalgal oil. Bioresour Technol 97:841–846

Petrick I, Dombrowski L, Kröger M, Beckert T, Kuchling T, Kureti S (2013) Algae biorefinery – material and energy use of algae. DBFZ Report No. 16

Petrusevski B, Bolier G, Van Breemen AN, Alaerts GJ (1995) Tangential flow filtration: a method to concentrate freshwater algae. Water Res 29:1419–1424

Price CA, Mendila-Morgenthaler LR, Goldstein M, Breden EN, Guillard RRL (1974) Harvest of planktonic marine algae by centrifugation into gradients of silica in the CF-6 continuous-flow zonal rotor. Biol Bull 147:136–145

Prommuak C, Pavasant P, Quitain AT, Goto M, Shotipruk A (2012) Microalgal lipid extraction and evaluation of single-step biodiesel production. Eng J 16:158–166

Rizwan M, Lee JH, Gani R (2013) Optimal processing pathway for the production of biodiesel from microalgal biomass: a superstructure based approach. Comput Chem Eng 58:305–314

Sim TS, Goh A, Becker EW (1988) Comparison of centrifugation, dissolved air flotation and drum filtration techniques for harvesting sewage-grown algae. Biomass 16:51–62

Sin G, Gernaey KV, Neumann MB, van Loosdrecht MCM, Gujer W (2009) Uncertainty analysis in WWTP model applications: a critical discussion using an example from design. Water Res 43(11):2894–2906. doi:10.1016/j.watres.2009.03.048

Sirin S, Trobajo R, Ibancz C (2012) Harvesting the microalgae Phaeodactylum tricornutum with polyaluminium chloride, aluminium sulphate, chitosan and alkalinity-induced flocculation. J Appl Phycol 24:1067–1080

Vicente G, Martinez M, Aracil J (2004) Integrated biodiesel production: a comparison of different homogeneous catalysts systems. Bioresour Technol 92:294–305

Wu Z, Zhu Y, Huang W, Zhang C, Li T, Zhang Y, Li A (2012) Evaluation of flocculation induced by pH increase for harvesting microalgae and reuse of flocculated medium. Bioresour Technol 110:496–502

Zhang Y, Dube MA, McLean DD, Kates M (2003) Biodiesel production from waste cooking oil: 1. Process design and technological assessment. Bioresour Technol 89:1–16

Multi-Actor Life-Cycle Assessment of Algal Biofuels for the U.S. Airline Industry

Datu Buyung Agusdinata and Daniel DeLaurentis

Abstract The chapter describes a hybrid approach for assessing the environmental impacts resulting from the production and use of algae-derived biofuels. It addresses the issue of the level of emissions reductions that can be expected from the adoption of algal biofuels. The approach used for answering the question combines a standard life-cycle assessment (LCA) method with the perspectives of multiple actors along the life-cycle stages and supply chain network (e.g. algae growers, refiners, distributors, and users), hence the Multi-Actor Life-cycle Assessment. In addition to the accounting of life-cycle emissions, the decision-making behaviors of actors are considered. The result of the approach is a prediction of emission impacts that may actually take place. The fundamental concepts of multi-actor life-cycle assessment methodology will be described and then applied via a system-based simulation model to assess potential emissions reduction resulting from the adoption of algal biofuels in the U.S. airline industry.

Keywords Multi-actor life-cycle assessment • Algal aviation biofuels • U.S. airlines industry • System dynamics • Actors' decision flow chart • Aviation environmental policies • Algal biofuels profitability • Simulation model • Biofuels adoption scenarios • Fleet-wide emissions impact

1 Introduction

Under pressure to reduce greenhouse effect from the combustion of fossil fuels, a range of stakeholders have been advocating for the development of biofuels. In the U.S., for example, the transportation sector contributes about 27 % of the total CO_2 emissions. The contribution of the aviation sector to the overall carbon emissions is

D.B. Agusdinata (✉)
Industrial & Systems Engineering and Environmental, Sustainability & Energy (ESE) Institute, Northern Illinois University, 590 Garden Road EB 236, DeKalb, IL, USA
e-mail: bagusdinata@niu.edu

D. DeLaurentis
School of Aeronautics & Astronautics, Purdue University, 701 West Stadium Avenue, West Lafayette, IN 47907, USA
e-mail: ddelaure@purdue.edu

A. Prokop et al. (eds.), *Algal Biorefineries*, DOI 10.1007/978-3-319-20200-6_18

relatively small at 2.9 % (FAA 2005). However, as the aviation industry is expected to grow between 2 and 4 % per annually, finding sustainable technological solutions including alternative fuels would help curb GHG emissions.

Liquid fuels derived from renewable feedstocks (i.e. biofuels) represent promising candidates for alternative fuels. Currently biofuels account for about 3 % of total transportation fuel consumed, mainly in the form of ethanol derived from corn grain and used as a blend for ground vehicles (e.g. E10 or E85 which respectively represent 10 % and 85 % ethanol blended with gasoline by volume).

To be fully adopted by the industry, aviation biofuels (or bio-jet) have to meet fuel standards for properties such as fuel density and performance requirements such as low freezing point (ASTM 2011). The requirement for "drop-in" fuels demand that alternative jet fuels must have characteristics sufficiently similar to petroleum jet fuels regardless of the feedstock and refining process so that they are compatible with existing aircraft engines. In addition, biofuels must be compatible with the existing infrastructure for distribution and delivery. The bio-jet produced by current refinery processes do not contain aromatic compounds, which account for up to 25 % of petro-jet by volume and are needed for proper lubrication of mechanical systems and sealing of fuel tanks. To meet such requirements, bio-jet fuels must be blended with conventional fuels. Currently, a 50–50 blend by volume between bio-jet and petro-jet fuel is the norm for meeting fuel property and performance specifications.

1.1 Algal Biofuels Potential

Algae is one energy feedstock considered to be sustainable for several reasons. The net energy ratio (NER) for algal biofuels production is generally lower than one (e.g. (Batan et al. 2010)) although the exact NER is variable based on production technique and the state of the art of technology used. A $NER<1$ means that the energy needed to produce the fuels is less than the energy generated from the produced fuels. The life-cycle greenhouse gas emissions are lower than for conventional petroleum-based fuels, though they could be higher than those of other renewable feedstocks (Agusdinata et al. 2011) (Fig. 1). Algae has the advantages that it does not compete with food production resources because it is inedible and can be grown in marginal land and has low land usage (Lardon et al. 2009). In addition, the production of algae needs CO_2 inputs and it is known that algae has a high CO_2 absorption and uptake rate.

1.2 Algal Biofuels Life-Cycle Impacts

Aviation biofuels derived from algae have potential to reduce aviation GHG emissions. Figure 1 shows algal biofuels emissions compared with those of various biosources (i.e. camelina, corn stover, switchgrass, and woody crops) and

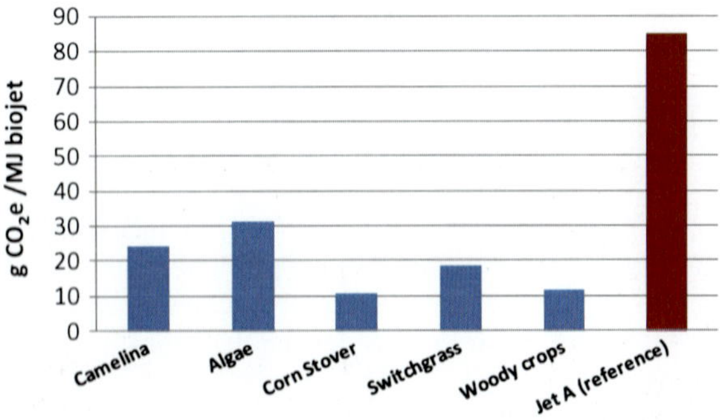

Fig. 1 Baseline unit life-cycle emissions of aviation biofuels (bio-jet)

petroleum-derived fuels (Hileman et al. 2008; Agusdinata et al. 2011). The GHG emissions are given as CO_2 emissions equivalent, (CO_2 e) per Mega Joule (MJ) energy produced. Algal biofuels produce about 30 g CO_2 e per M Joule energy. By contrast, convention jet fuels will produce about 84 g CO_2 e per M Joule energy. Compared to other biofuels, algal biofuels have higher unit emissions largely due to heat and electricity requirements in the production. The measure takes into account other greenhouse gases such as N_2O that are emitted during the use of fertilizer. Aviation biofuels' lower unit emissions are due to the fact that during feedstock cultivation, algae and other biomass plants absorb carbon dioxide from the atmosphere and during combustion carbon is converted back to carbon dioxide and returns to the atmosphere.

1.3 Algal Biofuels Challenges

To realize algae full potentials, the players in the sector need to address several challenges. One of them is to reduce the energy and fertilizer consumption to achieve positive energetic balance (Lardon et al. 2009). The other is production cost. It has been estimated that the unit cost of production range between $16–17 per gallon in 2015 (Agusdinata et al. 2011). This price is significantly higher than other biofuels, whose unit production costs range from $2–6 per gallon.

2 Multi-actor Life-Cycle Assessment (MA-LCA) Approach

The environmental sustainability of algal biofuels can be evaluated using life-cycle assessment (LCA) method. The method used to account for how technologies, processes, and products consume resources (i.e. energy and materials) and generate pollution (GHG emissions, air pollutants, and other toxic waste materials) to the environment. Using LCA one can look at all stages from raw material extraction, through manufacturing to disposal including all transportation needs throughout.

For transportation fuels, the LCA has also evolved to consider wider impacts such as human and ecological health, water consumption, land-use changes (e.g. deforestation), and bio-diversity. LCA implementation is guided by internationally accepted standards in terms of principles, frameworks, requirements, and methods (i.e. ISO 14040 and ISO 14044).

However, the standard LCA is technology centered. By itself, it cannot answer questions about the degree of adoption of a technology and the overall environmental benefits that can potentially be achieved as a result. To answer these types of questions we, have to consider the decisions made by actors involved in the entire life-cycle.

Multi-Actor Life-cycle Assessment (MA-LCA) is an approach to holistically account for the considerations of multi-actors in the development of a product, in our case aviation biofuels. The approach is based on system perspective couched within a standard life-cycle assessment (LCA) methodology (i.e. based on ISO 14040 and ISO 14044) (Fig. 2a). The approach combines a standard LCA method with the perspectives of multiple actors along the life-cycle stages and supply chain network (e.g. algae growers, refiners, distributors, and users). The difference between conventional and multi-actor LCA in the aviation application is illustrated in Fig. 2.

The MA-LCA is based on the idea that in each life-cycle stage, one can identify decision-making entities (individual humans or organizations). We call these entities actors. For algal aviation biofuels production, some of these actors are identified (Fig. 2b outer loop) across the algal biofuel life-cycle stages: algae growers, biorefineries, fuel distributors, airlines, and policymakers. While each influences a segment directly, impacts on adjacent segments are also often important.

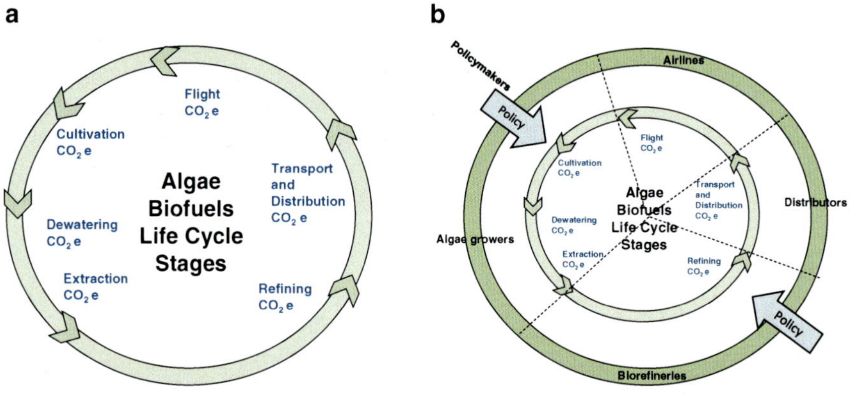

Standard LCA   Multi-Actor LCA

Fig. 2 Standard vs. multi-actor life-cycle assessment (**a**) Standard LCA (**b**) Multi-actor LCA

2.1 Factors Relevant to Life-Cycle Actors

Figure 3 shows the variables and their relationships that are relevant for some actors in life-cycle stages. Algae growers' costs, for example, are comprised of labor, electricity, capital, and fertilizer, which take a large proportion of the total growing costs. Feedstock cost, in turn, comprises a major operating cost for bio-refineries. The unit revenue is derived from the unit selling price of bio-jet and, in smaller proportion, from the feedstock by-product selling price.

The life-cycle emissions of bio-jet are influenced by two or three kinds of yield factor. For algae, three factors are involved: (1) feedstock yield factor: unit mass of feedstock produced per unit area of land, (2) feedstock oil yield factor: unit mass of feedstock needed to produce a unit mass of feedstock oil, and (3) bio-jet factor: unit mass of feedstock oil needed to produce a unit mass bio-jet.

Bio-jet demand, in turn, is driven by incentives given to airlines. The relative difference between the price of petro-jet and bio-jet determine the extra fuel cost incurred to airlines. Given the CO_2 credits, airlines sell or buy the carbon credits depending on whether their actual emissions exceed or are lower than the amount

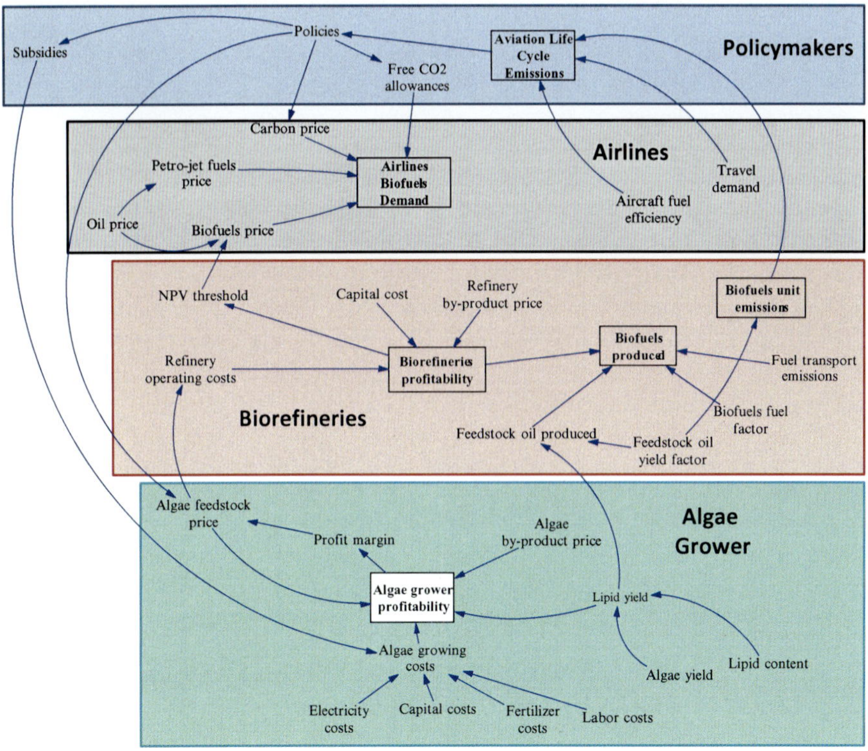

Fig. 3 A conceptual model of the inter-relationships of key model variables related to actors

permitted. Airlines can have lower life-cycle emissions by adopting bio-jet. The amount of carbon penalties or savings is determined by the price of carbon.

Lastly, as an overall system performance indicator, the level of emissions is comprised of some major factors: aircraft efficiency in terms of fuel consumption, the unit life-cycle emissions per unit mass fuel consumed and the level of airline operations and the aircraft fleet mix operated (these latter two driven by the growth of air travel demand). These factors and their relationships form a rationale that actors use to make their decisions.

2.2 *Actors' Decision Flowchart*

A decision flow chart captures the relationship and sequence that underlies actor decisions and drive impact of their policies on aviation life-cycle emissions. The flow chart in Fig. 4 shows simplified bio-jet supply and demand logic and provides access points for policies that can be directed toward influencing actor's decisions.

For airlines, the decision to use bio-jet fuel is mainly influenced by the degree of savings in carbon costs. To influence this decision, policymakers have leverage on establishing a carbon marketplace to regulate the CO_2 price. The earliest possible commercial use of bio-jets is expected to be 2015, the time when the fuel is certified. By this time, it is also assumed that the carbon cap and trade for aviation would have been implemented.

For bio-refineries, the response to bio-jet demand depends on whether a response provides viable return on investment. Thus, the net present value (NPV) and internal rate of return (IRR) are used as decision criteria. For instance, if the NPV > 0, IRR > 15 %, and there is enough bio-jet demand, the bio-refinery plant will be built creating demand for feedstock.

Feedstock producers will satisfy the demand only if they get a certain profit margin from producing feedstock. For instance, a threshold value of 10 % can be set, above which the feedstock will be produced, provided that there is demand and land available.

3 A Case Study in the U.S. Airline Industry

Algae-derived jet fuels blended with conventional fuels have been tested in commercial flights by airlines such as Alaska Airlines and United Airlines (Rahmes et al. 2009). The success of the flight tests provides a validation to the technical feasibility of biofuels. Suppose that instead of just a few individual flights, all commercial flights in the U.S. were to use algal biofuels, what would be the impact on greenhouse gas emissions?

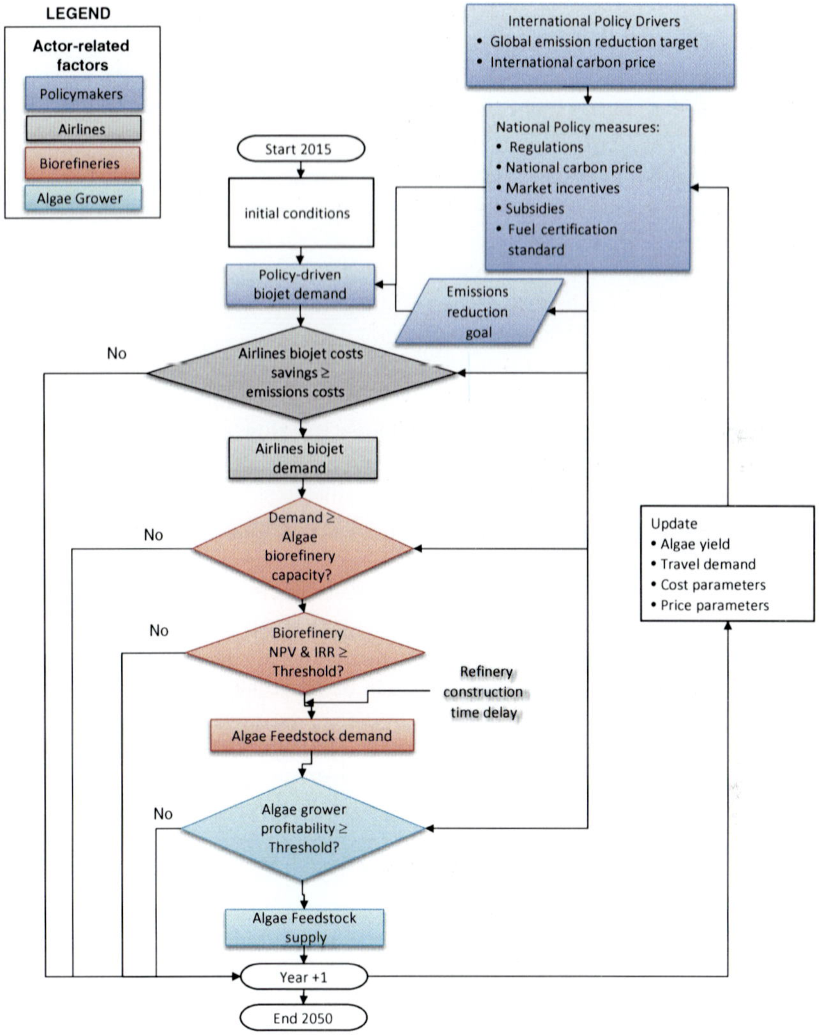

Fig. 4 Flowchart to operationalize MA-LCA (Modified from (Agusdinata et al. 2011))

3.1 Airlines Fleet-Wide Impacts of Algal Biofuels Model

To evaluate the impact of algal jet fuels on aviation emissions, a simulation model has been developed (Zhao et al. 2010). The Fleet-level Environmental Evaluation Tool (FLEET) was developed by researchers at Purdue University under sponsorship by the National Aeronautics and Space Administration (NASA) to assess environmental impact of aviation (emissions and noise) under scenarios of market demand and aircraft technology availability.

The simulation model integrates two major components: system dynamics and fleet allocation optimization. Figure 5 depicts how the two are integrated. The variety of dynamic relationships project flows of projected passenger demand and available aircraft in a fleet that airlines operate given some environmental constraints. The resource allocation then allocates aircraft to demanded trips in a 1 year period given the airline route network, resulting in an optimal fleet mix to satisfy demand. Outputs from the resource allocation optimization in the forms of environmental impact metrics may trigger environmental policies that feedback directly to both the optimization and the system dynamic loop. The system dynamics components are discussed and elaborated below.

As the influence diagram shows, the convoluted interactions among variables form many chains as well as feedback loops. For example, Demand Growth Rate affects Projected Demand, which affects the Desired Aircraft Fleet Size and Mix, which affects Aircraft Orders, which affects Order Backlog, and so forth. Appropriate system dynamics modeling of this "influence net" enables incorporation of the impact of a variety of time-varying events on the fleet forecasting. These events include (but are not limited to) quasi-steady events such as growth in yearly passenger demand for air transportation, volatility in jet fuel costs, and seasonal fluctuations that exist on top of the yearly demand increase. Introduction of new technology aircraft and retirement of aging and inefficient aircraft is another type of event. Other one-of-a-kind events can have profound impacts within a short period of time. These could include a sudden drop in demand due to economic downturn,

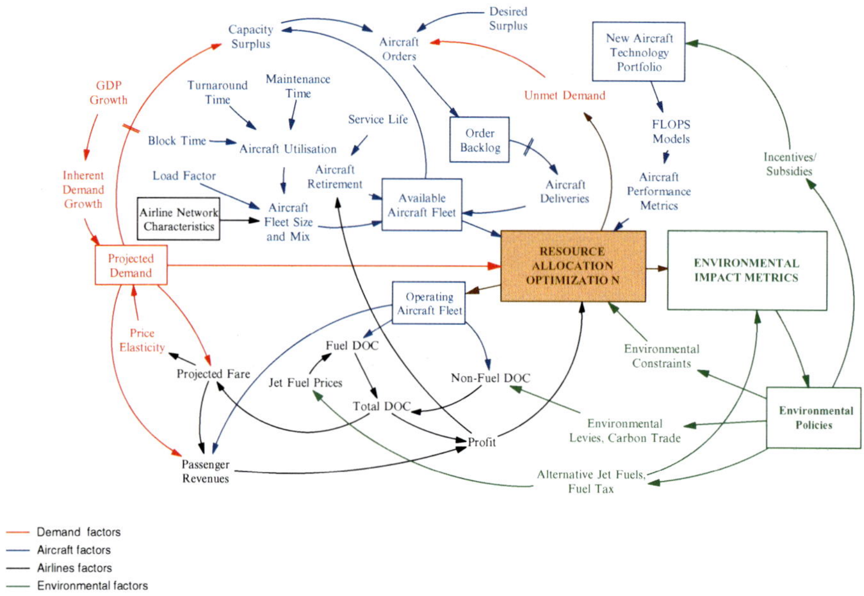

Fig. 5 FLEET (Zhao et al. 2010), An integrated system dynamics and resource allocation optimization approach

a rapid spike in jet fuel prices that raises operation cost drastically, or new regulation that limits aircraft noise and emissions.

A variety of policy options to mitigate emissions and noise from aviation have been discussed in the literature. These options include, for example, emission trading schemes (Scheelhaase and Grimme 2007). Furthermore, in addition to curfew hours, a levy modulated per aircraft type can be charged to compensate the population living nearby a particular airport affected by noise nuisance and emissions (Lu and Morrell 2006). Other policy includes incentives to facilitate the development of alternative jet fuel at its early stages (Agusdinata and Delaurentis 2011). All these measures are meant to discourage the inefficient use of fuel and noisy aircraft and accelerate the adoption of better aircraft and fuel, hence reduce environmental impacts.

The air transportation network modeled consists of only those routes that connect the WWLMINET 257 airports including international routes with either the origin or destination in the US. In 2005, approximately 65 % of all passenger air traffic – 80 % of international passengers traveling to and from the US and domestic passengers – had as origin or destination one of these airports. The 2005 passenger demand between these 257 airports is obtained from data provided by the Bureau of Transportation Statistics DB1B database (Bureau of Transportation Statistics (BTS) 2008).

3.2 The Resource Allocation Simulation Model

In a simplified formulation, the resource allocation model minimizes the airline's direct operating cost, which is the cost per flight (*cost*) multiplied by the number of flights (*flight*) across the number of aircraft type (*ac type*), *i*, and number of routes (*no. routes*), *j*. The model is subject to a condition that the supply (i.e. aircraft seat capacity, *cap*, multiplied by the number of flights) meets passenger demand (*dmd*) on each route. Additionally, the number of flights of each aircraft type should not exceed the aircraft trip limitations that aircraft can make per day (*trip*). Lastly, the number of flights should be a positive integer.

The RA model is formulated as:

Minimize

$$\sum_{j}^{no.routes} \sum_{i}^{ac\,type} cost_{ij} . flight_{ij}, \tag{1}$$

Subject to

$$\sum_{i}^{ac\,type} cap_i . flight_{ij} \geq dmd_j, \tag{2}$$

$$\sum_{i}^{ac\ type} flight_i \leq trip_i, \tag{3}$$

$$flight_{ij} \geq 0, \tag{4}$$

$$flight_{ij} = \text{integer}. \tag{5}$$

For all the flights of deployed aircraft resulting from the RA model, the total aviation fuel life cycle carbon emissions come from the contribution of alternative and conventional type. They are calculated using the formula given in Eq. 6 where $share_k$ is the proportion of fuel type k in the fuel mix. For an alternative fuel, the fuel share over time depends on its initial value and the growth. The *fuel_burn* term is the amount of fuel consumed in a flight. The *emission_factor* converts the amount of fuel consumed to the amount of CO_2 emissions.

$$\text{Total aviation life-cycle emissions} = \sum_{k}^{fuel\ type} \sum_{j}^{no.routes} \sum_{i}^{ac\ type} share_k.fuel\_burn_{ij}.flight_{ij}.emission\_factor.fuel\_factor_k, \tag{6}$$

When the emissions from fuel production and combustion are accounted for (i.e. fuel life cycle emissions), the *emission_factor is* set to be 3.16/0.861 = 3.67 kg CO_2 per kg fuel burn. Note that the 3.16 factor accounts for emissions from combustion only. The *fuel_factor* is the relative emission factor of fuel types to the reference conventional Jet fuel. The *fuel_factor* for algal based jet fuel is 30/84 = 0.36. To illustrate, for a 50–50 blend between *algal biofuels* and conventional fuel, every 1 kg of fuel consumed in a flight generates: Total fuel life cycle emissions =

$$[(0.5 \times 1 \times 3.67 \times 0.36) + ((1 - 0.5) \times 1 \times 3.67)] = 2.5\,\text{kg}\,CO_2\text{e}.$$

The resource allocation optimization problem above is solved using GAMS/ Cplex solver (www.ilog.com) and is implemented in MATLAB (www.mathworks.com). The detailed mathematical formulation of the simulation model can be found in (Zhao et al. 2010).

3.3 Algal Biofuels Adoption Level Scenarios and Cases

To estimate the impact on life-cycle emissions, three adoption level scenarios of algal bio-jet fuels are considered:

1. Scenario: "Baseline" – No algal biofuels, only conventional petro-jet fuels are used.
2. Scenario: "Constant Algae" – Constant 5 % share algae based fuel from the period of 2016–2050.

3. Scenario: "Increasing Algae" – Initial 1 % share starting 2016 that will increase linearly to 50 % in 2050.

Another important variable that needs to be considered is fuel price. As indicated in Fig. 5, the price that airlines pay for algal biofuels will impact its operating costs and will be passed to the passengers through ticket fare. When airlines increase the fare, demand may decrease due to price elasticity. In this case, the industry will fly fewer flights to accommodate reduced demand. The result is fewer carbon emissions.

Two cases are considered on the fuel price:

1. Case 1: fuel price parity – It is assumed that the price of algal biofuels is equal to that of conventional jet fuel. This assumption is necessary so that we can separate between reduction due to change in passenger demand and reduction due to biofuel adoptions.
2. Case 1: fuel price disparity – It is assumed that the price of algal biofuels is ten times that of conventional jet fuels. In this case, the reduction of emissions will result from both the reduction in passenger demand as well as adoption of algal biofuels.

4 Simulation Model Results and Analysis: Carbon Life-Cycle Emission Evolution

4.1 *Case 1: Fuel Price Parity*

Figure 6a shows demand evolution for all scenarios from 2005 to 2050. The data points are normalized by demand in 2005. Demand evolution exhibits similar patterns with an annual growth rate of approximately 1 % per year. This observation can be attributed to the pricing strategy where the airline does not lower price when operating cost is reduced due to technology improvement. Thus, technology improvement will not be reflected by demand.

Starting the year 2015, when algal biofuels are introduced, the trajectories of emissions start to diverge. From the period of 2005–2050, passenger demand increases by 250 % (or increases 3.5 times). During that period, the emissions go up by 165 % ("Baseline" scenario). The fact that emissions do not rise proportionately with the passenger demand is largely due to the improvement in aircraft fuel efficiency.

When algal biofuels are used by airlines at the share of 5 % of the total fuels consumed ("Constant Algae" scenario), the emissions will increase by 156 % (Fig. 6b). In the "Increasing Algae" scenario, in which the share of algae scenario increases linearly from 1 to 50 % over the period, life-cycle carbon emissions only increase by 81 %. This means that in this scenario, the total life-cycle emissions are 84 % lower than the "Baseline" scenario.

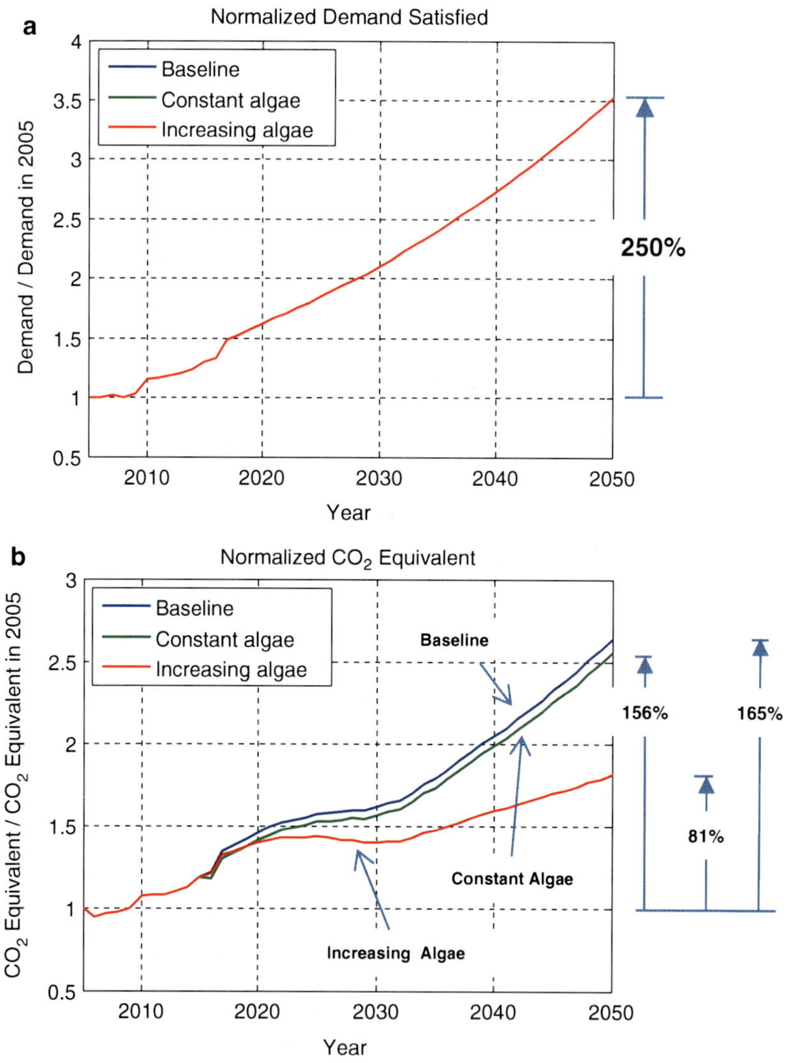

Fig. 6 The impact of algal biofuels to the normalized life-cycle CO2 emissions of the simulated U.S. Airline Industry (Case 1: fuel price parity) (**a**) Passenger demand evolution (**b**) Life-cycle carbon emissions evolution

4.2 Case 2: Fuel Price Disparity

Figure 7 shows the impact of algal biofuels adoption on passenger demand (Fig. 7a) and life-cycle CO_2 emissions (Fig. 7b). When algal biofuels price is ten times higher than that of conventional fuels, passenger demand will fall. In the "Constant Algae" scenario, passenger demand will increase by 215 % in 2050 instead of 250 % in the

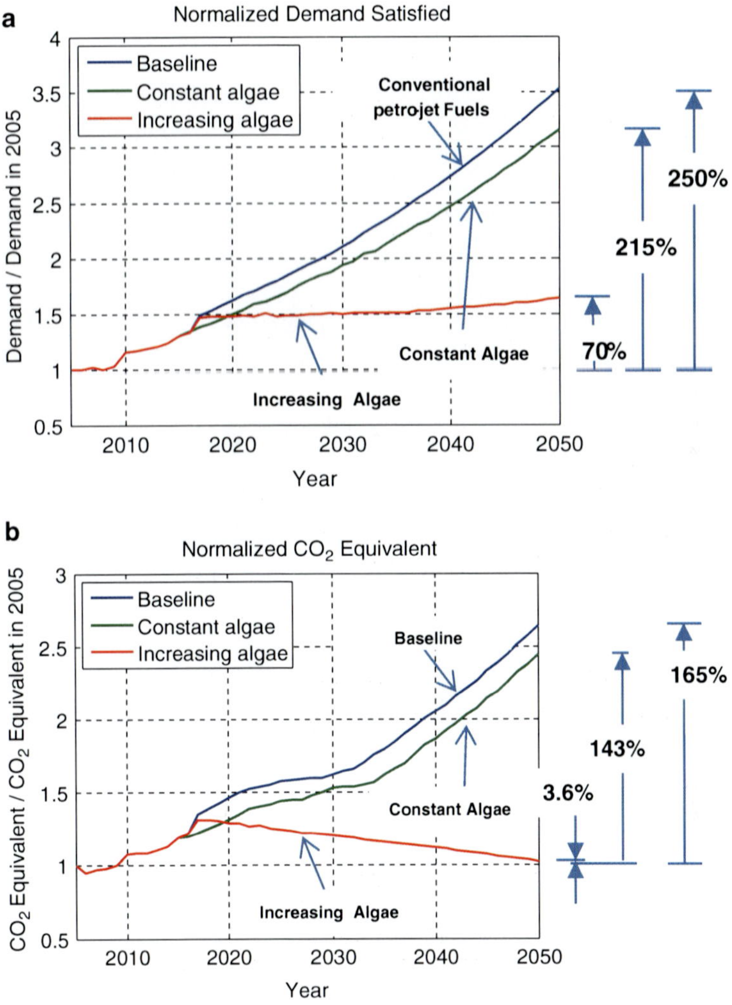

Fig. 7 The impact of algal biofuels to the normalized life-cycle CO_2 emissions of the simulated U.S. Airline Industry (Case 2: fuel price disparity) (**a**) Passenger demand evolution (**b**) Life-cycle carbon emissions evolution

"Baseline" scenario. The passenger demand will only increase by 70 % in 2050 in the "Increasing Algae" scenario.

The adoption of algal biofuels combined with the change in passenger demand has a considerable impact on life-cycle emissions. In 2050, the emissions in the "Constant Algae" scenario are 22 % lower than the "Baseline Scenario". The combined factors have so profound impact in the "Increasing Algae" scenario that they can bring down the emissions back almost to the 2005 level.

5 Concluding Remarks

The introduction of alternative jet fuel from algae has a considerable effect in reducing the life-cycle emissions of the U.S. airline industry. The multi-actor life-cycle assessment approach provides a more realistic adoption level of algal biofuels by considering actors' decision criteria. The FLEET simulation tool contributes to a better and more comprehensive understanding of the environmental impacts of air transportation by considering not only the emissions individual aircraft but their fleet-wide effect on the environment.

The simulation model presented in this chapter shows that algal biofuels, when adopted by the U.S. airline industry, will indeed reduce the life-cycle carbon emissions. In the most aggressive penetration simulated in the chapter (i.e. the "Increasing Algae" scenario) and assuming price parity, algal biofuels produce 85 % lower life-cycle CO_2 emissions than the "Baseline" scenario in 2050.

For the algal potentials to be reached, however, there is a mounting challenge to overcome. The production costs of algal biofuels should be low enough to make it at least price parity with the conventional jet fuels. Algal biofuels must also be competitive compared to other biofuels derived from other feedstocks such as lignocellulosic ones (e.g. switchgrass and woody trees).

Acknowledgement The authors would like to thank Kushal Moolchandani at Purdue University for his support in setting up and running the FLEET simulation model.

List of Acronyms

BTS	Bureau of Transportation Statistics
CO_2 e	Carbon dioxide equivalent
FAA	Federal Aviation Administration
FLEET	Fleet-level Environmental Evaluation Tool
GAMS	General Algebraic Modeling System
IRR	Internal Rate of Return
LCA	Life-cycle Assessment
LMINET	U.S. national airspace system network model, developed by the LMI company for NASA
MA-LCA	Multi-Actor Life-cycle Assessment
NASA	National Aeronautics and Space Administration
NER	Net Energy Ratio
NPV	Net Present Value
ISO	International Organization for Standardization
WWLMINET	Worldwide version of the LMINET

References

Agusdinata DB, Delaurentis DA (2011) Addressing equity issue in multi-actor policymaking via a system-of-systems approach: aviation emissions reduction case study. J Syst Sci Syst Eng 20(1):1–24

Agusdinata DB, Zhao F, Ileleji K, DeLaurentis DA (2011) Life cycle assessment of potential biojet fuel production in the United States. Environ Sci Technol 45(21):9133–9143

ASTM (2011) Standard specification for aviation turbine fuel containing synthesized hydrocarbons. ASTM D7566. doi:10.1520/D7566-14A

Batan L, Quinn J, Willson B, Bradley T (2010) Net energy and greenhouse gas emission evaluation of biodiesel derived from microalgae. Environ Sci Technol 44(20):7975–7980

Bureau of Transportation Statistics (BTS) (2008) www.bts.gov. Retrieved Aug 2008

FAA (2005) Aviation & emissions: a primer. Federal Aviation Administration, Office of Environment and Energy, Washington, DC

Hileman JI, Wong HM, Ortiz D, Brown N, Maurice L, Rumizen M (2008) The feasibility and potential environmental benefits of alternative fuels for commercial aviation. In: The 26th International Congress of the aeronautical Sciences (ICAS), Anchorage

Lardon L, Helias A, Sialve B, Stayer JP, Bernard O (2009) Life-cycle assessment of biodiesel production from microalgae. Environ Sci Technol 43(17):6475–6481

Lu C, Morrell P (2006) Determination and applications of environmental costs at different sized airports – aircraft noise and engine emissions. Transportation 33(1):45–61

Rahmes T, Kinder J, Crenfeldt G (2009) Sustainable Bio-Derived Synthetic Paraffinic Kerosene (Bio-SPK) jet fuel flights and engine tests program results. In: The 9th AIAA Aviation Technology, Integration, and Operations Conference (ATIO) and Aircraft Noise and Emissions Reduction Symposium (ANERS), Hilton Head, South Carolina

Scheelhaase J, Grimme W (2007) Emissions trading for international aviation – an estimation of the economic impact on selected European airlines. Air Transp Manag 13(5):253–263

Zhao J, Tetzloff IJ, Tyagi A, Dikshit P, Mane M, Agusdinata DB, Crossley WA, DeLaurentis DA (2010) Assessing new aircraft and technology impacts on fleet-wide environmental metrics including future scenarios. In: The 48th AIAA Aerospace Sciences meeting including the new horizons forum and aerospace exposition, Orlando, Florida

Index

A. Prokop et al. (eds.), *Algal Biorefineries*, DOI 10.1007/978-3-319-20200-6